Encyclopaedia

of

DISCRETE MATHEMATICS

Encyclopaedia of DISCRETE MATHEMATICS

Volume 1

Dr. Rakesh Kumar Pandey

ANMOL PUBLICATIONS PVT. LTD.
NEW DELHI - 110 002 (INDIA)

ANMOL PUBLICATIONS PVT. LTD.
Regd. Office: 4360/4, Ansari Road, Daryaganj,
New Delhi-110 002 (India)
Ph.: 23278000, 23261597
Branch Office: No. 1015, Ist Main Road, BSK IIIrd Stage
IIIrd Phase, IIIrd Block,
Bangalore-560 085 (India)
Tel.: 080-41723429
Visit us at: www.anmolpublications.com

Encyclopaedia of Discrete Mathematics

First Edition, 2009
ISBN 978-81-261-4108-1 (Set)

PRINTED IN INDIA

Printed at Mehra Offset Press, Delhi.

Contents

Preface

Discrete Mathematics stands among the *Leading Disciplines of Mathematics and Theoretical Computer Science*. It handles the objects of discrete values. This Encyclopaedia has more than one purpose. Students should learn a particular set of Mathematical Facts and know how to apply them more importantly. It also teaches students how to think Logically and Mathematically. To achieve these goals, the chapters are stressed on *Logical Reasoning, Graph Theory, Set Theory, Binary Relations, Recurrence Relations, Combinatorics, Functions, Matrices*, etc.

Encyclopaedia of Discrete Mathematics has been written for the students who are preparing for an *Advance Degree Programme*. With the background necessary for further study in Theoretical Computer Science. Sufficient topics have been included to accommodate students with varied background. The subject matter has been discussed in such a way that students will find no difficulty to understand it.

Author

List of Symbols

$\pm$	$\not\subset$	$\subset$	$\supset$	$\subseteq$	$\in$	$\notin$	∂
$\neq$	$\equiv$	$\approx$	$\angle$	$\vee$	Σ	Π	Ω
ϕ	α	β	χ	ε	γ	η	λ
μ	ν	$\perp$	π	θ	ρ	σ	τ
$\Re$	ω	ξ	ψ	ζ	Υ	$\exists$	$\int$
$\iint$	$\iiint$	$\oint$	$\times$	$\sqrt{\ }$	$\therefore$	$\because$	Δ
$\geq$	$\leq$	$\Rightarrow$	$\Downarrow$	$\Leftrightarrow$	∞	$\propto$	$\mathbb{R}$
$\mathbb{Z}$	$\mathbb{N}$	$\mathbb{C}$	$\mathbb{Q}$				

Chapter 1

Introduction to Discrete Mathematics

Discrete mathematics is mathematics that deals with discrete objects. Discrete objects are those which are separated from (not connected to/distinct from) each other. Integers (aka whole numbers), rational numbers (ones that can be expressed as the quotient of two integers), automobiles, houses, people etc. are all discrete objects. On the other hand real numbers which include irrational as well as rational numbers are not discrete.

As you know between any two different real numbers there is another real number different from either of them. So they are packed without any gaps and can not be separated from their immediate neighbors. In that sense they are not discrete. In this course we will be concerned with objects such as integers, propositions, sets, relations and functions, which are all discrete. We are going to learn concepts associated with them, their properties, and relationships among them among others.

Let us first see why we want to be interested in the formal/theoretical approaches in computer science.

Some of the major reasons that we adopt formal approaches are:

1. We can handle infinity or large quantity and indefiniteness with them.
2. Results from formal approaches are reusable.

As an example, let us consider a simple problem of investment. Suppose that we invest $1,000 every year with expected return of 10% a year. How much are we going to have after 3 years, 5 years, or 10 years ? The most naive way to find that out would be the brute force calculation.

Let us see what happens to $1,000 invested at the beginning of each year for three years. First let us consider the $1,000 invested at the beginning of the first year. After one year it produces a return of $100. Thus at the beginning of the second year, $1,100, which is equal to $1,000 * (1 + 0.1), is invested. This $1,100 produces $110 at the end of the second year. Thus at the beginning of the third year we have $1,210, which is equal to

\$1,000 * (1 + 0.1)(1 + 0.1), or \$1,000 * $(1 + 0.1)^2$. After the third year this gives us \$1,000 * $(1 + 0.1)^3$.

Similarly we can see that the \$1,000 invested at the beginning of the second year produces \$1,000 * $(1 + 0.1)^2$ at the end of the third year, and the \$1,000 invested at the beginning of the third year becomes \$1,000 * (1 + 0.1).

Thus the total principal and return after three years is \$1,000 * (1 + 0.1) + \$1,000 * $(1 + 0.1)^2$ + \$1,000 * $(1 + 0.1)^3$, which is equal to \$3,641.

One can similarly calculate the principal and return for 5 years and for 10 years. It is, however, a long tedious calculation even with calculators. Further, what if you want to know the principal and return for some different returns than 10%, or different periods of time such as 15 years ? You would have to do all these calculations all over again.

We can avoid these tedious calculations considerably by noting the similarities in these problems and solving them in a more general way. Since all these problems ask for the result of invesing a certain amount every year for certain number of years with a certain expected annual return, we use variables, say A, R and n, to represent the principal newly invested every year, the return ratio, and the number of years invested, respectively.

With these symbols, the principal and return after n years, denoted by S, can be expressed as

$$S = A(1 + R) + A(1 + R)^2 + ... + A(1 + R)^n.$$

As well known, this S can be put into a more compact form by first computing $S - (1 + R)S$ as

$$S = A\ ((1 + R)^{n+1} - (1 + R))/\ R.$$

Once we have it in this compact form, it is fairly easy to compute S for different values of A, R and n, though one still has to compute $(1 + R)^{n+1}$. This simple formula represents infinitely many cases involving all different values of A, R and n. The derivation of this formula, however, involves another problem. When computing the compact form for S, $S - (1 + R)S$ was computed using $S = A(1 + R) + A(1 + R)^2 + ... + A(1 + R)^n$. While this argument seems rigorous enough, in fact practically it is a good enough argument, when one wishes to be very rigorous, the ellipsis... in the sum for S is not considered precise. You are expected to interpret it in a certain specific way. But it can be interpreted in a number of different ways. In fact it can mean anything. Thus if one wants to be rigorous, and absolutely sure about the correctness of the formula, one needs some other way of verifying it than using the ellipsis. Since one needs to verify it for infinitely many cases (infinitely many values of A, R and n), some kind of formal approach, abstracted away from actual numbers, is required.

Suppose now that somehow we have formally verified the formula successfully and we are absolutely sure that it is correct. It is a good idea to write a computer program to compute that S, especially with $(1 + R)^{n+1}$ to be computed. Suppose again that we have written a program to compute S. How can we know that the program is correct ? As we know, there are infinitely many possible input values (that is, values of A, R and n). Obviously we can not test it for infinitely many cases. Thus we must take some formal approach.

Related to the problem of correctness of computer programs, there is the well known "Halting Problem". This problem, if put into the context of program correctness, asks whether or not a given computer program stops on a given input after a finite amount of time. This problem is known to be unsolvable by computers. That is, no one can write a computer program to answer that question. It is known to be unsolvable. But, how can we tell it is unsolvable ?. How can we tell that such a program can not be written? You can not try all possible solution methods and see they all fail. You can not think of all (candidate) methods to solve the Halting Problem. Thus you need some kind of formal approaches here to avoid dealing with a extremely large number (if not infinite) of possibilities.

Discrete mathematics is the foundation for the formal approaches. It discusses languages used in mathematical reasoning, basic concepts, and their properties and relationships among them. Though there is no time to cover them in this course, discrete mathematics is also concerned with techniques to solve certain types of problems such as how to count or enumerate quantities.

The kind of counting problems includes: How many routes exist from point A to point B in a computer network ? How much execution time is required to sort a list of integers in increasing order ? What is the probability of winning a lottery? What is the shortest path from point A to point B in a computer network ?

The subjects covered in this course include propositional logic, predicate logic, sets, relations, and functions, in particular growth of function.

The first subject is logic. It is a language that captures the essence of our reasoning, and correct reasoning must follow the rules of this language. We start with logic of sentences called propositional logic, and study elements of logic, (logical) relationships between propositions, and reasoning. Then we learn a little more powerful logic called predicate logic. It allows us to reason with statements involving variables among others. Earlier we also study sets, relations between sets, and operations on sets. Just about everything is described based on sets, when rigor is required. It is the basis of every theory in computer science and mathematics. We learn mathematical reasoning, in particular recursive definitions and mathematical induction. There are sets, operations and functions that can be defined precisely by recursive definitions.

Properties of those recursively defined objects can be established rigorously using proof by induction. They are an abstraction of relations we are familiar with in everyday life such as husband–wife relation, parent–child relation and ownership relation. They are also one of the key concepts in the discussion of many subjects on computer and computation.

For example, a database is viewed as a set of relations and database query languages are constructed based on operations on relations and sets. Graphs are also covered briefly here. They are an example of discrete structures and they are one of the most useful models for computer scientists and engineers in solving problems. Functions are a special type of relation and basically the same kind of concept as the ones we see in calculus.

Before we start the study of discrete structures, we briefly learn general framework of problem solving. If you are a good problem solver, you may skip that and go to logic.

PROBLEM SOLVING

Everyone must have felt at least once in his or her life how wonderful it would be if we could solve a problem at hand preferably without much difficulty or even with some difficulties. Unfortunately the problem solving is an art at this point and there are no universal approaches one can take to solving problems. Basically one must explore possible avenues to a solution one by one until one comes across a right path to a solution. Thus generally speaking, there is guessing and hence an element of luck involved in problem solving. However, in general, as one gains experience in solving problems, one develops one's own techniques and strategies, though they are often intangible. Thus the guessing is not an arbitrary guessing but an educated one.

A FRAMEWORK FOR PROBLEM SOLVING

The following four phases can be identified in the process of solving problems:

1. Understanding the problem
2. Making a plan of solution
3. Carrying out the plan
4. Looking back i.e. verifying

Each of the first two phases is going to be explained below a little more in detail. Phases (3) and (4) should be self–explanatory.

Understanding the Problem

Needless to say that if you do not understand the problem you can never solve it. It is also often true that if you really understand the problem, you can see a solution.

Below are some of the things that can help us understand a problem.

1. Extract the principal parts of the problem.

The principal parts are:

For "find" type problems, such as "find the principal and return for a given investment", UNKNOWNS, DATA and CONDITIONS, and for "proof" type problems HYPOTHESIS and CONCLUSION.

1. Be careful about hidden assumptions, data and conditions.
2. Consult definitions for unfamiliar (often even familiar) terminologies.
3. Construct one or two simple example to illustrate what the problem says.

Devising a Solution Plan

Start with the consideration of the principal parts: unknowns, data, and conditions for "find" problems, and hypothesis, and conclusion for "prove" problems. Once you identify the principal parts and understand them, the next thing you can do is to consider the problem from various angles and seek contacts with previously acquired knowledge. The first thing you should do is to try to find facts that are related to the problem at hand. Relevant facts usually involve words that are the same as or similar to those in the given problem. It is also a good idea to try to recall previously solved similar problems.

Look for a helpful idea that shows you the way to the end. Even an incomplete idea should be considered. Go along with it to a new situation, and repeat this process. There is no single method that works for all problems. However, there is a wealth of heuristics we can try. Following are some of the often used heuristics. Add your own heuristics as you gain experience.

1. Have I seen it before ?

That is, do I know similar or related problems ? Similar/related problems are ones with the same or a similar unknown or unknown may be different but the settings are the same or similar.

1. Do a little analysis on relationships among data, conditions and unknowns, or between hypothesis and conclusion.

3. What facts do I know related to the problem on hand ?

These are facts on the subjects appearing in the problem. They often involve the same or similar words. It is very important that we know inference rules.

Definitions: Make sure that you know the meaning of technical terms. This is obviously crucial to problem solving at any level. But especially at this level, if you know their meaning and understand the concepts, you can see a solution to most of the problems without much difficulty.

1. Compose a wish list of intermediate goals and try to reach them.

2. Have you used all the conditions/hypotheses ?: When you are looking for paths to a solution or trying to verify your solution, it is often a good idea to check whether or not you have used all the data/hypotheses. If you haven't, something is often amiss.
3. *Divide into cases*: Sometimes if you divide your problem into a number of separate cases based on a property of objects appearing in the problem, it simplifies the problem and clear your mind. For example if the problem concerns integers, then you may want to divide it into two cases: one for even numbers and the other for odd numbers.
4. *Proof by contradiction*: If you make an assumption, and that assumption produces a statement that does not make sense, then you must conclude that your assumption is wrong. For example, suppose that your car does not start. A number of things could be wrong. Let us assume for simplicity's sake that either the battery is dead or the starter is defective. So you first assume that the battery is dead and try to jump start it. If it doesn't start, you have a situation that does not make sense under your assumption of dead battery. That is, a good battery should start the car but it doesn't. So you conclude that your assumption is wrong. That is the battery is not the cause. Proof by contradiction follows that logic.

In this method we first assume that the assertion to be proven is not true and try to draw a contradiction i.e. something that is always false. If we produce a contradiction, then our assumption is wrong and therefore the assertion we are trying to prove is true. When you are stuck trying to justify some assertion, proof by contradiction is always a good thing to try.

1. Transform/Restate
2. *Working backward*: In this approach, we start from what is required, such as conclusion or final (desired) form of an equation etc., and assume what is sought has been found. Then we ask from what antecedent the desired result could be derived. If the antecedent is found, then we ask from what antecedent that antecedent could be obtained.... We repeat this process until either the data/ hypotheses are reached or some easy to solve problem is reached.
3. *Simplify the problem if possible*: Take advantage of symmetries which often exist.

Keep in mind that your first try may not work. But don't get discouraged. If one approach doesn't work, try another. You have to keep trying different approaches, different ideas. As you gain experience, your problem solving skills improve and you tend to find the right approach sooner.

Let us now look at some examples to illustrate the topics discussed above.

Example: This is an example in which you can find a solution once you analyze and understand the unknowns and data.

Problem: A survey of TV viewers shows the following results:

To the question "Do you watch comedies ?", 352 answered "Yes".,

To the question "Do you watch sports ?", 277 answered "Yes", and

To the question "Do you watch both comedies and sports ?", 129 answered "Yes".

Given these data, find, among people who watch at least one of comedies and sports, percentages of people who watch at least one of comedies and sports watch only comedies, only sports, and both comedies and sports.

Let us try to solve this problem following the framework presented above.

Understanding the Problem: This is a "find" type problem. So we try to identify unknowns, data and conditions.

The unknowns are the percentage of people who watch only comedies, the percentage of people who watch only sports, and the percentage of people who watch both comedies and sports. The data are the three numbers: 352, 277 and 129, representing the number of people who watch comedies, sports, and both comedies and sports, respectively. Note that 352 includes people who watch both comedies and sports as well as people who watch only comedies. Similarly for 277.

The conditions are not explicitly given in the problem statement. But one can see that the percentages must add up to 100, and they must be nonnegative.

Devising a Solution Plan: Here we first examine the principal parts in detail.

First let us consider the unknowns in more detail. To calculate the percentage of the people who watch only comedies, for example, we need the number of people who watch at least one of comedies and sports, and the number of people who watch only comedies. Thus actually two unknowns are involved in each of the required percentages, and the real unknowns are the number of people in each of the categories, and the number of people who watch at least one of comedies and sports.

Next let us look at the data. First the number 352 is the number of people who watch comedies. But that is not necessarily that of the people who watch only comedies. It includes that and the number of people who watch both comedies and sports. Similarly for the second number 277. Let us use symbols to represent each of the unknowns: Let C represent the number of people who watch only comedies, S that of the people who watch only sports, and T that of the people who watch at least one of those programs.

Then we have the following relationships among the unknowns:

$$C + 129 = 352$$

$$S + 129 = 277$$

$$C + S + 129 = T$$

From these equations we can easily obtain $C = 223$, $S = 148$, and $T = 500$.

Thus the required percentages are 44.6%, 29.6%, and 25.8%, respectively.

All we had to do to solve this problem is to analyze relationships between the data and the unknowns, that is, nothing much beyond "understanding the problem".

Example: This is a problem which you can solve using similar known results.

Problem: Find the (length of) diagonal of a rectangular parallelepiped given its length, width and height.

Again let us try to solve this problem following the framework presented above.

Understanding the Problem: This is a "find" type problem. So we try to identify unknowns, data and conditions.

The unknown is the diagonal of a rectangular parallelepiped, and the data are its length, width and height. Again there are no explicitly stated conditions. But the unknown and data must all be a positive number.

Before proceeding to the next phase, let us make sure that we understand the terminologies.

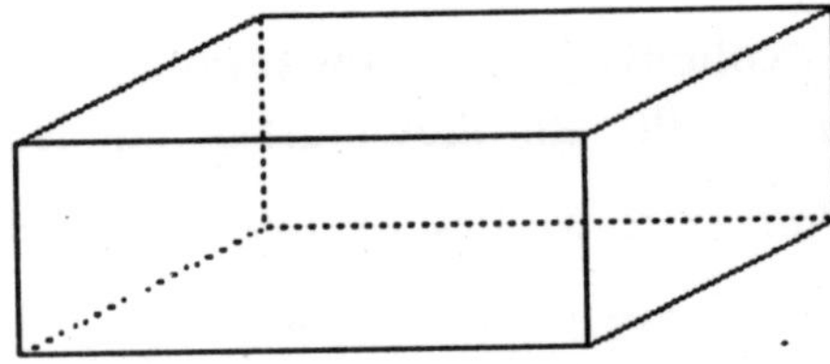

Rectangular Parallelpiped

First a rectangular parallelepiped is a box with rectangular faces like a cube except that the faces are not necessarily a square but a rectangle as shown below.

Next a diagonal of a rectangular parallelepiped is the line that connects its two vertices (corner points) that are not on the same plane. It is shown in the figure below.

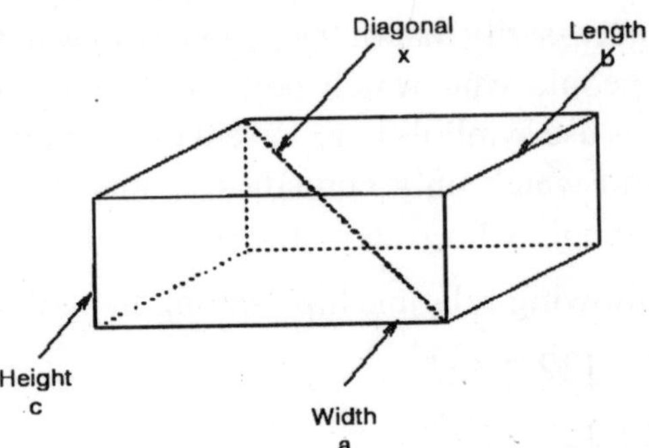

Devising a Solution Plan: Here we first try to find relevant facts. Relevant facts often involve the same or similar words or concepts. Since the unknown is a diagonal, we look for facts concerning diagonal. Note that drawing figures here is quite helpful.

One of the facts that immediately comes to our mind in this problem is Pythagoras' theorem. It has to do with right triangles and is shown below.

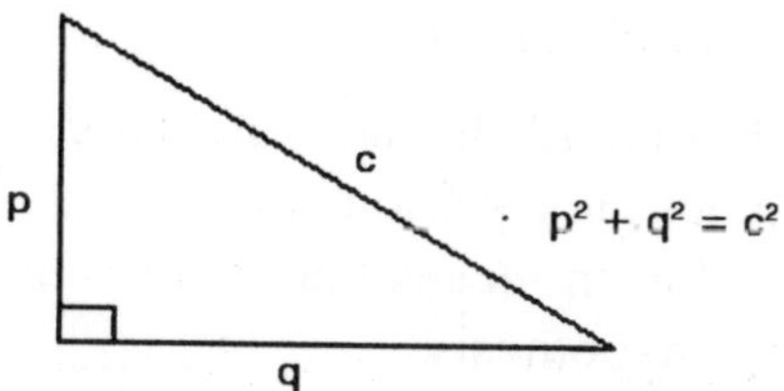

Let us try to see whether or not this theorem helps. To use this theorem, we need a right triangle involving a diagonal of a parallelepiped. As we can see below, there is a right triangle with a diagonal x as its hypotenuse.

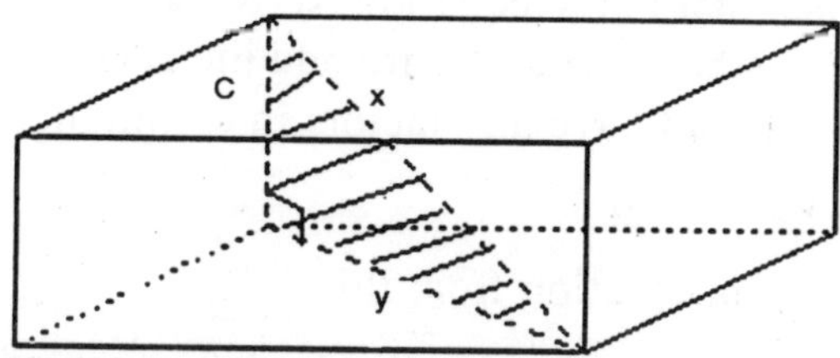

However, the triangle here involves two unknowns: x and y. Since x is what we are looking for, we need to find the value of y. To find y, we note another right triangle shown below.

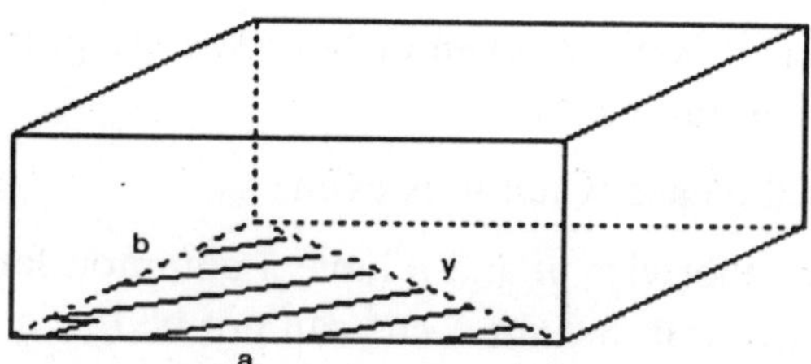

Applying Pythagoras' theorem again, we can obtain the value of y.

Thus $\quad y^2 = a^2 + b^2$

is obtained from the second triangle, and

$$x^2 = c^2 + y^2$$

is derived from the first triangle.

From these two equations, we can find that x is equal to the positive square root of $a^2 + b^2 + c^2$.

Example: This is a proof type problem and "proof by contradiction" is used.

Problem: Given that a, b, and c are odd integers, prove that equation $ax^2 + bx + c = 0$ can not have a rational root.

Understanding the Problem: This is a "prove" type problem.

The hypothesis is that a, b, and c are odd integers, and the conclusion is that equation $ax^2 + bx + c = 0$ can not have a rational root.

The hypothesis is straightforward. In the conclusion, "rational root" means a root, that is, the value of x that satisfies the equation, and that can be expressed as m/n, where m and n are integers. So the conclusion means that there is no number of the form m/n that satisfies the equation under the hypothesis.

Devising a Solution Plan: For this problem, let us try "proof by contradiction". When you are asked to prove the impossibility of an event or non–existence of certain things, this approach often is quite helpful. Following the "proof by contradiction", let us assume that the conclusion is false, that is the equation $ax^2 + bx + c = 0$ has a rational root m/n, where m and n are integers, when a, b, and c are odd integers. We can assume without loss of generality that m and n do not have any factors in common. Then

$$a(m/n)^2 + b(m/n) + c = 0.$$

Let us try to derive a contradiction from this.

First let us make this equation simpler, that is, let us get rid of fractions.

Since n is not equal to 0, multiplying the both sides of (1) by n^2, we get

$$am^2 + bmn + cn^2 = 0.$$

Since m is an integer, it is either even or odd. We are going to consider those cases one by one. That is "divide into cases".

Let us first consider the case when m is even.

Then n is odd, since otherwise m and n have a common factor 2. Now $am^2 + bmn$ is even, and cn^2 is odd. Hence $am^2 + bmn + cn^2$ can not be 0.

Next let us consider the case when m is odd.

By an argument similar to the previous case, we can see that n is also odd.

If m and n are odd, then am^2, bmn, and cn^2 are all odd, since a, b, and c are odd integers. However, the sum of three odd numbers can not be equal to 0.

Thus by assuming that the conclusion is false, we have arrived at a contradiction, that is m/n does not satisfy the equation. Hence our assumption must be wrong, and therefore the conclusion is correct.

Example: This is another proof type problem and "working backward" technique is used.

Problem: Prove that $(a + b + c)^2 \leq 4(ab + bc + ca)$, if a, b, c are three sides of a triangle.

Understanding the Problem: This is a "prove" type problem.

The hypothesis is that a, b, and c are three sides of a triangle, and the conclusion is that inequality $(a + b + c)^2 \leq 4(ab + bc + ca)$ holds.

Devising a Solution Plan: Here we try "Working Backward" heuristic. That is manipulate the conclusion possibly using the hypothesis and reduce it into something that is obviously true.

First by multiplying out the left hand side of the inequality, $(a + b + c)^2 = a^2 + b^2 + c^2 \leq 2(ab + bc + ca)$.

Hence if $a^2 + b^2 + c^2$ $2(ab + bc + ca)$, then the conclusion holds.

Next, to see what we can try, note that we have not used the hypothesis yet; and see if it can help here.

It is well known that the sum of two sides of a triangle is greater than the third side.

Hence $a + b > c$, $b + c > a$, and $c + a > b$ hold.

From these we can obtain $c(a + b) > c^2$, $a(b + c) > a^2$, and $b(c + a) > b^2$.

By adding these three inequalities, we get

$a^2 + b^2 + c^2 < a(b + c) + b(c + a) + c(a + b) = 2(ab + bc + ca)$.

Hence $a^2 + b^2 + c^2 < 2(ab + bc + ca)$.

Hence $a^2 + b^2 + c^2$ $2(ab + bc + ca)$.

Hence $(a + b + c)^2$ $4(ab + bc + ca)$ holds.

Example: This is a find type problem and "working backward" technique is used.

Problem: Given a 4 quart pail and a 9 quart pail, obtain 6 quarts of water in the 9 quart pail using these two pails. You can fill or empty the pails and you can have as much water as you want.

Understanding the Problem: This is a "find" type problem.

The problem is to obtain 6 quarts of water in the 9 quart pail using 4 quart and 9 quart pails as measures. You can fill either pail from the water source or from the other pail, and you can empty the pails any time.

Devising a Solution Plan: You can solve this in a number of different ways. Here we try "Working Backward" heuristic. It starts with the desired solution and work backward step by step. At each step we try to find a state that immediately precedes the current state such that we can reach the current state from that state with one simple operation such as filling a pail or emptying a pail in this problem. We repeat this process until we reach some easily reachable state such as empty pails, full pails, one pail full and the other empty, etc.

Our solution to the original problem is obtained by traversing this process backward to the desired state.

Let us denote the 9 quart pail by A and the 4 quart pail by B for simplicity.

In this problem, the desired state is to have 6 qts in A.

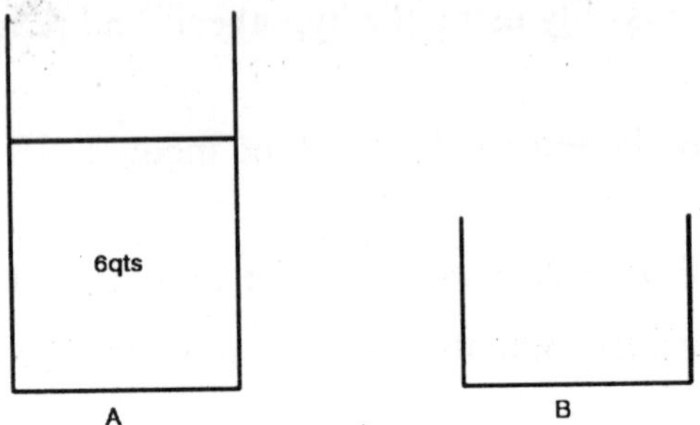

Thus in the first step of "working backward", we ask how we could get to the desired state with one operation. As one can easily see if we could dump 3 qts from 9 qts in A, then we would have 6 qts in A. To be able to dump 3 qts from A we need 1 qt in B. Thus the state immediately preceding the current state is the one where A is full ,and B has 1 qt in it.

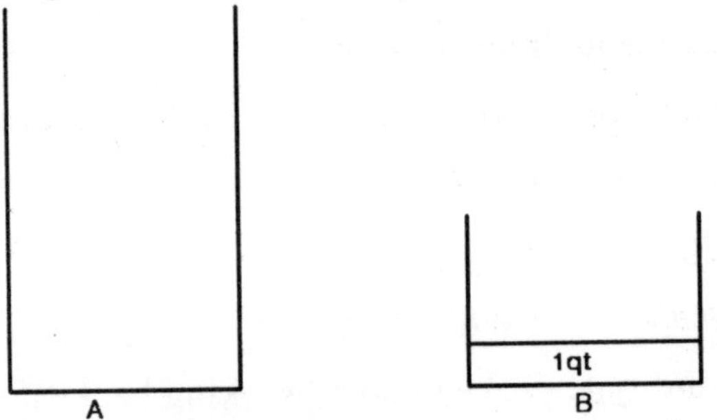

In the second step, the question we ask is how to get 1 qt in B.

It does not look easy to get 1 qt in B. So let us see whether or not we can get 1 qt in A. If we have 1 qt in A, then we can certainly get 1 qt in B without any trouble. thus we might say that the third state is to have 1 qt in A.

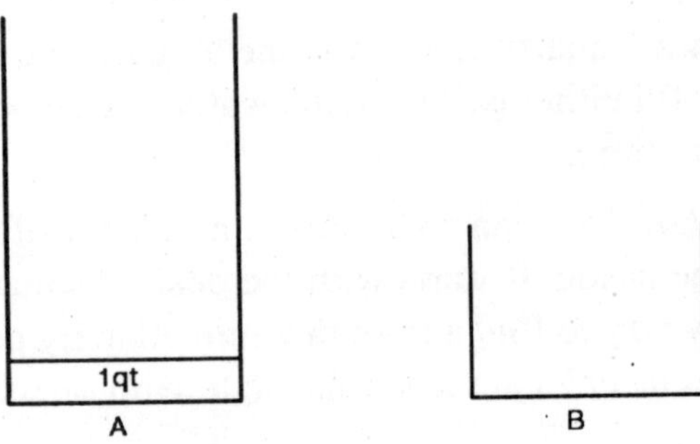

In the third step, the question we ask is how to get 1 qt in A.

This is relatively easy to accomplish because all you have to do is to get rid of 8 qts from a full A, which can be done by emptying A twice into B.

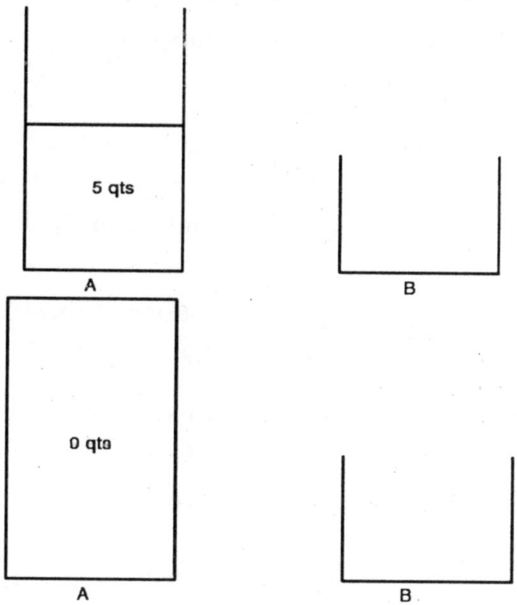

Since this state can be easily reached (all you have to do to get to this state is to fill A with water), we stop here. Our solution to the original problem is now obtained by going this process backward.

Thus first we fill up A. Then dump A into B leaving 5 qts in A. Then dump A into B again. This gives us 1 qt in A. Pour that into B. Then fill A and empty it into B. We now have 6 qts in A, which is what is required.

EXTRAPOLATION

One of the most important thing, if not the most important, in the university education is the cultivation of the ability to extrapolate. Obviously we can not anticipate all the problems you are going to encounter in the future and prepare you for them. So what we try to do is to teach you the essentials and hope that you can deal with deviations or new problems successfully. Here several examples of extrapolation are going to be given to illustrate what I mean by "extrapolation" and how to deal with new problems with the knowledge and experiences you presently have.

Example:

Problem: A survey of TV viewers shows the following results:

To the question "Do you watch comedies ?", 374 replied "Yes".

To the question "Do you watch sports ?", 360 replied "Yes".

To the question "Do you watch detective stories ?", 350 replied "Yes".

To the question "Do you watch comedies and sports ?", 134 replied "Yes".

To the question "Do you watch comedies and detectives ?", 96 replied "Yes".

To the question "Do you watch detectives and sports ?", 241 replied "Yes".

To the question "Do you watch all three ?", 37 replied "Yes".

Find the percentages of people who watch only comedies, only sports, only detective stories, comedies and sports, comedies and detectives, sports and detectives, and all three.

By doing the same kind of analyses on the unknowns and data, one can find these percentages.

The total number of people who watch at least one of these programs is

$$374 + 360 + 350 - 134 - 96 - 241 + 37 = 650,$$

because each of 134, 96 and 241 is counted twice into 374 + 360 + 350, and 37 is counted in three times and subtracted out three times in 374 + 360 + 350 – 134 – 96 – 241.

Similarly the number of those watching only comedies is

$$374 - 134 - 96 + 37 = 181,$$

the number of those watching only detectives is

$$350 - 241 - 96 + 37 = 50,$$

the number of those watching only sports is

$$360 - 134 - 241 + 37 = 22,$$

the number of those watching only comedies and detectives is

$$96 - 37 = 59,$$

the number of those watching only comedies and sports is

$$350 - 37 = 313,$$

the number of those watching only detectives and sports is

$$241 - 37 = 204.$$

The calculation of the percentages are omitted.

Example: 300 people were surveyed on TV programs they watch and all 300 responded. 75 of them say they watch a regular sports program, 210 say they watch a regular comedy program, and 36 say they watch both.

There is a special detective story program which conflicts with the regular sports and comedy programs. If those who watch regular programs do not want to miss them, how many people can watch the special ?

To answer the question, all you have to do is to find the total number of people who watch at least one of the regular sports and comedy programs. Then 300 minus that number is the answer. Thus this problem is essentially the same as Example,

and the answer is 300 – (75 + 210 – 36) = 51.

Example: 300 people were surveyed. 50 of them watch sports, 140 watch comedies, and 134 do not watch either of them. Then how many of them watch both comedies and sports ?

Since 300 were surveyed and 134 do not watch either sports or comedies, 166 watch at least one. Hence 50 + 140 – 166 = 24 watch both. Thus 50 – 24 = 26 watch only sports, and 140 – 24 = 116 watch only comedies.

Chapter 2

Logical Reasoning

Logic is a language for reasoning. It is a collection of rules we use when doing logical reasoning. Human reasoning has been observed over centuries from at least the times of Greeks, and patterns appearing in reasoning have been extracted, abstracted, and streamlined. The foundation of the logic we are going to learn here was laid down by a British mathematician George Boole in the middle of the 19th century, and it was further developed and used in an attempt to derive all of mathematics by Gottlob Frege, a German mathematician, towards the end of the 19th century.

A British philosopher/mathematician, Bertrand Russell, found a flaw in basic assumptions in Frege's attempt but he, together with Alfred Whitehead, developed Frege's work further and repaired the damage. The logic we study today is more or less along this line.

In logic we are interested in true or false of statements, and how the truth/falsehood of a statement can be determined from other statements. However, instead of dealing with individual specific statements, we are going to use symbols to represent arbitrary statements so that the results can be used in many similar but different cases. The formalization also promotes the clarity of thought and eliminates mistakes. There are various types of logic such as logic of sentences (propositional logic), logic of objects (predicate logic), logic involving uncertainties, logic dealing with fuzziness, temporal logic etc. Here we are going to be concerned with propositional logic and predicate logic, which are fundamental to all types of logic.

INTRODUCTION TO PROPOSITIONAL LOGIC

Propositional logic is a logic at the sentential level. The smallest unit we deal with in propositional logic is a sentence. We do not go inside individual sentences and analyze or discuss their meanings. We are going to be interested only in true or false of sentences, and major concern is whether or not the truth or falsehood of a certain sentence follows from those of a set of sentences, and if so, how. Thus sentences considered in this logic are not arbitrary sentences but are the ones that are true or false. This kind of sentences are called propositions.

PROPOSITION

Sentences considered in propositional logic are not arbitrary sentences but are the ones that are either true or false, but not both. This kind of sentences are called propositions.

If a proposition is true, then we say it has a truth value of "true"; if a proposition is false, its truth value is "false".

For example, "Grass is green", and "2 + 5 = 5" are propositions.

The first proposition has the truth value of "true" and the second "false".

But "Close the door", and "Is it hot outside ?"are not propositions.

Also "x is greater than 2", where x is a variable representing a number, is not a proposition, because unless a specific value is given to x we can not say whether it is true or false, nor do we know what x represents.

Similarly "x = x" is not a proposition because we don't know what "x" represents hence what "=" means. For example, while we understand what "3 = 3" means, what does "Air is equal to air" or "Water is equal to water" mean ? Does it mean a mass of air is equal to another mass or the concept of air is equal to the concept of air ? We don't quite know what "x = x" mean. Thus we can not say whether it is true or not. Hence it is not a proposition.

ELEMENTS OF PROPOSITIONAL LOGIC

Simple sentences which are true or false are basic propositions. Larger and more complex sentences are constructed from basic propositions by combining them with connectives. Thus propositions and connectives are the basic elements of propositional logic. Though there are many connectives, we are going to use the following five basic connectives here:

NOT, AND, OR, IF_THEN (or IMPLY), IF_AND_ONLY_IF.

They are also denoted by the symbols:

$$\neg, \wedge, \vee, \rightarrow, \leftrightarrow,$$

respectively.

Often we want to discuss properties/relations common to all propositions. In such a case rather than stating them for each individual proposition we use variables representing an arbitrary proposition and state properties/relations in terms of those variables. Those variables are called a propositional variable. Propositional variables are also considered a proposition and called a proposition since they represent a proposition hence they behave the same way as propositions.

A proposition in general contains a number of variables. For example (P ∨ Q) contains variables P and Q each of which represents an arbitrary proposition. Thus a proposition takes different values depending on the values of the constituent variables. This relationship of the value of a proposition and those of its constituent variables can be represented by a table. It tabulates the value of a proposition for all possible values of its variables and it is called a truth table. For example the following table shows the relationship between the values of P, Q and P ∨ Q:

OR		
P	**Q**	**(P ∨ Q)**
F	F	F
F	T	T
T	F	T
T	T	T

In the table, F represents truth value false and T true.

This table shows that P ∨ Q is false if P and Q are both false, and it is true in all the other cases.

MEANING OF THE CONNECTIVES

Let us define the meaning of the five connectives by showing the relationship between the truth value (i.e. true or false) of composite propositions and those of their component propositions. They are going to be shown using truth table. In the tables P and Q represent arbitrary propositions, and true and false are represented by T and F, respectively.

NOT	
P	P
T	F
F	T

This table shows that if P is true, then (P) is false, and that if P is false, then (P) is true.

AND		
P	Q	(P Ω Q)
F	F	F
F	T	F
T	F	F
T	T	T

This table shows that (P ∧ Q) is true if both P and Q are true, and that it is false in any other case.

Similarly for the rest of the tables.

OR		
P	Q	(P Q)
F	F	F
F	T	T
T	F	T
T	T	T

IMPLIES		
P	Q	(P Q)
F	F	T
F	T	T
T	F	F
T	T	T

When P Q is always true, we express that by P Q. That is P Q is used when proposition P always implies proposition Q regardless of the value of the variables in them. Also see a note on the truth value of IMPLIES when P is False.

IF AND ONLY IF		
P	Q	(P ↔ Q)
F	F	T
F	T	F
T	F	F
T	T	T

When P Q is always true, we express that by P ⇔ Q. That is ⇔ is used when two propositions always take the same value regardless of the value of the variables in them.

CONSTRUCTION OF COMPLEX PROPOSITIONS

SYNTAX OF PROPOSITIONS

First it is informally shown how complex propositions are constructed from simple ones. Then more general way of constructing propositions is given.

In everyday life we often combine propositions to form more complex propositions without paying much attention to them. For example combining "Grass is green", and "The sun is red" we say something like "Grass is green and the sun is red", "If the sun is red, grass is green", "The sun is red and the grass is not green" etc. Here "Grass is green", and "The sun is red" are propositions, and form them using connectives "and", "if... then..." and "not" a little more complex propositions are formed. These new propositions can in turn be combined with other propositions to construct more complex propositions. They then can be combined to form even more complex propositions. This process of obtaining more and more complex propositions can be described more generally as follows:

Let X and Y represent arbitrary propositions. Then

$[\neg X]$, $[X \wedge Y]$, $[X \vee Y]$, $[X \rightarrow Y]$, and $[X \leftrightarrow Y]$
are *propositions.*

Note that X and Y here represent an arbitrary proposition. This is actually a part of more rigorous definition of proposition which we see later.

Example: [P → [Q V R]] is a proposition and it is obtained by first constructing [Q V R] by applying [X V Y] to propositions Q and R considering them as X and Y, respectively, then by applying [X → Y] to the two propositions P and [Q V R] considering them as X and Y, respectively.

Note: Rigorously speaking X and Y above are place holders for propositions, and so they are not exactly a proposition. They are called a *propositional variable*, and propositions formed from them using connectives are called a *propositional form*. However, we are not going to distinguish them here, and both specific propositions such as "2 is greater than 1" and propositional forms such as $(P \vee Q)$ are going to be called a proposition.

CONVERSE AND CONTRAPOSITIVE

For the proposition $P \rightarrow Q$, the proposition $Q \rightarrow P$ is called its converse, and the proposition $\neg Q \rightarrow \neg P$ is called its contrapositive.

For example for the proposition "If it rains, then I get wet",

***Converse*:** If I get wet, then it rains.

Contrapositive: If I don't get wet, then it does not rain.

The converse of a proposition is not necessarily logically equivalent to it, that is they may or may not take the same truth value at the same time.

On the other hand, *the contrapositive of a proposition is always logically equivalent to the proposition.* That is, they take the same truth value regardless of the values of their constituent variables. Therefore, "If it rains, then I get wet." and "If I don't get wet, then it does not rain." are logically equivalent. If one is true then the other is also true, and vice versa.

IF THEN VARIATIONS

If–then statements appear in various forms in practice. The following list presents some of the variations. These are all *logically* equivalent, that is as far as true or false of statement is concerned there is no difference between them. Thus if one is true then all the others are also true, and if one is false all the others are false.

1. If p, then q.
2. p implies q.
3. If p, q.
4. p only if q.
5. p is sufficient for q.
6. q if p.
7. q whenever p.
8. q is necessary for p.
9. It is necessary for p that q.

For instance, instead of saying "If she smiles then she is happy", we can say "If she smiles, she is happy", "She is happy whenever she smiles", "She smiles only if she is happy" etc. without changing their truth values.

"Only if" can be translated as "then". For example, "She smiles only if she is happy" is equivalent to "If she smiles, then she is happy".

Note that "She smiles only if she is happy" means "If she is not happy, she does not smile", which is the contrapositive of "If she smiles, she is happy".

You can also look at it this way: "She smiles only if she is happy" means "She smiles only when she is happy". So any time you see her smile you know she is happy. Hence "If she smiles, then she is happy". Thus they are logically equivalent.

Also "If she smiles, she is happy" is equivalent to "It is necessary for her to smile that she is happy". For "If she smiles, she is happy" means "If she smiles, she is *always* happy". That is, she never fails to be happy when she smiles. "Being happy" is inevitable consequence/necessity of "smile". Thus if "being happy" is missing, then "smile" can not be there either. "Being happy" is necessary "for her to smile" or equivalently "It is necessary for her to smile that she is happy".

FROM ENGLISH TO PROPOSITION

As we are going to see in the next section, reasoning is done on propositions using inference rules. For example, if the two propositions "if it snows, then the school is closed", and "it snows" are true, then we can conclude that "the school is closed" is true. In everyday life, that is how we reason.

To check the correctness of reasoning, we must check whether or not rules of inference have been followed to draw the conclusion from the premises. However, for reasoning in English or in general for reasoning in a natural language, that is not necessarily straightforward and it often encounters some difficulties. Firstly, connectives are not necessarily easily identified as we can get a flavor of that from the previous topic on variations of if then statements. Secondly, if the argument becomes complicated involving many statements in a number of different forms twisted and tangled up, it can easily get out of hand unless it is simplified in some way.

One solution for that is to use symbols (and mechanize it). Each sentence is represented by symbols representing building block sentences, and connectives. For example, if P represents "it snows" and Q represents "the school is closed", then the previous argument can be expressed as

$[[P \rightarrow Q] \wedge P] \rightarrow Q$,

or

$P \rightarrow Q$

P

....

Q

This representation is concise, much simpler and much easier to deal with. In addition today there are a number of automatic reasoning systems and we can verify our arguments in symbolic form using them. One such system called TPS is used for reasoning exercises in this course. For example, we can check the correctness of our argument using it.

To convert English statements into a symbolic form, we restate the given statements using the building block sentences, those for which symbols are given, and the connectives of propositional logic (not, and, or, if then, if and only if), and then substitute the symbols for the building blocks and the connectives. For example, let P be the proposition "It is snowing", Q be the proposition "I will go the beach", and R be the proposition "I have time".

Then first "I will go to the beach if it is not snowing" is restated as "If it is not snowing, I will go to the beach". Then symbols P and Q are substituted for the respective sentences to obtain $\sim P \rightarrow Q$. Similarly, "It is not snowing and I have time only if I will go to the beach" is restated as "If it is not snowing and I have time, then I will go to the beach", and it is translated as $(\sim P \wedge R) \rightarrow Q$.

REASONING WITH PROPOSITIONS

INTRODUCTION TO REASONING

Logical reasoning is the process of drawing conclusions from premises using rules of inference. Here we are going to study reasoning with propositions. Later we are going to see reasoning with predicate logic, which allows us to reason about individual objects. However, *inference rules of propositional logic are also applicable to predicate logic* and reasoning with propositions is fundamental to reasoning with predicate logic.

These inference rules are results of observations of human reasoning over centuries. Though there is nothing absolute about them, they have contributed significantly in the scientific and engineering progress the mankind have made. Today they are universally accepted as the rules of logical reasoning and they should be followed in our reasoning.

Since inference rules are based on identities and implications, we are going to study them first. We start with three types of proposition which are used to define the meaning of "identity" and "implication".

TYPES OF PROPOSITION

Some propositions are always true regardless of the truth value of its component propositions.

For example $(P \vee \neg P)$ is always true regardless of the value of the proposition P.

A proposition that is always true called a tautology.

There are also propositions that are always false such as $(P \wedge \neg P)$.

Such a proposition is called a contradiction.

A proposition that is neither a tautology nor a contradiction is called a contingency.

For example $(P \vee Q)$ is a contingency.

These types of propositions play a crucial role in reasoning. In particular every inference rule is a tautology as we see in identities and implications.

IDENTITIES

From the definitions(meaning) of connectives, a number of relations between propositions which are useful in reasoning can be derived. Below some of the often encountered pairs of logically equivalent propositions, also called *identities*, are listed. These identities are used in logical reasoning. In fact we use them in our daily life, often more than one at a time, without realizing it.

If two propositions are logically equivalent, one can be substituted for the other in any proposition in which they occur without changing the logical value of the proposition. Below $\Leftrightarrow$ corresponds to $\rightarrow$ and it means that the equivalence is always true (a tautology), while $\rightarrow$ means the equivalence may be false in some cases, that is in general a contingency.

That these equivalences hold can be verified by constructing truth tables for them. First the identities are listed, then examples are given to illustrate them.

List of Identities:

1. $P \Leftrightarrow (P \vee P)$......... idempotence of $\vee$
2. $P \Leftrightarrow (P \wedge P)$......... idempotence of $\wedge$
3. $(P \vee Q) \Leftrightarrow (Q \wedge P)$......... commutativity of $\vee$
4. $(P \wedge Q) \Leftrightarrow (Q \wedge P)$......... commutativity of $\vee$
5. $[(P \vee Q) \vee R] \Leftrightarrow [P (Q R)]$......... associativity of $\vee$
6. $[(P \wedge Q) \wedge R] \Leftrightarrow [P \wedge (Q \wedge R)]$......... associativity of $\wedge$
7. $\neg(P \vee Q) \Leftrightarrow (\neg P \wedge \neg Q)$......... DeMorgan's Law
8. $\neg(P \wedge Q) \Leftrightarrow (\neg P \vee \neg Q)$......... DeMorgan's Law
9. $[P \vee (Q \vee R] \Leftrightarrow [(P \wedge Q) \vee (P \wedge R)]$......... distributivity of $\wedge$ over $\vee$
10. $[P \vee (Q \wedge R] \Leftrightarrow [(P \vee Q) \wedge (P \vee R)]$......... distributivity of $\vee$ over $\wedge$
11. $(P \vee \text{True}) \Leftrightarrow \sigma\text{True}$
12. $(P \wedge \text{False}) \Leftrightarrow \text{False}$
13. $(P \vee \text{False}) \Leftrightarrow P$
14. $(P \wedge \text{True}) \Leftrightarrow P$
15. $(P \vee \neg P) \Leftrightarrow \text{True}$
16. $(P \wedge \neg P) \Leftrightarrow \text{False}$
17. $P \Leftrightarrow \neg(\neg P)$......... double negation
18. $(P \rightarrow Q) \Leftrightarrow (\neg P \vee Q)$......... implication
19. $(P \rightarrow Q) \Leftrightarrow [(P \rightarrow Q) \wedge (Q \rightarrow P)]$......... equivalence
20. $[(P \wedge Q) \rightarrow R] \Leftrightarrow [P \rightarrow (Q \rightarrow R)]$.......... exportation
21. $[(P \rightarrow Q) \wedge (P \rightarrow \neg Q)] \Leftrightarrow\rightarrow P$......... absurdity
22. $(P \rightarrow Q) \Leftrightarrow (\neg Q \rightarrow \neg P)$......... contrapositive

Let us see some example statements in English that illustrate these identities.

Examples:

1. $P \Leftrightarrow (P \vee P)$.......... idempotence of $\vee$

   What this says is, for example, that "Tom is happy." is equivalent to "Tom is happy or Tom is happy". This and the next identity are rarely used, if ever, in everyday life. However, these are useful when manipulating propositions in reasoning in symbolic form.

2. $P \Leftrightarrow (P \wedge P)$.......... idempotence of $\wedge$

Similar to 1. above.

3. $(P \vee Q) \Leftrightarrow (Q \vee P)$.......... commutativity of $\vee$

What this says is, for example, that "Tom is rich or (Tom is) famous." is equivalent to "Tom is famous or (Tom is) rich".

4. $(P \wedge Q) \Leftrightarrow (Q \vee P)$.......... commutativity of $\vee$

What this says is, for example, that "Tom is rich and (Tom is) famous." is equivalent to "Tom is famous and (Tom is) rich".

5. $[(P \vee Q) \vee R] \Leftrightarrow [P \vee (Q \vee R)]$.......... associativity of $\vee$

What this says is, for example, that "Tom is rich or (Tom is) famous, or he is also happy." is equivalent to "Tom is rich, or he is also famous or (he is) happy".

6. $[(P \wedge Q) \wedge R] \Leftrightarrow [P \wedge (Q \wedge R)]$.......... associativity of $\wedge$

Similar to 5. above.

7. $\neg(P \vee Q) \Leftrightarrow (\neg P \wedge \neg Q)$.......... DeMorgan's Law byby

For example, "It is not the case that Tom is rich or famous." is true if and only if "Tom is not rich and he is not famous."

8. $\neg(P \wedge Q) \Leftrightarrow (\neg P \vee \neg Q)$.......... DeMorgan's Law

For example, "It is not the case that Tom is rich and famous." is true if and only if "Tom is not rich or he is not famous."

9. $[P \wedge (Q \vee R] \Leftrightarrow [(P \wedge Q) \vee (P \wedge R)]$.......... distributivity of $\wedge$ over $\vee$

What this says is, for example, that "Tom is rich, and he is famous or (he is) happy." is equivalent to "Tom is rich and (he is) famous, or Tom is rich and (he is) happy".

10. $[P \vee (Q \wedge R] \Leftrightarrow [(P \vee Q) \wedge (P \vee R)]$.......... distributivity of $\vee$ over $\wedge$

Similarly to 9. above, what this says is, for example, that "Tom is rich, or he is famous and (he is) happy." is equivalent to "Tom is rich or (he is) famous, and Tom is rich or (he is) happy".

11. $(P \vee \text{True}) \Leftrightarrow$ True. Here True is a proposition that is always true. Thus the proposition $(P \vee \text{True})$ is always true regardless of what P is.

This and the next three identities, like identities 1 and 2, are rarely used, if ever, in everyday life. However, these are useful when manipulating propositions in reasoning in symbolic form.

12. $(P \wedge \text{False}) \Leftrightarrow$ False

13. $(P \vee \text{False}) \Leftrightarrow P$

14. (P ∧ True) ⇔ P

15. (P ∨¬P) ⇔ True

What this says is that a statement such as "Tom is 6 foot tall or he is not 6 foot tall." is always true.

16. (P ∧¬P) ⇔ False

What this says is that a statement such as "Tom is 6 foot tall and he is not 6 foot tall." is always false.

17. P ⇔¬(¬P).......... double negation

What this says is, for example, that "It is not the case that Tom is not 6 foot tall." is equivalent to "Tom is 6 foot tall."

18. (P → Q) ⇔ (P ¬ Q).......... implication

For example, the statement "If I win the lottery, I will give you a million dollars." is not true, that is, I am lying, if I win the lottery and don't give you a million dollars. It is true in all the other cases. Similarly, the statement "I don't win the lottery or I give you a million dollars." is false, if I win the lottery and don't give you a million dollars. It is true in all the other cases. Thus these two statements are logically equivalent.

19. (P ↔ Q) ⇔ [(P → Q) ∧ (Q → P)].......... equivalence

What this says is, for example, that "Tom is happy if and only if he is healthy." is logically equivalent to ""if Tom is happy then he is healthy, and if Tom is healthy he is happy."

20. [(P ∧ Q) → R] ⇔ [P → (Q → R)].......... exportation

For example, "If Tom is healthy, then if he is rich, then he is happy." is logically equivalent to "If Tom is healthy and rich, then he is happy."

21. [(P → Q) ∧ (P →¬Q)] ⇔¬ P.......... absurdity

For example, if "If Tom is guilty then he must have been in that room." and "If Tom is guilty then he could not have been in that room." are both true, then there must be something wrong about the assumption that Tom is guilty.

22. (P → Q) ⇔ (¬Q →¬P).......... contrapositive

For example, "If Tom is healthy, then he is happy." is logically equivalent to "If Tom is not happy, he is not healthy." The identities 1 ~ 16 listed above can be paired by duality relation, which is defined below, as 1 and 2, 3 and 4,..., 15 and 16. That is 1 and 2 are dual to each other, 3 and 4 are dual to each other,.... Thus if you know one of a pair, you can obtain the other of the pair by using the duality.

DUAL OF PROPOSITION

Let X be a proposition involving only ¬, ∧, and as a connective. Let X^* be the proposition obtained from X by replacing ∨ with ∨, ∨with ∧, T with F, and F with T. Then X^* is called the dual of X.

For example, the dual of [P ∧ Q] ∨P is [P ∨ Q] ∧P, and the dual of [¬P ∧ Q] ∨¬ [T ∧¬ R] is [¬P ∧¬Q] ∧¬[F ∨¬ R].

Property of Dual: If two propositions P and Q involving only,, and as connectives are equivalent, then their duals P^* and Q^* are also equivalent.

EXAMPLES OF USE OF IDENTITIES

Here a few examples are presented to show how the identities in Identities can be used to prove some useful results.

1. ¬(P → Q) ⇔ (P ∧¬Q)

What this means is that the negation of "if P then Q" is "P but not Q". For example, if you said to someone "If I win a lottery, I will give you $100,000." and later that person says "You lied to me." Then what that person means is that you won the lottery but you did not give that person $100,000 you promised.

To prove this, first let us get rid of → using one of the identities: (P → Q) ⇔ (¬P ∨ Q).

That is, ¬(P → Q) ⇔¬(¬P ∨ Q).

Then by De Morgan, it is equivalent to ¬¬P ∧¬ Q, which is equivalent to P ∧¬Q, since the double negation of a proposition is equivalent to the original proposition as seen in the identities.

2. P ∨ (P ∧ Q) ⇔ P — Absorption

What this tells us is that P ∨ (P ∧ Q) can be simplified to P, or if necessary P can be expanded into P ∨ (P ∧ Q).

To prove this, first note that P ⇔ (P ∧ T).
Hence

P ∨ (P ∧ Q)

⇔ (P ∧ T) ∨ (P ∧ Q)

⇔ P ∧ (T ∨ Q), by the distributive law.

⇔ (P ∧ T), since (T ∨ Q) ⇔ T.

⇔ P, since (P ∧ T) ⇔ P.

Note that by the duality

P ∧ (P ∨ Q) ⇔ P also holds.

IMPLICATIONS

The following implications are some of the relationships between propositions that can be derived from the definitions(meaning) of connectives.

$\Rightarrow$ below corresponds to $\rightarrow$ and it means that the implication always holds. That is it is a tautology.

These implications are used in logical reasoning. When the right hand side of these implications is substituted for the left hand side appearing in a proposition, the resulting proposition is implied by the original proposition, that is, one can deduce the new proposition from the original one. First the implications are listed, then examples to illustrate them are given.

List of Implications:

1. $P \Rightarrow (P \vee Q)$.......... addition
2. $(P \wedge Q) \Rightarrow P$.......... simplification
3. $[P \wedge (P \rightarrow Q] \Rightarrow Q$.......... modus ponens
4. $[(P \rightarrow Q) \wedge \neg Q] \Rightarrow \neg P$.......... modus tollens
5. $[\neg P \wedge (P \vee Q] \Rightarrow Q$.......... disjunctive syllogism
6. $[(P \rightarrow Q) \wedge (Q \rightarrow R)] \Rightarrow (P \rightarrow R)$.......... hypothetical syllogism
7. $(P \rightarrow Q) \Rightarrow [(Q \rightarrow R) \rightarrow (P \rightarrow R)]$
8. $[(P \rightarrow Q) \wedge (R \rightarrow S)] \Rightarrow [(P \wedge R) \rightarrow (Q \wedge S)]$
9. $[(P \leftrightarrow Q) \wedge (Q \leftrightarrow R)] \Rightarrow (P \leftrightarrow R)$

Examples:

1. $P \Rightarrow (P \vee Q)$.......... addition

   For example, if the sun is shining, then certainly the sun is shining or it is snowing. Thus "if the sun is shining, then the sun is shining or it is snowing." "If $0 < 1$, then $0 \leq 1$ or a similar statement is also often seen.

2. $(P \wedge Q) \Rightarrow P$.......... simplification

   For example, if it is freezing and (it is) snowing, then certainly it is freezing. Thus "If it is freezing and (it is) snowing, then it is freezing."

3. $[P \wedge (P \rightarrow Q] \Rightarrow Q$.......... modus ponens

   For example, if the statement "If it snows, the schools are closed" is true and it actually snows, then the schools are closed.

   This implication is the basis of all reasoning. Theoretically, this is all that is necessary for reasoning. But reasoning using only this becomes very tedious.

4. $[(P \to Q) \wedge \neg Q] \Rightarrow \neg P$.......... modus tollens

   For example, if the statement "If it snows, the schools are closed" is true and the schools are not closed, then one can conclude that it is not snowing.

   Note that this can also be looked at as the application of the contrapositive and modus ponens. That is, $(P \to Q)$ is equivalent to $(\neg Q) \to (\neg P)$. Thus if in addition Q holds, then by the modus ponens, $\neg P$ is concluded.

5. $[\neg P \wedge (P \vee Q] \Rightarrow Q$.......... disjunctive syllogism

   For example, if the statement "It snows or (it) rains." is true and it does not snow, then one can conclude that it rains.

6. $[(P \to Q) \wedge (Q \to R)] \Rightarrow (P \to R)$.......... hypothetical syllogism

   For example, if the statements "If the streets are slippery, the school buses can not be operated." and "If the school buses can not be operated, the schools are closed." are true, then the statement "If the streets are slippery, the schools are closed." is also true.

7. $(P \to Q) \Rightarrow [(Q \to R) \to (P \to R)]$

   This is actually the hypothetical syllogism in another form. For by considering $(P \to Q)$ as a proposition S, $(Q \to R)$ as a proposition T, and $(P \to R)$ as a proposition U in the hypothetical syllogism above, and then by applying the "exportation" from the identities, this is obtained.

8. $[(P \to Q) \wedge (R \to S)] \Rightarrow [(P \wedge R) (Q \wedge S)]$

   For example, if the statements "If the wind blows hard, the beach erodes." and "If it rains heavily, the streets get flooded." are true, then the statement "If the wind blows hard and it rains heavily, then the beach erodes and the streets get flooded." is also true.

9. $[(P \leftrightarrow Q) \wedge (Q \leftrightarrow R)] \Rightarrow (P \leftrightarrow R)$

This just says that the logical equivalence is transitive, that is, if P and Q are equivalent, and if Q and R are also equivalent, then P and R are equivalent.

REASONING WITH PROPOSITIONS

Logical reasoning is the process of drawing conclusions from premises using rules of inference. The basic inference rule is modus ponens. It states that if both $P \to Q$ and P hold, then Q can be concluded, and it is written as P.

$$\frac{P \to Q}{Q}$$

Here the lines above the dotted line are premises and the line below it is the conclusion drawn from the premises. For example if "if it rains, then the game is not played" and "it rains" are both true, then we can conclude that the game is not played. In addition to modus ponens, one can also reason by using identities and implications. If the left(right) hand side of an identity appearing in a proposition is replaced by the right(left) hand side of the identity, then the resulting proposition is logically equivalent to the original proposition. Thus the new proposition is deduced from the original proposition. For example in the proposition $P \wedge (Q \rightarrow R)$, $(Q \rightarrow R)$ can be replaced with $(\neg Q \vee R)$ to conclude $P \wedge (\neg Q \vee R)$, since $(Q \rightarrow R) \Leftrightarrow (\neg Q \vee R)$

Similarly if the left(right) hand side of an implication appearing in a proposition is replaced by the right(left) hand side of the implication, then the resulting proposition is logically implied by the original proposition. Thus the new proposition is deduced from the original proposition. The tautologies listed as "implications" can also be considered inference rules as shown below.

Rules of Inference	*Tautological Form*	*Name*
P / $\overline{P \vee Q}$	$P \Rightarrow (P \vee Q)$	addition
$P \wedge Q$ / $\overline{P}$	$(P \wedge Q) \Rightarrow P$	simplification
P / $P \rightarrow Q$ / $\overline{Q}$	$[P \wedge (P \rightarrow Q)] \Rightarrow Q$	modus ponens
$\neg Q$ / $P \rightarrow Q$ / $\overline{\neg P}$	$[\neg Q \wedge (P \rightarrow Q)] \Rightarrow \neg P$	modus tollens
$P \vee Q$ / $\neg P$ / $\overline{Q}$	$[(P \vee Q) \wedge \neg P] \Rightarrow Q$	disjunctive syllogism
$P \rightarrow Q$ / syllogism / $Q \rightarrow R$ / $\overline{P \rightarrow R}$	$[(P \rightarrow Q) \wedge (Q \rightarrow R)] \Rightarrow [P \rightarrow R]$	h y p o t h e t i c a l
P / Q / $\overline{P \wedge Q}$		conjunction
$(P \rightarrow Q) \wedge (R \rightarrow S)$ / $P \vee R$ / $\overline{Q \vee S}$	$[(P \rightarrow Q) \wedge (R \rightarrow S) \wedge (P \vee R)] \Rightarrow [Q \vee S]$	constructive dilemma
$(P \rightarrow Q) \wedge (R \rightarrow S)$ / $\neg P \vee \neg R$ / $\overline{\neg P \vee \neg R}$	$[(P \rightarrow Q) \wedge (R\ S) \wedge (\neg Q \vee \neg S)] \Rightarrow [\neg P \vee \neg R]$	destructive dilemma

EXAMPLE OF INFERENCING

Consider the following argument:

1. Today is Tuesday or Wednesday.
2. But it can't be Wednesday, since the doctor's office is open today, and that office is always closed on Wednesdays.
3. Therefore today must be Tuesday.

This sequence of reasoning (inferencing) can be represented as a series of application of modus ponens to the corresponding propositions as follows.

The modus ponens is an inference rule which deduces Q from $P \to Q$ and P.

T: Today is Tuesday.

W: Today is Wednesday.

D: The doctor's office is open today.

C: The doctor's office is always closed on Wednesdays.

The above reasoning can be represented by propositions as follows.

1. $T \vee W$
2. $\dfrac{D}{C}$
3. $\dfrac{\sim W}{T}$

To see if this conclusion T is correct, let us first find the relationship among C, D, and W:

C can be expressed using D and W. That is, restate C first as the doctor's office is always closed if it is Wednesday. Then $C \leftrightarrow (W \to \sim D)$ Thus substituting $(W \to \sim D)$ for C, we can proceed as follows.

$$\begin{array}{c} D \\ W \to \sim d \\ \hline \sim W \end{array}$$

which is correct by modus tollens.

From this $\sim W$ combined with $T \vee W$ of 1. above,

$$\begin{array}{c} \sim W \\ W \vee W \\ \hline T \end{array}$$

which is correct by disjunctive syllogism.

Thus we can conclude that the given argument is correct.

To save space we also write this process as follows eliminating one of the ~*W*'s:

$$\frac{\begin{array}{c} D \\ W \to \sim D \end{array}}{\sim W}$$

$$\frac{T \vee W}{T}$$

PRO OF OF IDENTITIES

All the identities in Identities can be proven to hold using truth tables as follows. In general two propositions are logically equivalent if they take the same value for each set of values of their variables.

Thus to see whether or not two propositions are equivalent, we construct truth tables for them and compare to see whether or not they take the same value for each set of values of their variables.

For example consider the commutativity ∨ of:

$(P \vee Q) \Leftrightarrow (Q \vee P)$.

To prove that this equivalence holds, let us construct a truth table for each of the proposition $(P \vee Q)$ and $(Q \vee P)$.

A truth table for $(P \vee Q)$ is, by the definition of ∨,

P	*Q*	*(P ∨ Q)*
F	F	F
F	T	T
T	F	T
T	T	T

A truth table for $(Q \vee P)$ is, by the definition of ∨,

P	*Q*	*(Q ∨ P)*
F	F	F
F	T	T
T	F	T
T	T	T

A truth table for $(Q \vee P)$ is, by the definition of ∨,

P	*Q*	*(Q ∨ P)*
F	F	F
F	T	T
T	F	T
T	T	T

As we can see from these tables (P ∨ Q) and (Q ∨ P) take the same value for the same set of value of P and Q. Thus they are (logically) equivalent. We can also put these two tables into one as follows:

P	*Q*	*(P ∨ Q)*	*(Q ∨ P)*
F	F	F	F
F	T	T	T
T	F	T	T
T	T	T	T

Using this convention for truth table we can show that the first of De Morgan's Laws also holds.

P	*Q*	*¬(P ∨ Q)*	*¬P ∧¬ Q*
F	F	T	T
F	T	F	F
T	F	F	F
T	T	F	F

By comparing the two right columns we can see that ¬(P ∨ Q) and ¬P ∧¬Q are equivalent.

PROOF OF IMPLICATIONS

1. All the implications in Implications can be proven to hold by constructing truth tables and showing that they are always true. For example consider the first implication "addition": P ⇒ (P ∨ Q).

To prove that this implication holds, let us first construct a truth table for the proposition P ∨ Q.

P	*Q*	*(P ∨ Q)*
F	F	F
F	T	T
T	F	T
T	T	T

Then by the definition of $\rightarrow$, we can add a column for $P \rightarrow (P \vee Q)$ to obtain the following truth table.

P	Q	$(P \vee Q)$	$P \rightarrow (P \vee Q)$
F	F	F	T
F	T	T	T
T	F	T	T
T	T	T	T

The first row in the rightmost column results since P is false, and the others in that column follow since $(P \vee Q)$ is true.

The rightmost column shows that $P \rightarrow (P \vee Q)$ is always true.

2. Some of the implications can also be proven by using identities and implications that have already been proven.

For example suppose that the identity "exportation":

$$[(X \wedge Y) \rightarrow Z] \Leftrightarrow [X \rightarrow (Y \rightarrow Z)],$$

and the implication "hypothetical syllogism":

$$[(P \rightarrow Q) \wedge (Q \rightarrow R)] \Rightarrow (P \rightarrow R)$$

have been proven. Then the implication No. 7:

$$(P \rightarrow Q) \Rightarrow [(Q \rightarrow R) \rightarrow (P \rightarrow R)]$$

can be proven by applying the "exportation" to the "hypothetical syllogism" as follows:

Consider $(P \rightarrow Q)$, $(Q \rightarrow R)$, and $(P \rightarrow R)$ in the "hypothetical syllogism" as X, Y and Z of the "exportation", respectively.

Then since $[(X \wedge Y) \rightarrow Z] \Leftrightarrow [X \rightarrow (Y \rightarrow Z)]$ implies $[(X \wedge Y) \rightarrow Z] \Rightarrow [X \rightarrow (Y \rightarrow Z)]$, the implication of No. 7 follows.

Similarly the modus ponens (implication No. 3) can be proven as follows:

Noting that $(P \rightarrow Q) \Leftrightarrow (\neg P \vee Q)$,

$P \wedge (P \rightarrow Q)$

$\Leftrightarrow P \wedge (\neg P \vee Q)$

$\Leftrightarrow (P \wedge \neg P) \vee (P \wedge Q)$ by the distributive law

$\Leftrightarrow F \vee (P \wedge Q)$

$\Leftrightarrow (P \wedge Q)$

$\Rightarrow Q$

Also the exportation (identity No. 20), $(P \rightarrow (Q \rightarrow R)) \Leftrightarrow (P \wedge Q) \rightarrow R)$ can be proven using identities as follows:

$(P \rightarrow (Q \rightarrow R)) \Leftrightarrow \neg P \vee (Q \rightarrow R)$

$\Leftrightarrow \neg P \vee (\neg Q \vee R)$

$\Leftrightarrow (\neg P \vee \neg Q) \vee R$

$\Leftrightarrow \neg(P \wedge Q) \vee R$

$\Leftrightarrow (P \wedge Q) \rightarrow R$

3. Some of them can be proven by noting that a proposition in an implication can be replaced by an equivalent proposition without affecting its value.

For example by substituting $(\neg Q \rightarrow \neg P)$ for $(P \rightarrow Q)$, since they are equivalent being contrapositive to each other, modus tollens (the implication No. 4): $[(P \rightarrow Q) \wedge \neg Q] \Rightarrow \neg P$, reduces to the modus ponens: $[X \wedge (X \rightarrow Y)] \Rightarrow Y$. Hence if the modus ponens and the "contrapositive" in the "Identities" have been proven, then the modus tollens follows from them.

PREDICATE LOGIC

The propositional logic is not powerful enough to represent all types of assertions that are used in computer science and mathematics, or to express certain types of relationship between propositions such as equivalence.

For example, the assertion "x is greater than 1", where x is a variable, is not a proposition because you can not tell whether it is true or false unless you know the value of x. Thus the propositional logic can not deal with such sentences. However, such assertions appear quite often in mathematics and we want to do inferencing on those assertions.

Also the pattern involved in the following logical equivalences can not be captured by the propositional logic:

"Not all birds fly" is equivalent to "Some birds don't fly".

"Not all integers are even" is equivalent to "Some integers are not even".

"Not all cars are expensive" is equivalent to "Some cars are not expensive",

....

Each of those propositions is treated independently of the others in propositional logic. For example, if P represents "Not all birds fly" and Q represents "Some integers are not even", then there is no mechanism in propositional logic to find out tha P is equivalent to Q. Hence to be used in inferencing, each of these equivalences must be listed individually rather than dealing with a general formula that covers all these equivalences collectively and instantiating it as they become necessary, if only propositional logic is used.

Thus we need more powerful logic to deal with these and other problems. The predicate logic is one of such logic and it addresses these issues among others.

PREDICATE

A predicate is a verb phrase template that describes a property of objects, or a relationship among objects represented by the variables.

For example, the sentences "The car Tom is driving is blue", "The sky is blue", and "The cover of this book is blue" come from the template "is blue" by placing an appropriate noun/noun phrase in front of it.

The phrase "is blue" is a predicate and it describes the property of being blue. Predicates are often given a name. For example any of "is_blue", "Blue" or "B" can be used to represent the predicate "is blue" among others. If we adopt B as the name for the predicate "is_blue", sentences that assert an object is blue can be represented as "B(x)", where x represents an arbitrary object. B(x) reads as "x is blue".

Similarly the sentences "John gives the book to Mary", "Jim gives a loaf of bread to Tom", and "Jane gives a lecture to Mary" are obtained by substituting an appropriate object for variables *x*, *y*, and *z* in the sentence "*x* gives *y* to *z*". The template "... gives... to..." is a predicate and it describes a relationship among three objects. This predicate can be represented by Give(x, y, z) or G(x, y, z), for example.

Note: The sentence "John gives the book to Mary" can also be represented by another predicate such as "gives a book to". Thus if we use B(x, y) to denote this predicate, "John gives the book to Mary" becomes B(John, Mary). In that case, the other sentences, "Jim gives a loaf of bread to Tom", and "Jane gives a lecture to Mary", must be expressed with other predicates.

QUANTIFICATION

FORMING PROPOSITIONS FROM PREDICATES

A predicate with variables is not a proposition. For example, the statement $x > 1$ with variable *x* over the universe of real numbers is neither true nor false since we don't know what *x* is. It can be true or false depending on the value of *x*.

For $x > 1$ to be a proposition either we substitute a specific number for *x* or change it to something like "There is a number *x* for which $x > 1$ holds", or "For every number *x*, $x > 1$ holds".

More generally, a predicate with variables (called an atomic formula) can be made a proposition by applying one of the following two operations to each of its variables:

1. Assign a value to the variable
2. Quantify the variable using a quantifier.

For example, $x > 1$ becomes $3 > 1$ if 3 is assigned to x, and it becomes a true statement, hence a proposition.

In general, a quantification is performed on formulas of predicate logic (called wff), such as $x > 1$ or $P(x)$, by using quantifiers on variables. There are two types of quantifiers: universal quantifier and existential quantifier.

The universal quantifier turns, for example, the statement $x > 1$ to "for every object x in the universe, $x > 1$", which is expressed as "$\forall x\ x > 1$". This new statement is true or false in the universe of discourse. Hence it is a proposition once the universe is specified.

Similarly the existential quantifier turns, for example, the statement $x > 1$ to "for some object x in the universe, $x > 1$", which is expressed as "$\exists x\ x > 1$." Again, it is true or false in the universe of discourse, and hence it is a proposition once the universe is specified.

UNIVERSE OF DISCOURSE

The universe of discourse, also called universe, is the set of objects of interest. The propositions in the predicate logic are statements on objects of a universe. The universe is thus the domain of the (individual) variables. It can be the set of real numbers, the set of integers, the set of all cars on a parking lot, the set of all students in a classroom etc. The universe is often left implicit in practice. But it should be obvious from the context.

THE UNIVERSAL QUANTIFIER

The expression: $\forall x\ P(x)$, denotes the universal quantification of the atomic formula P(x). Translated into the English language, the expression is understood as: "*For all x, P(x)* holds", "*for each x, P(x)* holds" or "*for every x, P(x)* holds". $\forall$ is called the universal quantifier, and $\forall x$ means all the objects x in the universe. If this is followed by $P(x)$ then the meaning is that $P(x)$ is true for every object x in the universe. For example, "All cars have wheels" could be transformed into the propositional form, $\forall x\ P(x)$, where:

1. $P(x)$ is the predicate denoting: x has wheels, and
2. The universe of discourse is only populated by cars.

Universal Quantifier and Connective AND

If all the elements in the universe of discourse can be listed then the universal quantification $\forall x\ P(x)$ is equivalent to the conjunction: $P(x_1)) \wedge P(x_2) \wedge P(x_3) \wedge ... \wedge P(x_n)$.

For example, in the above example of $\forall x\ P(x)$, if we knew that there were only 4 cars in our universe of discourse (c_1, *c2, c3* and *c4*) then we could also translate the statement as: $P(c1) \wedge P(c2) \wedge P(c3) \wedge P(c4)$

THE EXISTENTIAL QUANTIFIER

The expression: $\exists x P(x)$, denotes the existential quantification of $P(x)$. Translated into the English language, the expression could also be understood as: "There exists an x such that *P(x)*" or "There is at least one x such that *P(x)*" is called the existential quantifier, and $\exists x$ means at least one object x in the universe. If this is followed by *P(x)* then the meaning is that *P(x)* is true for at least one object x of the universe. For example, "*Someone loves you*" could be transformed into the propositional form, $\exists x\ P(x)$, where:

•*P(x)* is the predicate meaning: x loves you,

•The universe of discourse contains (but is not limited to) all living creatures.

Existential Quantifier and Connective OR

If all the elements in the universe of discourse can be listed, then the existential quantification $\exists x P(x)$ is equivalent to the disjunction: $P(x_1) \vee P(x_2) \vee P(x_3) \vee ... \vee P(x_n)$.

For example, in the above example of $\exists x\ P(x)$, if we knew that there were only 5 living creatures in our universe of discourse (say: me, he, she, rex and fluff), then we could also write the statement as: *P(me)* ∨*P(he)* ∨*P(she)* ∨*P(rex)* ∨*P(fluff)*

An appearance of a variable in a wff is said to be bound if either a specific value is assigned to it or it is quantified. If an appearance of a variable is not bound, it is called free. The extent of the application(effect) of a quantifier, called the scope of the quantifier, is indicated by square brackets []. If there are no square brackets, then the scope is understood to be the smallest wff following the quantification.

For example, in $\exists x\ P(x, y)$, the variable x is bound while y is free. In $\forall x\ [\exists y\ P(x, y) \vee Q(x, y)]$, x and the y in $P(x, y)$ are bound, while y in $Q(x, y)$ is free, because the scope of $\exists y$ is $P(x, y)$. The scope of $\forall x$ is $[\exists y\ P(x, y) \vee Q(x, y)]$.

HOW TO READ QUANTIFIED FORMULAS

When reading quantified formulas in English, read them from left to right. $\forall x$ can be read as "for every object x in the universe the following holds" and $\exists x$ can be read as "there erxists an object x in the universe which satisfies the following" or "for some object x in the universe the following holds". Those do not necessarily give us good English expressions. But they are where we can start. Get the correct reading first then polish your English without changing the truth values.

For example, let the universe be the set of airplanes and let $F(x, y)$ denote "x flies faster than y". Then

$\forall x \forall y\ F(x, y)$ can be translated initially as "For every airplane x the following holds: x is faster than every (any) airplane y". In simpler English it means "Every airplane is faster than every airplane (including itself !)".

$\forall x\ \forall y\ F(x, y)$ can be read initially as "For every airplane x the following holds: for some airplane y, x is faster than y". In simpler English it means "Every airplane is faster than some airplane".

$\exists x \forall y\ F(x, y)$ represents "There exist an airplane x which satisfies the following: (or such that) for every airplane y, x is faster than y". In simpler English it says "There is an airplane which is faster than every airplane" or "Some airplane is faster than every airplane".

$\exists x \exists y\ F(x, y)$ reads "For some airplane x there exists an airplane y such that x is faster than y", which means "Some airplane is faster than some airplane".

ORDER OF APPLICATION OF QUANTIFIERS

When more than one variables are quantified in a wff such as $\exists y\ \forall x\ P(x, y)$, they are applied from the inside, that is, the one closest to the atomic formula is applied first. Thus $\exists y\ \forall x\ P(x, y)$ reads $\exists y\ [\forall x\ P(x, y)]$, and we say "there exists an y such that for every x, $P(x, y)$ holds" or "for some y, $P(x, y)$ holds for every x".

The positions of the same type of quantifiers can be switched without affecting the truth value as long as there are no quantifiers of the other type between the ones to be interchanged.

For example $\exists x\ \exists y\ \exists z\ P(x, y, z)$ is equivalent to $\exists y\ \exists x\ \exists z\ P(x, y, z)$, $\exists z\ \exists y\ \exists x\ P(x, y, z)$, etc. It is the same for the universal quantifier.

However, the positions of different types of quantifiers can not be switched.

For example $\forall x\ \exists y\ P(x, y)$ is not equivalent to $\exists y\ \forall x\ P(x, y)$. For let $P(x, y)$ represent $x < y$ for the set of numbers as the universe, for example. Then $\forall x\ \exists y\ P(x, y)$ reads "for every number x, there is a number y that is greater than x", which is true, while $\exists y\ \forall x\ P(x, y)$ reads "there is a number that is greater than every (any) number", which is not true.

WELL–FORMED FORMULA FOR FIRST ORDER PREDICATE LOGIC SYNTAX RULES

Not all strings can represent propositions of the predicate logic. Those which produce a proposition when their symbols are interpreted must follow the rules given below, and they are called wffs(well–formed formulas) of the first order predicate logic.

RULES FOR CONSTRUCTING WFFS

A predicate name followed by a list of variables such as $P(x, y)$, where P is a predicate name, and x and y are variables, is called an atomic formula.

Wffs are constructed using the following rules:

1. *True* and *False* are wffs.
2. Each propositional constant (i.e. specific proposition), and each propositional variable (i.e. a variable representing propositions) are wffs.
3. Each atomic formula (i.e. a specific predicate with variables) is a wff.
4. If A, B, and C are wffs, then so are $\neg A$, $(A \wedge B)$, $(A \vee B)$, $(A \rightarrow B)$, and $(A \leftrightarrow B)$.

5. If x is a variable (representing objects of the universe of discourse), and A is a wff, then so are $\forall x\, A$ and $\exists x\, A$.

Note: More generally, arguments of predicates are something called a term. Also variables representing predicate names (called predicate variables) with a list of variables can form atomic formulas. But we do not get into that here.)

For example, "The capital of Virginia is Richmond." is a specific proposition. Hence it is a wff by Rule 2.

Let B be a predicate name representing "being blue" and let x be a variable. Then $B(x)$ is an atomic formula meaning "x is blue". Thus it is a wff by Rule 3. above. By applying Rule 5. to $B(x)$, $\forall x B(x)$ is a wff and so is $\exists x B(x)$. Then by applying Rule 4. to them $\forall x\, B(x) \wedge \exists x\, B(x)$ is seen to be a wff. Similarly if R is a predicate name representing "being round". Then $R(x)$ is an atomic formula. Hence it is a wff. By applying Rule 4 to $B(x)$ and $R(x)$, a wff $B(x) \wedge R(x)$ is obtained.

In this manner, larger and more complex wffs can be constructed following the rules given above.

Note, however, that strings that can not be constructed by using those rules are not wffs. For example, $\forall x B(x) R(x)$, and $B(\exists x)$ are NOT wffs, NOR are $B(R(x))$, and $B(\exists x\, R(x))$.

One way to check whether or not an expression is a wff is to try to state it in English. If you can translate it into a correct English sentence, then it is a wff.

More examples: To express the fact that Tom is taller than John, we can use the atomic formula *taller*(Tom, John), which is a wff. This wff can also be part of some compound statement such as *taller*(Tom, John) $\wedge$–*taller*(John, Tom), which is also a wff.

If x is a variable representing people in the world, then *taller(x*,Tom), $\forall x$ *taller(x*,Tom), $\exists x$ *taller(x*,Tom), $\exists x\ \forall y$ *taller(x,y)* are all wffs among others.

However, *taller(*$\exists$x,John) and *taller*(Tom $\wedge$ Mary, Jim), for example, are NOT wffs.

FROM WFF TO PROPOSITION

INTERPRETATION

A wff is, in general, not a proposition. For example, consider the wff $\forall x\, P(x)$. Assume that $P(x)$ means that x is non–negative (greater than or equal to 0). This wff is true if the universe is the set $\{1, 3, 5\}$, the set $\{2, 4, 6\}$ or the set of natural numbers, for example, but it is not true if the universe is the set $\{-1, 3, 5\}$, or the set of integers, for example. Further more the wff $\forall x\, Q(x, y)$, where $Q(x, y)$ means x is greater than y, for the universe $\{1, 3, 5\}$ may be true or false depending on the value of y.

As one can see from these examples, the truth value of a wff is determined by the universe, specific predicates assigned to the predicate variables such as P and Q, and the values assigned to the free variables. The specification of the universe and predicates, and an assignment of a value to each free variable in a wff is called an interpretation for the wff.

For example, specifying the set $\{1, 3, 5\}$ as the universe and assigning 0 to the variable y, for example, is an interpretation for the wff $\forall x\ Q(x, y)$, where $Q(x, y)$ means x is greater than y. $\forall x\ Q(x, y)$ with that interpretation reads, for example, "Every number in the set $\{1, 3, 5\}$ is greater than 0".

As can be seen from the above example, a wff becomes a proposition when it is given an interpretation.

There are, however, wffs which are always true or always false under any interpretation. Those and related concepts are discussed below.

SATISFIABLE, UNSATISFIABLE AND VALID WFFS

A wff is said to be satisfiable if there exists an interpretation that makes it true, that is if there are a universe, specific predicates assigned to the predicate variables, and an assignment of values to the free variables that make the wff true.

For example, $\forall x\ N(x)$, where $N(x)$ means that x is non–negative, is satisfiable. For if the universe is the set of natural numbers, the assertion $\exists x\ N(x)$ is true, because all natural numbers are non–negative. Similarly $\forall x\ N(x)$ is also satisfiable.

However, $\forall x\ [N(x) \wedge \neg N(x)]$ is not satisfiable because it can never be true. A wff is called invalid or unsatisfiable, if there is no interpretation that makes it true.

A wff is valid if it is true for every interpretation*.

For example, the wff $\forall x\ P(x) \vee \exists \neg\ x\ P(x)$ is valid for any predicate name P, because $\exists x\ P(x)$ is the negation of $x\ P(x)$.

However, the wff $\forall x\ N(x)$ is satisfiable but not valid.

Note that a wff is not valid iff it is unsatisfiable for a valid wff is equivalent to true. Hence its negation is false.

EQUIVALENCE

Two wffs W_1 and W_2 are equivalent if and only if $W_1 \leftrightarrow W_2$ is valid, that is if and only if $W_1 \leftrightarrow W_2$ is true for all interpretations.

For example $\forall x\ P(x)$ and $\neg\exists x\ \neg P(x)$ are equivalent for any predicate name P. So are $\forall x[P(x) \wedge Q(x)]$ and $[\forall x\ P(x) \wedge \forall x\ Q(x)]$ for any predicate names P and Q.

* To be precise, it is not for every interpretation but for the ones that "make sense". For example you don't consider the universe of set of people for the predicate $x > 1$ as an interpretation.

Also an interpretation assingns a specific predicate to each predicate varialbe. A rigorous definition of interpretation etc. are, however, beyond the scope of this course.

TRANSCRIBING ENGLISH TO PREDICATE LOGIC WFFS

English sentences appearing in logical reasoning can be expressed as a wff. This makes the expressions compact and precise. It thus eliminates possibilities of misinterpretation of sentences. The use of symbolic logic also makes reasoning formal and mechanical, contributing to the simplification of the reasoning and making it less prone to errors.

Transcribing English sentences into wffs is sometimes a non–trivial task. In this course we are concerned with the transcription using given predicate symbols and the universe.

To transcribe a proposition stated in English using a given set of predicate symbols, first restate in English the proposition using the predicates, connectives, and quantifiers. Then replace the English phrases with the corresponding symbols.

Example: Given the sentence "Not every integer is even", the predicate "$E(x)$" meaning x is even, and that the universe is the set of integers,

first restate it as "It is not the case that every integer is even" or "It is not the case that for every object x in the universe, x is even."

Then "it is not the case" can be represented by the connective "", "every object x in the universe" by "$\forall x$", and "x is even" by $E(x)$.

Thus altogether wff becomes $\neg\forall x\ E(x)$.

This given sentence can also be interpreted as "Some integers are not even". Then it can be restated as "For some object x in the universe, x is not integer". Then it becomes $\exists x \neg\ E(x)$.

More examples: A few more sentences with corresponding wffs are given below. The universe is assumed to be the set of integers, $E(x)$ represents x is even, and $O(x)$, x is odd.

"Some integers are even and some are odd" can be translated as

$\exists x\ E(x) \wedge \exists x\ O(x)$

"No integer is even" can go to

$\forall x\ \neg E(x)$

"If an integer is not even, then it is odd" becomes

$\forall x\ [\neg E(x) \rightarrow O(x)]$

"2 is even" is

$E(2)$

More difficult translation: In these translations, properties and relationships are mentioned for certain type of elements in the universe such as relationships between integers in the universe of numbers rather than the universe of integers. In such a case the element type is specified as a precondition using if_then construct.

Examples: In the examples that follow the universe is the set of numbers including real numbers, and complex numbers. *I(x)*, *E(x)* and *O(x)* representing "x is an integer", "x is even", and "x is odd", respectively.

"All integers are even" is transcribed as

$\forall x\ [I(x) \to E(x)]$

It is first restated as "For every object in the universe (meaning for every numnber in this case) if it is integer, then it is even". Here we are interested in not any arbitrary object(number) but a specific type of objects, that is integers. But if we write $\forall x$ it means "for any object in the universe". So we must say "For any object, if it is integer.." to narrow it down to integers.

"Some integers are odd" can be restated as "There are objects that are integers and odd", which is expressed as

$\exists x\ [I(x) \wedge E(x)]$

For another interpretation of this sentence see a note

"A number is even only if it is integer" becomes

$\forall x\ [E(x) \to I(x)]$

"Only integers are even" is equivalent to "If it is even, then it is integer". Thus it is translated to

$\forall x\ [E(x) \to I(x)]$

REASONING WITH PREDICATE LOGIC REASONING

Predicate logic is more powerful than propositional logic. It allows one to reason about properties and relationships of individual objects. In predicate logic, one can use some additional inference rules, which are discussed below, as well as those for propositional logic such as the equivalences, implications and inference rules. The following four rules describe when and how the universal and existential quantifiers can be added to or deleted from an assertion.

1. *Universal Instantiation*:

$$\frac{\forall \times P(x)}{P(c)}$$

where c is some arbitrary element of the universe.

Go to Universal Instantiation for further explanations and examples.

2. *Universal Generalization*:

$$\frac{P(c)}{\forall \times P(x)}$$

where *P(c)* holds for every element c of the universe of discourse.

Go to Universal Generalization for further explanations and examples.

3. *Existential Instantiation*:

$$\frac{\exists \times P(x)}{P(c)}$$

where *c* is some element of the universe of discourse. It is not arbitrary but must be one for which P(c) is true.

Go to Existential Instantiation for further explanations and examples.

4. *Existential Generalization*:

$$\frac{P(c)}{\exists \times P(x)}$$

where *c* is an element of the universe.

Go to Existential Generalization for further explanations and examples.

NEGATION OF QUANTIFIED STATEMENT

Another important inference rule is the following:

$\neg\exists x\ P(x) \Leftrightarrow \forall\ x\ \neg P(x)$

This, for example, shows that if P(x) represnts x is happy and the universe is the set of people, then "There does not exist a person who is happy" is equivalent to "Everyone is not happy".

Thus the left side can be substituted for the right side whenever necessary in reasoning and vice versa.

Example: As an example of inference using these rules, let us consider the following reasoning:

A check is void if it has not been cashed for 30 days. This check has not been cashed for 30 days. Therefore this check is void. You can not cash a check which is void. Therefore you can not cash this check. We now have a check which can not be cashed.

This can be put into symbolic form using the following predicates assuming the universe is the set of all objects:

$C(x)$: x is a check.

$T(x)$: x has been cashed within 30 days.

$V(x)$: x is void.

$S(x)$: x can be cashed.

This check represents a specific object in the universe which corresponds to "this check".

$\forall x\ [[C(x) \wedge \neg\ T(x)] \rightarrow V(x)]$

$\neg T(This\_check)$

....

$V(This\_check)$

$\forall x\ [[C(x) \wedge V(x)] \rightarrow \neg S(x)]$

....

$S(This\_check)$

....

$\exists x\ [C(x) \wedge \neg\ S(x)]$.

Here the reasoning proceeds as follows:

From $\forall x\ [[C(x) \wedge \neg T(x)] \rightarrow V(x)]$ by Universal Instantiation

$[[C(This\_check) \wedge \neg\ T(This\_check)] \rightarrow V(This\_check)]$

Since This_check is a check and $T(This\_check)$,

$[C(This\_check)\ T(This\_check)]$ holds.

Hence

$[[C(This\_check) \wedge \neg\ T(This\_check)] \rightarrow V(This\_check)]$

$[C(This\_check) \wedge \neg\ T(This\_check)]$

....

$V(This\_check)$

by Modus Ponens.

Then from $\forall x\ [[C(x) \wedge V(x)] \rightarrow \neg\ S(x)]$ by Universal Instantiation,

$[[C(This\_check) \wedge V(This\_check)] \rightarrow \neg\ S(This\_check)]$

Since $V(This\_check)$, and $C(This\_check)$,

$[[C(This\_check) \wedge V(This\_check)] \rightarrow \neg\ S(This\_check)]$

$[C(This\_check) \wedge V(This\_check)]$

....

$\neg S(This\_check)$

by Modus Ponens.

Since $C(This\_check)$ also holds,

$\neg S(This\_check)$

$C(This\_check)$

....

$C(This\_check) \wedge \neg S(This\_check)$

Then by Existential Generalization $\exists x\ [C(x) \wedge \neg\ S(x)]$.

QUANTIFIERS AND CONNECTIVES

Quantifiers and Connectives 1

There are following four important relationships between quantifiers and connectives. They are used frequently in reasoning.

1. $\forall x\ [P(x) \wedge Q(x)] \Leftrightarrow [\forall x\ P(x) \wedge \forall x\ Q(x)]$
2. $[\forall x\ P(x) \vee x\ Q(x)] \Rightarrow \forall\ x\ [P(x) \vee Q(x)]$
3. $\exists x\ [P(x)\ Q(x)] \Leftrightarrow [\exists x\ P(x) \vee \exists\ x\ Q(x)]$
4. $\exists x\ [P(x) \wedge Q(x)] \Rightarrow [\exists x\ P(x) \wedge \exists\ x\ Q(x)]$

Let us see what these mean with examples.

Let *P(x)* represent "x is rich", and *Q(x)* "x is happy", and let the universe be a set of three people.

Also let *LHS* (*RHS*) denote the left (right) hand side of the implication or equivalence. Then

1. $\forall x\ [P(x) \wedge Q(x)] \Leftrightarrow [\forall x\ P(x) \wedge \forall\ x\ Q(x)]$ can be illustrated as follows:

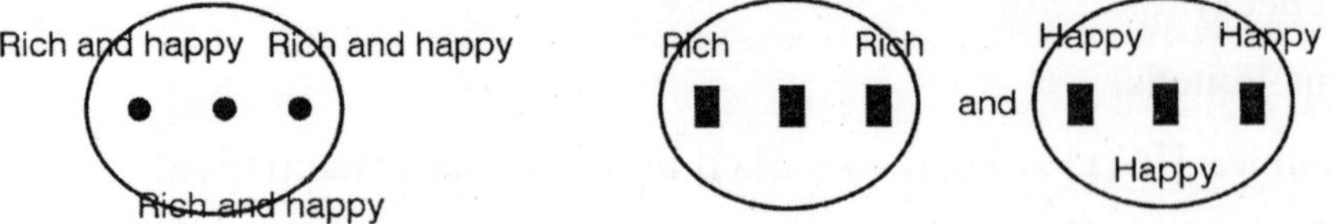

That is, LHS says everyone is rich and happy, and RHS says everyone is rich and everyone is happy.

Thus LHS and RHS describe the same situation. That is, LHS is true if and only if RHS is true.

2. $[\forall x\ P(x) \vee \forall\ x\ Q(x)] \Rightarrow \forall\ x\ [P(x) \vee Q(x)]$ for the same interpretation as 1. above can be shown as follows:

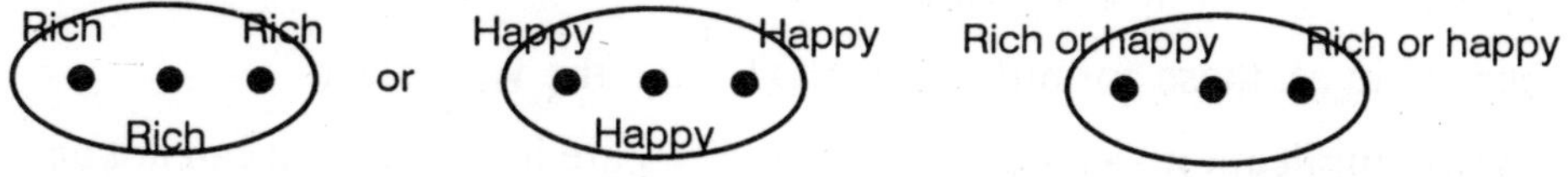

LHS RHS

That is, LHS says everyone is rich or everyone is happy, and RHS says everyone is rich or happy. Thus if LHS is true, then RHS is certainly true. However on RHS it can happen that two people are rich but the third is not rich but happy. In that case LHS is not true while RHS is true. Thus RHS does not necessarily imply LHS.

3. $\exists x\ [P(x) \vee Q(x)] \Leftrightarrow [\exists x\ P(x)\ x \vee \exists Q(x)]$, again for the same example, can be shown as follows:

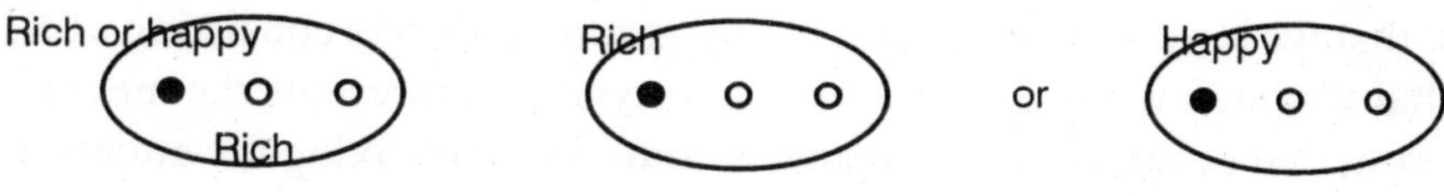

LHS RHS

LHS says someone is rich or happy, and RHS says someone is rich or someone is happy. Thus clearly LHS implies RHS. Also if someone is rich then that person is certainly rich or happy. Thus RHS implies LHS.

4. $\exists x\ [P(x) \wedge Q(x)] \Rightarrow [\exists x\ P(x) \wedge \exists x\ Q(x)]$, for the same example, can be shown as follows:

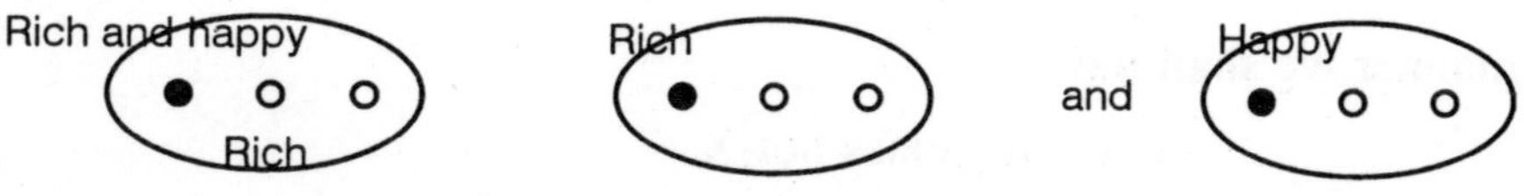

LHS RHS

LHS say someone is rich and happy. Hence there is someone who is rich and there is someone who is happy. Hence LHS implies RHS. However, since RHS can be true without anyone being rich and happy at the same time, RHS does not necessarily imply LHS.

QUANTIFIERS AND CONNECTIVES 2

If a wff (Q below) in the scope of a quantifier does not have the variable (x below) that is quantified by that quantifier, then that wff can be taken out of the scope of that quantifier. That is,

1. $\forall x\, [P(x) \wedge Q] \Leftrightarrow [\forall x\, P(x) \wedge Q]$
2. $[\forall x\, P(x) \vee Q] \Leftrightarrow x\, [P(x) \vee Q]$
3. $\exists x\, [P(x) \vee Q] \Leftrightarrow [\exists x\, P(x) \vee Q]$
4. $\exists x\, [P(x) \wedge Q] \Leftrightarrow [\exists x\, P(x) \wedge Q]$,

   where Q in all these formulas DO NOT have the variable x.

Note: When implication $\rightarrow$ and/or equivalence $\leftrightarrow$ are involved, you can not necessarily take Q outside the scope. To see what happens, express $\rightarrow$ and $\leftrightarrow$ using $\vee$ and $\wedge$, and apply the above formulas. For example $\forall x\, [P(x) \rightarrow Q]$ is NOT equivalent to $\forall x\, P(x) \rightarrow Q$. Rather it is equivalent to $\exists x\, P(x) \rightarrow Q$.

CALCULUS OF LOGIC

The universe of conceivable objects is represented by 1 or unity. This is assumed as the primary and subject conception. All subordinate conceptions of class are understood to be formed from it by limitation, according to the following scheme.

Suppose that we have the conception of any group of objects consisting of Xs, Ys, and others, and that x, which we shall call an elective symbol, represents the mental operation of selecting from that group all the Xs which it contains, or of fixing the attention upon the Xs to the exclusion of all which are not Xs, y the mental operation of selecting the Ys, and so on; then, 1 or the universe being the subject conception, we shall have

$$x\,1 \text{ or } x = \text{the class X},$$

$$y\,1 \text{ or } y = \text{the class Y},$$

$$xy\,1 \text{ or } xy = \text{the class each member of which is both } X \text{ and Y},$$

and so on.

In like manner we shall have

$$1-x = \text{the class not-X},$$

$$1-y = \text{the class not-Y},$$

$$x(1-y) = \text{the class whose members are Xs but not-Ys},$$

$$(1-x)(1-y) = \text{the class whose members are neither Xs nor Ys},$$

Furthermore, from consideration of the nature of the mental operation involved, it will appear that the following laws are satisfied. Representing by x, y, z any elective symbols whatever,

$$x(y+z) = xy + xz$$

$$xy = yx, \text{ \&c.}$$

$$x^{\pi} = x, \text{ \&c.}$$

From the first of these it is seen that elective symbols are distributive in their operation; from the second that they are *commutative*. The third I have termed the index law; it is peculiar to elective symbols. The truth of these laws does not at all depend upon the nature, or the number, or the mutual relations, of the individuals included in the different classes. There may be but one individual in a class, or there may be a thousand. There may be individuals common to different classes, or the classes may be mutually exclusive. All elective symbols are distributive, and commutative, and all elective symbols satisfy the law expressed. These laws are in fact embodied in every spoken or written language. The equivalence of the expressions "good wise man" and "wise good man," is not a mere truism, but an assertion of the law of commutation exhibited. And there are similar illustrations of the other laws.

With these laws there is connected a general axiom. We have seen that algebraic operations performed with elective symbols represent mental processes. Thus the connexion of two symbols by the sign + represents the aggregation of two classes into a single class, the connexion of two symbols *xy* as in multiplication, represents the mental operation of selecting from a class *Y* those members which belong also to another class X, and so on. By such operations the conception of a class is modified. But beside this the mind has the power of perceiving relations of equality among classes. The axiom in question, then, is that *if a relation of equality is perceived between two classes, that relation remains unaffected when both subjects are equally modified by the operations above described.* (A). This axiom, and not "Aristotle's dictum," is the real foundation of all reasoning, the form and character of the process being, however, determined by the three laws already stated.

It is not only true that every elective symbol representing a class satisfies the index law (3), but it may be rigorously demonstrated that any combination of elective symbols $\phi(xyz..)$, which satisfies the law $\phi(xyz..)^{\pi} = \phi xyz..)$, represents an intelligible conception,– a group or class defined by a greater or less number of properties and consisting of a greater or less number of parts.

The four categorical propositions upon which the doctrine of ordinary syllogism is founded, are

All Ys are Xs.	A,
No Ys are Xs.	E,
Some Ys are Xs.	I,
Some Ys are not Xs.	O.

We shall consider these with reference to the classes among which relation is expressed.

A. The expression All Ys represents the class *Y* and will therefore be expressed by *y*, the copula are by the sign =, the indefinite term, Xs, is equivalent to Some Xs. It is a convention of language, that the word Some is expressed in the subject, but not in the predicate of a proposition. The term Some Xs will be expressed by vx, in which v is an elective symbol appropriate to a class V, some members of which are Xs, but which is in other respects arbitrary. Thus the proposition A will be expressed by the equation

$$y = vx$$

E. In the proposition, No Ys are Xs, the negative particle appears to be attached to the subject instead of to the predicate to which it manifestly belongs. We do not intend to say that those things which are not-Ys are Xs, but that things which are Ys are not-Xs. Now the class not-Xs is expressed by $1-x$; hence the proposition No Ys are Xs, or rather All Ys are not-Xs, will be expressed by

$$y = v(1-x)$$

I. In the proposition Some Ys are Xs, or Some Ys are Some Xs, we might regard the Some in the subject and the Some in the predicate as having reference to the same arbitrary class V, and so write

$$vy = vx,$$

but it is less of an assumption to refrain from doing this. Thus we should write

$$vy = v'x$$

v' referring to another arbitrary class V'.

O. Similarly, the proposition Some Ys are not-Xs, will be expressed by the equation

$$vy = v'(1-x).$$

It will be seen from the above that the forms under which the four categorical propositions A, E, I, O are exhibited in the notation of elective symbols are analogous with those of pure language, *i.e.* with the forms which human speech would assume, were its rules entirely constructed upon a scientific basis. In a vast majority of the propositions which can be conceived by the mind, the laws of expression have not been modified by usage, and the analogy becomes more apparent, *e.g.* the interpretation of the equation

$$z = x(1-y) + y(1-x),$$

is, the class Z consists of all Xs which are not-Ys and of all Ys which are not-Xs.

GENERAL THEOREMS RELATING TO ELECTIVE FUNCTIONS

We have now arrived at this step,–that we are in possession of a class of symbols x, y, z, &c. satisfying certain laws, and applicable to the rigorous expression of any categorical proposition whatever. It will be our next business to exhibit a few of the general theorems of the calculus which rest upon the basis of those laws, and these theorems we shall afterwards apply to the discussion of particular examples.

Of the general theorems I shall only exhibit two sets: those which relate to the development of functions, and those which relate to the solution of equations.

Theorems of Development

1. If x be any elective symbol, then

   $\phi(x) = \phi(1)\, x + \phi(0)\, (1-x)$

   the coefficients $\phi(1)$, $\phi(0)$, which are quantitative or common algebraic functions, are called the moduli, and x and $1-x$ the constituents.

2. For a function of two elective symbols we have

   $\phi(xy) = \phi(11)xy + \phi(10)\, x(1-y)$

   $+ \phi(01)(1-x)y + \phi(00)(1-x)(1-y)$

   in which $\phi(11)$, $\phi(10)$, &c. are quantitative, and are called the moduli, and xy, $x(1-y)$, &c. the constituents.

3. Functions of three symbols:

   $\phi(xyz) - \phi(111)xyz + \phi(110)xy(1-z)$

   $+ \phi(101)x(1-y)z + \phi(100)x(1-y)(1-z)$

   $+ \phi(011)(1-x)yz + \phi(010)(1-x)y(1-z)$

   $+ \phi(001)xy(1-z) + \phi(000)(1-x)(1-y)(1-z)$

   in which $\phi(111)$, $\phi(110)$, &c. are the moduli, and xyz, $xy(1-z)$, &c. the constituents.

From these examples the general law of development is obvious. And I desire it to be noted that this law is a mere consequence of the primary laws which have been expressed.

Theorem: *If we have any equation* $\phi(xyz..) = 0$, and fully expand the first member, then every constituent whose modulus does not vanish may be equated to 0. This enables us to interpret any equation by a general rule.

Rule: Bring all the terms to the first side, expand this in terms of all the elective symbols involved in it, and equate to 0 every constituent whose modulus does not vanish.

As an example, let us take the equation

$$x + 2y - 3xy = 0$$

Here $\phi(xy) = x + 2y - 3xy$, whence the values of the moduli are

$\phi(11) = 0$, $\phi(10) = 1$, $\phi(01) = 2$, $\phi(00) = 0$,

so that the expansion gives

$x(1-y) + 2y(1-x) = 0$,

which is in fact only another form. We have, then, by the Rule

$x(1-y) = 0$

$y(1-x) = 0$

The former implies that there are no Xs which are not-Ys, the latter that there are no Ys which are not-Xs, these together expressing the full significance of the original equation.

We can, however, often recombine the constituents with a gain of simplicity. In the present instance, we have $x-y=0$, or $x=y$, that is, the class X is identical with the class Y. This proposition is equivalent to the two former ones. All equations are thus of equal significance which give, on expansion, the same series of constituent equations, and *all are interpretable*.

GENERAL SOLUTION OF ELECTIVE EQUATIONS

1. The general solution of the equation $(xy) = 0$, in which two elective symbols only are involved, y being the one whose value is sought, is

$$y = \frac{\phi(10)}{\phi(10)-\phi(11)}x + \frac{\phi(100)}{\phi(100)-\phi(01)}(1-x)$$

The coefficients

$$\frac{\phi(10)}{\phi(10)-\phi(11)}, \frac{\phi(00)}{\phi(00)-\phi(01)}$$

are here the moduli.

2. The general solution of the equation $\phi(xyz) = 0$, z being the symbol whose value is to be determined, is

$$z = \frac{\phi(110)}{\phi(110)-\phi(111)}xy + \frac{\phi(100)}{\phi(100)-\phi(101)}x(1-y)$$
$$+\frac{\phi(010)}{\phi(010)-\phi(011)}(1-x)y + \frac{\phi(000)}{\phi(000)-\phi(001)}(1-x)(1-y)$$

the coefficients of which we shall still term the moduli. The law of their formation will readily be seen, so that the general theorems which have been given for the solution of elective equations of two and three symbols, may be regarded as examples of a more general theorem applicable to all elective equations whatever. in applying these results it is to be observed, that if a modulus assume the form 0/0 it is to be replaced by an arbitrary elective symbol w, and that if a modulus assume any numerical value except 0 or 1, the constituent of which it is a factor must be separately equated to 0. Although these conditions are deduced solely from the laws to which the symbols are obedient, and without any reference to interpretation, they nevertheless render the solution of every equation interpretable in logic. To such formulae also *every question upon the relations of classes may be referred*. One or two very simple illustrations may suffice.

3. Given $yx = yz + x(1-z)$.

The Ys which are Xs consist of the Ys which are Zs and the Xs which are not-Zs. Required the class Z.

Here $\phi(xyz) = yx - yz - x(1-z)$,

$\phi(111) = 0$, $\phi(110) = 0$, $\phi(101) = 0$,

$\phi(100) = -1$, $\phi(011) = -1$, $\phi(010) = 0$,

$\phi(001) = 0$, $\phi(000) = 0$;

and substituting in (14), we have

$$z = 0/0\ xy + x(1-y) + 0/0\ (1-x)(1-y)$$
$$= x(1-y) + wxy + w'(1-x)(1-y).$$

Hence the class Z includes all Xs which are not-Ys, an indefinite number of Xs which are Ys, and an indefinite number of individuals which are neither Xs nor Ys. The classes w and w' being quite arbitrary, the indefinite remainder is equally so; it may vanish or not.

Since $1-z$ represents a class, not-Z, and satisfies the index law

$$(1-z)^{\pi} = 1-z,$$

as is evident on trial, we may, if we choose, determine the value of this element just as we should determine that of z.

Let us take, in illustration of this principle, the equation $y = vx$, (All Ys are Xs), and seek the value of $1-x$, the class not-X.

Put $1-x = z$ then $y = v(1-z)$, and if we write this in the form $y - v(1-z) = 0$ and represent the first member by $\phi(vyz)$, v here taking the place of x, we shall have

$\phi(111) = 1$, $\phi(110) = 0$, $\phi(101) = 0$, $\phi(100) = -1$,

$\phi(011) = 1$, $\phi(010) = 1$, $\phi(001) = 0$, $\phi(000) = 0$;

the solution will thus assume the form

$$z = \frac{0}{0-1}vy + \frac{-1}{-1-0}v(1-y) + \frac{1}{1-1}(1-v)y + \frac{0}{0-0}(1-v)(1-y),$$

or

$$1-x = v(1-y) + 1/0\ (1-v)y + 0/0\ (1-v)(1-y).$$

The infinite coefficient of the second term in the second member permits us to write

$$y(1-v) = 0,$$

the coefficient 0/0 being then replaced by w, an arbitrary elective symbol, we have

$$1-x = v(1-y) + w(1-v)(1-y),$$

or

$$1-x = \{v + w(1-v)\}\ (1-y).$$

We may remark upon this result that the coefficient $v + w(1-v)$ in the second member satisfies the condition

$$\{v + w(1-v)\}^{\pi} = v + w(1-v),$$

as is evident on squaring it. It therefore represents a *class*. We may replace it by an elective symbol u, we have then

$$1-x = u(1-y),$$

the interpretation of which is

All not-Xs are not-Ys.

This is a known transformation in logic, and is called conversion by contraposition, or negative conversion. But it is far from exhausting the solution we have obtained. Logicians have overlooked the fact, that when we convert the proposition All Ys are (some) Xs into All not-Xs are (some) not-Ys there is a relation between the two (*somes*), understood in the predicates.

The equation shews that whatever may be that condition which limits the Xs in the original proposition,–the not-Ys in the converted proposition consist of all which are subject to the same condition, and of an arbitrary remainder which are not subject to that condition. The equation further shews that there are no Ys which are not subject to that condition. We can similarly reduce the equation $y = v(1-x)$, No Ys are Xs, to the form $x = v'(1\ y)$ No Xs are Ys, with a like relation between v and v'.

If we solve the equation $y = vx$ All Ys are Xs, with reference to v, we obtain the subsidiary relation $y(1-x) = 0$ No Ys are not-Xs, and similarly from the equation $y = v(1-x)$ (No Ys are Xs) we get $xy = 0$.

These equations, which may also be obtained in other ways, I have employed in the original treatise. All equations whose interpretations are connected are similarly connected themselves, by solution or development.

ON SYLLOGISM

The forms of categorical propositions already deduced are

$y = vx$,	All Ys are Xs,
$y = v(1-x)$,	No Ys are Xs,
$vy = v'x$,	Some Ys are Xs,
$vy = v'(1-x)$,	Some Ys are not-Xs,

whereof the two first give, by solution, $1-x = v'(1-y)$. All not-Xs are not-Ys, $x = v'(1-y)$, No Xs are Ys. To the above scheme, which is that of Aristotle, we might annex the four categorical propositions

$1-y = vx$,	All not-Ys are Xs,
$1-y = v(1-x)$,	All not-Ys are not-Xs,
$v(1-y) = v'x$,	Some not-Ys are Xs,
$v(1-y) = v'(1-x)$,	Some not-Ys are not-Xs,

the first two of which are similarly convertible into

$1-x = v'y$,	All not-Xs are Ys,
$x = v'y$,	All Xs are Ys,
or	No not-Xs are Ys,

If now the two premises of any syllogism are expressed by equations of the above forms, the elimination of the common symbol y will lead us to an equation expressive of the conclusion.

Ex. 1. All Ys are Xs, $-y = vx$,

All Zs are Ys, $-z = v'y$,

The elimination of y gives

$$z = vv'x,$$

the interpretation of which is

All Zs are Xs,

the form of the coefficient vv' indicates that the predicate of the conclusion is limited by both the conditions which separately limit the predicates of the premises.

Ex. 2. All Ys are Xs, $-y = vx$,

All Ys are Zs, $-y = v'z$.

The elimination of y gives

$$v'z = vx,$$

which is interpretable into Some Zs are Xs. It is always necessary that one term of the conclusion should be interpretable by means of the equations of the premises. In the above case both are so.

Ex. 3. All Xs are Ys, $-x = vy$,

No Zs are Ys, $-z = v'(1 - y)$.

Instead of directly eliminating y let either equation be transformed by solution. The first gives

$$1 - y = u(1 - x),$$

u being equivalent to $v + w(1 - v)$, in which w is arbitrary. Eliminating $1 - y$ between this and the second equation of the system, we get

$$z = v'u(1 - x),$$

the interpretation of which is

No Zs are Xs.

Had we directly eliminated y, we should have had

$$vz = v'(v - x),$$

the reduced solution of which is

$$z = v'\{v + (1 - v)\}(1 - x),$$

in which w is an arbitrary elective symbol. This exactly agrees with the former result.

These examples may suffice to illustrate the employment of the method in particular instances. But its applicability to the demonstration of general theorems is here, as in other cases, a more important feature. I subjoin the results of a recent investigation of the Laws of Syllogism. While those results are characterized by great simplicity and bear, indeed, little trace of their mathematical origin, it would, I conceive, have been very difficult to arrive at them by the examination and comparison of particular cases.

LAWS OF SYLLOGISM DEDUCED FROM THE ELECTIVE CALCULUS

We shall take into account all propositions which can be made out of the classes X, Y, Z, and referred to any of the forms embraced in the following system,

A,	All Xs are Zs.	A',	All not-Xs are Zs.
E,	No Xs are Zs.	E'	No not-Xs are Zs, or
	(All not-Xs are not-Zs.)		
I,	Some Xs are Zs.	I',	Some not-Xs are Zs.
O,	Some Xs are not-Zs.	O',	Some not-Xs are not-Zs.

It is necessary to recapitulate that quantity (universal and particular) and quality (affirmative and negative) are understood to belong to the *terms* of propositions which is indeed the correct view.

Thus, in the proposition All Xs are Ys, the subject All Xs is universal-affirmative, the predicate (some) Ys particular-affirmative.

In the proposition, Some Xs are Zs, both terms are particular-affirmative.

The proposition No Xs are Zs would in philosophical language be written in the form All Xs are not-Zs. The subject is universal-affirmative, the predicate particular-negative.

In the proposition Some Xs are not-Zs, the subject is particular-affirmative, the predicate particular-negative. In the proposition All not-Xs are Ys the subject is universal-negative, the predicate particular-affirmative, and so on.

In a pair of premises there are four terms, viz. two subjects and two predicates; two of these terms, viz. those involving the Y or not-Y may be called the middle terms, the two others the extremes, one of these involving X or not-X, the other Z or not-Z.

The following are then the conditions and the rules of inference.

Case 1st. The middle terms of like quality.

Condition of Inference. One middle term universal.

Rule: Equate the extremes.

Case 2nd. The middle terms of opposite qualities.

1st. Condition of Inference: One extreme universal.

Rule: Change the quantity and quality of that extreme, and equate the result to the other extreme.

2nd. Condition of inference. Two universal middle terms.

Rule: Change the quantity and quality of either extreme, and equate the result to the other extreme.

I add a few examples,

1st All Ys are Xs.

All Zs are Ys.

This belongs to Case 1. All Ys is the universal middle term. The extremes equated give All Zs are Xs, the stronger term becoming the subject.

2nd	All Xs are Ys	= All Xs are Ys
	No Zs are Ys	No Zs are not-Ys

This belongs to Case 2, and satisfies the first condition. The middle term is particular-affirmative in the first premise, particular-negative in the second. Taking All-Zs as the universal extreme, we have, on changing its quantity and quality, Some not-Zs, and this equated to the other extreme gives

All Xs are (some) not-Zs = No Xs are Zs.

If we take All Xs as the universal extreme we get

No Zs are Xs.

3rd All Xs are Ys.

Some Zs are not-Ys.

This also belongs to Case 2, and satisfies the first condition. The universal extreme All Xs becomes, some not-Xs, whence

Some Zs are not-Xs.

4th All Ys are Xs.

All not-Ys are Zs.

This belongs to Case 2, and satisfies the second condition. The extreme Some Xs becomes All not-Xs, ——

∴ All not-Xs are Zs. ——

The other extreme treated in the same way would give

All not-Zs are Xs, ——

which is an equivalent result.

If we confine ourselves to the Aristotelian premises A, E, I, O, the second condition of inference in Case 2 is not needed. The conclusion will not necessarily be confined to the Aristotelian system.

Ex.	Some Ys are not-Xs = Some Ys are not-Xs
No Zs are Ys	All Zs are not-Ys

This belongs to Case 2, and satisfies the first condition. The result is

Some not-Zs are not-Xs.

These appear to me to be the ultimate laws of syllogistic inference. They apply to every case, and they completely abolish the distinction of figure, the necessity of conversion, the arbitrary and partial rules of distribution, &c. If all logic were reducible to the syllogism these might claim to be regarded as the rules of logic. But logic, considered as the science of the relations of classes has been shewn to be of far greater extent. Syllogistic inference, in the elective system, corresponds to elimination. But this is not the highest in the order of its processes. All questions of elimination may in that system be regarded as subsidiary to the more general problem of the solution of elective equations. To this problem all questions of logic and of reasoning, without exception, may be referred.

For the fuller illustrations of this principle I must however refer to the original work. The theory of hypothetical propositions, the analysis of the positive and negative elements, into which all propositions are ultimately resolvable, and other similar topics are also there discussed. Undoubtedly the final aim of speculative logic is to assign the conditions which render reasoning possible, and the laws which determine its character and expression. The general axiom and the laws, appear to convey the most definite solution that can at present be given to this question. When we pass to the consideration of hypothetical propositions, the same laws and the same general axiom which ought perhaps also to be regarded as a law, continue to prevail; the only difference being that the subjects of thought are no longer classes of objects, but cases of the coexistent truth or falsehood of propositions.

Those relations which logicians designate by the terms conditional, disjunctive, &c., are referred by Kant to distinct conditions of thought. But it is a very remarkable fact, that the expressions of such relations can be deduced the one from the other by mere analytical process. From the equation $y = vx$, which expresses the *conditional* proposition, "If the proposition Y is true the proposition X is true," we can deduce

$$yx + (1 - y)x + (1 - y)(1 - x) = 1,$$

which expresses the *disjunctive* proposition, "Either Y and X are together true, or X is true and Y is false, or they are both false," and again the equation $y(1 - x) = 0$, which expresses a relation of coexistence, *viz.* that the truth of Y and the falsehood of X do not coexist. The distinction in the mental regard, which has the best title to be regarded as fundamental, is, I conceive, that of the affirmative and the negative. From this we deduce the direct and the inverse in operations, the true and the false in propositions, and the opposition of qualities in their terms.

The view which these enquiries present of the nature of language is a very interesting one. They exhibit it not as a mere collection of signs, but as a system of expression, the elements of which are subject to the laws of the thought which they represent. That those laws are as rigorously mathematical as are the laws which govern the purely quantitative conceptions of space and time, of number and magnitude, is a conclusion which I do not hesitate to submit to the exactest scrutiny.

Chapter 3

Graph Theory

A graph is a symbolic representation of a network and of its connectivity. It implies an abstraction of the reality so it can be simplified as a set of linked nodes.

The development of graph theory is very similar the development of probability theory, where much of the original work was motivated by efforts to understand games of chance. The large portions of graph theory have been motivated by the study of games and recreational mathematics. Generally speaking, we use graphs in two situations. Firstly, since a graph is a very convenient and natural way of representing the relationships between objects we represent objects by vertices and the relationship between them by lines. In many situations (problems) such a pictorial representation may be all that is needed. Example of such applications are chemical molecule and map coloring as shown in the diagrams below. Other examples are signal-flow graphs, Konigsberg bridges, tracing maze etc.

H H
H — C — C — O — H
H H

Secondly, we take the graph as mathematical model, solve the appropriate graph-theoretic problem, and then interpret the solution in terms of the original problem.

BASIC GRAPH DEFINITION

Graph theory is a branch of mathematics concerned about how networks can be encoded and their properties measured.

In transport geography most networks have an obvious spatial foundation, namely road, transit and rail networks, which tend to be defined more by their links than by their nodes. This it is not necessarily the case for all transportation networks. For instance, maritime and air networks tend to be more defined more by their nodes than by their links since links are often not clearly defined.

A telecommunication system can also be represented as a network, while its spatial expression can have limited importance and would actually be difficult to represent. Mobile telephone networks or the Internet, possibly to most complex graphs to be considered, are relevant cases of networks having a structure that can be difficult to symbolize.

However, cellular phones and antennas can be represented as nodes while the links could be individual phone calls. Servers, the core of the Internet, can also be represented as nodes within a graph while the physical infrastructure between them, namely fiber optic cables, can act as links. Consequently, all transport networks can be represented by graph theory in one way or the other.

The following Elements are Fundamental at Understanding Graph Theory

Graph. A graph G is a set of vertex (nodes) v connected by edges (links) e. Thus $G=(v, e)$.

Vertex (Node). A node v is a terminal point or an intersection point of a graph. It is the abstraction of a location such as a city, an administrative division, a road intersection or a transport terminal (stations, terminuses, harbors and airports).

Edge (Link). An edge e is a link between two nodes. The link (i, j) is of initial extremity i and of terminal extremity j. A link is the abstraction of a transport infrastructure supporting movements between nodes. It has a direction that is commonly represented as an arrow. When an arrow is not used, it is assumed the link is bi-directional.

Sub-Graph. A sub-graph is a subset of a graph G where p is the number of sub-graphs. For instance $G' = (v', e')$ can be a distinct sub-graph of G. Unless the global transport system is considered in its whole, every transport network is in theory a sub-graph of another. For instance, the road transportation network of a city is a sub-graph of a regional transportation network, which is itself a sub-graph of a national transportation network.

Buckle. A link that makes a node correspond to itself is a buckle.

Planar Graph. A graph where all the intersections of two edges are a vertex. Since this graph is located within a plane, its topology is two-dimensional.

Non-planar Graph. A graph where there are no vertex at the intersection of at least two edges. This implies a third dimension in the topology of the graph since there is the possibility of having a movement "passing over" another movement such as for air transport. A non-planar graph has potentially much more links than a planar graph.

LINKS AND THEIR STRUCTURES

A transportation network enables flows of people, freight or information, which are occurring along its links. Graph theory must thus offer the possibility of representing movements as linkages, which can be considered over several aspects:

Connection. A set of two nodes as every node is linked to the other. Considers if a movement between two nodes is possible, whatever its direction. Knowing connections makes it possible to find if it is possible to reach a node from another node within a graph.

Path. A sequence of links that are traveled in the same direction. For a path to exist between two nodes, it must be possible to travel an uninterrupted sequence of links. Finding all the possible paths in a graph is a fundamental attribute in measuring accessibility and traffic flows.

Chain. A sequence of links having a connection in common with the other. Direction does not matter.

Length of a Link, Connection or Path. Refers to the label associated with a link, a connection or a path. This label can be distance, the amount of traffic, the capacity or any attribute of that link. The length of a path is the number of links (or connections) in this path.

Cycle. Refers to a chain where the initial and terminal node is the same and that does not use the same link more than once is a cycle.

Circuit. A path where the initial and terminal node corresponds. It is a cycle where all the links are traveled in the same direction. Circuits are very important in transportation because several distribution systems are using circuits to cover as much territory as possible in one direction (delivery route).

BASIC STRUCTURAL PROPERTIES

The organization of nodes and links in a graph conveys a structure that can be described and labeled. The basic structural properties of a graph are:

Symmetry and Asymmetry. A graph is symmetrical if each pair of nodes linked in one direction is also linked in the other. By convention, a line without an arrow represents a link where it is possible to move in both directions. However, both directions have to be defined in the graph. Most transport systems are symmetrical but asymmetry can often occur as it is the case for maritime (pendulum) and air services. Asymmetry is rare on road transportation networks, unless one-way streets are considered.

Completeness. A graph is complete if two nodes are linked in at least one direction. A complete graph has no sub-graph.

Connectivity. A complete graph is described as connected if for all its distinct pairs of nodes there is a linking chain. Direction does not have importance for a graph to be connected, but may be a factor for the *level* of connectivity. If $p>1$ the graph is not connected because it has more than one sub-graph. There are various levels of connectivity, depending on the degree at which each pair of nodes is connected.

Complementary. Two sub graphs are complementary if their union results in a complete graph. Multimodal transportation networks are complementary as each sub-graph benefits from the connectivity of other sub-graphs.

Root. A node *r* where every other node is the extremity of a path coming from *r* is a root. Direction has an importance. A root is generally the starting point of a distribution system, such as a factory or a warehouse.

Trees. A connected graph without a cycle is a tree. A tree has the same number of links than nodes plus one. ($e = v\text{-}1$). If a link is removed, the graph ceases to be connected. If a new link between two nodes is provided, a cycle is created. A branch of root *r* is a tree where no links are connecting any node more than once.

Articulation Node. In a connected graph, a node is an articulation node if the sub-graph obtained by removing this node is no longer connected. It therefore contains more than one sub-graph ($p > 1$). An articulation node is generally a port or an airport, or an important hub of a transportation network, which serves as a bottleneck.

Isthmus. In a connected graph, an isthmus is a link that is creating, when removed, two sub-graphs having at least one connection.

HISTORY OF GRAPH THEORY

The origin of graph theory can be traced back to Euler's work on the Konigsberg bridges problem, which subsequently led to the concept of an Eulerian graph. The study of cycles on polyhedra by the Thomas P. Kirkman and William R. Hamilton led to the concept of a Hamiltonian graph.

The concept of a tree, a connected graph without cycles, appeared implicitly in the work of Gustav Kirchhoff, who employed graph-theoretical ideas in the calculation of currents in electrical networks or circuits. Later, Arthur Cayley, James J. Sylvester George Polya, and others use 'tree' to enumerate chemical molecules.

The study of planar graphs originated in two recreational problems involving the complete graph K_5 and the complete bipartite graph $K_{3,3}$. These graphs proved to be planarity, as was subsequently demonstrated by Kuratowski. First problem was presented by A. F. Mobius around the year 1840 as follows:

Once upon a time, there was a king with five sons.

In his will he stated that after his death the sons

should divide the kingdom into five provinces so

that the boundary of each province should have
a frontiers line in common with each of the other
four provinces.

Here the problem is whether one can draw five mutually neighboring regions in the plane.

The king further stated that all five brothers should
join the provincial capital by roads so that no two
roads intersect.

Here the problem is that deciding whether the graph K_5 is planar.

The origin of second problem is unknown but it is first mentioned by H. Dudeney in 1913 in its present form.

The puzzle is to lay a water, gas, and electricity
to each of the three houses without any pipe
crossing another.

This problem is that of deciding whether the graph $K_{3,3}$ is planar.

The celebrated four-color problem was first posed by Francis Guthrie in 1852. and a celebrated incorrect "proof" by appeared in 1879 by Alfred B. Kempe. It was proved by Kenneth Appel and Wolfgang Haken in 1976 and a simpler and more systematic proof was produced by Neil Roberton, Daniel Sanders, Paul Seymour, and Robin Thomas in 1994.

DEFINITIONS AND EXAMPLES

Informally, a graph is a diagram consisting of points, called vertices, joined together by lines, called edges; each edge joins exactly two vertices. A graph G is a triple consisting of a vertex set of V(*G*), an edge set E(G), and a relation that associates with each edge two vertices (not necessarily distinct) called its endpoints.

Definition of Graph

A graph $G = (V, E)$ consists of a (finite) set denoted by V, or by $V(G)$ if one wishes to make clear which graph is under consideration, and a collection E, or $E(G)$, of unordered pairs $\{u, v\}$ of distinct elements from V. Each element of V is called a vertex or a point or a node, and each element of E is called an edge or a line or a link.

Formally, a graph G is an ordered pair of dsjoint sets (V, E), where $E \subseteq V \times V$. Set V is called the vertex or node set, while set E is the edge set of graph G. Typically, it is assumed that self-loops (i.e. edges of the form (u, u), for some $u \in V$) are not contained in a graph.

Directed and Undirected Graph

A graph G = (V, E) is directed if the edge set is composed of ordered vertex (node) pairs. A graph is undirected if the edge set is composed of unordered vertex pair.

Vertex Cardinality

The number of vertices, the cardinality of V, is called the order of graph and devoted by |V|. We usually use n to denote the order of G. The number of edges, the cardinality of E, is called the size of graph and denoted by |E|. We usually use m to denote the size of G.

Neighbor Vertex and Neighborhood

We write $v_iv_j \in E(G)$ to mean $\{v_i, v_j\} \in E(G)$, and if $e = v_i\, v_j \in E(G)$, we say v_i and v_j are adjacent.

Formally, given a graph G = (V, E), two vertices $v_i, v_j \in V$ are said to be neighbors, or adjacent nodes, if $(v_i, v_j) \in E$. If G is directed, we distinguish between incoming neighbors of v_i (those vertices $v_j \in V$ such that $(v_j, v_i) \in E$) and outgoing neighbors of v_i (those vertices $v_j \in V$ such that $(v_i, v_j) \in E$).

The open neighborhood $N(v)$ of the vertex v consists of the set vertices adjacent to v, that is, $N(v) = \{w \in v: vw \in E\}$. The closed neighborhood of v is $N[v] = N(v) \cup \{v\}$. For a set S N V, the open neighborhood N(S) is defined to be $U_{v \in S}N(v)$, and the closed neighborhood of S is $N[S] = N(S) \cup S$.

Vertex Degree: The degree deg(v) of vertex v is the number of edges incident on v or equivalently, $\deg(v) = |N(v)|$. The degree sequence of graph is $(\deg(v_1), \deg(v_2), ..., \deg(v_n))$, typically written in nondecreasing or nonincreasing order. The minimum and maximum degree of vertices in V(G) are denoted by $d(G)$ and $\delta(G)$, respectively. If $d(G) = \Delta(G) = r$, then graph G is said to be regular of degree r, or simply r-regular.

Formally, given a graph G = (V, E), the degree of a vertex $v \in V$ is the number of its neighbors in the graph. That is,

$\deg(v) = |\{u \in V: (v, w) \in E\}|$.

If G is directed, we distinguish between in-degree (nimber of incoming neighbors) and out-degree (number of outgoing neighbors) of a vertex.

Loop and Multiple Edges: A loop is an edge whose endpoints are equal i.e., an edge joining a vertex to it self is called a loop. We say that the graph has multiple edges if in the graph two or more edges joining the same pair of vertices.

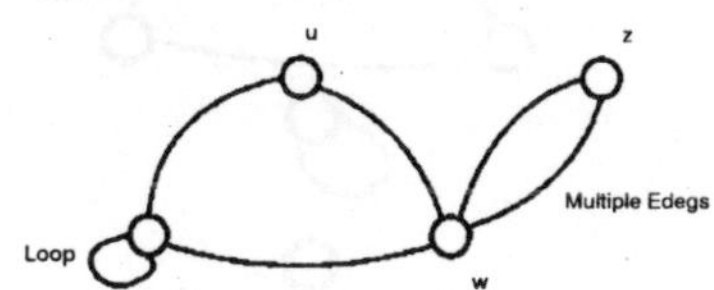

Simple Graph: A graph with no loops or multiple edges is called a simple graph. We specify a simple graph by its set of vertices and set of edges, treating the edge set as a set of unordered pairs of vertices and write $e = uv$ (or $e = vu$) for an edge e with endpoints u and v.

When u and v are endpoints of an edge, they are adjacent and are neighbors.

Connected Graph

A graph that is in one piece is said to be connected, whereas one which splits into several pieces is disconnected.

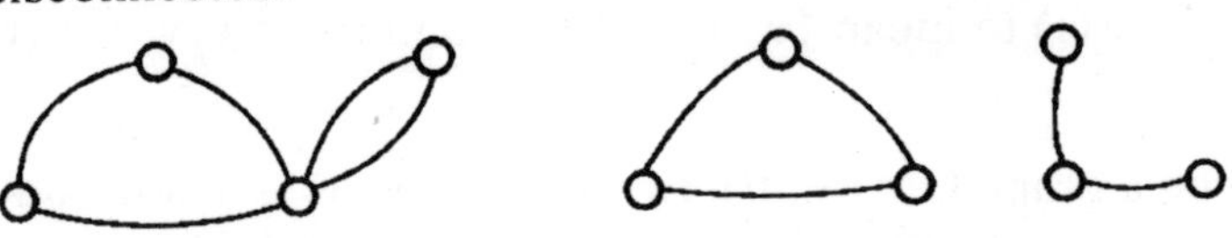

Connected Non-Simple Graph　　　Disconnected Simple Graph

A graph G is connected if there is a path in G between any given pair of vertices, otherwise it is disconnected. Every disconnected graph can be split up into a number of connected subgraphs, called components.

Subgraph

Let G be a graph with vertex set $V(G)$ and edge-list E(G). A subgraph of G is a graph all of whose vertices belong to $V(G)$ and all of whose edges belong to $E(G)$. For example, if G is the connected graph below:

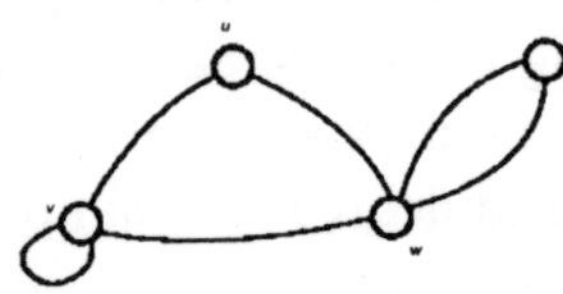

where $V(G) = \{u, v, w, z\}$ and $E(G) = (uv, uw, vv, vw, wz, wz\}$ then the following four graphs are subgraphs of G.

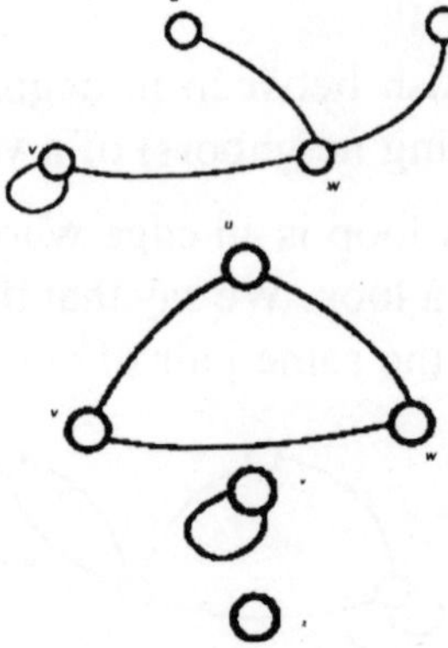

Degree (or Valency)

Let G be a graph with loops, and let v be a vertex of G. The degree of v is the number of edges meeting at v, and is denoted by $deg(v)$.

For example, consider, the following graph G

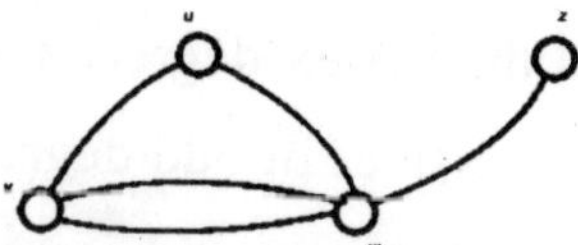

The graph G has $deg(u) = 2$, $deg(v) = 3$, $deg(\underline{w}) = 4$ and $deg(z) = 1$.

Regular Graph

A graph is regular if all the vertices of G have the same degree. In particular, if the degree of each vertex is r, the G is regular of degree r.

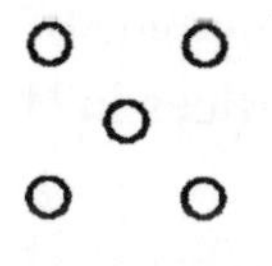

Graph with degree $r = 0$

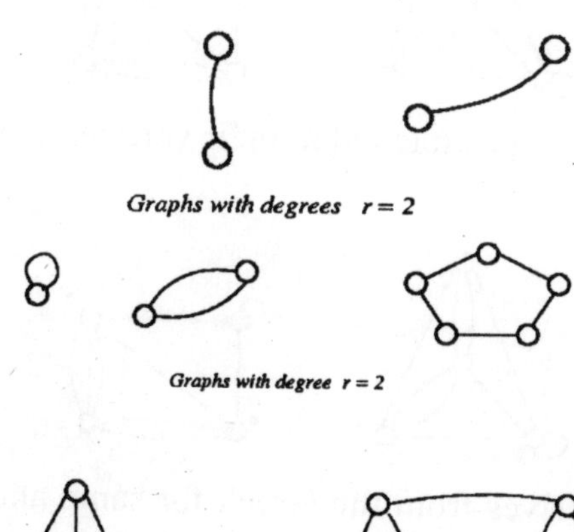

Graphs with degrees $r = 2$

Graphs with degree $r = 2$

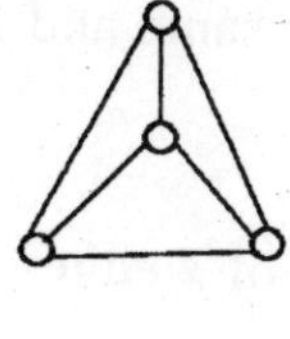

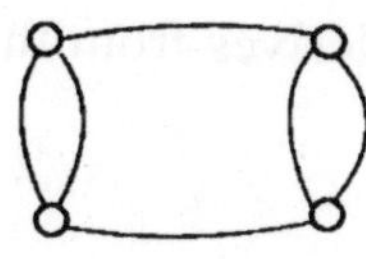

Graphs with degrees $r = 3$

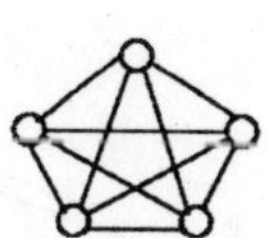

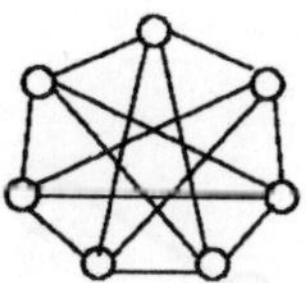

Graphs with degrees $r = 4$

The Handshaking Lemma: In any graph, the sum of all the vertex-degree is equal to twice the number of edges.

Proof: Since each edge has two ends, it must contribute exactly 2 to the sum of the degrees. The result follows immediately.

The Following are the consequences of the Handshaking lemma.

1. In any graph, the sum of all the vertex-degree is an even number.
2. In any graph, the number of vertices of odd degree is even.
3. If G is a graph which has n vertices and is regular of degree r, then G has exactly 1/2 nr edges.

Isomorphic Graphs

Two graph G and H are isomorphic if H can be obtained from G by relabeling the vertices – that is, if there is a one-to-one correspondence between the vertices of G and those of H, such that the number of edges joining any pair of vertices in G is equal to the number of edges joining the corresponding pair of vertices in H. For example, two unlabeled graphs, such as

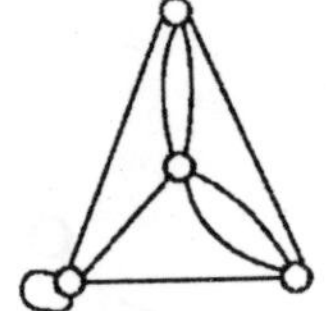

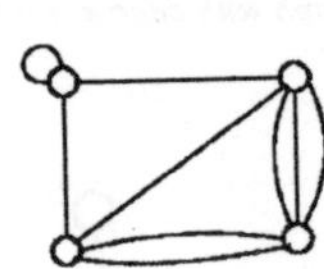

are isomorphic if labels can be attached to their vertices so that they become the same graph.

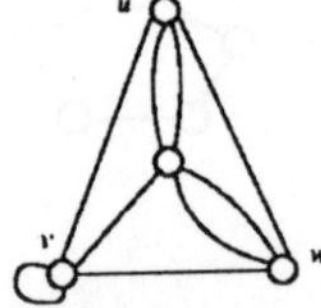

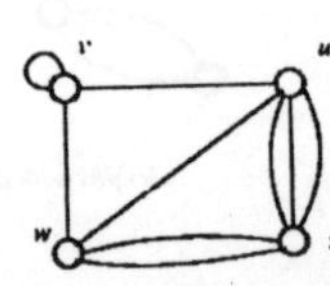

The word isomorphic derives from the Greek for same and form.

WALK

A walk of length k in a graph G is a succession of k edges of G of the form uv, vw, wx,..., yz.

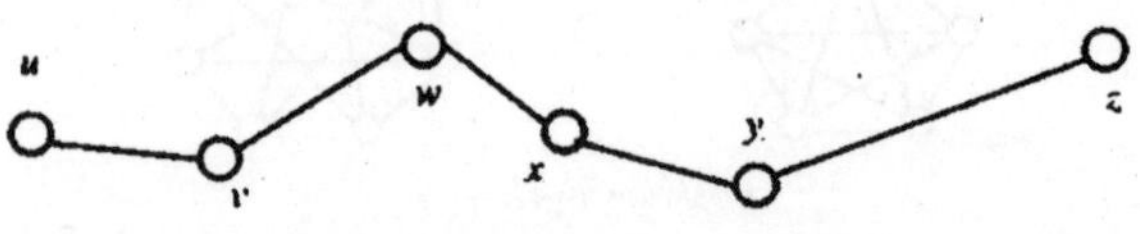

We denote this walk by *uvwx.. yz* and refer to it as a walk between *u* and *z*.

Trail and Path

If all the edges (but no necessarily all the vertices) of a walk are different, then the walk is called a trail. If, in addition, all the vertices are difficult, then the trail is called path.

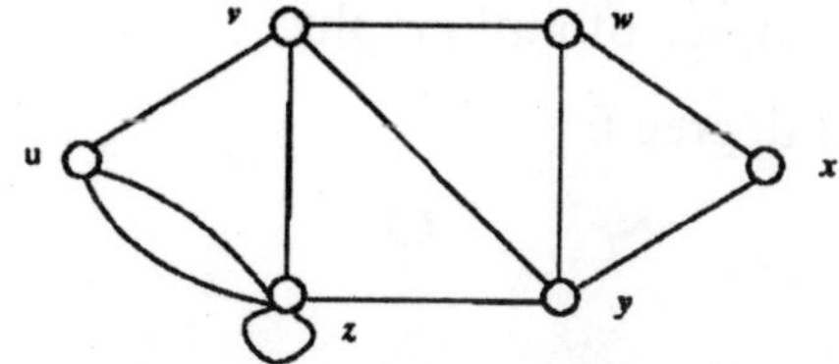

1. The walk *vzzywxy* is a trail since the vertices y and z both occur twice.
2. The walk *vwxyz* is a path since the walk has no repeated vertices.

Complete Graphs

A computer graph is a graph in which every two distinct vertices are joined by exactly one edge. The complete graph with *n* vertices is denoted by K_n.

The following are the examples of complete graphs.

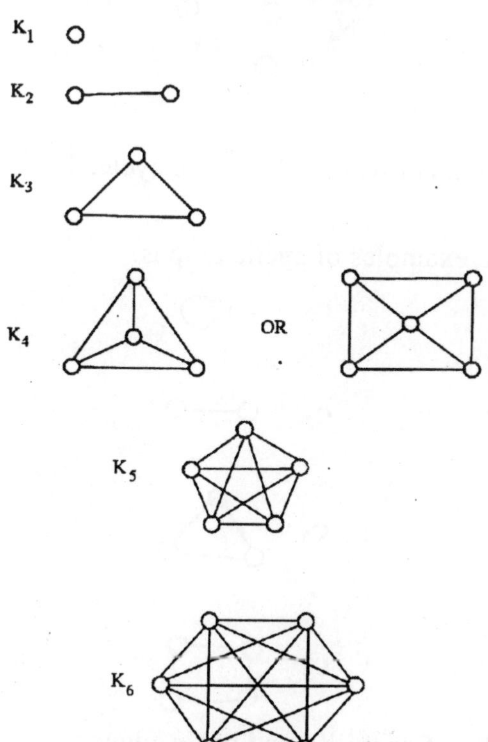

The graph K_n is regular of degree n-1, and therefore has $1/2n(n-1)$ edges, by consequence 3 of the handshaking lemma.

Null Graphs

A null graphs is a graph containing no edges. The null graph with n vertices is denoted by N_n.

The following are the examples of null graphs.

Note that N_n is regular of degree 0.

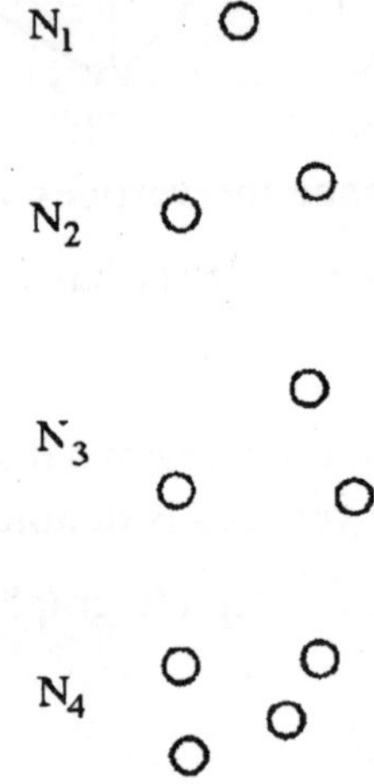

Cycle Graphs

A cycle graph is a graph consisting of a single cycle. The cycle graph with n vertices is denoted by C_n.

The following are the examples of cyclic graphs.

C_1

C_2

C_3

C_4

Note that C_n is regular of degree 2, and has n edges.

Path Graphs

A path graph is a graph consisting of a single path. The path graph with n vertices is denoted by P_n.

The following are the examples of path graphs.

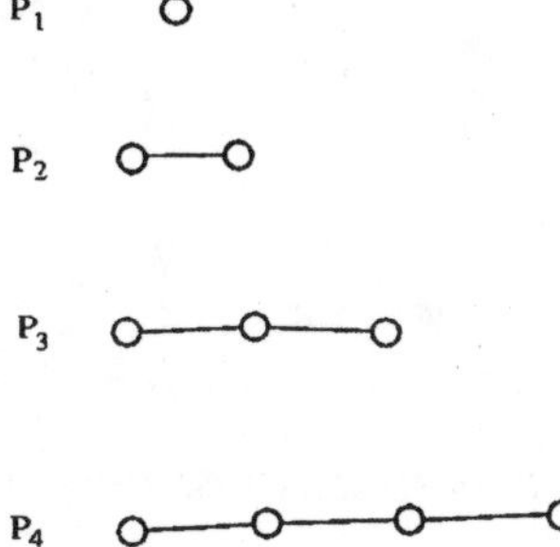

Note that path graph, P_n, has n-1 edges, and can be obtained from cycle graph, C_n, by removing any edge.

Bipartite Graphs

A bipartite graph is a graph whose vertex-set can be split into two sets in such a way that each edge of the graph joins a vertex in first set to a vertex in second set. The examples of bipartite graphs are:

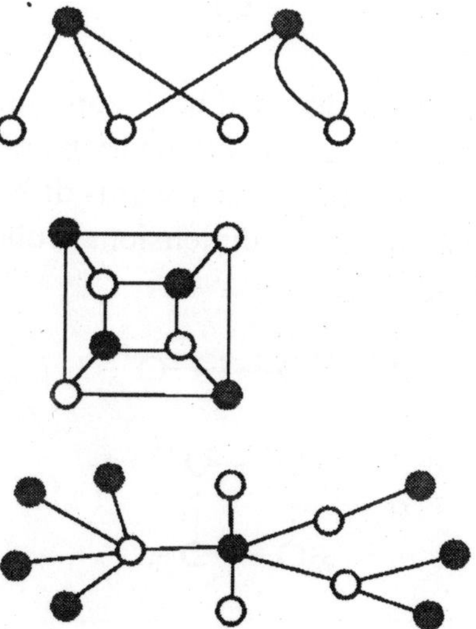

Complete Bipartite Graph

A complete bipartite graph is a bipartite graph in which each vertex in the first set is joined to each vertex in the second set by exactly one edge. The complete bipartite graph with r vertices and 3 vertices is denoted by $K_{r,s}$.

The following are some examples.

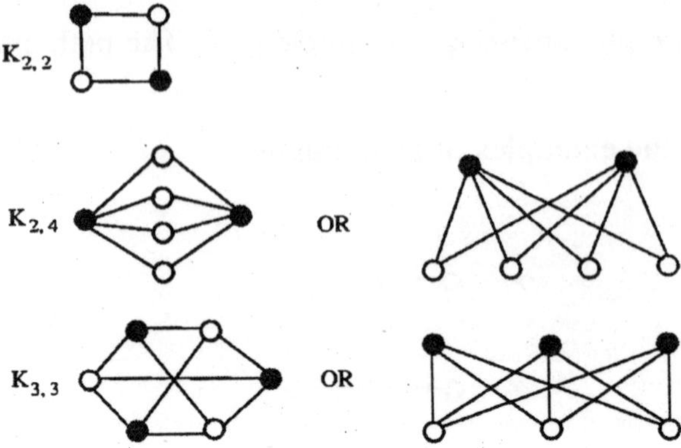

Note that $K_{r,s}$ has r+s vertices (r vertices of degrees, and s vertices of degree r), and rs edges. Note also that $K_{r,s} = K_{s,r}$.

An Important Note: A complete bipartite graph of the form $K_{r,s}$ is called a star graph.

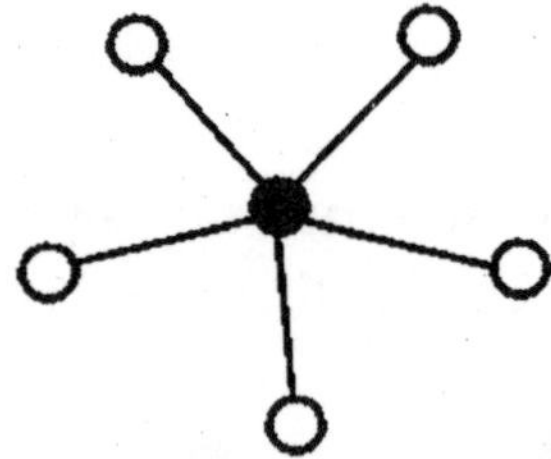

Cube Graph

The cube graphs is a bipartite graphs and have appropriate in the coding theory. The cube graphs constructed by taking as vertices all binary words of a given length and joining two of these vertices if the corresponding binary words differ in just one place. The binary words of length *k* is called the *k*-cube (or *k*-dimensional cube) graph and is denoted by Q_k. The following are some examples.

Q_1 0 1

Q_2 10 11 00 01

Q_3 100 101 110 111 010 011 000 001

Note that Q_k has 2^k vertices and is regular of degree k. It follows from consequence 3 of the handshaking lemma that Q_k has $k * 2^{k-1}$ edges.

The Peterson Graph

This graph is named after a Danish mathematician, Julius Peterson(1839-1910), who discovered the graph in a paper of 1898.

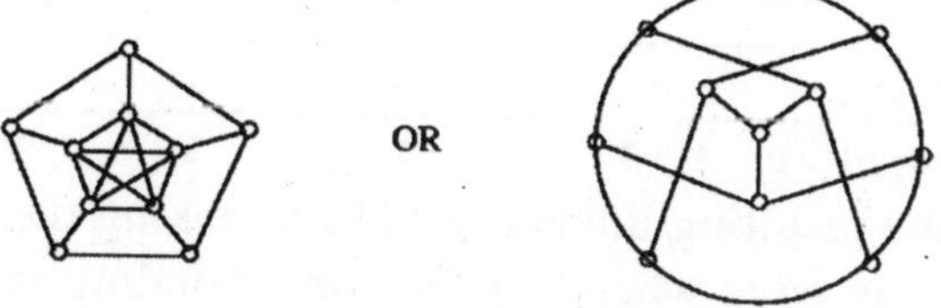

Tree Graph

A tree is a connected graph which has no cycles.

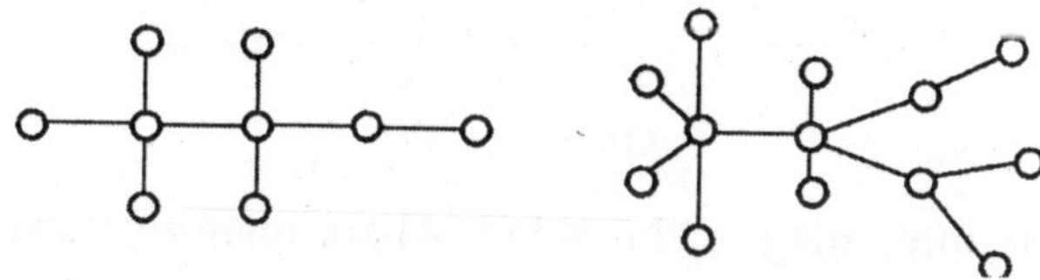

Spanning Tree

If G is a connected graph, the spanning tree in G is a subgraph of G which includes every vertex of G and is also a tree.

Consider the following graph

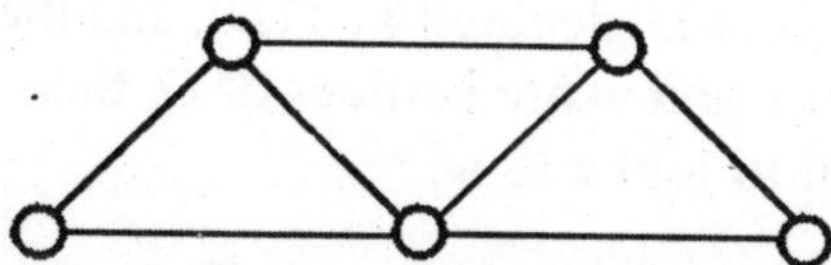

The following are the three of its spanning trees:

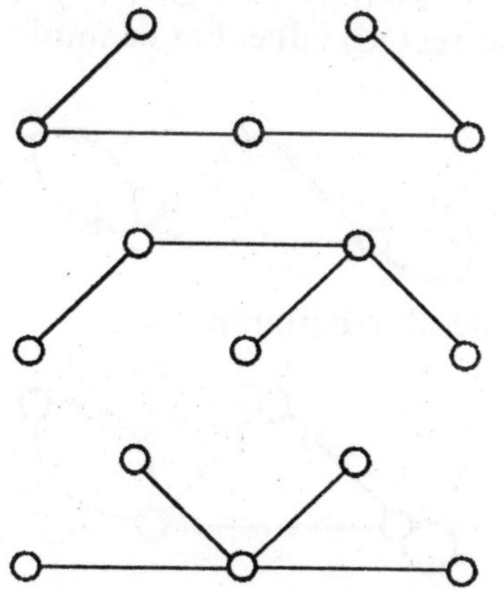

Interval Graphs

Consider the intervals (0, 3), (2, 7), (-1, 1), (2, 3), (1, 4), (6, 8) which may be illustrated as

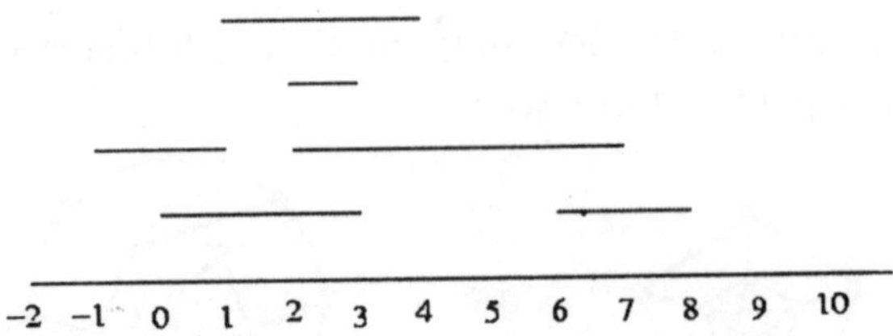

We can construct the resulting interval graphs by taking the interval as vertices, join two of these vertices by an edge whenever the corresponding intervals have at least one point in common.

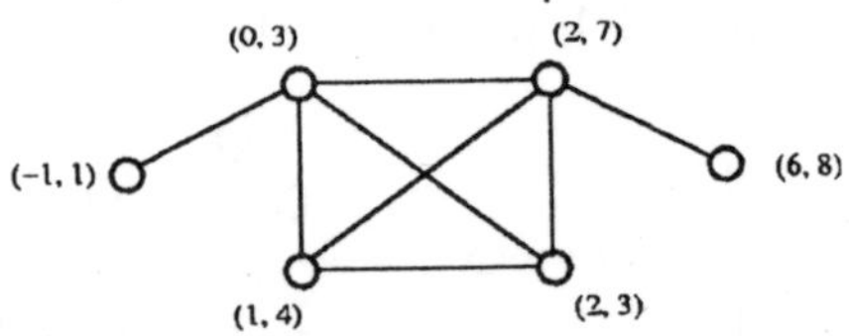

Note that since the intervals (–1, 1) and (1, 4) are open intervals, they do not have a point in common.

Definition of Diagraph

A directed graph or diagraph *D* consists of a set of elements, called vertices, and a list of ordered pairs of these elements, called arcs. The set of vertices is called arcs. The set of vertices is called the vertex-set of *D*, denoted by *V*(*D*), and the list of arcs is called the arc-list of *D*, denoted by *A*(*D*). If v and w are vertices of *D*, then an arc of the form *vw* is said to be directed from v to *w*, or to join *v* to *w*.

The underlying graph of diagraph is the graph obtained by replacing each arc of diagraph by corresponding (undirected) edge. For example, consider the following digraph

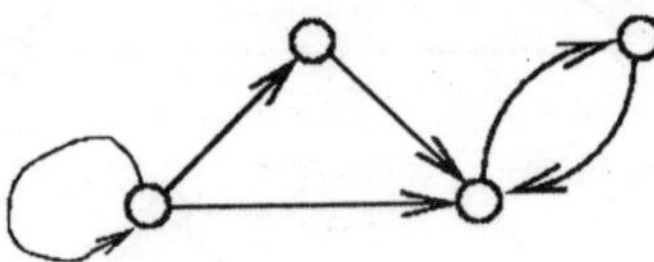

The underlying graph of the above digraph is

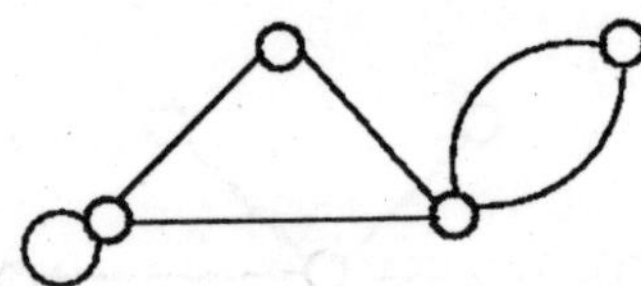

The Platonic Graphs

The following regular solids are called the Platonic solids:

Tetrahedron

Hexahedron (cube)

Octahedron

Dodecahedron

Icosahedron

The name Platonic arises from the fact that these five solids were mentioned in Plato's Timaeus. A Platonic graph is obtained by projecting the corresponding solid on to a plane.

Tetrahedron

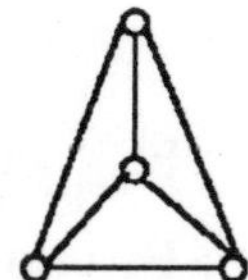

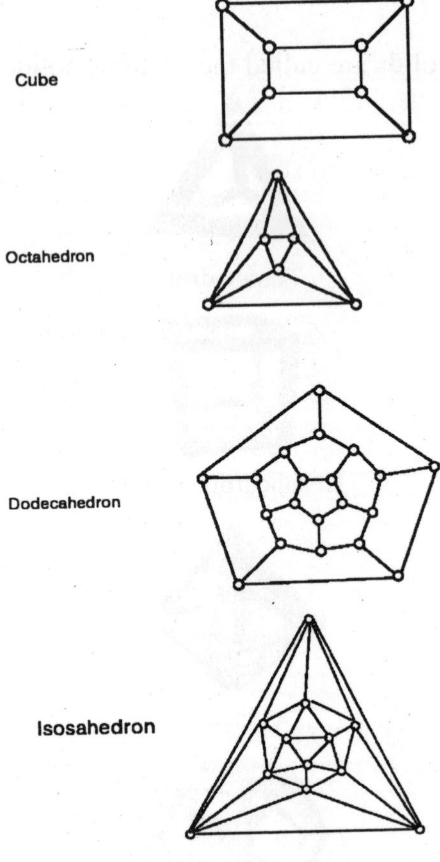

EULER GRAPHS

Consider the following road map

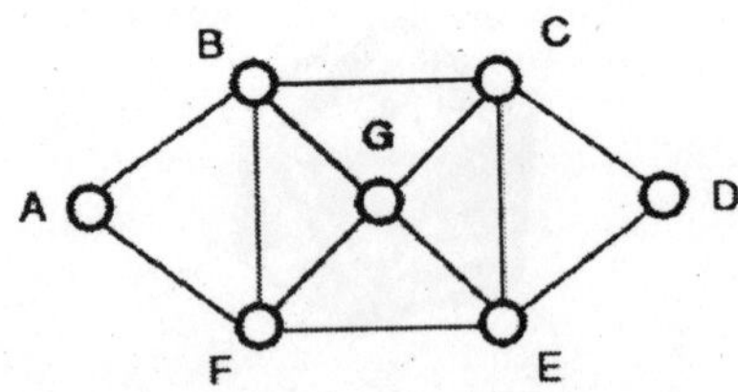

The explorer's Problem: An explorer wants to explore all the routes between a number of cities. Can a tour be found which traverses each route only once? Particularly, find a tour which starts at A, goes along each road exactly once, and ends back at A.

Examples of such tour are

A B C D E F B G C E G F A

A F G C D E G B C E F B A

The Explorer travels along each road (edges) just once but may visit a particular city (vertex) several times.

THE TRAVELER'S PROBLEM

A traveler wants to visit a number of cities. Can a tour be found which visits each city only once? Particularly, find a tour which starts at A, goes to each city exactly once, and ends back at A.

Examples of such tour are

A B C D E G F A

A F E D C G B A

The travelers visits each city (vertex) just once but may omit several of the roads (edges) on the way.

Eulerian Trail

A connected graph G is Eulerian if there is a closed trail which includes every edge of G, such a trail is called an Eulerian trail.

Hamiltonian Cycle

A connected graph G is Hamiltonian if there is a cycle which includes every vertex of G; such a cycle is called a Hamiltonian cycle.

Consider the following examples:

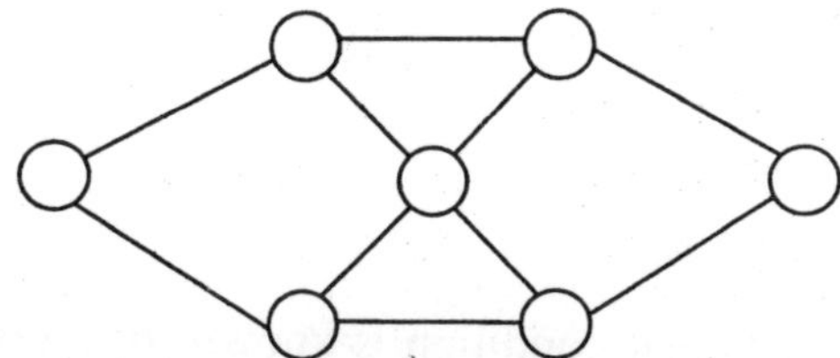

This graph is BOTH Eulerian and Hamiltonian.

This graph is Eulerian, but NOT Hamiltonian.

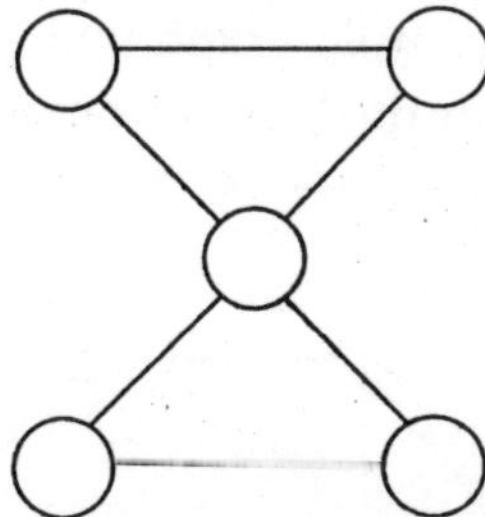

This graph is an Hamiltionian, but NOT Eulerian.

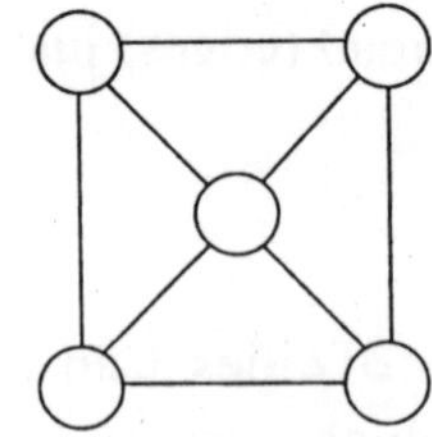

This graph is NEITHER Eulerian NOR Hamiltionian

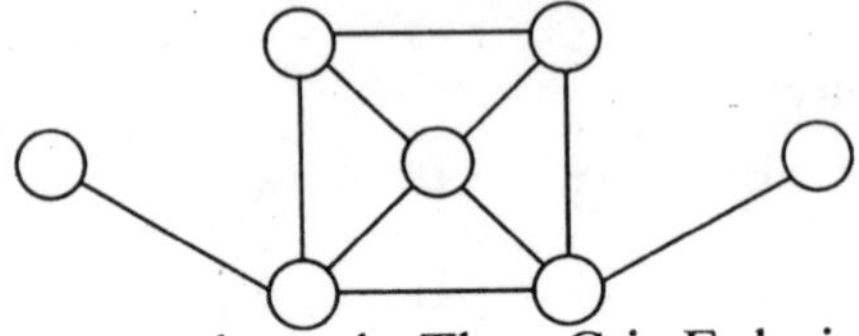

Theorem: Let G be a connected graph. Then G is Eulerian if and only if every vertex of G has even degree.

1. Necessary Condition
   - If G is Eulerian, then every vertex of G has even degree.
2. Sufficient Condition
   - If every vertex of G has even degree, then G is Eulerian.

Eulerian-Type Problems

1. Edge-traceable graphs.
2. Diagrams-Tracing Puzzles.
3. Dominoes.
4. Mazes and labyrinths,
5. The Chinese Postman Problem.
6. The Rotating Drum Problem.

Neither necessary nor sufficient condition is known for a graph to be Hamiltonian. The search for necessary or sufficient conditions is a major area of study in graph theory today.

Sufficient Condition

Dirac's Theorem: Let G be a simple graph with n vertices where $n \geq 3$ If $\deg(v) \geq 1/2\, n$ for each vertex v, then G is Hamiltonian. For example,

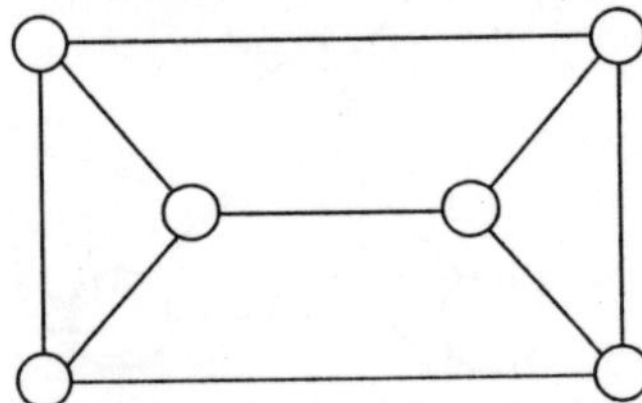

$n = 6$ and $\deg(v) = 3$ for each vertex, so this graph is Hamiltonian by Dirac's theorem.

Ore's Theorem: *Let G be a simple graph with n vertices where n ≥ 2 if deg(v) + deg(w) ≥ n for each pair of non-adjacent vertices v and w, then G is Hamiltonian.*

For example,

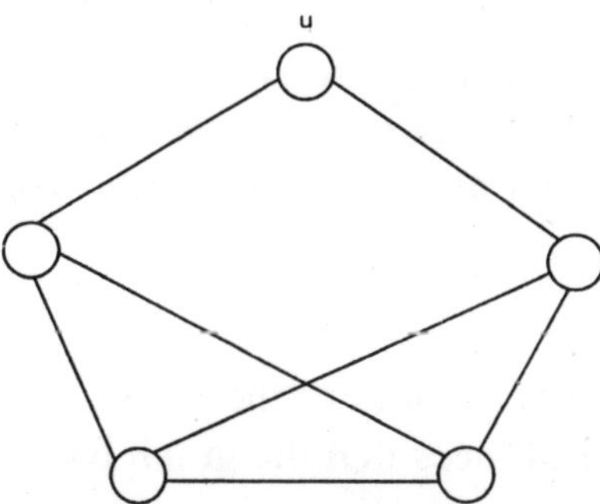

$n = 5$ but $\deg(u) = 2$, so Dirac's theorem does not apply. However, $\deg(v) + \deg(w) \geq 5$ for all pairs of vertices v and w (infact, for all pairs of vertices v and w), so this graph is Hamiltonian by Ore's theorem.

Note that if $\deg(v) \geq 1/2\ n$ for each vertex, then $\deg(v) + \deg(w) \geq n$ for each pair of vertices v and w. It follows that Dirac's theorem can be deduced from Ore's theorem, so we prove only Ore's threoem.

GRAPHS FROM AN ABSTRACT VIEWPOINT

Pictures of graphs showing all of the edges and vertices are great ways to represent a graph. It is immediately clear which vertices are joined by edges, and you can often recognize the graph from its picture. For example, if you were given a very abstract description of the cube graph like: *The graph has eight vertices and 12 edges. Every vertex meets exactly three vertices.* It might take you some time and effort to recognize that this *is* the cube graph. If you see a picture that looks exactly like a cube, then you know immediately that you are dealing with the cube graph.

However, pictorial representations of graphs have their limitations, too. For example, if you wanted to prove a theorem about graphs, and all you had were pictorial representations, then your proof might have to consist of drawing every single, possible graph, and then demonstrating the truth of your theorem for each picture that you have drawn. A pretty time-consuming process, to say the least.

To help us find other ways of describing graphs, consider the following definitions:

Definition: A graph G consists of a set of vertices and a set of edges, where each edge joins an unordered pair of vertices. The set of vertices of G is denoted by $V(G)$ and the set of edges is denoted by $E(G)$.

Definition: A loop is an edge that joins a vertex to itself. A multiple edge occurs when there is more than one edge joining a pair of vertices.

Definition: A simple graph is a pair ($V(G)$, $E(G)$) where $V(G)$ is a non-empty finite set of elements, and $E(G)$ is a finite set of unordered pairs of distinct elements of $V(G)$.

ABSTRACT DEFINITIONS AND EXAMPLES OF GRAPHS

Pictorial representations of graphs are really useful representations, but are sometimes inconvenient to use when proving theorems about combinatorial graphs. For example, planar graphs (graphs that can be drawn on a page so that no edges cross) obey the following formula:

$$v - e + f = 2$$

where v represents the number of vertices of the graph, e is the number of edges that the graph has, and f is the number of faces that the graph has.

1. How could you prove this formula?
2. If you took the summer course *Geometry and the Imagination*, then you might remember that we proved a similar formula:

$$v - e + f = 2k$$

where $k = 1$ for a sphere, $k = 0$ for a Euclidean plane, and $0k = 1$ for a hyperbolic plane. We did the case of the sphere in class. Remember that we put a 'grid' of triangles on the sphere, and then counted up the number of faces, edges and vertices. The formula we got for the *sphere* was:

$$v - e + f = 2$$

which is exactly the same as the formula for the *planar* graph (i.e. graph that you can draw in the *Euclidean plane*, a.k.a. a piece of paper). Why is the formula for a *planar* graph the same as the geometry formula for the *sphere* (and not the same as the geometry formula for the *Euclidean plane*) ?

PICTURE OF GRAPHS VERSUS ABSTRACT DEFINITIONS

One of the approaches that you might have tried in the thinking opportunity above is to draw some examples of planar graphs, and count the numbers of edges, vertices and faces to check to make sure that the formula is correct. Here are some examples:

THE TETRAHEDRAL GRAPH

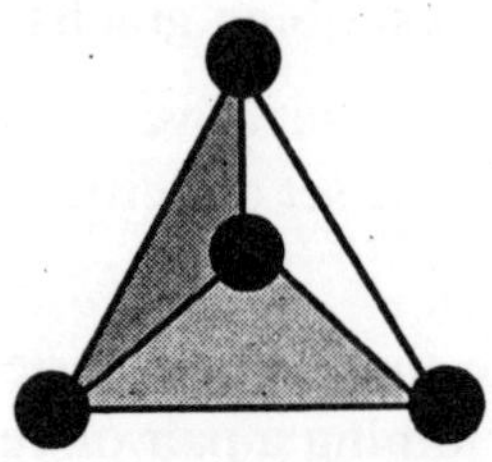

There are 4 *vertices*. $v = 4$.

There are 6 *edges*. $e = 6$.

There are 4 *faces*. $f = 4$.

So: $v - e + f = 4 - 6 + 4 = 2$.

THE CUBE GRAPH

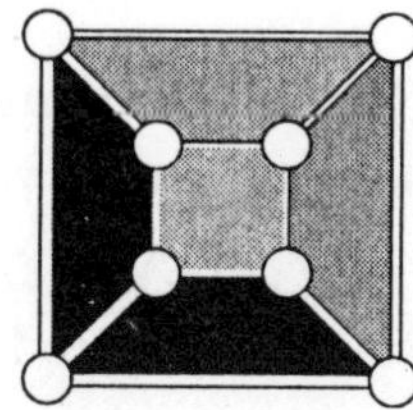

There are 8 *vertices*. $v = 8$.

There are 12 *edges*. $e = 12$.

There are 6 *faces*. $f = 6$.

So: $v - e + f = 8 - 12 + 6 = 2$.

THE DODECAHEDRAL GRAPH

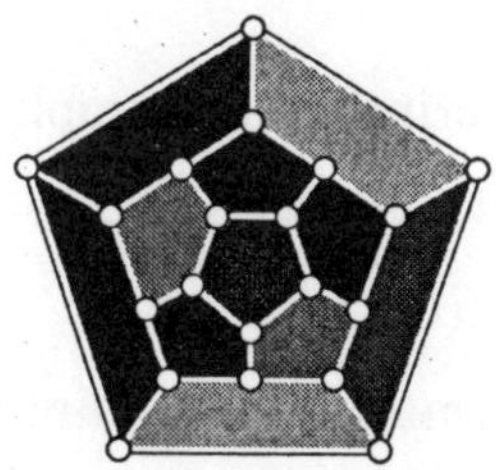

There are 20 *vertices*. $v = 20$.

There are 30 *edges*. $e = 30$.

There are 12 *faces*. $f = 12$.

So: $v - e + f = 20 - 30 + 12 = 2$.

These examples verify the formula, and the pictorial representations of the graphs make it very easy to identify and count the numbers of edges, vertices and faces for each graph. However, if we were to try and prove the formula:

$$v - e + f = 2$$

using only these pictorial representations of the graphs, then we would have to draw out *every* single planar graph, and verify the formula for each picture that we drew. Since there are a *lot* of possible planar graphs, this would be an impractical approach. One answer to this is to make *abstract* definitions. These definitions are useful because they let us prove things more easily, but they are usually harder to understand than the pictorial representations of graphs.

Abstract Definitions

Definition: A graph G consists of a set of vertices and a set of edges, where each edge joins an unordered pair of vertices. The set of vertices of G is denoted by $V(G)$ and the set of edges is denoted by $E(G)$.

Definition: A loop is an edge that joins a vertex to itself. A multiple edge occurs when there is more than one edge joining a pair of vertices.

Definition: A simple graph is a pair ($V(G)$, $E(G)$) where $V(G)$ is a non-empty finite set of elements, and $E(G)$ is a finite set of unordered pairs of distinct elements of $V(G)$.

New for this lesson:

Definition: Two vertices are said to be adjacent if there is an edge joining them. The vertices are said to be incident to the edge. Similarly, two distinct edges are adjacent if they have at least one vertex in common.

USING MATRICES TO REPRESENT GRAPHS

The ideas of adjacency and incidence can be used to come up with different ways of representing a graph.

Let's take a pretty abstract description of a graph G:

1. We are given a vertex set $V(G) = \{v_{1,}, v_n\}$
2. We are given an edge set $E(G) = \{e_{1,}, e_m\}$

One way to represent this information is to point out when a particular edge meets a particular vertex (i.e. when a particular vertex is incident to a particular edge).

For example with the graph:

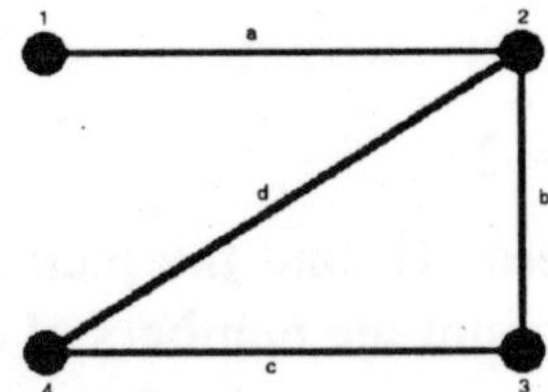

1. Vertex 1 is incident with side '*a*.'
2. Vertex 2 is incident with side '*a*,' with side '*b*,' and with side '*d*.'
3. Vertex 3 is incident with side '*b*,' and side '*c*.'
4. Vertex 4 is incident with side '*c*,' and side '*d*.'

This information could be represented by drawing a grid like:

	Side 'a'	**Side 'b'**	**Side 'c'**	**Side 'd'**
Vertex 1				
Vertex 2				
Vertex 3				
Vertex 4				

The table is filled in by writing '1' in a box if that vertex is incident with that edge, and writing '0' in a box if that vertex is not incident with that edge. The completed table would look like:

	Side 'a'	**Side 'b'**	**Side 'c'**	**Side 'd'**
Vertex 1	1	0	0	0
Vertex 2	1	1	0	1
Vertex 3	0	1	1	0
Vertex 4	0	0	1	1

Getting rid of the vertex and side labels, the table becomes a matrix known as the incidence matrix:

$$\textit{Incidence Matrix} = \begin{bmatrix} 1 & 0 & 0 & 0 \\ 1 & 1 & 0 & 1 \\ 0 & 1 & 1 & 0 \\ 0 & 0 & 1 & 1 \end{bmatrix}$$

More abstractly:

Definition: Suppose G is a graph with a vertex set $V(G) = \{v_{1,}, v_n\}$ and an edge set $E(G) = \{e_{1,}, e_n\}$. We will write $v_i \sim e_j$ if and only if the ith vertex of G is incident with the jth edge of G. Then the nm matrix $B = [b_{ij}]$ whose ij^{th} entry, b_{ij}, is defined by:

$$b_{ij} = \begin{cases} 1 & , v_i \sim e_j \\ 0 & , otherwise \end{cases}$$

is called the incidence matrix of the graph G.

The idea of adjacency can also be used to describe a graph using a matrix. Again, let's take a pretty abstract description of a graph G:

1. We are given a vertex set $V(G) = \{v_{1,}, v_n\}$

2. We are given an edge set $E(G) = \{e_1,, e_m\}$

This information could be represented by pointing out when a particular vertices are joined by edges (i.e. pointing out when vertices are adjacent). Recycling the previous example:

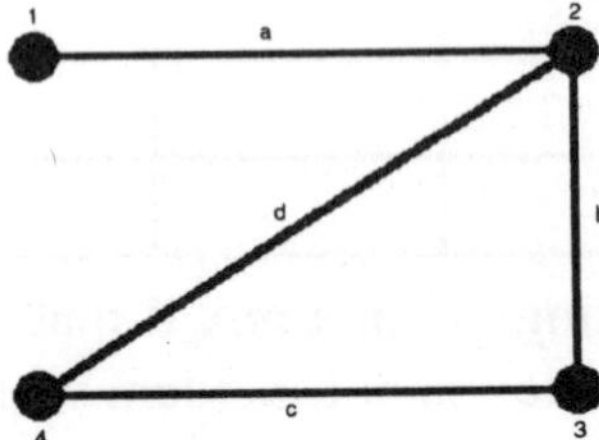

1. Vertex 1 is adjacent to vertex 2.
2. Vertex 2 is adjacent to vertex 1, to vertex 3, and to vertex 4.
3. Vertex 3 is adjacent to vertex 2 and to vertex 4.
4. Vertex 4 is adjacent to vertex 2 and vertex 3.

In a similar spirit to the table made to record the incidence information, we could make a table to record the adjacency information:

	Vertex 1	*Vertex 2*	*Vertex 3*	*Vertex 4*
Vertex 1	0	1	0	0
Vertex 2	1	0	1	1
Vertex 3	0	1	0	1
Vertex 4	0	1	1	0

1. The adjacency matrix made from this table would look like:

$$\textit{Adjacency Matrix} = \begin{bmatrix} 0 & 1 & 0 & 0 \\ 1 & 0 & 1 & 1 \\ 0 & 1 & 0 & 1 \\ 0 & 1 & 1 & 0 \end{bmatrix}$$

Is it possible for two different graphs to have the same adjacency matrix? Conversely, if two graphs have the same adjacency matrix, are they guaranteed to be the same graph? You might like to try making up some examples to test your hypotheses.

1. How could the adjacency matrix be modified to guarantee that two graphs have the same adjacency matrix if and only if they are the same graph? You might like to think about what information a graph contains, and how much of it is recorded in the above definition of the adjacency matrix.
2. What algebraic properties or patterns can you see in the adjacency matrix? Are

these patterns just a fluke, caused by the particular graph used in the example, or are they genuine, mathematical patterns? If you think that they are real patterns, why do they arise? For example, notice that all the entries on the main diagonal of the matrix are zero. Will this be the case for every possible graph?

More Definitions

Definition: The degree or valency of a vertex of G is the number of edges incident with that vertex.

For example, the degrees of the vertices of the following graph are:

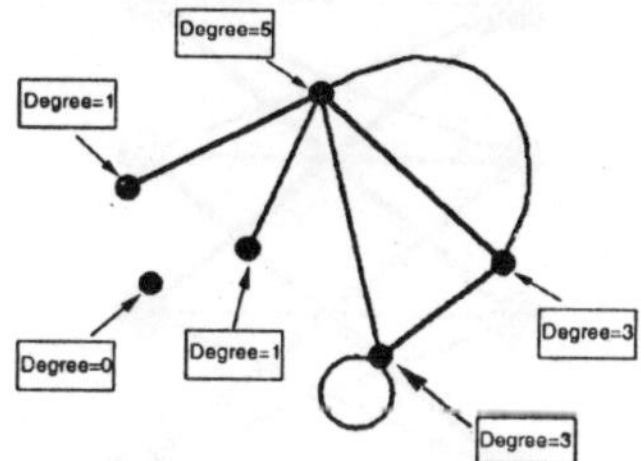

Definition: If all the vertices of a simple graph have the same degree, and if k is that common degree then the graph is said to be k regular (alternatively, regular of degree k).

Definition: Suppose G is a graph. A subgraph of G is a graph whose vertices all belong to $V(G)$ and whose edges all belong to $E(G)$.

For example, if we a graph resembling:

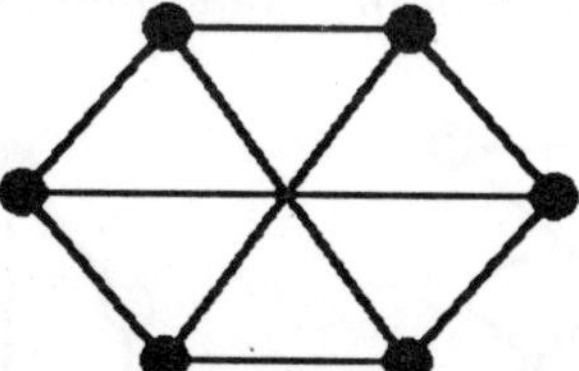

then *some* of the subgraphs of this graph (each is drawn using a different color) include:

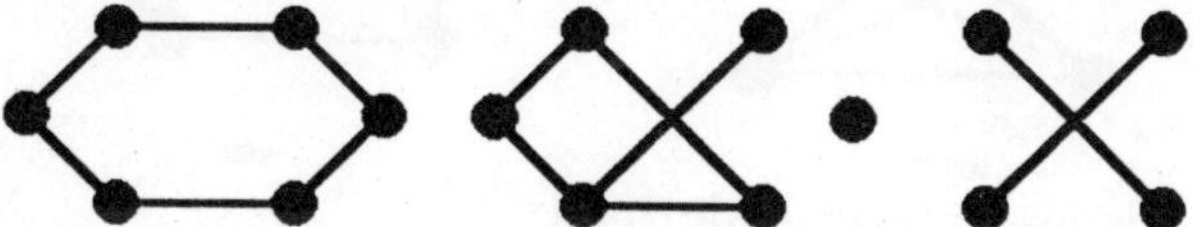

WHEN ARE TWO GRAPHS THE SAME?

Finally in this lesson, let us address the question of what it means for two graphs to be the same. As you can imagine, there are many different ways of drawing the same graph. For example, both of the following are the cube graph: But they don't look all that similar. The question of when graphs are the same has been implicit in some of the thinking opportunities.

For example, to decide whether the incident and adjacency matrices are different for different graphs, you have to decide what it means for graphs to be different in the first place, and this can sometimes be very hard to tell from the pictures.

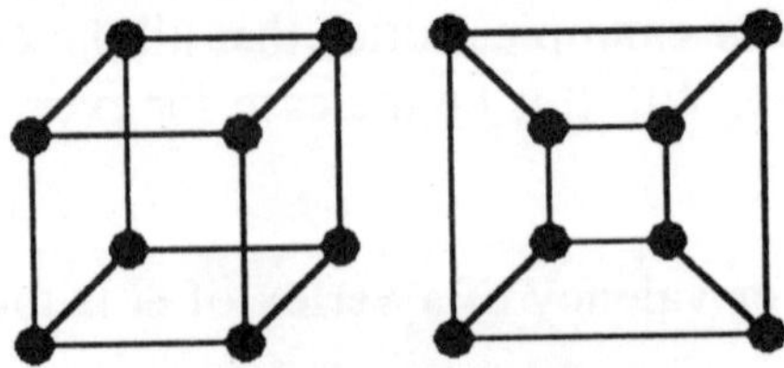

For example, believe it or not, the following graph:

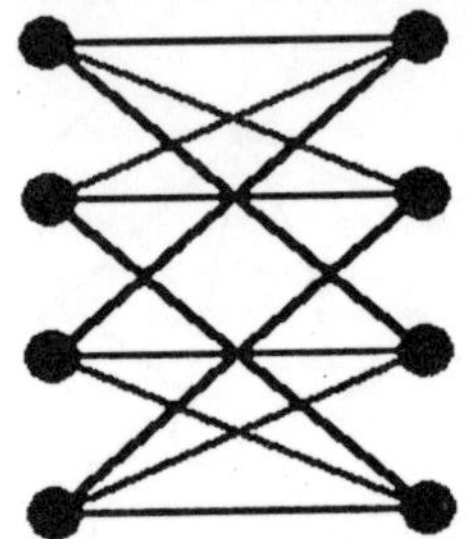

is the cube graph too. In graph theory, two graphs which are really the same are called isomorphic. This means that one pictorial representation of one graph can be re-drawn (preserving all of the vertices and edges) to resemble the other graph. If you attended the summer course *Geometry and the Imagination*, you might remember "schmooshing" operations. Two graphs are isomorphic if you can schmoosh one into the other. As an example, let's schmoosh this funny looking graph into the cube graph (colors are added to show how the schmooshing goes).

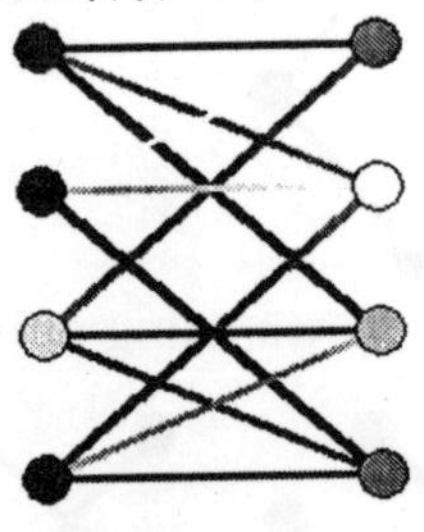

Before...

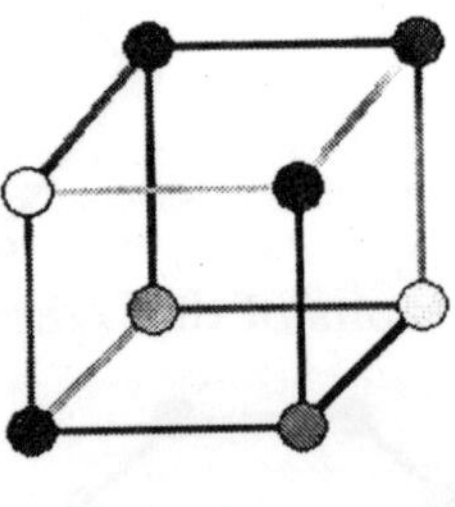

... After

In abstract terms, isomorphism is defined:

Definition: Two graphs, G_1 and G_2, are isomorphic if there is a one-to-one correspondence between the vertices of G_1 and those of G_2 with the property that the number of edges joining any two vertices of G_1 is equal to the number of edges joining the corresponding vertices of G_2.

Definition: If G is a graph with vertex set $V(G) = \{v_1,, v_n\}$ then the nn matrix

$A = [a_{ij}]$ whose ij^{th} entry is the number of edges joining vertex I to vertex j is called the adjacency matrix.

PATHS AND CIRCUITS

So far, much of this course has consisted of advancing definitions and examples that try to capture the meaning if mathematical concepts in abstract language. Let's try to see how some of these definitions might be used to decide theoretical and practical questions.

If you are interested in the theoretical side of mathematics, then you might like to consider the following proposition.

Proposition: Suppose that G is a graph, with $n > 0$ vertices. Let $d_1, \ldots, d_n$ denote the degrees of the vertices of the graph G (see lesson 2 for the definition of the degree of a vertex). Then the sum of the degrees is an even number, and can be written in the form:

$$\sum_{j=1}^{n} d_j = 2m$$

1. This proposition is *not true*. There are graphs that satisfy the hypotheses of the proposition but do not obey the conclusion. See if you can find an example of a graph that does not satisfy the proposition.
2. The proposition can be fixed to be mathematically true by adding hypotheses that the graph must satisfy. Can you find hypotheses that make the proposition true? You might like to consider the example that you came up with, and try to identify what features of your example mess up the proposition.
3. The number $m >= 0$ is not the same for every graph. You might like to try some examples of graphs that satisfy your hypotheses to get some ideas for what m might be.
4. See if you can use the definitions and concepts outlined in the first two lessons to develop a mathematical proof of the proposition (the version that includes your hypotheses).

If you are interested in applying mathematics to solve practical problems, you might like to think about the following statement:

In a gathering of six people, either there are three people who all know each other, or else there are three people none of whom knows the other two.

1. How could you use a graph to represent the gathering of six people, and how could you represent who knows who.
2. The Latin phrase, "Nosce te ipsum," can be translated as, "Know thyself." How would you represent the possibility of people at the gathering 'knowing themselves' on your graph? If people at the gathering are allowed to 'know themselves,' is the statement about who knows who at the gathering still valid?
3. If you assume that there are three people at the gathering who all know each other,

how would you represent this using a graph? Let's call a graph like this a 'Type A' graph.

4. If, instead, you assume that there are three people at the gathering, none of whom knows the other two, how would you represent this using a graph? Let's call a graph like this a 'Type B' graph.
5. How could you show that any possible graph that represents the gathering of six people has to be a 'Type A' of a 'Type B' graph?
6. You might like to consider the following line of reasoning, and try to apply it to the questions posed in this thinking opportunity.

We could represent the people at the gathering by vertices on a graph. A gathering of six people would be represented by a graph with six vertices. If two people at the gathering know each other, then we could draw an edge between the two vertices. (Here, the possibility of people knowing themselves is ignored.) For example, a gathering might resemble:

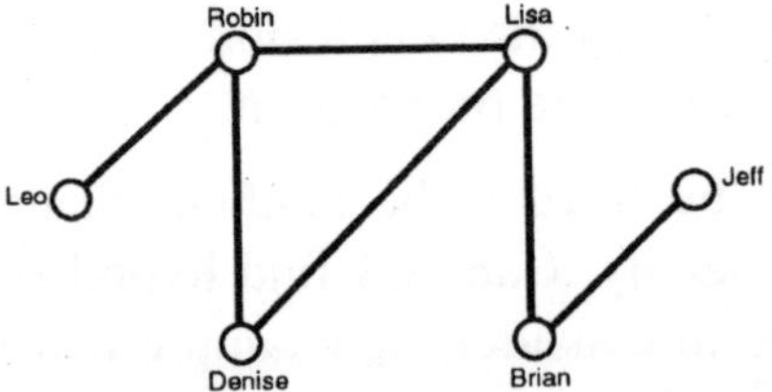

At this gathering, Robin, Lisa and Denise all know each other. Graphically, the edges linking Robin's, Lisa's and Denise's vertices forms a triangle.

Another gathering might look like:

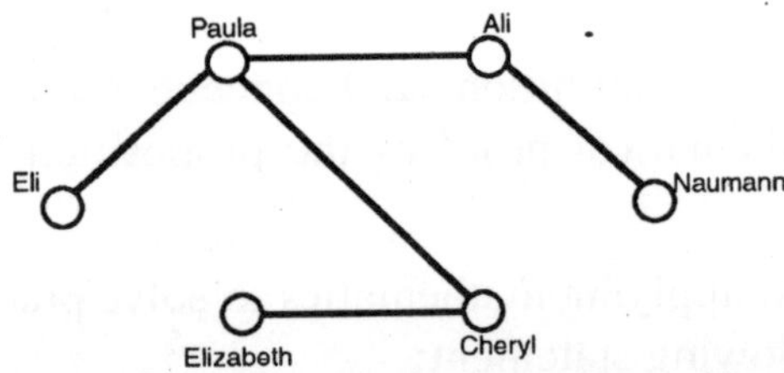

Here, Ali, Eli and Elizabeth do not know each other. There is a graph, based on the pink and green graph, that has an edge between two vertices if the two vertices *don't* have an edge between them in the pink and green graph. This is called the *conjugate* of the pink and green graph, and resembles:

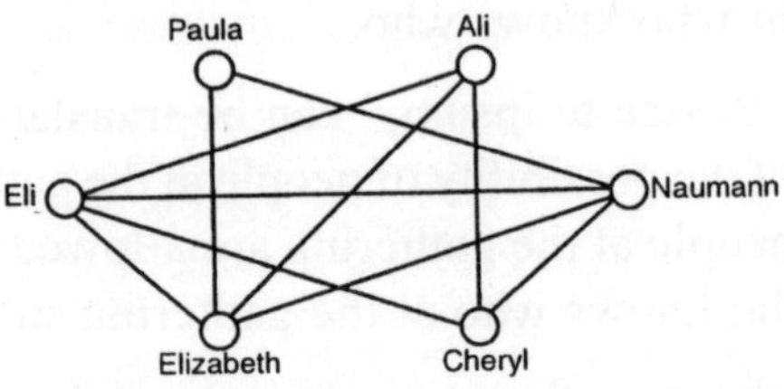

(Here, both yellow and black edges are part of the conjugate graph). In the *conjugate* graph, there is a triangle between Ali, Eli, and Elizabeth (it has been highlighted in yellow). There are other triangles in the conjugate graph, for example, between Paula, Elizabeth and Naumann. There is another triangle between Ali, Eli and Cheryl, and another between Ali, Naumann and Cheryl. Each of these triangles represents a trio of people at the gathering, none of who know the other two in the trio.

So, proving statement: *In a gathering of six people, either there are three people who all know each other, or else there are three people none of whom knows the other two.*

Is equivalent to proving the statement: *On a simple graph with exactly six vertices, either the graph has a triangle, or the conjugate of the graph has a triangle.*

Incidentally, a collection of edges that form a triangle (or a square, pentagon, or any other closed figure) is called a *circuit*, one of the themes of the current lesson.

WALKS

One of the very first problems in graph theory was the "Seven Bridges of Kφnigsberg." The problem goes something like this:

During the eighteenth century the city of Kφnigsberg (in East Prussia) was divided into four sections (including the island of Kneiphof) by the Pregel river. Seven bridges connected these regions, as shown in the picture below.

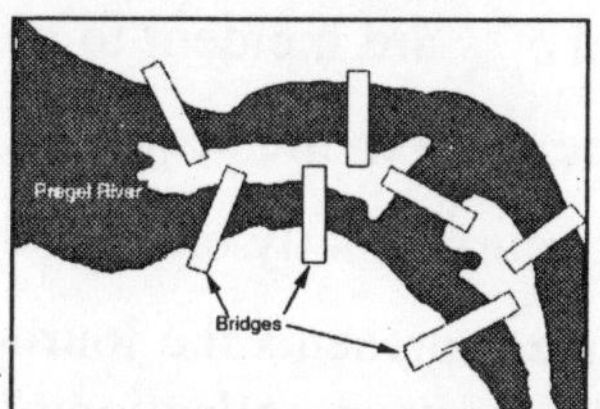

It was said that people spent their Sundays walking around, trying to find a starting point so that they could walk about the city, cross each bridge exactly once, and then return to their starting point. Can you find a starting point, and a path around the city that allow you to do this?

The bridges of the city of Kφnigsberg could be represented more abstractly as pairs of vertices joined by edges. For example:

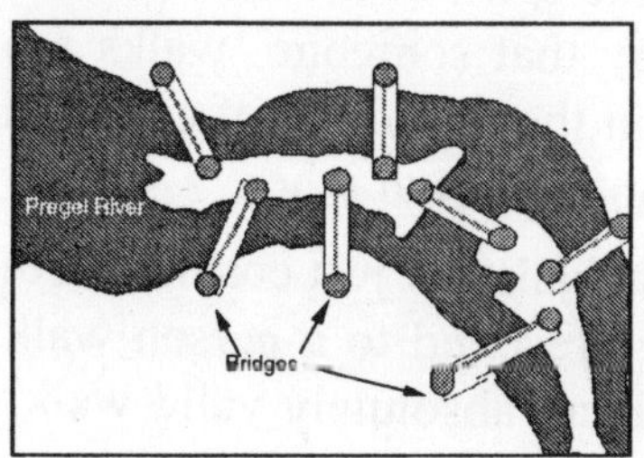

(The vertices and edges of the graph are drawn in pink). Completing the graph by drawing in the main streets of Kφnigsberg might give:

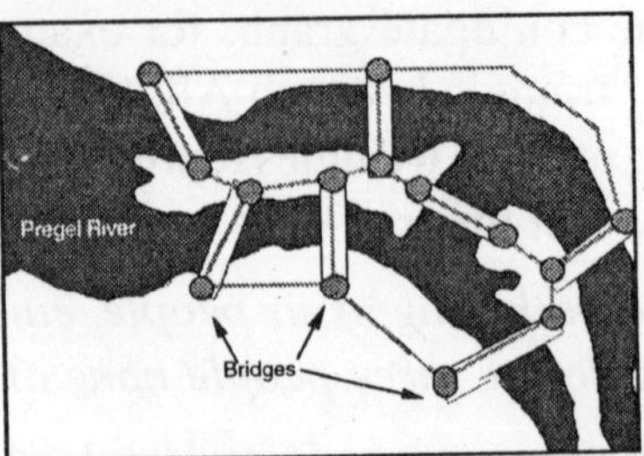

Making things more abstract, and dispensing with all but the vertices and the edges,

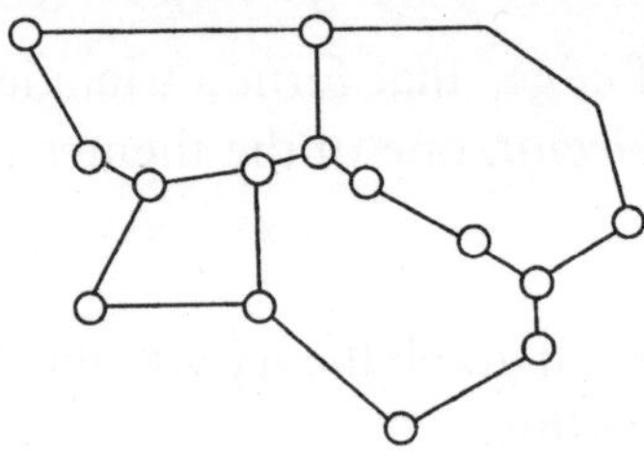

(the edges representing the bridges are colored blue). Then the problem of the Seven Bridges of Kφnigsberg could be re-stated as:

Is there a sequence of edges $e_1,\dots, e_m$ in the above pink and blue graph such that:

1. Consecutive edges e_j and e_{j+1} are incident to a common vertex, and,
2. Each of the blue edges belongs to the sequence, and,
3. Each of the blue edges occurs exactly once in the sequence?

Because this kind of sequence represents the journey that the people of Kφnigsberg could take while *walking* around the city, a collection of edges from *any* graph with these kinds of properties is called a *walk.* Precisely and abstractly:

Definition: Let G denote a graph. A walk in G is a finite sequence of edges $e_1,\dots, e_m$ with each edge, e_j, incident to vertices v_{j-1} and v_j such that any two consecutive edges e_j and e_{j+1} are incident to a common vertex v_j. The vertex v_0 is called the initial vertex and the vertex v_m is called the final vertex. The number of edges in the walk is called the length of the walk. Although the spirit of the definition of a walk seems pretty clear (it is supposed to represent the way that someone 'walks around' a graph), there is a little ambiguity. The point is that, in the definition of a walk, there is no requirement that the edges in the walk are different from each other.

So, it is possible to have a walk that just consists of the very same edge repeated over and over again. This might correspond to a person walking up and down the same old street again and again. This is an absolutely valid walk to go on (although the scenery might get repetitive after a while!), but it can be a technically difficult situation to handle.

It is sometimes very helpful, when proving technical results, to be able to assume that all of the edges in the walk are different. This leads to the definition of a *path*:

Definition: Let G denote a graph. A path in G is a walk in G in which all of the edges are distinct. A path in which all of the *vertices* are distinct is called a simple path. A closed walk is a walk from a vertex to itself (i.e. the initial and final vertices are the same vertex). A circuit is a closed, simple path.

1. Most of the terminology that we have encountered (words such as *path* and *circuit*) is quite suggestive. In my mind, the word *path* conjures up an image of a track through the woods which leads you from one place to another. Sometimes when hiking, you see signs that help to mark the path you want to take. Usually, to prevent yourself from getting lost, you hike from one marker sign to the next. This could be represented by vertices for the marker signs, and edges for the hikes between them. So, a 'walk in which all the edges are distinct' captures what happens when you hike along a path quite accurately for me. If it were up to you to find names for the concepts that we have defined in this lesson, what names would you choose? How do these names capture the essence of the mathematics in ways that appeal to your intuition?
2. To develop concrete examples of the concepts defined so far, try to identify all of the:
3. Walks,
4. Paths,
5. Simple paths,
6. Closed walks, and,
8.. Circuits

In the following graph (it will be helpful to label the edges):

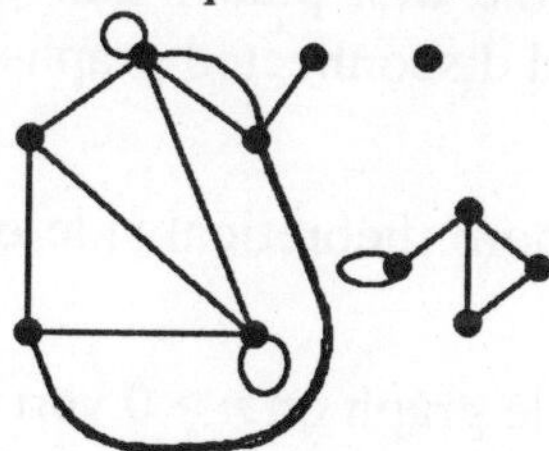

Connected Graphs

In the supplement to lesson 2 ('Examples of Graphs'), the *union* of two graphs was defined. When viewed from the perspective of the edge and vertex sets, the idea of the union of two graphs is just like the union of sets. To clarify, if G_1 and G_2 are two graphs, and a graph G_3 is defined to be their union, $G_3 = G_1 \cup G_2$, then the vertex and edge sets of G_3 ($V(G_3)$ and $E(G_3)$ respectively) are given by:

$$V(G_3) = V(G_1) \cup V(G_2)$$
$$E(G_3) = E(G_1) \cup E(G_2)$$

However, in terms of pictorial representations of graphs, the union of two graphs seems rather pointless.

G_1 $\cup$ G_2 = $G_1 \cup G_2$

To get the pictorial representation of the union, you just draw the two graphs next to one another! From the perspective of pictorial representations, the concept of the union of two graphs doesn't seem to be terribly useful. However, the concept of the union of a graph allows for an elegant definition of what it means for a graph to be 'connected.'

Definition: A graph is said to be connected if it cannot be expressed as the union of two graphs. Otherwise, the graph is said to be disconnected.

Definition: The connected component of a graph G containing a given vertex v an element of $V(G)$ is the largest sub-graph of G that contains v and is a connected graph.

1. Connected and disconnected graphs can be illustrated using the examples of gatherings of people mentioned earlier. The graphs of all the gatherings, such as,

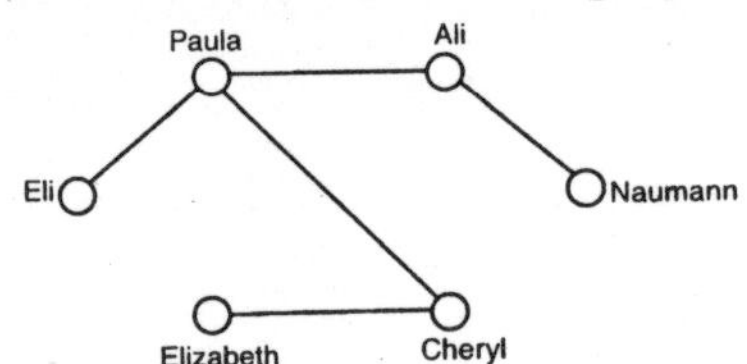

are connected graphs, because they can't be expressed as two separate graphs drawn side-by-side. This sort of makes sense - a person would have to know at least one other person to be invited to the gathering in the first place. Can you think of other examples from everyday life where connected and disconnected graphs arise because of the nature of the situation that you describe?

If you are interested in the more theoretical side of graph theory, you might like to consider following proposition.

Proposition: Let G be a simple graph on $n > 0$ vertices. If the graph G has more than: $\frac{1}{2}(n-1)(n-2)$ edges, then the graph G is connected.

1. Do you think that proposition is true? You might like to try the $n = 1$, $n = 2$ and $n = 3$ cases to test the predictions of the proposition.
2. If you are convinced that the proposition is true, how would you go about proving it?

3. If you think that the proposition is *not* true, how is it flawed? Can you add some hypotheses or restrictions on the graph G that will 'fix' the proposition?

MEASURES OF DISTANCE AND SPREAD

Sometimes, it is helpful to have a notion of how 'far apart' two vertices of a graph are, and 'big' a graph is. For us, these concepts are useful when considering a special kind of path that a graph may have called a *Hamiltonian* cycle (presumably in honor of William Hamilton, who left the '4-color Problem' alone for 25 years). The measures of distance and spread make the most sense in the context of connected graphs (or for the connected components of disconnected graphs), which is why they are defined now.

The two graphs:

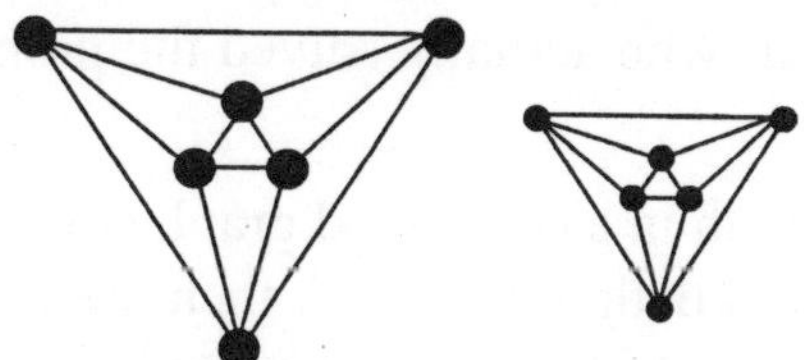

are very different sizes when size is meant in the normal, everyday way. The graph on the left is clearly much bigger than the graph on the right. However, if you look closely at the vertices and edges present in each of these graphs, you can see that they are *exactly the same graph*.

1. The definitions of size and distance that will be given for graphs do not depend on the lengths that edges or the size that vertices are drawn in pictorial representations of graphs. Can you think of applications of graphs to phenomena in the world where this viewpoint is appropriate?
2. Can you think of other applications of graphs to phenomena in the world where it *would* make sense to take the length that edges and the sizes of vertices into account when considering notions of distance and size? Can you think of a way to define these new, more elaborate mathematical objects in an abstract way? Can any of the definitions, propositions and theorems.that we have thought about be modified to make sense with the new class of mathematical objects that you have defined?

Definition The distance between any two vertices v and w in the same connected component of a graph is the length of the shortest simple path from v to w, and is denoted by $d(v, w)$.

Remember that a simple path is a special case of a walk, and the length of a walk between two vertices is the number of edges that are 'walked across.'

Definition: The diameter of a connected graph is the maximum distance between two of its vertices. The girth of a graph is the length of the shortest circuit.

EULERIAN GRAPHS

Remember how the problem of the Seven Bridges of Kφnigsberg could be re-stated as:

Is there a sequence of edges $e_1, \ldots, e_m$ in the pink and blue graph such that:

1. Consecutive edges e_j and e_{j+1} are incident to a common vertex, and,
2. Each of the blue edges belongs to the sequence, and,
3. Each of the blue edges occurs exactly once in the sequence?

By defining paths, we have been able to characterize the first requirement of the problem in abstract terms. To characterize the next two requirements, we will study more specialized kinds of paths, namely Eulerian paths (named for the famous Swiss mathematician Leonhard Euler, who actually solved the problem of the 'Seven Bridges') and Hamiltonian paths.

Definition: An Eulerian path in a connected graph *G* is a path in which every edge in *G* occurs exactly once. If such a path exists in *G*, then the graph is called semi-Eulerian.

Definition: An Euler tour in a connected graph *G* is a closed path which includes every edge of *G* exactly once. A graph that has an Euler tour is called an Eulerian graph.

1. Is every connected graph Eulerian? Try to give an argument for why this might be true, or an example if you don't think that this is true.
2. Is it possible for a graph to be semi-Eulerian without being Eulerian? Conversely, is it possible for a graph to be Eulerian without being semi-Eulerian? Again, try to produce arguments or examples to justify your conclusions.
3. How could concepts such as Euler paths or Euler tours be applied to help solve the problem of the Seven Bridges of Kφnigsberg?
4. How do you think that the requirements of the problem motivated the definitions given? Do you think that this is a valid way to do mathematics? Can you think of other examples (from mathematics or more broadly) where practical considerations have lead to abstract or general definitions that are rich enough to be studied for their own sake?

If you are interested in the theoretical side of graph theory, you might like to consider the following proposition:

Proposition: If *G* is a graph in which the degree of each vertex is at least two, then *G* must contain a circuit.

Remember the definition of a circuit: a circuit is a closed, simple path.

1. Can you find an example of a graph that makes this proposition fail? What

specific problems or properties would a graph have to have in order to make the proposition fail? You might like to think of a list of possible ways that you could make the proposition fail, and treat each possibility in turn. If you are able to make the proposition fail, are there any hypotheses that you can add for *G* that will save the proposition?

2. If you were unable to find a way of making the proposition fail, or if you have added hypotheses, and feel that the proposition might actually be true after all, you might like to try and find an argument that proves the proposition.

You might have decided that not every connected graph is Eulerian, an example of a connected graph that is *not* Eulerian is the following.

Eulerian graphs are quite special. If you took the time to verify whether the above example was actually Eulerian or not, you might have found that it is actually quite time consuming and difficult to decide whether a graph *is* Eulerian or not. Eulerian graphs can be characterized as follows.

Theorem: A finite, connected, loop-free graph is Eulerian if and only if the degree of each vertex in the graph is even.

Note: The statement of this theorem includes a phrase, "If and only if." This makes the theorem kind of 'double-barreled.' It means that *if* all the vertices of a finite, connected, loop-free graph, *G*, have even degrees, then *G* is Eulerian. The theorem *also* means that if a finite, connected, loop-free graph, *G*, is Eulerian, then all the vertices of *G* have even degrees. Because the theorem is 'double-barreled,' the proof has to be 'double-barreled' as well.

Proof: First to be established is the 'barrel' of the theorem that says:

If a finite, connected, loop-free graph, G, *is Eulerian, then all of the vertices of* G *have even degrees.*

To prove this, suppose that *G* is a finite, connected, loop-free Eulerian graph. Then *G* has an Euler tour. Each time that the Euler tour passes through a vertex, it uses two edges incident to that vertex. Each edge occurs exactly once in the Euler tour, so every vertex has even degree.

Next to be established is the 'barrel' of the theorem that says:

If all the vertices of a finite, connected, loop-free graph, G, *have even degree, then* G *is Eulerian.*

To prove this, suppose that *G* is a finite, connected, loop-free where every vertex has even degree. Firstly, dispense with the somewhat pathological case where *G* has no edges (i.e. each vertex has degree zero). Then the result is holds, because if the graph has no edges, then the 'empty path' consisting of no edges 'includes' every edge of the graph once. This sounds kind of nuts, but the reason that it works is that there is no requirement that a path actually contain any edges (check the definition). So, a path that contains no edges whatsoever is a mathematical possibility, although it might not occur as an intuitive possibility to everyone.

Next, to the more intuitively obvious situation where the graph G actually has some edges. By the proposition that you might have proved in the last 'thinking opportunity,' the graph G contains a circuit. Remove this circuit (just the edges, not the vertices) from G, leaving a subgraph H that is a union of connected components. By induction on the number of edges, each of these connected components has an Euler tour. An Euler tour in G can now be constructed by starting at any one of the vertices of the original circuit, following this circuit around, and every time a new connected component of H is encountered, an Euler tour in this component is made. This ends the proof.

The proof of the theorem referred to a mathematical technique called 'Proof by Induction.'

1. If you have studied mathematical induction, try to explain how induction is used to make the proof of the theorem work.
2. One of my teachers once described mathematical induction with a little story.

You check into a motel. The desk clerk informs you that you will be staying in room number 'N.' However, the desk clerk isn't very helpful, and instead of giving you the key to room 'N,' he gives you the key to room 1. You ask him what the deal is, and he tells you that the key to open room 'k + 1' is located in room 'k' How can you get to your room?

If you haven't studied mathematical induction, you might like to have a look at the proof of the following theorem (which uses mathematical induction). Try to figure out why the argument works.

Theorem: For any integer $n > 0, \sum_{j=1}^{n} j = \frac{n(n+1)}{2}$.

Proof: Firstly, do the $n = 1$ case. When $n = 1$, the left hand side of the formula gives,

$$L.H.S. = \sum_{j=1}^{1} j = 1.,$$

while the right hand side of the formula gives,

$$R.H.S. = \frac{1(1+1)}{2} = \frac{2}{2} = 1.$$

Since the left hand side and the right hand side of the formula agree for $n = 1$, we conclude that the theorem holds for the case $n = 1$.

Next, the inductive step. Assume that the theorem holds true for some integer $k > 0$.

We will show, on the basis of this assumption, that the theorem also holds true for the integer $k + 1$.

The assumption that the theorem holds for k means that we can assume that,

$$\sum_{j=1}^{k} j = \frac{1(k+1)}{2}$$

For the integer $k + 1$, the theorem predicts that the right hand side should be

$$R.H.S. = \frac{(k+1)(k+2)}{2}.$$

For the integer $k + 1$, the left hand side is:

$$R.H.S. = \left(\sum_{j=1}^{k+1} j\right) + (k+1) = \frac{k(k+1)}{2} + (k+1) \cdot$$

More algebra gives that,

$$L.H.S. = \frac{k(k+1) + 2(k+1)}{2} = \frac{(k+1)(k+2)}{2} = R.H.S.$$

Having shown that the left and right hand sides of the theorem agree for the integer $k + 1$, when it has been assumed that the theorem works for the integer k, completes the proof.

The following proposition can be proved using the theorem characterizing Eulerian graphs. See if you can come up with an argument to explain why it is a true statement. (Observe that this proposition is 'double-barreled' as it contains the 'if and only if' phrase. Make sure your argument is 'double-barreled' too.)

Proposition: A finite, connected, loop-free graph is semi-Eulerian if and only if it has at most two vertices of odd degree.

1. The theorem about Eulerian graphs is very specific; only finite, connected, loop-free graphs. Can the theorem be adapted to handle graphs that are infinite, disconnected, or that have loops? You might like to try and find some examples or develop arguments to support your conclusions.

HAMILTON CYCLES

Euler paths and Euler tours were mainly concerned with the *edges* of a graph. An Euler path, for example, included every *edge* of the graph exactly once. Another special type of path focuses more on the *vertices* of the graph. These are called *Hamiltonian* paths.

Definition: Any walk in a connected graph G that passes through each vertex of the graph exactly once is called a Hamiltonian path. If such a path exists, then the graph is called semi-Hamiltonian.

Definition: A Hamilton cycle is a closed Hamiltonian path, and any graph that has a Hamilton cycle is said to be Hamiltonian.

1. Before, there was a theorem that gave criteria for a graph G to have an Euler tour. These criteria were:
2. G had to be finite,
3. G had to be loop-free,
4. G had to be connected, and,
5. Each vertex of G had to have even degree.

What sort of criteria would a graph have to satisfy in order to have a Hamiltonian cycle? What about a Hamiltonian path? Try to give examples of graphs that show that your criteria are essential.

1. It is possible to have graphs that are *not* Hamiltonian or semi-Hamiltonian. It is also possible to have graphs that are semi-Hamiltonian, but not Hamiltonian. Classify the following examples of graphs.

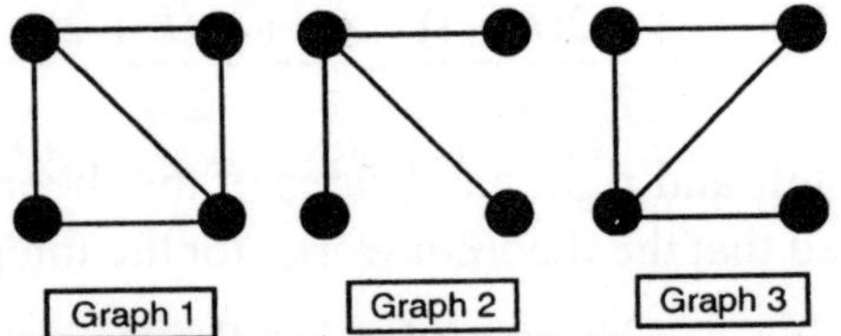

Graph 1 Graph 2 Graph 3

As with Eulerian graphs, checking graphs 'by hand' to see if they are Hamiltonian or not is very time-consuming, and actually quite difficult (especially for very large graphs with lots of vertices and edges).

It would be nice to develop another theorem that would provide simpler criteria for deciding whether a graph was Hamiltonian or not. There is a result (known as Orı's Theorem) that goes part of the way towards this. Unfortunately, Orı's Theorem is not a 'double-barreled' theorem. Orı's Theorem gives a *sufficient* condition for a graph to be Hamiltonian. Unfortunately, it does not give a *necessary* condition; there are examples of Hamiltonian graphs that do not fulfill all of the requirements of Orı's Theorem.

Theorem: If G is a finite, connected, loop-free graph with n vertices, with the property that the degrees of any two adjacent vertices is greater than or equal to n, then G is Hamiltonian.

Note: As mentioned, there are examples of Hamiltonian graphs that do not satisfy the criteria of Orı's Theorem. For example, the cube graph is Hamiltonian - you might like to try and find a Hamiltonian cycle.

However, the cube graph is a graph on 8 vertices ($n = 8$), but the degrees of adjacent vertices always sums to 6.

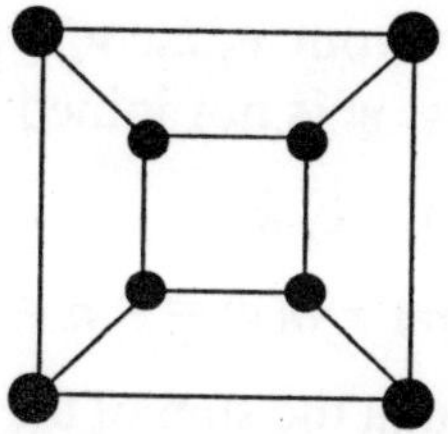

Proof: Assume that G has no Hamiltonian cycle. Add as few new edges to G as are necessary to produce a Hamiltonian cycle. (Note that the complete graph on n vertices has a Hamiltonian cycle, so eventually, you will add enough edges to produce a Hamiltonian cycle).

Suppose that the last edge that you added is between two vertices v and w, and that this is the final link in what becomes a Hamiltonian cycle. Schematically, the Hamiltonian cycle might look like this:

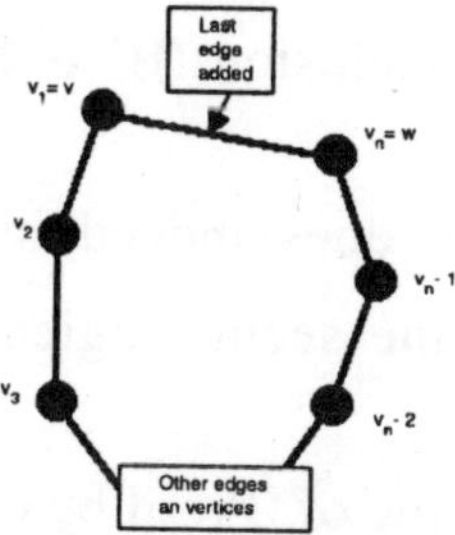

Label the vertices as indicated in the schematic. Now consider the 'neighbors' of v and w in G (i.e. the vertices adjacent to v and w in G.)

Firstly, v is not adjacent to itself and is not adjacent to w in G (remember that the edge joining v and w was the last edge *added* to G), but is adjacent to, say k of the vertices v_2, $v_3, \ldots, v_{n-1}$. So, $d(v)$ = degree of v in $G = k$.

Secondly, w is not adjacent to itself, and is not adjacent to v in G, but is adjacent to, say, l of the vertices $v_2, v_3, \ldots, v_{n-1}$. So, $d(w)$ = degree of w in $G = l$.

Suppose that v is adjacent to v_j $(2 =< j =< n - 1)$, then w cannot be joined to v_{j-1} or else:

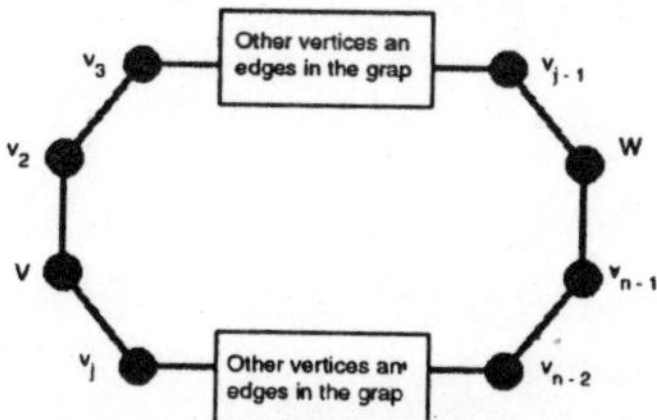

would have been a Hamiltonian cycle in G before the edge linking v and w was added (remember that it has been assumed that the fewest possible edges have been added to the graph G in order to produce a Hamiltonian cycle).

Thus, for each of the k vertices from $v_2,..., v_{n-1}$ that v is adjacent to, there is one vertex that w is not adjacent to. Thus, w is not joined to at least k vertices. Therefore,

$$l = d(w) = <(n-1)\ k,$$

and,

$$l + k = d(w) + d(v) = < n - 1 < n.$$

This contradicts the assumption that the sum of the degrees of any two adjacent vertices must be at least n. So, G must have had a Hamiltonian cycle. This concludes the proof.

1. The proof of Orı's Theorem used a technique of proof called 'Proof by Contradiction.' This technique can be described as something like:
2. Suppose you want to prove that hypothesis 'A' implies conclusion 'B.'
3. Start by assuming the hypothesis 'A.'
4. Next, assume that conclusion 'B' is actually false.
5. Show that assuming that conclusion 'B' is false leads to a contradiction with hypothesis 'A.'
6. Conclude that hypothesis 'A' does, indeed, imply conclusion 'B.'

Do you think that this technique seems logically valid? Can you identify what premise(s) it relies upon?

Can you explain how the technique of 'Proof by Contradiction' is used to prove Orı's Theorem? There is a second, less explicit use of 'Proof by Contradiction' in the body of the proof. Can you identify it?

There is another result called Dirac's Theorem. Dirac's Theorem may be stated:

Theorem: If G is a simple graph on n vertices with the property that all of its vertices have degree at least $\frac{n}{2}$, then G is Hamiltonian.

1. Can you use Orı's Theorem to prove Dirac's Theorem?
2. Can you find a way to prove Dirac's Theorem without using Orı's Theorem? (A proof by contradiction might work here).

Chapter 4

Graph Connectivity

In mathematics and computer science, connectivity is one of the basic concepts of graph theory. It is closely related to the theory of network flow problems. The connectivity of a graph is an important measure of its robustness as a network.

DEFINITIONS OF COMPONENTS, CUTS AND CONNECTIVITY

In an undirected graph G, two vertices u and v are called connected if G contains a path from u to v. Otherwise, they are called disconnected. A graph is called connected if every pair of distinct vertices in the graph is connected (directly or indirectly). A connected component is a maximal connected subgraph of G.

Each vertex belongs to exactly one connected component, as does each edge. A directed graph is called weakly connected if replacing all of its directed edges with undirected edges produces a connected (undirected) graph. It is strongly connected or strong if it contains a directed path from u to v for every pair of vertices u,v. The strong components are the maximal strongly connected subgraphs. A cut or vertex cut of a connected graph G is a set of vertices whose removal renders G disconnected.

The connectivity or vertex connectivity $\kappa(G)$ is the size of a smallest vertex cut. A graph is called k-connected or k-vertex-connected if its vertex connectivity is k or greater. A complete graph with n vertices has no cuts at all, but by convention its connectivity is n-1. A vertex cut for two vertices u and v is a set of vertices whose removal from the graph disconnects u and v. The local connectivity $\kappa(u,v)$ is the size of a smallest vertex cut separating u and v. Local connectivity is symmetric for undirected graphs; that is, $\kappa(u,v)=\kappa(v,u)$. Moreover, $\kappa(G)$ equals the minimum of $\kappa(u,v)$ over all pairs of vertices u,v.

2-connectivity is also called "biconnectivity" and 3-connectivity is also called "triconnectivity". Analogous concepts can be defined for edges. Thus an edge cut of G is a set of edges whose removal renders the graph disconnected. The edge-connectivity $\lambda(G)$ is the size of a smallest edge cut, and the local edge-connectivity $\lambda(u,v)$ of two vertices u,v

is the size of a smallest edge cut disconnecting u from v. Again, local edge-connectivity is symmetric. A graph is called k-edge-connected if its edge connectivity is k or greater.

MENGER'S THEOREM

One of the most important facts about connectivity in graphs is Menger's theorem, which characterizes the connectivity and edge-connectivity of a graph in terms of the number of independent paths between vertices. If u and v are vertices of a graph G, then a collection of paths between u and v is called independent if no two of them share a vertex (other than u and v themselves). Similarly, the collection is edge-independent if no two paths in it share an edge. The greatest number of independent paths between u and v is written as $\kappa'(u,v)$, and the greatest number of edge-independent paths between u and v is written as $\lambda'(u,v)$. Menger's theorem asserts that $\kappa(u,v) = \kappa'(u,v)$ and $\lambda(u,v) = \lambda'(u,v)$ for every pair of vertices u and v. This fact is actually a special case of the max-flow min-cut theorem.

BRIDGE

A bridge is a single edge whose removal disconnects a graph.

The above graph G1 can be split up into two components by removing one of the edges bc or bd. Therefore, edge bc or bd is a bridge.

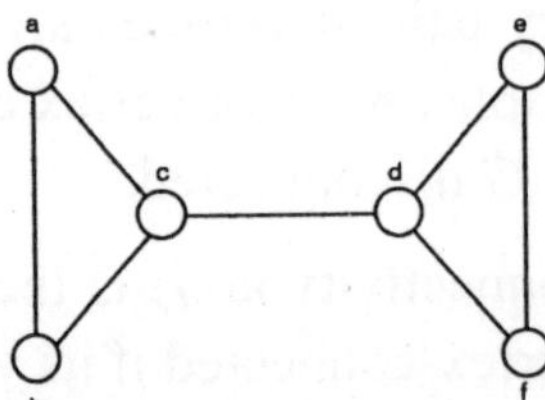

The above graph G2 can be disconnected by removing a single edge, cd. Therefore, edge cd is a bridge.

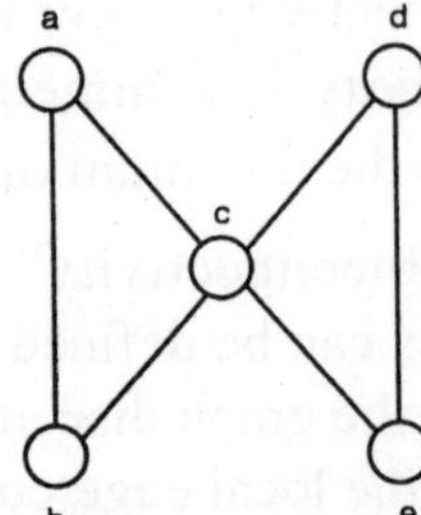

The above graph G3 cannot be disconnected by removing a single edge, but the removal of two edges (such as ac and bc) disconnects it.

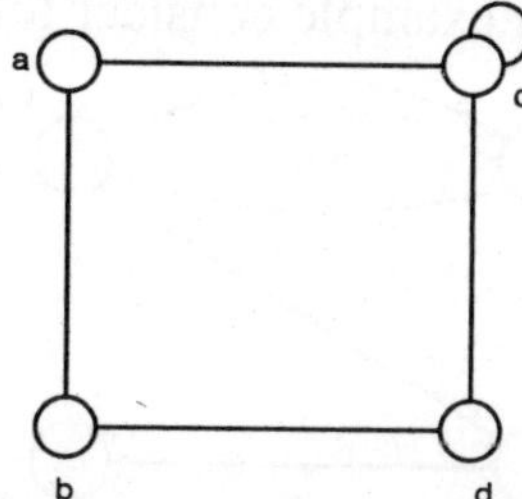

The above graph G4 can be disconnected by removing two edges such as ac and dc.

Edge Connectivity

The edge-connectivity $\lambda(G)$ of a connected graph G is the smallest number of edges whose removal disconnects G. When $\lambda(G) \geq k$, the graph G is said to be k-edge-connected.

For example, the edge connectivity of the above four graphs G1, G2, G3, and G4 are as follows:

1. *G1* has edge-connectivity 1.
2. *G2* has edge connectivity 1.
3. *G3* has edge connectivity 2.
4. *G4* has edge connectivity 2.

Cut Set

A cut set of a connected graph G is a set S of edges with the following properties

1. The removal of all edges in S disconnects G.
2. The removal of some (but not all) of edges in S does not disconnects G.

As an example consider the following graph

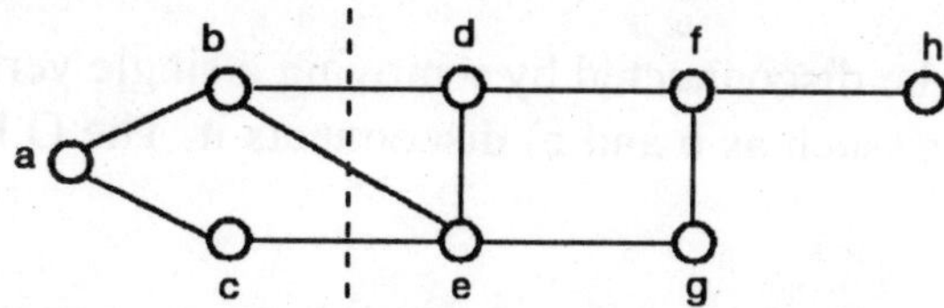

We can disconnect G by removing the three edges bd, bc, and ce, but we cannot disconnect it by removing just two of these edges. Note that a cut set is a set of edges in which no edge is redundant.

Vertex Connectivity

The connectivity (or vertex connectivity) K(G) of a connected graph G (other than a complete graph) is the minimum number of vertices whose removal disconnects *G*. When $K(G) \geq k$, the graph is said to be *k*-connected (or *k*-vertex connected). When we remove a vertex, we must also remove the edges incident to it. As an example consider following graphs.

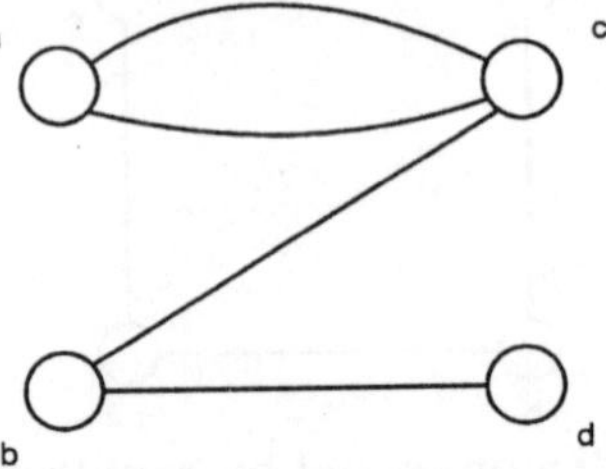

The above graph G can be disconnected by removal of single vertex (either b or c). The G has connectivity 1.

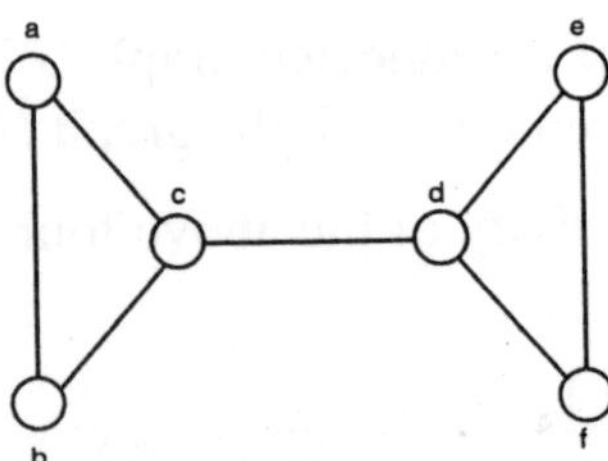

The above graph G can be disconnected by removal of single vertex (either c or d). The vertex c or d is a cut-vertex. The G has connectivity 1. The above G can be disconnected by removing just one vertex i.e., vertex c. The vertex c is the cut-vertex. The G has connectivity 1.

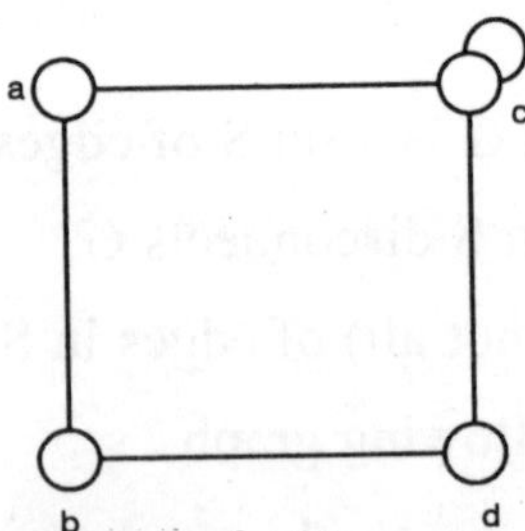

The above G cannot be disconnected by removing a single vertex, but the removal of two non-adjacent vertices (such as b and c) disconnects it. The G has connectivity 2.

Cut-Vertex

A cut-vertex is a single vertex whose removal disconnects a graph. It is important to note that the above definition breaks down if G is a complete graph, since we cannot then disconnects G by removing vertices. Therefore, we make the following definition.

Connectivity of Complete Graph

The connectivity $k(k_n)$ of the complete graph k_n is n-1. When $n\text{-}1 \geq k$, the graph k_n is said to be k-connected.

Vertex-Cut set. A vertex-cut set of a connected graph G is a set S of vertices with the following properties.

1. The removal of all the vertices in S disconnects G.
2. The removal of some (but not all) of vertices in S does not disconnects G.

Consider the following graph

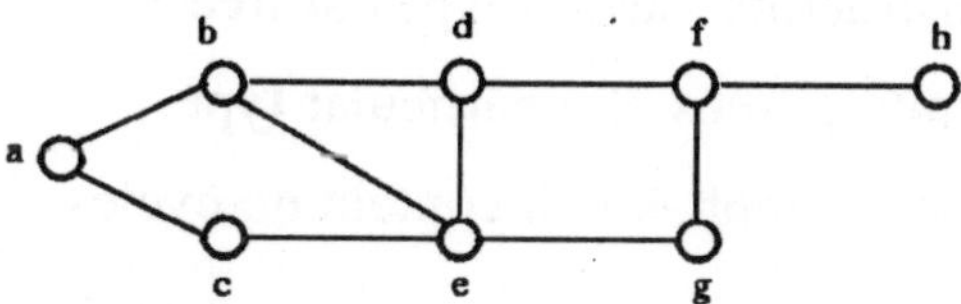

We can disconnects the graph by removing the two vertices b and e, but we cannot disconnect it by removing just one of these vertices. the vertex-cutset of G is {b, e}.

Note that the connectivity k(G) does not exceed the edge-connectivity λ(G). This inequality holds for all connected graph. Formally, for any connected graph G we have

$$K(G) \leq \lambda(G) \leq \delta(G)$$

where δ(G) is the smallest vertex-degree in G. But it is certainly possible for both inequality in above theorem to be strict inequalities (that is, $k(G) < \lambda(G) < \delta(G)$) For example, in the following graph,

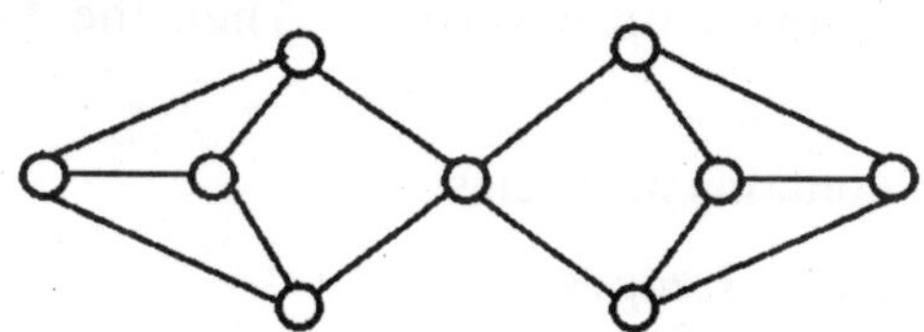

K(G)=1, λ(G) = 2, and δ(G) = 3.

TREES GRAPHS

An acyclic graph (also known as a forest) is a graph with no cycles. A tree is a connected acyclic graph. Thus each component of a forest is tree, and any tree is a connected forest.

Theorem: The following are equivalent in a graph G with n vertices.

1. G is a tree.
2. There is a unique path between every pair of vertices in G.
3. G is connected, and every edge in G is a bridge.
4. G is connected, and it has (n – 1) edges.
5. G is acyclic, and it has (n – 1) edges.
6. G is acyclic, and whenever any two arbitrary nonadjacent vertices in G are joined by and edge, the resulting enlarged graph G' has a unique cycle.

7. G is connected, and whenever any two arbitrary nonadjacent vertices in G are joined by an edge, the resulting enlarged graph has a unique cycle.

   Generally speaking, algorithms associated with trees can be divided into three types.

8. Algorithms for searching and labeling a given tree.
9. Algorithms for constructing various types of tree.
10. Algorithms for counting trees of a particular type.

Tree A tree is a connected graph which contain no cycles.

For example,

n = 1

n = 2

n = 3

n = 4 OR

n = 5

OR OR

Theorem: Let T be a graph with n vertices. Then the following statements are equivalent.

1. T is connected and contains no cycles.
2. T is connected and has n-1 edges.
3. T has n-1 edges and contains no cycles.
4. T is connected and each edge is abridge.
5. Any two vertices of T are connected by exactly one path.
6. T contains no cycles, but the addition of any new edge creates exactly one cycles.

SPANNING TREES

Let G be a connected graph. A spanning tree in G is a subgraph of G that includes all the vertices of G and is also a tree. The edges of the trees are called branches. For example, consider the following graph G

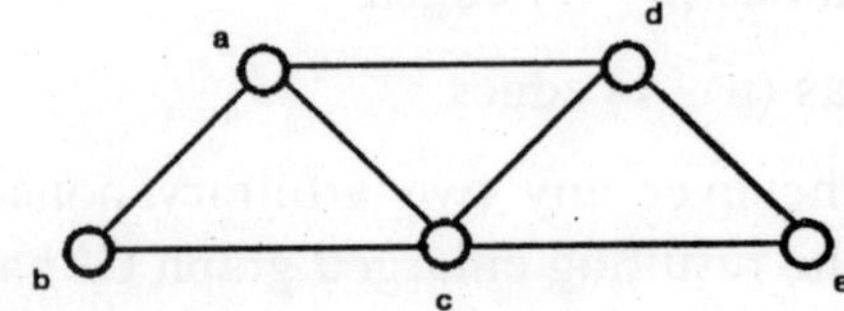

The three spanning trees G are:

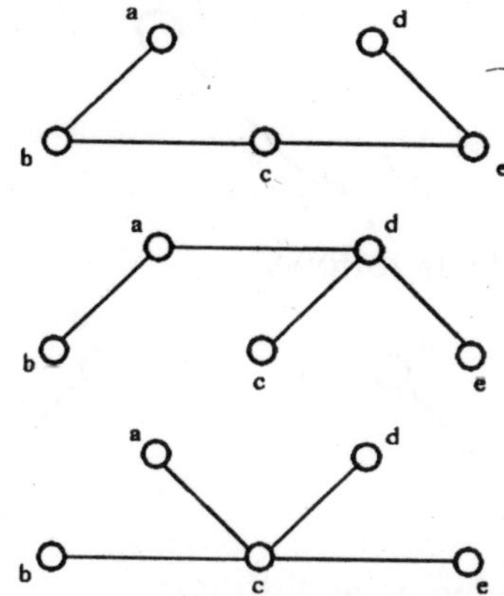

We can find a spanning tree systematically by using either of two methods.

1. Cutting-down Method
   - Start choosing any cycle in G.
   - Remove one of cycle's edges.
   - Repeat this procedure until there are no cycle left.

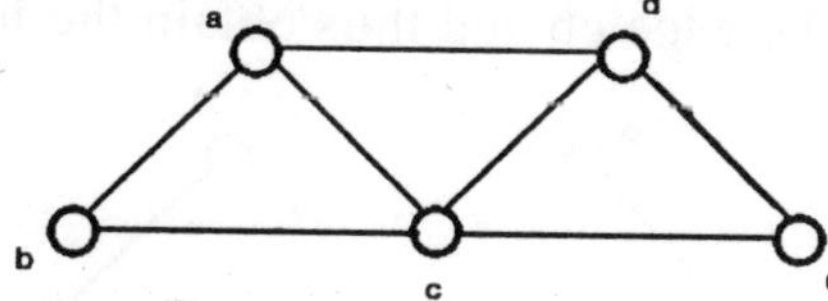

For example, given the graph G

1. We remove the edge ac which destroy the cycle adca in the above graph and we get
2. We remove the edge cb, which destroy the cycle adcba in the above graph and we get
3. We remove the edge ec, which destroy the cycle decd in the above graph and thus obtained the following spanning tree.
4. Building-up Method
   - Select edges of G one at a time. in such a way that no cycles are created.
   - Repeat this procedure until all vertices are included.

For example, for the following graph G

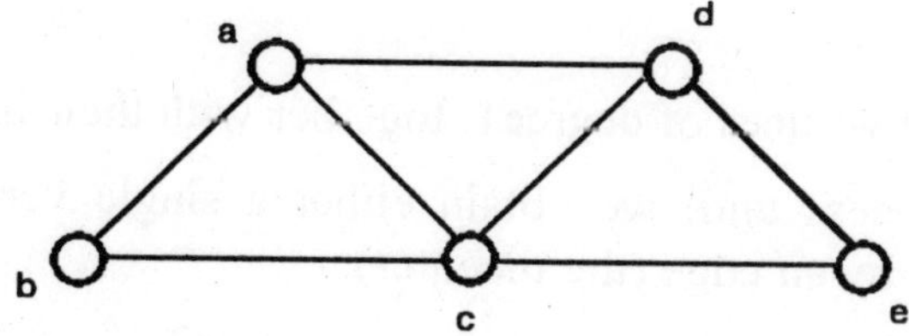

1. Choose the edge ab

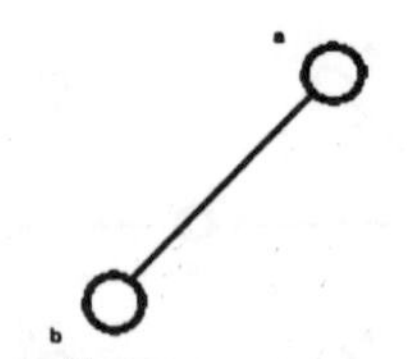

2. Next choose the edge de as follows:

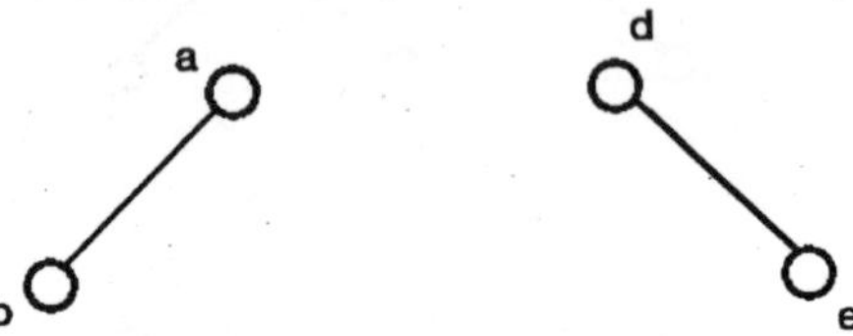

3. After that choose the edge ec as follows:

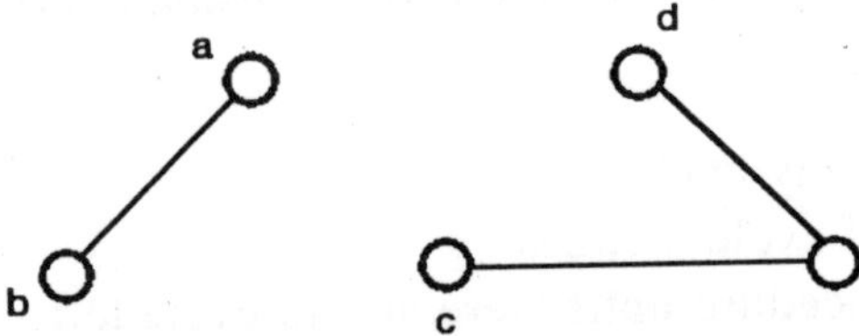

4. Finally, we choose the edge cb and thus obtain the following spanning tree.

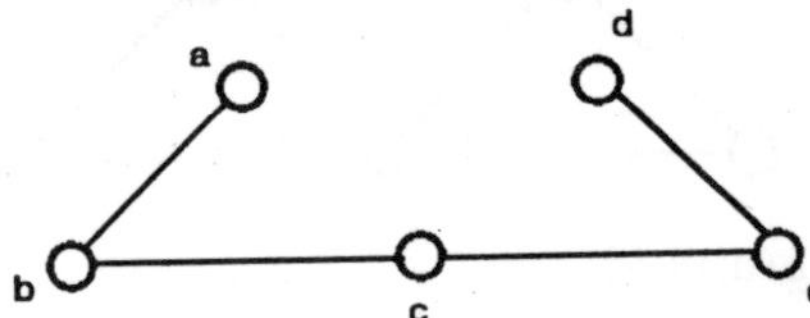

Theorem: A graph is connected if and only if it has a spanning tree.

Proof: Let G be a connected graph. Delete edges from G that are not bridges until we get a connected subgraph H in which each edge is a bridge. Then H is a spanning tree. On the other hand, if there is a spanning tree in G, there is a path between any pair of vertices in G; thus G is connected.

Centers and Bicenters

It is convenient to start building up the tree at the middle of a tree and move outwards. This was the approach used by Arthur Cayley when he counted the number of chemical molecules by building them step by step. But what do we mean by the "middl≥ of a tree?

There are two straight forward ways to compute centers and bicenters.

Algorithm1

1. Remove all the vertices of degree1, together with their incident edges.
2. Repeat the process until we obtain either a single vertex (the center) or two vertices joined by an edge (the bicenter).

A tree with a center is called a central tree, and a tree with a bicenter is called a bicentral tree. Note that every tree is either central or bicentral, but not both.

For example, given a following tree.

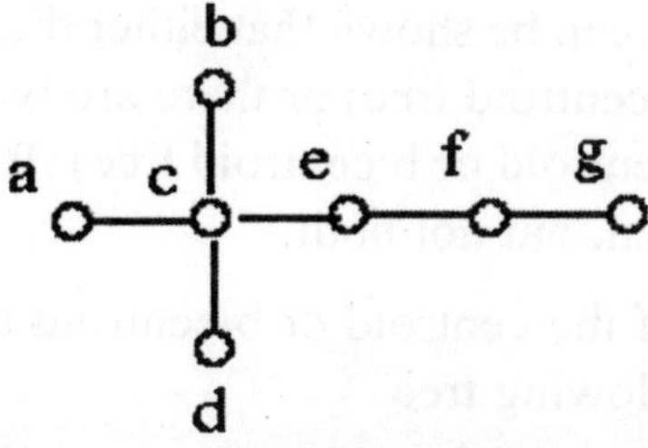

Remove all vertices of degree 1.

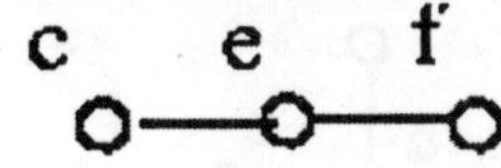

Remove all vertices of degree 1.

Therefore, a tree is central with center e.

Another example, given a following tree.

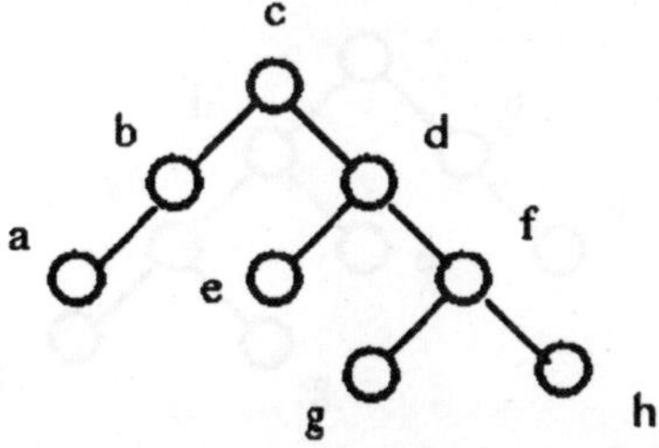

Remove all vertices of degree 1.

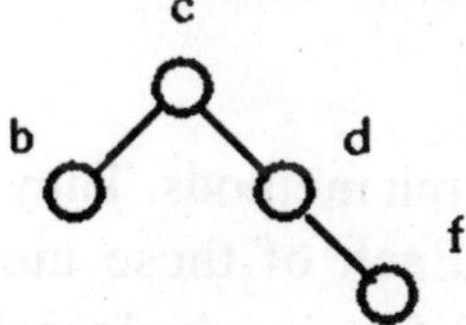

Remove all vertices of degree 1.

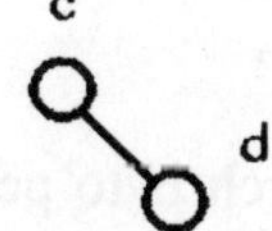

Therefore, a given tree is bicentral with bicenter cd.

Algorithm 2

For each vertex v of the degree 2 or more, count the number of vertices in each of the subtrees emanating from v, and let n_v be the maximum of these numbers.

If the tree has n vertices it can be shown that either there is just one vertex v for which $n_v \leq 1/2(n-1)$ (the centroid or centroid tree) or there are two adjacent vertices v and w for which $n_v = n_w = 1/2n$ (the bicentroid or bicentroid tree). It is easy to see that every tree is either centroidal or bicentroidal, but not both.

Note that we can think of the centroid or bicentroid as the 'center of gravity' of the tree.·For example, given a following tree

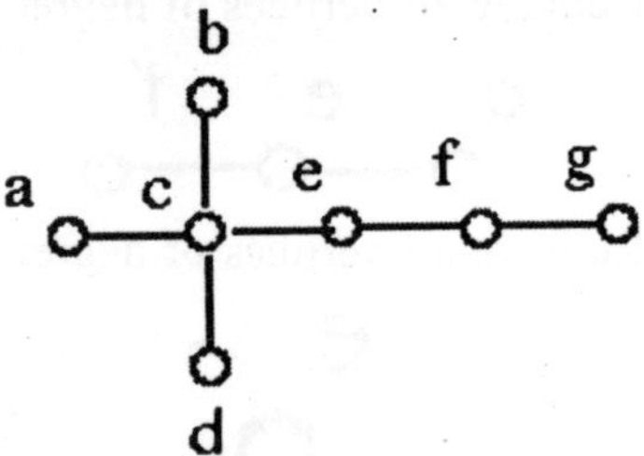

Since $n_c = 4$, $n_e = 4$, $n_f = 5$ and $n_g = 6$. Therefore, we have a bicentroidal tree with bicentroid *ce*. Another example, given a following tree.

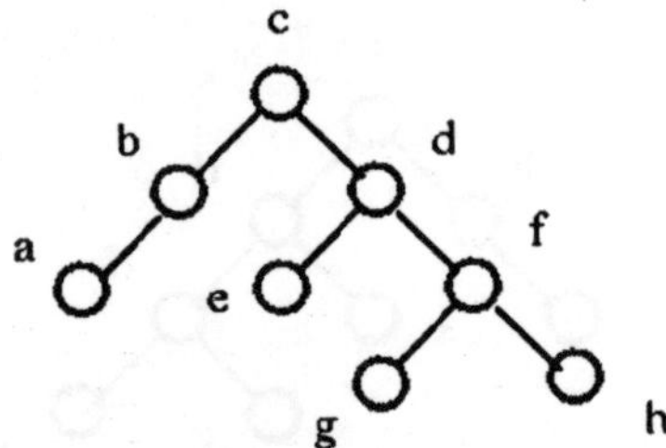

Since $n_b = 6$, $n_c = 5$, $n_d = 3$ and $n_f = 5$. Therefore, we have a centroidal tree with centroid *d*.

Tree Searching

There are two well-known search methods. They are brown as depth-first search(DFS) and breadth-first search (BFS). Each of these methods lists the vertices as they are encountered, and indicates the direction in which each edge is first traversed. The methods differ only in the way in which the vertex-lists are constructed.

Depth-First Search (DFS)

The basic idea of depth-first search is to penetrate as deeply as possible into a tree before fanning out to other vertices. This idea may be depicted as follows:

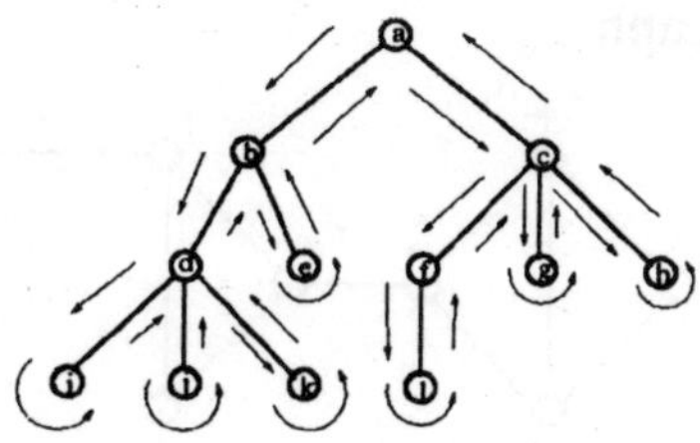

$a \to b \to d \to i \to j \to k \to e \to c \to f \to i \to g \to h.$

Another examplc, the expression $a + \{(b - c) \times d\}$ can represented by following tree:

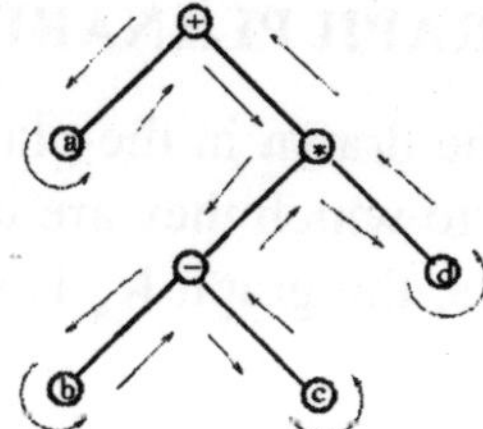

Example: Consider the graph

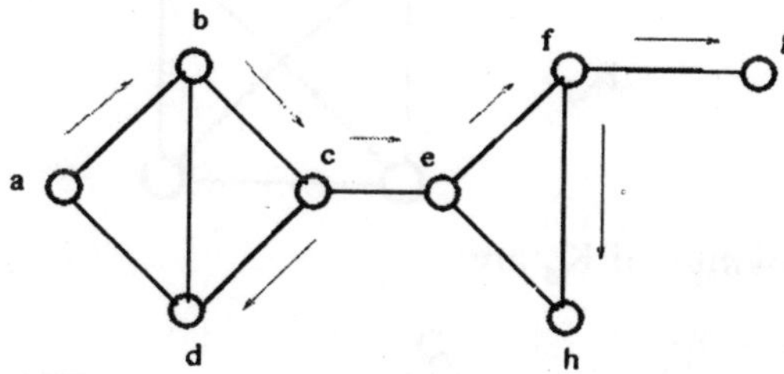

Note that we have marked those edges we used when going to new vertex. These edges form a spanning tree, called a DFS spanning tree.

Breadth-First Search (BFS)

The basic idea of breadth-first search is to fan out to as many vertices as possible before penetrating deep into a tree. This idea may be depicted as follows:

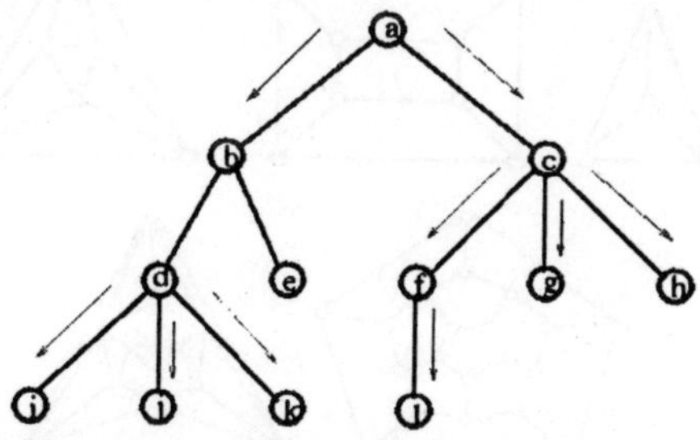

$a \to b \to c \to d \to e \to f \to g \to h \to i \to j \to k \to e$

above example clearly shows that breadth-first search must complete each level before proceeding to the next one.

Example: Consider the graph

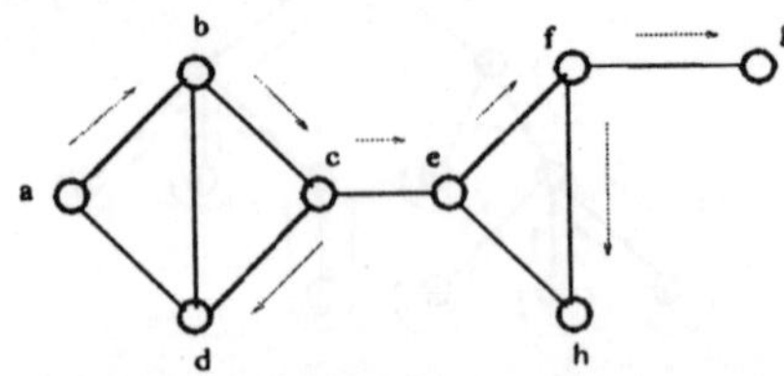

Note that we have marked those edges we used when going to new vertex. These edges form a spanning tree, called a BFS spanning tree.

GRAPH PLANARITY

A graph G is planar if it can be drawn in the plane in such a way that no two edges meet each other except at a vertex to which they are incident. Any such drawing is called a plane drawing of G. For example, the graph K_4 is planar, since it can be drawn in the plane without edges crossing.

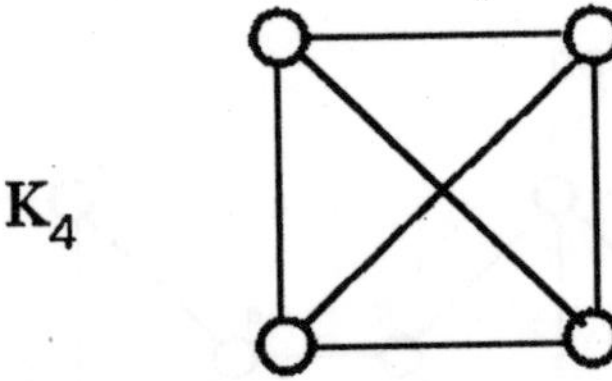

The three plane drawings of K_4 are:

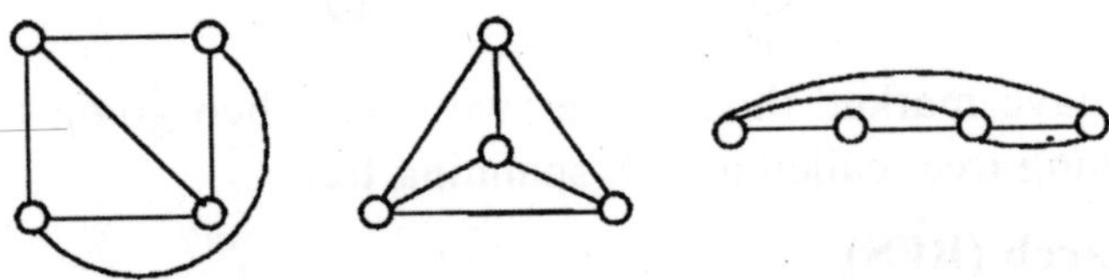

The five Platonic graphs are all planar.

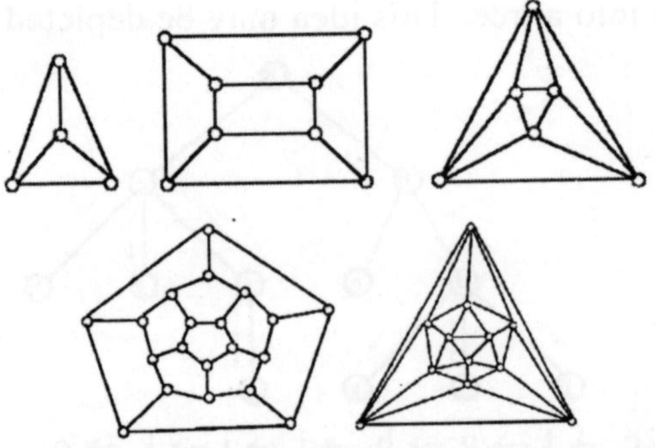

On the other hand, the complete bipartite graph $K_{3,3}$ is not planar, since every drawing of $K_{3,3}$ contains at least one crossing. why? because $K_{3,3}$ has a cycle which must appear in any plane drawing.

To study planar graphs, we restrict ourselves to simple graphs.

1. If a planar graph has multiple edges or loops.
   - Collapse the multiple edges to a single edge.
   - Remove the loops.
2. Draw the resulting simple graph without crossing.
3. Insert the loops and multiple edges.

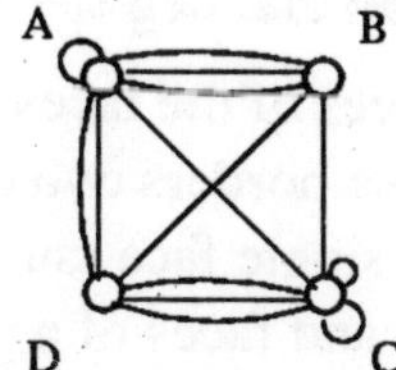

Remove loops and multiple edge.

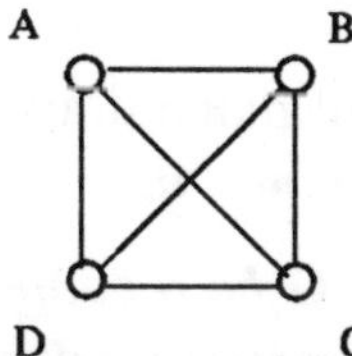

Draw without multiple edge.

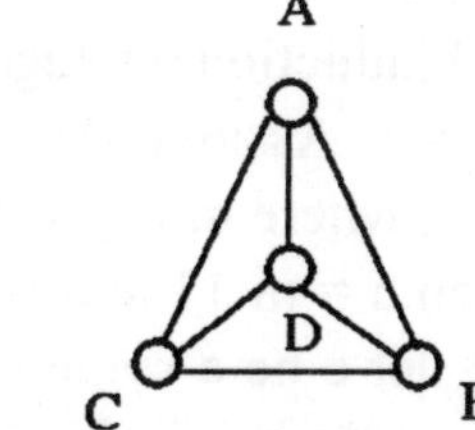

Insert loops and multiple edges.

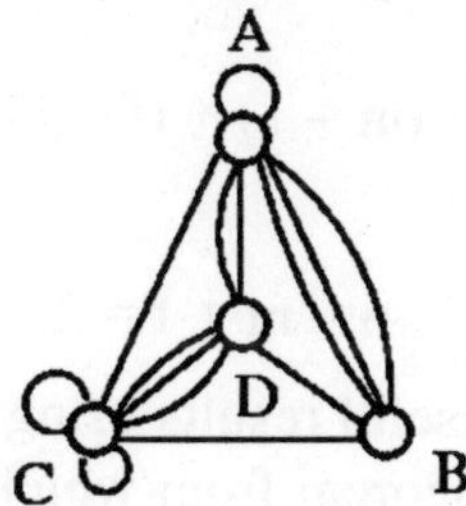

Euler's Formula

If G is a planar graph, then any plane drawing of G divides the plane into regions, called faces. One of these faces is unbounded, and is called the infinite face. If f is any face, then the degree of f (denoted by deg f) is the number of edges encountered in a walk around the boundary of the face f. If all faces have the same degree (g, say), the G is face-regular of degree g.

For example, the following graph G has four faces, f_4 being the infinite face.

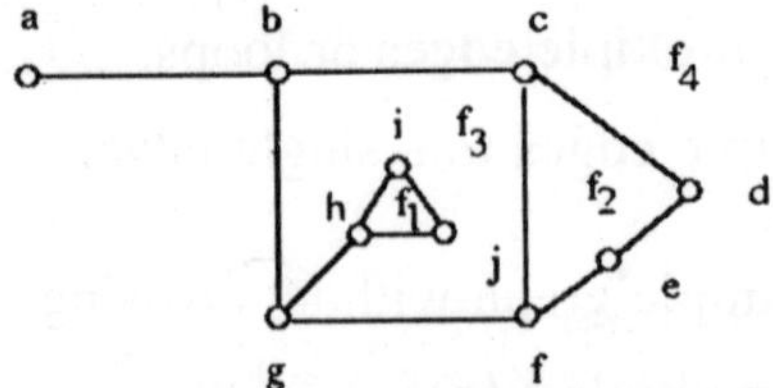

It is easy to see from above graph that deg f_1= 3, deg f_2= 4, deg f_3= 9, deg f_4= 8.

Note that the sum of all the degrees of the faces is equal to twice the number of edges in the the graph, since each edge either borders two different faces (such as bg, cd, and cf) or occurs twice when walk around a single face (such as ab and gh). The Euler's formula relates the number of vertices, edges and faces of a planar graph. If n, m, and f denote the number of vertices, edges, and faces respectively of a connected planar graph, then we get $n - m + f = 2$.

The Euler formula tells us that all plane drawings of a connected planar graph have the same number of faces namely, $2 + m - n$.

Theorem: (Euler's Formula) *Let G be a connected planar graph, and let n, m and f denote, respectively, the numbers of vertices, edges, and faces in a plane drawing of G. Then* $n - m + f = 2$.

Proof: We employ mathematical induction on edges, m. The induction is obvious for m = 0 since in this case n = 1 and f = 1. Assume that the result is true for all connected plane graphs with fewer than m edges, where m is greater than or equal to 1, and suppose that G has m edges. If G is a tree, then n = m +1 and f =1 so the desired formula follows. On the other hand, if G is not a tree, let e be a cycle edge of G and consider G – e. The connected plane graph G – e has n vertices, m –1 edges, and f –1 faces so that by the inductive hypothesis,

$$n - (m - 1) + (f - 1) = 2$$

which implies that

$$n - m + f = 2.$$

We can obtains a number of useful results using Euler's formula. (A "corollary" is a theorem associated with another theorem from which it can be easily derived.)

Corollary: *Let G be a connected planar simple graph with n vertices, where* $n \geq 3$ *and m edges. Then* $m \leq 3n - 6$.

Proof: For graph G with f faces, it follows from the handshaking lemma for planar graph that $2m \geq 3f$ (why?) because the degree of each face of a simple graph is at least 3), so $f \leq 2/3$ m.

Combining this with Euler's formula

Since $n - m + f = 2$

We get $m - n + 2 \leq 2/3\ m$

Hence $m \leq 3n - 6$.

As an example of Corollary, show that K_5 is non-planar.

Proof: Suppose that K_5 is a planar graph. Since K_5 has 5 vertices and 10 edges it follows from Corollary 1 that 10 $(3 \times 5) - 6 = 9$. This contradiction shows that K_5 is non planar.

It is important to note that $K_{3,3}$ has 6 vertices and 9 edges, and it is true that $9 \leq (3 \times 6) - 6 = 12$. This fact simply shows that we cannot use Corollary to prove that $K_{3,3}$ is non-planar. This leads us to following corollary.

Corollary: *Let G be a connected planar simple graph with n vertices and m edges, and no triangles. Then $m \leq 2n - 4$.*

Proof: For graph *G* with *f* faces, it follows from the handshaking lemma for planar graphs that $2m \geq 4f$ (why because the degree of each face of a simple graph without triangles is at least 4), so that $f \leq 1/2\ m$.

Combining this with Euler's formula

Since $n - m + f = 2$

Implies $m - n + 2 = f$

We get m – n + 2 d" 1/2m

Hence $m \leq 2n - 4$

As an example of Corollary, show that $K_{3,3}$ is non-planar.

Proof. Suppose that $K_{3,3}$ is a planar graph. Since $K_{3,3}$ has 6 vertices and 9 edges and no triangles, it follows from Corollary that $9 \leq (2 \times 6) - 4 = 8$. This contradiction shows that $K_{3,3}$ is non-planar.

Corollary: Let G be a connected planar simple graph. Then G contains at least one vertex of degree 5 or less.

Proof: From Corollary, we get m d" 3n – 6. Suppose that every vertex in G has degree 6 or more. Then we have $2m \geq 6n$ (why? because 2m is the sum of the vertex-degree), and therefore $m \geq 3n$. This contradiction shows that at least one vertex has degree 5 or less.

Now we will show by using Euler's formula that there are only five regular convex polyhedra – namely, the tetrahedron, cube, octahedron, dodecahedron, and isosahedron.

Theorem: *There are only 5 regular convex polyhedra.*

Proof: We prove this theorem by showing that there are only 5 connected planar graph G with following properties.

1. G is regular of degree d, where $d \geq 3$.
2. Any plane drawing of G is face-regular of degree g where $g \geq 3$.

Let n, m and f be the numbers of vertices, edges, and faces of such a planar graph G. Then, from properties (i) and (ii), we ge

$m = 1/2$ dn

$= 1/2$ gf

This gives us $n = 2m/d$ and $f = 2m/g$

Here Euler's formula ($n - m + f = 2$) holds, since G is a planar graph.

Therefore, $2m/d - m + 2m/g = 2$

Which can be written as $1/d - 1/2 + 1/g = 1/m$

Since $1/m > 0$, it follows that $1/d + 1/g > 1/2$

Note that each of d and g is at least 3, so each of 1/d and 1/g is at most 1/3.

Therefore, $1/d > 1/2 - 1/3 = 1/6$ and $1/g > 1/2 - 1/3 = 1/6$.

and we conclude that $d < 6$ and $g < 6$.

This means that the only possible values of d and g are 3, 4, and 5. However, if both d and g are greater than 3, then

$1/d + 1/g \leq 1/4 + 1/4 = 1/2$

which is a contradiction. This leaves us with just five cases:

Case 1: When $d = 3$ and $g = 3$.

we get $1/m = 1/3 - 1/2 + 1/3 = 1/6$

Therefore $m = 6$

It follows that $n = 8$ and $f = 4$ and this gives the Tetrahedron.

Case 2: When $d = 3$ and $g = 4$.

we get $1/m = 1/3 - 1/2 + 1/4 = 1/12$

Therefore $m = 12$

It follows that $n = 8$ and $f = 6$ and this gives the Cube.

Case 3: When $d = 3$ and $g = 5$.

we get $1/m = 1/3 - 1/2 + 1/5 = 1/30$

Therefore m = 30

It follows that n = 20 and f = 12 and this gives the Dodecahedron.

Case 4: When d = 4 and g = 3.

we get 1/m = 1/4 – 1/2 + 1/3 = 1/12

Therefore m = 12

It follows that n = 6 and f = 8 and this gives the Octahedron.

Case 5: When d = 5 and g = 3.

we get 1/m = 1/5 – 1/2 + 1/3 = 1/30

Therefore m = 30

It follows that n = 12 and f = 20 and this gives the Icosahedron.

And this completes the proof.

Planarity Testing

The Corollaries 1, 2 and their generalization are often useful for showing that graph is not planar. Unfortunately, there are many graphs which satisfy these inequalities but are not planar. Therefore, we need other way to decide planarity.

Some important observations:

1. Observation 1
   - Not all graphs are planar.
   - For example, we know K_5 and $K_{3,3}$ are not planar.
2. Observation 2
   - If G is a planar graph, then every subgraph of G is planar;
   - We usually stated observation 2 as follows
   - Observation 2a
   - If G contains a nonplanar graph as a subgraph, then G is non-planar. For example, following graph is nonplanar

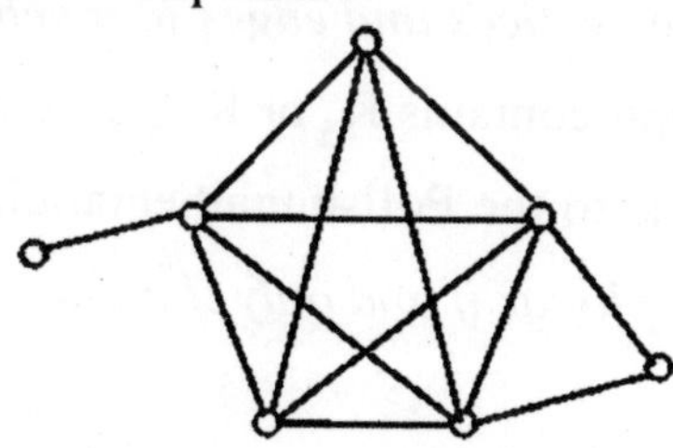

- Since it contains K5 as a subgraph.
- The following graph is also non-planar

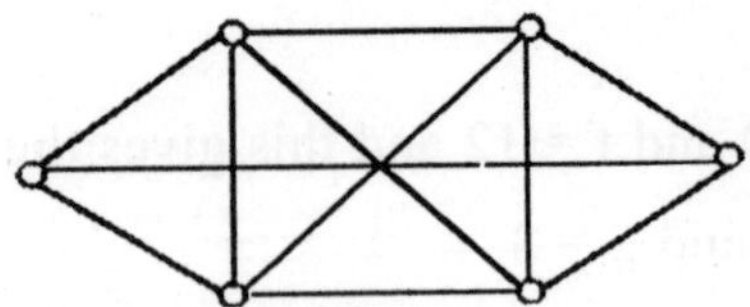

– Since the it contains $K_{3,3}$ as a subgraph.

3. Observation 3
   – If G is a planar graph, then every subdivsion of G is planar, we usually stated observation 3 in the following way.
   – Observation 3a
   – If G is a subdivision of a non-planar graph, then G is non-planar.
   – For example, following graph is non-planar,

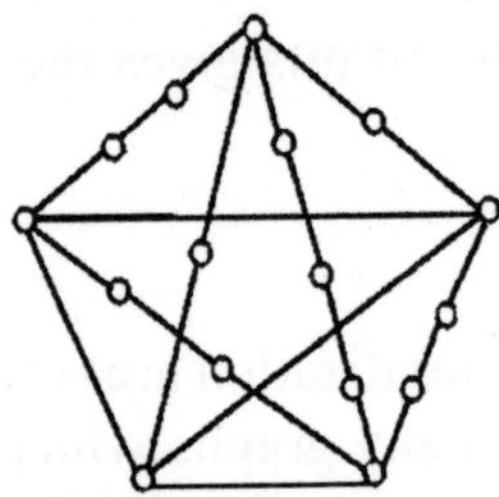

– Since it is a subdivision of K_5.

– Also, the following graph is non-planar,

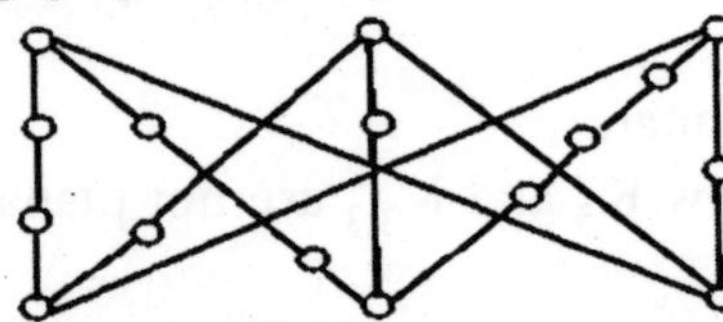

– Since it is a subdivision of $K_{3,3}$.

It follows from observations (2a) and (3a) that, if any graph G contains a subdivision of K_5 and $K_{3,3}$ as a subgraph, then G must be non-planar.

Why are we so obsessed with K_5 and $K_{3,3}$? *The reason is that all non-planar graphs can be obtained by adding vertices and edges to a subdivision of K_5 and $K_{3,3}$.*

Every non-planar graph contains K_5 or $K_{3,3}$ as a subgraph.

Following result is due to the Polish mathematician K. Kuratowski.

Theorem: *A graph is planar if and only if it does not contain a subdivision of K_5 and $K_{3,3}$ as a subgraph.*

Graph Contraction

In the following figure contradiction is done by bringing the vertex *w* closer and closer to *v* until *w* and *v* coincide and then coalescing multiple edges into a single edge.

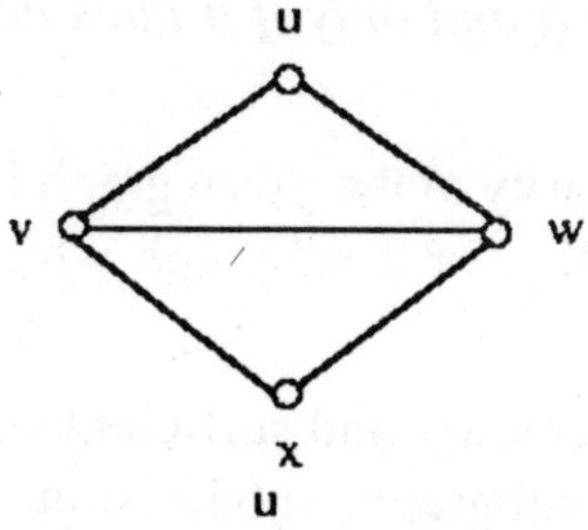

Bring vertex *w* closer to *v*.

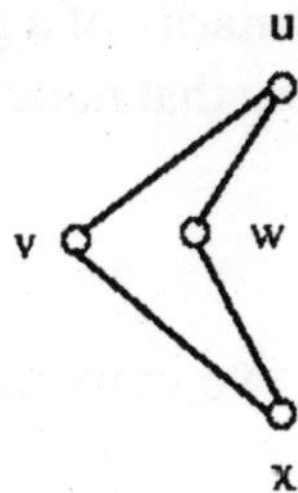

Coalesce vertex *v* and vertex *w*.

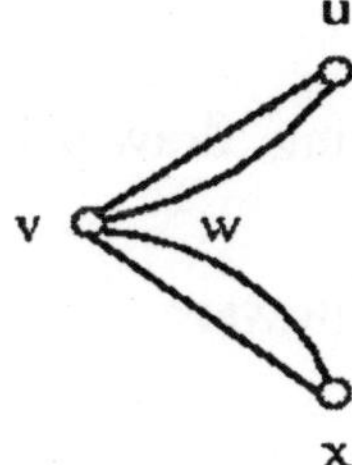

Finally, coalesce multiple edges and we have

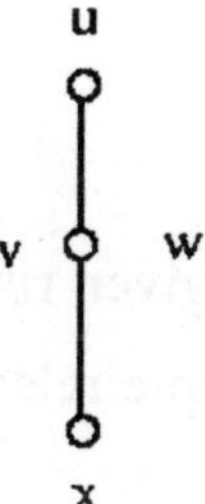

A contraction of a graph is the result of a sequence of edge-contractions. For example, K_5 is a contraction of the Petersen graph

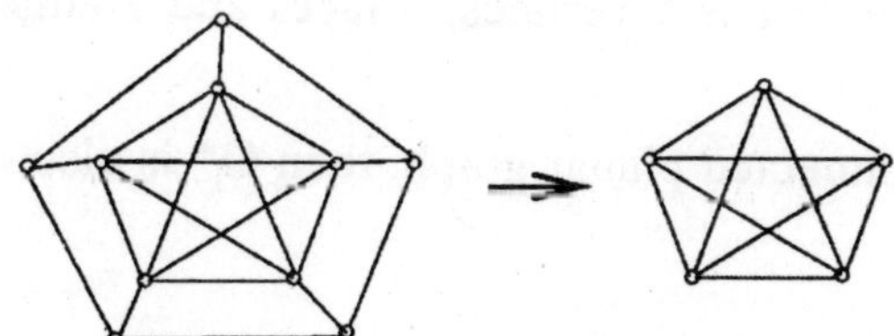

Theorem: *A graph is planar if and only if it does not contain a subgraph which has* K_5 *and* $K_{3,3}$ *as a contraction.*

The basic idea to test the planarity of the given graph is if we are able to spot a subgraph which is a subdivision of K_5 or $K_{3,3}$ or a subgraph which contracts to K_5 or $K_{3,3}$ then a given graph is non-planar.

Theorems 3 and 4 give us necessary and sufficient conditions for a graph to be planar in purely graph-theoretic sense (subgraph, subdivision, $K_{3,3}$, etc) rather than geometric sense (crossing, drawing in the plane, etc). This is the reason, why there exists no algorithm uses these two theorems for testing the planarity of a graph. Since this would involve looking at a large number of subgraph and verifying that none of them is a subdivision of, or contracts toK_5 or $K_{3,3}$.

Duality

Given a connected planar graph G, we construct dual graph G* in three stages.

1. Take a plane drawing of G.
2. Choose one point inside each face of the plane drawing – these points are the vertices of G*.
3. For each e of the plane drawing, draw a line connecting the vertices of G* on each side of e.

This procedure is illustrated as follows:

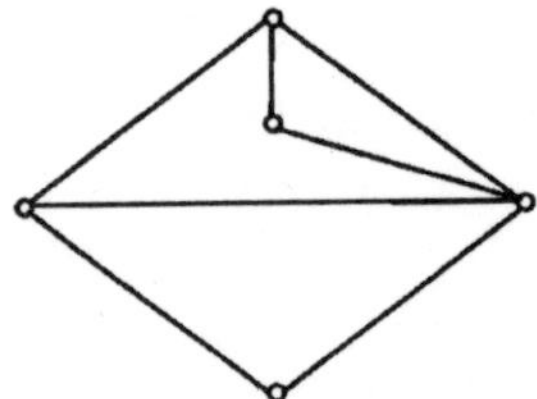

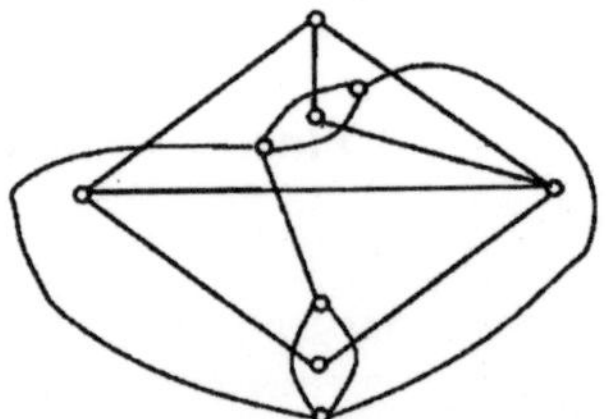

Note that each plane drawing of G given rise to just one dual graph G*.

Following theorem illustrates a simple relationship between the number of vertices, faces and edges of a graph and its dual.

Theorem: *If G is a connected planar graph with n vertices, f faces and m edges, then G* has f vertices, n faces and m edges.*

In the above example, G has 5 vertices, 4 faces and 7 edges, and G* has 4 faces, 5 faces, and seven edges.

Note that if G is a connected planar graph, then G* is also connected planar graph.

Chapter 5

Graph Embedding

In topological graph theory, an embedding of a graph G on a surface Σ is a representation of G on Σ in which points of Σ are associated to vertices and simple arcs (homeomorphic images of $[0,1]$) are associated to edges in such a way that:

1. The endpoints of the arc associated to an edge e are the points associated to the end vertices of e,
2. An arc include no points associated with other vertices,
3. Two arcs never intersect at a point which is interior to either of the arcs.

Here a surface is a compact, connected 2-manifold.

Often, an embedding is regarded as an equivalence class (under homeomorphisms of Σ) of representations of the kind just described. Informally, an embedding of a graph into a surface is a drawing of the graph on the surface in such a way that its edges may intersect only at their endpoints. If a graph G is embedded on a closed surface Σ, the complement of the union of the points and arcs associated to the vertices and edges of G is a family of regions (or faces). A 2-cell embedding is an embedding in which every face is homeomorphic to an open disk. A closed 2-cell embedding is an embedding in which the closure of every face is homeomorphic to a closed disk.

The genus of a graph is the minimal integer n such that the graph can be embedded in a surface of genus n. In particular, a planar graph has genus 0, because it can be drawn on a sphere without self-crossing. The non-orientable genus of a graph is the minimal integer n such that the graph can be embedded in a non-orientable surface of (non-orientable) genus n. Fixed-parameter tractable algorithms are known to check whether a graph can be embedded into a surface of a given fixed genus as well as to find the genus of a graph. That is, although the problem is NP-complete when the genus is unbounded, it can be solved in an amount of time that is linear in the graph size and doubly exponential in the genus.

EMBEDDINGS OF GRAPHS OF HIGHER-DIMENSIONAL MANIFOLDS

It is known that any graph can be embedded into a three-dimensional space. The idea is simple. Pick a line in the space and place the vertices of a graph along this line an any order. Draw as many planes through the line as there are edges in the graph and embed each edge into an individual plane. The above special case of embedding is called book embedding of the graph. This metaphor comes from imagining that each of the planes where an edge is drawn is like a page of a book. It was observed that in fact several edges may be drawn in the same "page".

In this section we introduce the best known parameter involving nonplanar graphs. On a sphere we placed a number of handles or equivalently, inserted a number of holes, so that we can draw a graph with edge-crossings. The number of handles (or holes) is referred to as the genus of the surface and denoted as of or g(G) or gen(G). By the genus, g, of a graph G is meant the smallest genus of all surfaces on which G can be embedded. Every graph has a genus; in fact a graph of size m can be embedded on a surface of genus m.

Genus of a Graph

The term "planar" comprises two conditions.

A graph is planar if

1. It is drawn without edge-crossings
2. It is drawn in a plane.

The concept "genus" includes the first condition but generalizes the seconds by considering other surfaces.

The first four members of an infinite family of surfaces i.e., sphere, one-hole torus, two-hole torus, and three-hole torus are shown below

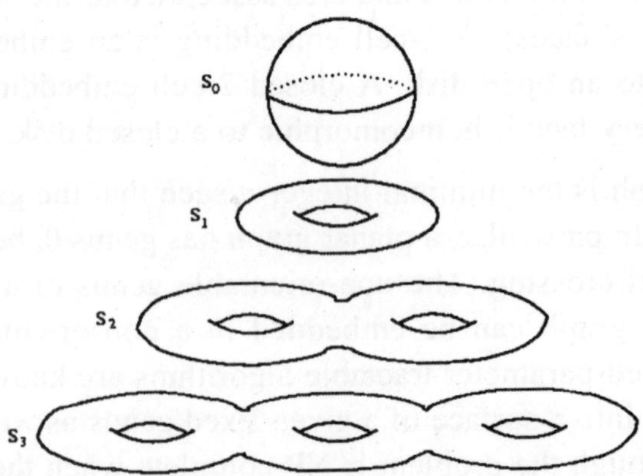

These surfaces are taken to be hollow and of negligible thickness. For example, we should think of S_0 as a beachball rather than a baseball and of S_1 as a valueless inner-tube

rather than a doughnut. In the names of the surfaces subscripts refer to the number of holes. Thus, S_0 is a "torus" with no holes, S_1 is a one-hole torus, S_{100} is a hundred-hole torus, etc.

Definition: The genus of a graph, denoted "g" is the subscript on the first surface among the family of surfaces, S_0, S_1,... on which the graph can be drawn without edge-crossings. In this definition we implicitly assumes that every graph has a geuns; That is, given any complicated nonplanar graph G, we can systematically search a sequence S_0, S_1, S_2,... that reveal some surfaces on which G can be drawn without edge-crossing. The genus, g, is just the subscript of the first surface.

Theorem: *The set of all planar graphs is equal to the set of all graphs with g = 0.*

Proof Idea: Here we have two statements to establish. First, "Every planar graph has g = o" and second "Every graph with g = 0 is planar." To prove that two sets are equal we have to show that each is a subset of the other.

This theorem shows that the "planar" concept is merely a special case of the "genus" concept. Planar graphs are the graphs of genus 0.

Note that despite of the fact that edges can go "around the back" of a sphere, we cannot avoid edge-crossings on spheres when they cannot be avoided in a plane.

Example: K_4 has g = 0 because it is a planar.

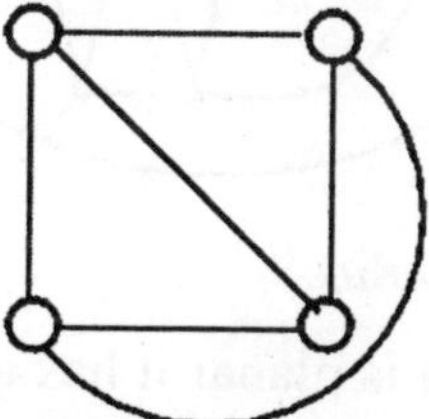

Example: Each cyclic graph, C_v, has g=0 because it is planar.

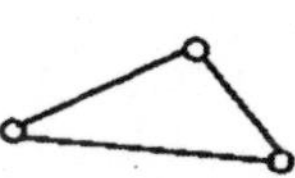

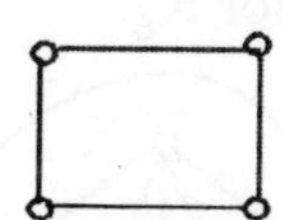

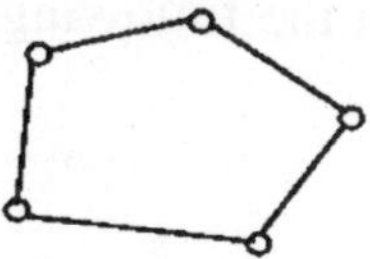

Example. The complete bipartite graph $K_{3,3}$ (utility graph) has g = 1 because it is nonplanar and so by theorem 1 cannot be drawn without edge-crossings on S_0;

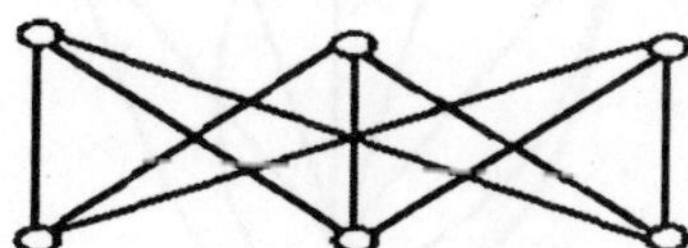

but it can be drawn without edge-crossings on S_1 (one-hole torus or doughnut).

Thus, the subscript 1 in the is S_1 the first surface (doughnut) in the family S_0, S_1,... on which $k_{3,3}$ can be drawn without edge-crossings. Similarly, K_5 has g = 1 for the same reasons see the following figure.

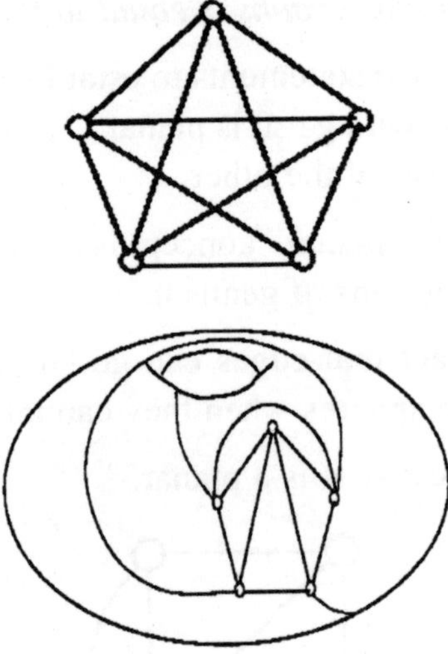

Theorem: *Every graph has a genus.*

Proof: Let G be any graph. If G is planar it has g = 0 by Theorem 1, so let us assume that G is nonplanar. Take a drawing of G in a plane and transfer to the surface of S_0. Add to S_0 enough "handl ≥ to serve as "overpass", thereby eliminating the edge-crossings. This has been shown in the following figure for K_6.

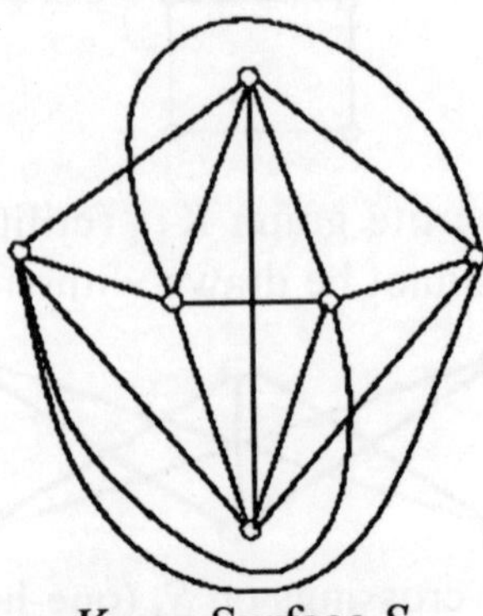

K_6 on Surface S_0

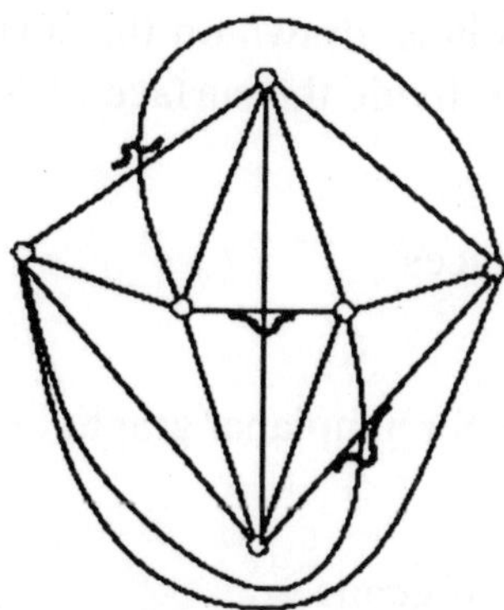

K_6 on Surface S_0 with three handles (or holes).

Let the number of handles be n. The number n is finite, as a graph can have only finite number of edges and hence only a finite number of edge-crossings. Topologically, the surface consisting S_0 of with n handles can be "continuously deformed" into S_n and the edges and vertices of G can be carried along with this continuous deformation. Following figure shows this for S_0 with three handles (k_6 have been omitted).

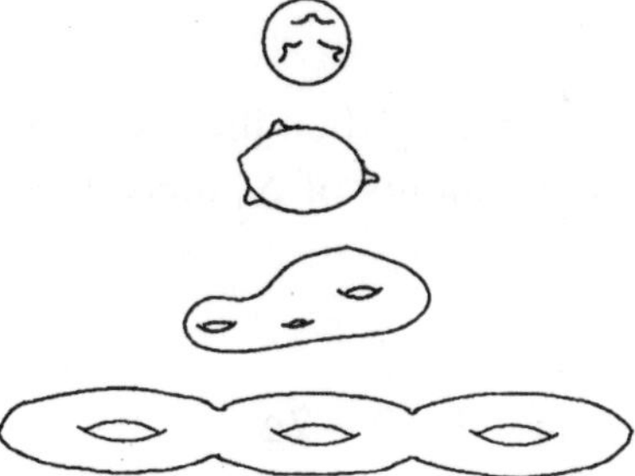

Thus G can be drawn on S_n without edge-crossings. Since there is at least one member, S_n, of the sequence of surfaces S_0, S_1, S_2 on which G can be drawn without edge-crossings, there must be a first such number S_g. Then by definition g is the genus of G and we have the theorem.

Theorem: *If a graph G has genus g then G can be drawn without edge-crossings on every surface S_n for which $n \geq g$.*

For specific graph G, its genus g cuts the sequence S_0, S_1,... into two pieces. The first part S_0, S_1,... S_{g-1} is finite and consists of all surfaces in the family on which G cannot be drawn without edge-crossings. The second part S_g, S_{g+1},... is infinite and consists of all surfaces on which G can be drawn without edge-crossings. The "genus" is use to specify the extent to which nonplanar graph is "nonplanar". For example a graph of genus 100 is much farther from planarity than a graph of genus 4. Up to now the term "fac≥ has been defined only for planar graphs. To speak of the "faces" of say, complete bipartite graph, would have been to speak nonsense. But with new family of surfaces it is now possible to define the term "fac≥ for any graph as follows:

If a graph G of genus g has been drawn on the surface of S_g without edge-crossings, then the edges and vertices of G divide the surface of S_g into regions faces of G.

Example: $k_{3,3}$ has $f = 3$.

< fig 5 → Edit → Include faces >

< fig 6 → Include faces >

Now that "fac≥ makes sense for nonplanar graphs as well, we revisit the Euler formula.

Euler's Second Formula

For every connected graph of genus g,

$$v + f - e = 2 - 2g$$

It is easy to see that for planar graphs this reduces to the

$$v + f - e = 2$$

Our proof of the Euler's second formula is based on the following assumption that we shall not prove.

ASSUMPTION

If G is a connected graph of genus g then there exists a crossing-free drawing of G on S_g such that through each of the g holes of S_g there is a ring composed of vertices and edges of G.

Euler's Second Formula

If G is connected then n – m + f = 2 – 2g.

Proof: Some Consequences of Euler's Second Formula

Lemma: If a connected graph G has n≥3 and genus g then

g ≥ (1/6)e – (1/2)(n – 2)

Proof: By the lemma we have $3f \leq 2n$. Since G is connected Euler's second formula applies and we have

n – m + f = 2 – 2g.

Rewritten f = – n + m + 2 – 2g

multiply by 3

3f = –3n + 3m + 6 -6g

Combining with the inequality, 3f d" 2n.

–3n + 3m + 6 – 6g ≤ 2n

which can be rewritten as

$-6g \leq - m + 3n - 6$

multiply by –1/6 given

$g \geq (1/6)m - (1/2)(n-2)$

Using the Theorem

We can find a lower bound for the genus of a connected graph, even if we know nothing more than its cardinality of vertices and edges. For example, consider the graph with 52 vertices and 201 edges. Then by the theorem we have

$g \geq (1/6)201 - (1/2)50 = 81/2.$

But g is an integer so we can conclude that the genus of G is at least 9.

The next theorem can be used to find an upper bound for the genus of a graph.

Lemma: If a graph H of genus g_H can be drawn on S_n without edge-crossing, then $g_H \geq n$.

Lemma: If H is a subgraph of G, then $g_H \geq g_g$.

Theorem: If n ≥ 3 the complete graph k_n has genus

g = left ceiling {(n-3)(n–4)}/12} right ceiling.

Corollary: If G has n ≥ 3 and genus g then

g ≤ left ceiling {(n – 3)(n – 4)}/12} right ceiling.

The lower and upper bounds for the genus of connected graph, even if we know only its number of vertices, m, and number of edges, n, is as follows

left ceiling (1/6)e – (1/2)(v – 2) right ceiling ≤ g ≤ left ceiling{(v – 3)(v – 4)}/12 right ceiling.

Conclusion

Euler formula

$n - m + f = 2 - 2g$

If G is a connected graph with n ≥ 3 then

$g \geq (1/6)m - (1/2)n+1$

If G is a connected graph with smallest cycle of length *k*, then

$g \geq m/2(1 - (2/k)) - (n/2) 1$

If G is a connected, triangle-free graph with n≥3, then

$g \geq (1/4)m - (1/2)n+1$

If G is a graph having blocks $B_1, B_2, \ldots, B_k$

then

$$g = {}^{k}\Sigma_{i=1}\ g(B_i)$$

A consequence of this result is: If G is a graph with component G1, G2,...Gk, then

$$g = {}^{k}\Sigma_{i=1}\ g(Gi)$$

Computing the genus of Graphs

The genus of the complete graph is given by

$g(k_n)$ = left ceiling {(n-3)(n-4)}/12 right ceiling, $n \geq 3$.

The genus of the complete bipartite graph is given by

$g(k_{r,s})$ = left ceiling {(r-2)(s-2)/4}right ceiling, $r,s \geq 2$.

The genus of the n-cube is given by

$$G(Q_n) = (n-4).\ 2^{n-3} + 1,\ n \geq 2$$

GRAPH COLORING

In this topic we look at ways of coloring things in a graph (where 'things' means vertices, edges or faces) in such a way that adjacent things are colored differently.

VERTEX COLORINGS

The vertices must be colored differentlv if thev are joined by an edge.

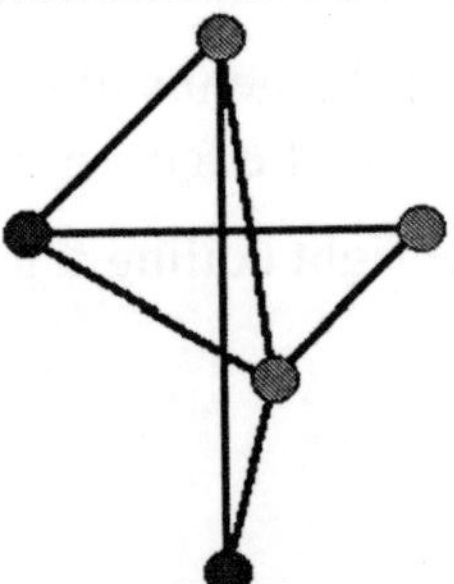

EDGE COLORINGS

Edges with vertces in common must be colored differently.

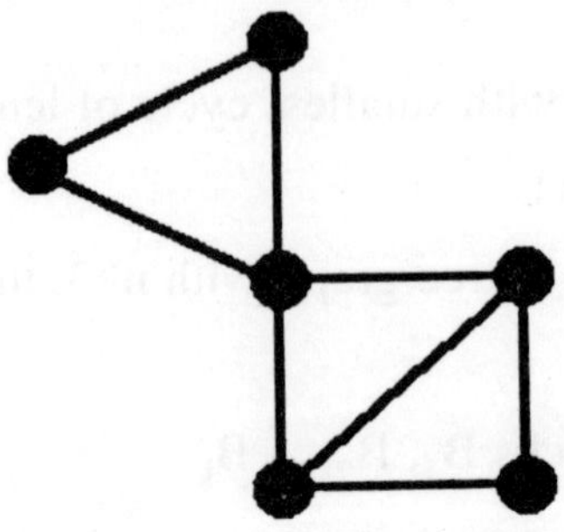

FACE COLORINGS

Faces with adjacent edges are always colored differently.

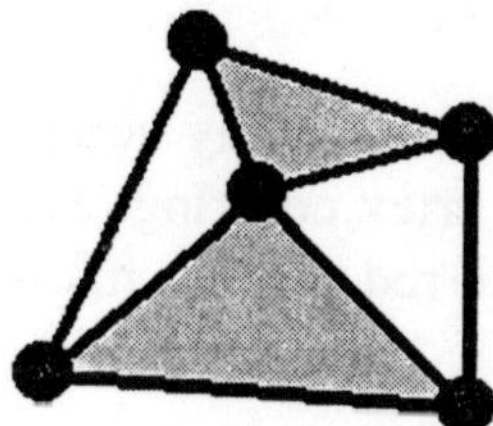

Note: We are usually interested in knowing the smallest number of colors that have to be used to color a graph.

Vertex Colorings

***Definition*:** A graph is said to be *k-vertex colorable* (or just *k-colorable*) if its vertices can be colored using up to *k* different colors in such a way that each vertex is of a single color and any two adjacent vertices have different colors (i.e. so that no two vertices of the same color are adjacent).

Definition: The *vertex chromatic number*, or simply the *chromatic number* of a fintite, loop-free graph *G* is the smallest positive number *k* such that *G* is *k*-colorable. The chromatic number of a graph *G* is usually denoted by c (*G*).

Note: The graph must be loop-free, because if vertices are allowed to have loops, then they can be adjacent to themselves. If a vertex is adjacent to itself, then it must be colored a different color from itself, which is impossible.

Examples: The triangle graph has chromatic number 3.

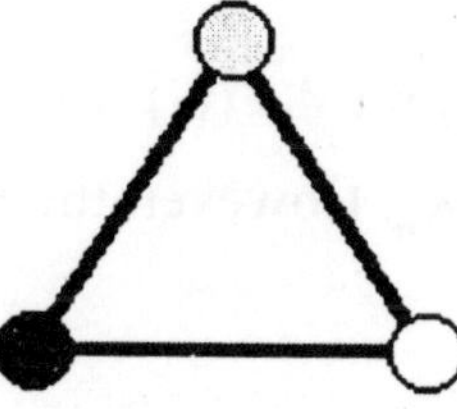

The cube graph has chromatic number 2.

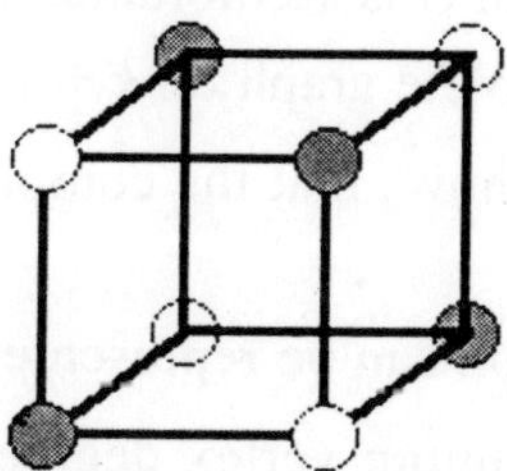

Lemma: For a finite, loop-free graph, *G*, c(*G*) = 2 if and only if *G* is a bipartite graph.

Proof: Firstly, suppose that G is a bipartite graph, with parts A and B, then color all vertices in A red, and all vertices in B blue. Then every edge joins a blue vertex to a red vertex. Therefore, G is 2-colorable, so c(G) = < 2. Now, (G) > 1, since once you have an edge, you need at least two colors.

Next, assume that c(G) = 2. Consider any vertex coloring of G with colors red and blue. Then, by the definition of a vertex coloring, every edge joins a red vertex to a blue vertex. Therefore, G is bipartite with red vertices in one part, and blue vertices in the other part. This completes the proof of the lemma.

Note: $\chi(K_n) = n$. Since any two vertices are joined by an edge and must therefore be colored differently.

Theorem: If every vertex of G has degree $d(v) < k$, then G is k-colorable.

Proof: Use induction on n, the number of vertices of the graph G. If $n = 1$ or $n = 2$, the assertion is easily seen to be true. Suppose $n > 2$, and assume that the proposition is valid for all graphs with fewer than n vertices.

Choose any vertex v of G and delete it and all the edges incident to v. This leaves a subgraph H of G with n - 1 vertices satisfying the given hypothesis (i.e. that every vertex has degree less than k). By the inductive hypothesis, $\chi(H)$ k, i.e. H is k-colorable.

Now, consider any particular k-coloring of H. Since $d(v) < k$, the vertices of H that were adjacent to v in G are colored with at most k-1 different colors. Thus, there's at least one color left with which we may color v, so that it is of a different color to each of its neighbors. This gives a coloring of G using the same colors as H. Therefore, G is k-colorable. This completes the proof of the theorem.

Note: The bound,

$$\chi(G) \leq \overset{\max}{v \in V(G)} \{d(v) + 1\},$$

cannot be improved. For example, K_n. However, the theorem can be improved to Brook's theorem.

Theorem: (Brook's Theorem).

If $\overset{\max}{v \in V(G)} \{d(v) + 1\} = k$, then G is k-colorable, unless it has a connected component isomorphic to K_{k+1} (i.e. the complete graph on k + 1 vertices.)

Remark: Brook's Theorem shows that the complete graphs are the 'worst cases' for vertex colorability.

Examples: The Peterson graph can be represented pictorially as,

The Peterson graph has maximum vertex degree 3, and does not have a connected component isomorphic to K_4. Therefore, c (*Peterson*) = 3.

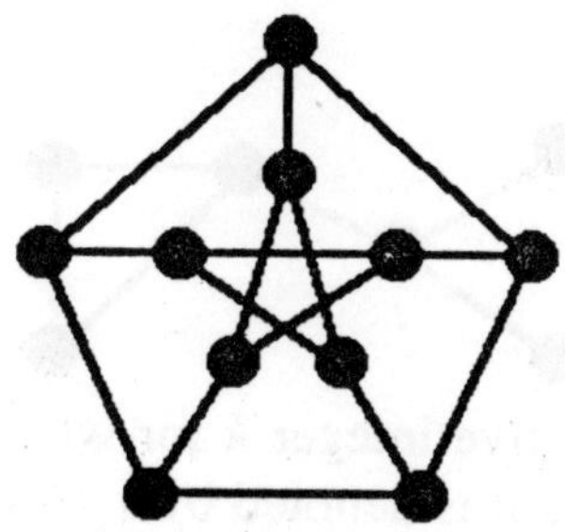

DUAL GRAPHS AND FACE COLORINGS

If G has m edges, n verteices and f faces, then the dual G^8 has f vertices, m edges and n faces.

Note that $(G^*)^*$ is isomorphic to G.

Theorem: (The Six Color Theorem for Finite Planar Graphs).

If G is a finite, simple planar graph, then G is 6-colorable (i.e. $\chi(G) = <6$).

Proof: If G has fewer than seven vertices, then the result is obvious. Suppose that G has n vertices with $n >= 7$. For induction, suppose that all such graphs on $n - 1$ vertices are 6-colorable.

Now, planarity and Euler's Polyhedron Formula implied that G has at least one vertex with degree less than 6, say vertex v. Remove this vertex and all edges incident to it, leaving a subgraph H on $n - 1$ vertices, which is 6-colorable (by the inductive hypothesis). Then v can be colored with any color different to the colors of its (fewer than 6) neighbors. This completes the proof of the Six Color Theorem.

Note: This theorem can be improved to the 'Five Color Theorem,' and ultimately to the 'Four Color Theorem.'

Face Colorings

Definition: A *map* is a simple, connected plane graph with no isthmus (i.e. an edge whose removal disconnects the graph into two components).

Definition: A map is said to be *k-colorable* if all of its faces can be colored with k colors (each face of one color) so that no two faces with a common edge have the same color.

Note: A map is k-colorable (faces) if and only if the vertices of the dual graph are k-colorable.

Edge Colorings

Definition: A graph G is said to be k-edge colorable if its edges can be colored using up to k-colors such that no two edges with a vertex in common have the same color.

Example:

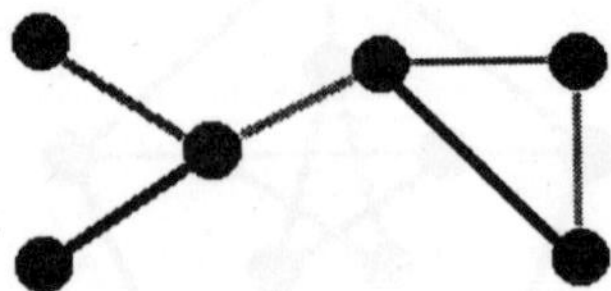

Definition: The smallest positive integer k for which G is k-colorable is called the *edge- chromatic number* of G, and it is denoted by $\chi_e(G)$.

Theorem: (Vizings Theorem).

If G is a finite, loop-free graph whose largest vertex degree is r, then:

$r =< \chi_e(G) =< r + 1$.

A GLOSSARY OF GRAPHS

Some graphs have special properties. For example, the graphs,

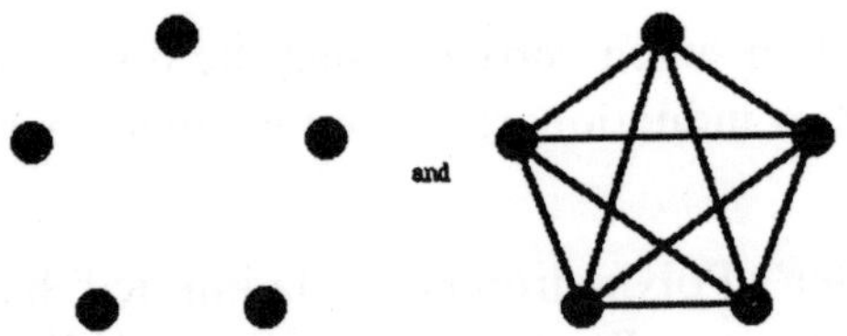

are both graphs on five vertices. In the graph on the left, no vertices are connected at all, whereas in the graph on the right, every vertex is joined to every other vertex by exactly one edge. This graph on the left is called the null graph on five vertices, and the graph on the right is called the complete graph on five vertices.

Definition: A graph whose edge set is empty is called a null graph. A null graph on n vertices is denoted by N_n.

Definition: A simple graph in which every pair of distinct vertices are adjacent is called a complete graph. The complete graph on n vertices is denoted by K_n.

Definition: A connected, 2-regular graph is called a circuit graph. The circuit graph on $n > 3$ vertices is denoted by C_n.

1. Is it possible to have circuit graphs with 1 or 2 vertices? If so, what would they look like. If not, what goes wrong with the definition?

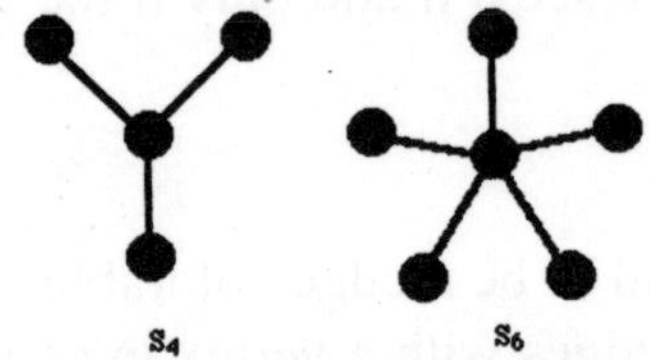

Definition: The star graph, S_n, has n vertices $v_1,\ldots, v_n$ with single edges joining v_1 to v_j for $2 \le j \le n$. For example,

Definition: The wheel graph, W_n, has n vertices $v_1,\ldots, v_n$ with single edges joining v_1 to v_j for $2 \le j \le n$. It also has single edges joining v_j to v_{j+1} for $2 \le j \le n$, and a single edge joining v_n to v_2. For example,

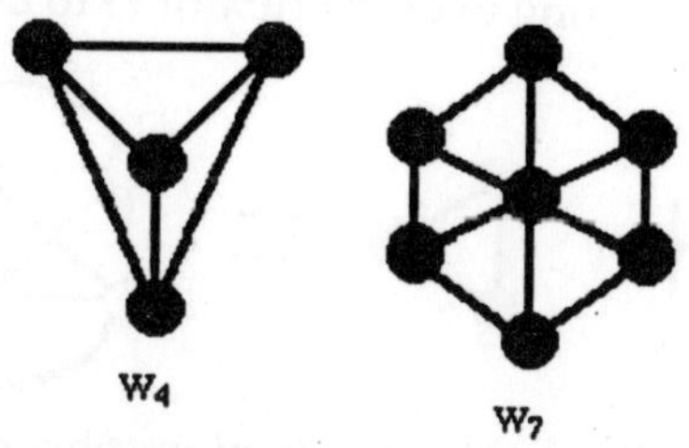

Definition: Suppose that the vertex set of a graph G can be split in two non-empty, disjoint sets, A and B in such a way that every edge of G joins a vertex of A to a vertex of B. The graph G is then said to be a bipartite graph.

1. The stipulation that the sets A and B are non-empty sets might seem a bit pedantic. However, if it is omitted, there are quite serious consequences. See if you can prove: If the requirement that sets A and B are non-empty is dropped, then every graph is a bipartite graph. If this was the case, there would be little sense in calling a graph a bipartite graph - we could just call it a graph, and save some typing.

· The following graphs are all bipartite graphs. Can you show how they satisfy the definition given above?

Are any of the special types of graph defined so far examples of bipartite graphs?

Definition Suppose that G is a bipartite graph, and that the vertex set can be split into non-empty disjoint sets A and B. If every vertex of A is joined to every vertex of B, then G is called a complete bipartite graph. Such a graph is denoted by $K_{m,\,n}$ where A contains m vertices, and B contains n vertices, and every vertex in A is joined to every vertex in B by a single edge. For example,

NEW GRAPHS FROM OLD

There are several constructions or operations that help to generate further examples of graphs from the examples we already have. The three discussed here are the sum, the union, and the complements of graphs.

Definition: If two graphs are taken to be $G_1 = (V(G_1), E(G_1))$ and $G_2 = (V(G_2), E(G_2))$. where $V(G_1)$ and $V(G_2)$ are assumed to be disjoint, then their union $G_1 \cup G_2$ is defined to be the graph with vertex set $V(G_1) \cup V(G_2)$, and edge set $E(G_1) \cup E(G_2)$. For example,

Definition: If G and H are graphs with disjoint vertex sets, then their sum, $G + H$, is the graph with vertex set $V(G)$ U $V(H)$ and edge set consisting of the union of $E(G)$ and $E(H)$ together with single edges joining every vertex of G to every vertex of H. For example,

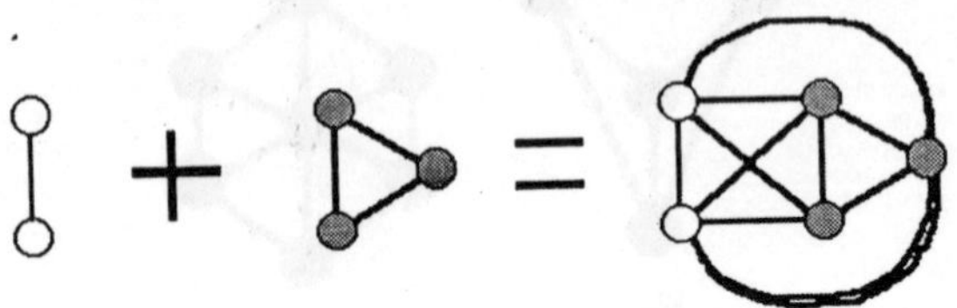

1. Can you find a relationship between wheel graphs, null graphs and circuit graphs?
2. Can you find a relationship between star graphs and null graphs?

Definition: If G is a simple graph with vertex set $V(G)$ then the complement, $\overline{G}$, of G is the graph with the same vertex set as G, and edge set given by the rule:

Two vertices v and w are adjacent in $\overline{G}$ if and only if they are not adjacent in G.

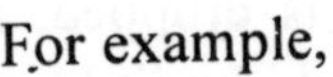

For example,

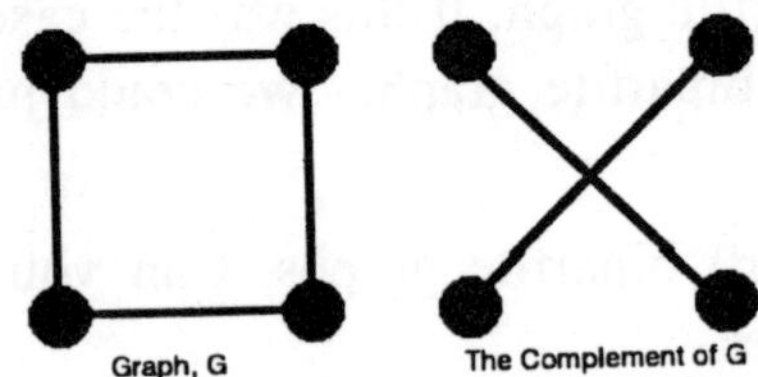

1. What is the complement of the complete graph on n vertices?
2. Can you find an argument that justifies the proposition:

Proposition: The complement of $K_{m,\,n}$ is K_m U K_n.

As you work on an argument, you might like to try some examples to see how the mathematics works.

1. Can you find an argument that justifies the following proposition:

Proposition: Suppose that G is a k-regular graph on n vertices. Then the complement of G is a $(n - k - 1)$-regular graph on n vertices.

Let G be a graph with no loops. A k-coloring of G is an assignment of k colors to the vertices of G in such a way that adjacent vertices are assigned different colors.

If G has a k-coloring, then G is said to be k-coloring, then G is said to be k-colorable. The chromatic number of G, denoted by X(G), is the smallest number k for which is k-colorable. For example,

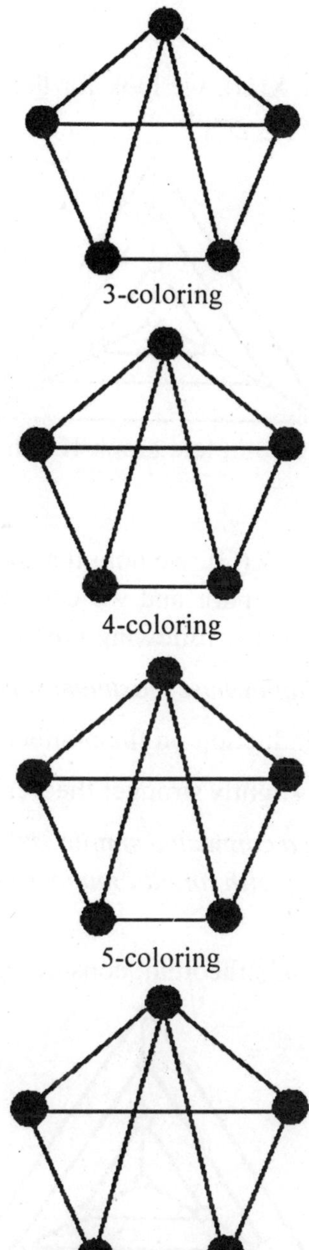

3-coloring

4-coloring

5-coloring

Not a permissible coloring, since one of the edge has color blue at both ends.

It is easy to see from above examples that chromatic number of G is at least 3. That is X(G) d" 3, since G has a 3-coloring in first diagram. On the other hand, X(G) ≥ 3, since G contains three mutually adjacent vertices (forming a triangle)., which must be assigned different colors. Therefore, we have X(G) = 3.7

LOWER BOUND OF X(G)

To obtain a lower bound for X(G), we look for the largest complete subgraph in G. For example, consider the following graph.

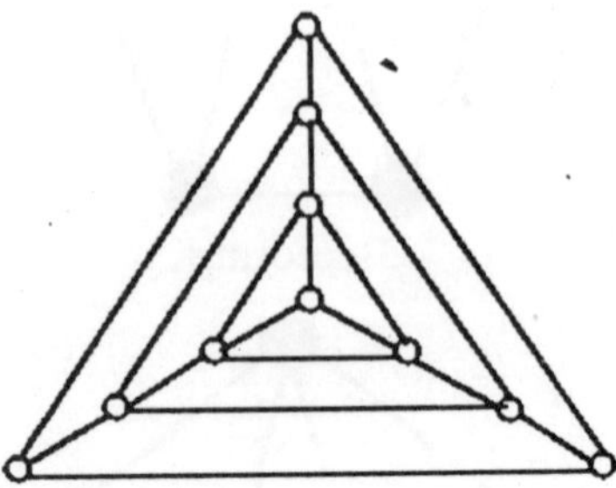

Since this graph contains the complete graph K_4, therefore X(G) ≥ 4.

UPPER BOUND OF X(G)

To obtain an upper bound for X(G), we note that if G has n vertices, then X(G) ≤ 4. However, this upper bound is very poor and we can improve it if we know the largest vertex-degree in G, which gives us the following theorem.

Theorem: *If G is a simple graph whose maximum vertex-degree is d, then X(G) ≤ d + 1.*

Proof Idea: Mathematical induction on the number of vertices of G.

We can prove the following slightly stronger theorem, which illustrates the same idea.

Brook's Theorem: *Let G be a connected simple graph whose maximum vertex-degree is d. If G is neither a cycle graph with an odd number of vertices, nor a complete graph, then X(G) ≤ d.*

To illustrate the use of Brook's theorem, consider graph G.

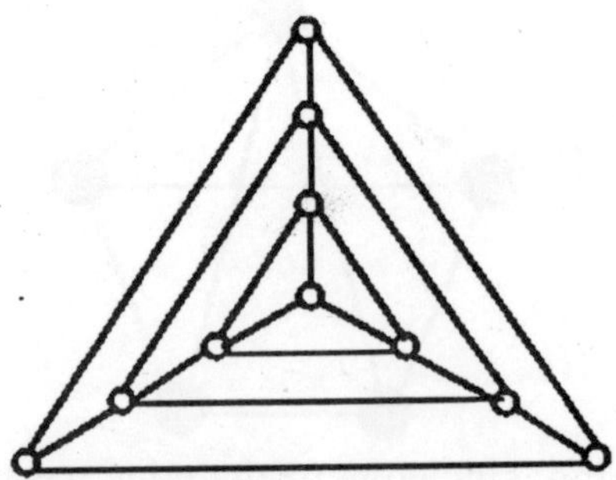

Since G contains the complete graph K_4, so X(G) ≤ 4. On the other hand, G satisfies the conditions of Brook's theorem with d = 4 and so X(G) ≤ 4. It follows from these two pieces of information that X(G) = 4. It is important to note that Brook's theorem does not always give a tight bound. For example, if G is the bipartite graph $k_{1,100}$, then X(G) = 2, whereas Brook's theorem gives us the upper bound X(G) ≤ 100.

CHROMATIC POLYNOMIALS

All known algorithms for finding the chromatic number of a graph are some what inefficient. A good estimation for the chromatic number of given graph involves the idea of a chromatic polynomials. Let G be a simple graph, and let $P_G(k)$ be the number of ways of coloring the vertices of G with k colors in such a way that no two adjacent vertices are assigned the same color. The function $P_G(k)$ is called the chromatic polynomial of G.

As an example, consider complete graph K_3 as shown in the following figure.

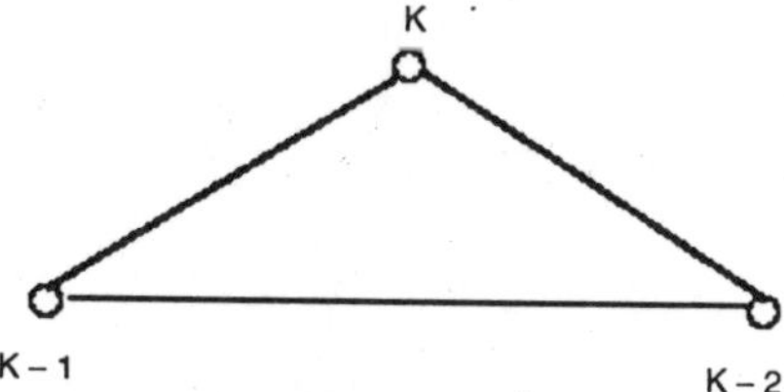

Then the top vertex can be assigned any of the k colors, the left vertex can be assigned any k – 1 colors, and right vertex can be assigned any of the k – 2 colors. The chromatic polynomial of K_3 is therefore k(k – 1)(k – 2). The extension of this immediately gives us the following result. If G is the complete graph K_n, then $P_G(k) = k(k-1)(k-2)\ldots(k-n+1)$.

EDGE COLORINGS

Let G be a graph with no loops. A *k*-edge-coloring of G is an assignment of *k* colors to the edges of G in such a way that any two edges meeting at a common vertex are assigned different colors,. If G has a *k*-edge coloring, then G is said to be *k*-edge colorable. The chromatic index of G, denoted by X'(G), is the smallest *k* for which G is *k*-edge-colorable. For example, consider the following graphs with eight edges:

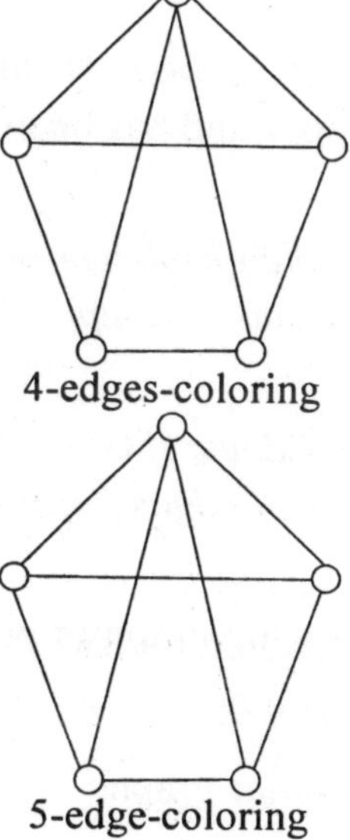

4-edges-coloring

5-edge-coloring

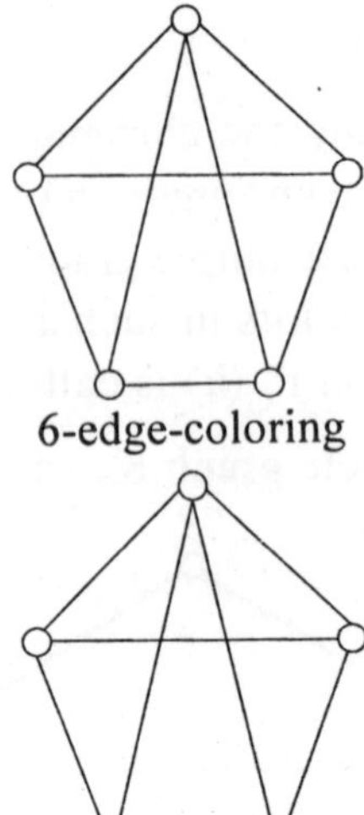

6-edge-coloring

Not a permissible coloring. Since two of the edges colored blue meet at a common vertex. From the above examples, it follows that $X'(G) \leq 4$, since G has a 4-edge-coloring in figure a (above).

On the other hand, $X'(G) \geq 4$, since G contains 4 edges meeting at a common vertex i.e., a vertex of degree 4, which must be assigned different colors. Therefore, $X'(G) = 4$.

LOWER BOUND OF X'(G)

To obtain a lower bound for $X'(G)$, we look for the largest vertex-degree, d, in given G, which gives us $X'(G) \geq d$.

UPPER BOUND OF X'(G)

To obtain an upper bound for $X'(G)$, we note that if G has *m* edges, then $X'(G) \leq m$. However, this upper bound is very poor and has been improved by V. G. Vizing and C. E. Shannon.

Vizing's Theorem: *If G is a simple graph whose maximum vertex-degree is d, then D* $\leq X'(G) \leq d+1$. *Following two theorems give upper bounds for the chromatic index of a graph with multiple edges.*

Vizing's Theorem for Multiple Edges: *If G is a graph whose maximum vertex-degree is d, and if h is the maximum number of edges joining a pair of vertices, then* $d \leq X'(G) \leq d+h$.

For example, consider the following graph in which d = 6 and h = 3.

Therefore bounds are $6 \leq X'(G) \leq 9$.

In fact, $X'(G) = 8$ for this particular graph.

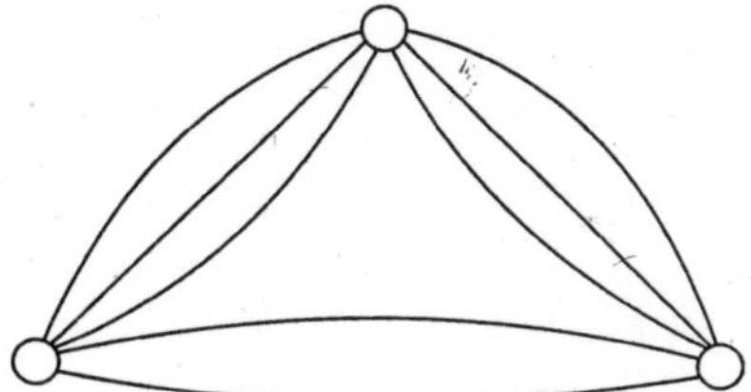

Shannon's Theorem: *If G is a graph whose maximum vertex-degree is d, then d ≤ X'(G) ≤ 3/2 d.*

For example, consider the following graph we have d = 6, and so bounds are 6 ≤ X'(G) ≤ 9. If d is odd, then (3/2)d is not an integer. In which case we can strengthen the bound to (3/2)d – 1/2.

We consider this section with an important theorem by Hungarian mathematician Denes Konig.

Konig's Theorem: *If G is n bipartite graph whose maximum vertex degree is d, then X'(G) = d.*

Proof Idea: Mathematical induction on the number of edge of G.

Konig's theorem tells us that every bipartite graph (not necessarily simple) with maximum vertex-degree d can be edge-colored with just d colors.

Chapter 6

Tree Structure

In graph theory, a tree is a graph in which any two vertices are connected by exactly one path. Alternatively, any connected graph with no cycles is a tree. A forest is a disjoint union of trees. Trees are widely used in computer science data structures such as binary search trees, heaps, tries, Huffman trees for data compression, etc. In some of the previous lessons, there have been thinking opportunities that relate mathematically defined words (such as path and circuit) to the everyday meaning of the words.

A tree is an undirected simple graph G that satisfies any of the following equivalent conditions:

1. G is connected and has no cycles.
2. G has no cycles, and a simple cycle is formed if any edge is added to G.
3. G is connected, and it is not connected anymore if any edge is removed from G.
4. G is connected and the 3-vertex complete graph K_3 is not a minor of G.
5. Any two vertices in G can be connected by a unique simple path.

If G has finitely many vertices, say n of them, then the above statements are also equivalent to any of the following conditions:

1. G is connected and has $n - 1$ edges.
2. G has no simple cycles and has $n - 1$ edges.

An undirected simple graph G is called a *forest* if it has no simple cycles.

A directed tree is a directed graph which would be a tree if the directions on the edges were ignored. Some authors restrict the phrase to the case where the edges are all directed towards a particular vertex, or all directed away from a particular vertex.

A tree is called a rooted tree if one vertex has been designated the *root*, in which case the edges have a natural orientation, *towards* or *away* from the root. The *tree-order* is the partial ordering on the vertices of a tree with *u* d" *v* if and only if the unique path from the root to *v* passes through *u*. A tree which is a subgraph of some graph *G* is a normal tree if the ends of every edge in *G* are comparable in this tree-order. Rooted trees, often with additional structure such as ordering of the neighbors at each vertex, are a key data structure in computer science; see tree data structure. In a context where trees are supposed to have a root, a tree without any designated root is called a free tree.

A polytree has at most one undirected path between any two vertices. In other words, a polytree is a directed acyclic graph (DAG) for which there are no undirected cycles either.

A labeled tree is a tree in which each vertex is given a unique label. The vertices of a labeled tree on *n* vertices are typically given the labels 1, 2, …, *n*. A recursive tree is a labeled rooted tree where the vertex labels respect the tree order (*i.e.*, if $u < v$ for two vertices *u* and *v*, then the label of *u* is smaller than the label of *v*). An irreducible (or *series-reduced*) tree is a tree in which there is no vertex of degree 2. An ordered tree is a tree for which an ordering is specified for the children of each node. An n-ary tree is a tree for which each node which is not a leaf has at most *n* children. *2-ary trees* (resp. *3-ary trees*) are sometimes called binary trees (resp. ternary trees) For example, a mathematical path might resemble:

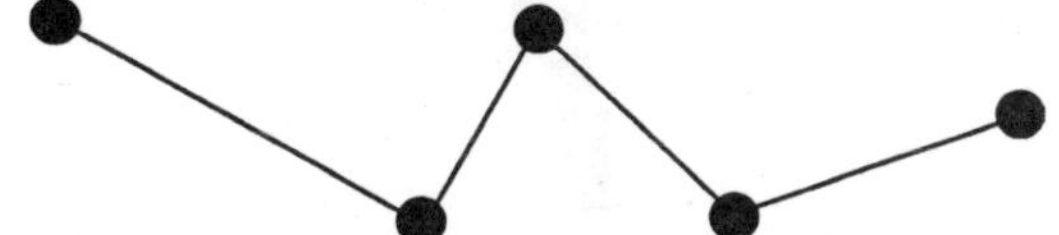

The everyday use of the word path might evoke an image resembling:

There is a correspondence between the mathematical meaning of *path*, and the image evoked by the everyday sense of the word *path*. In this instance, the vertices are given by the trail marker sings, and the edges are given by the sections of the trail joining the trail markers. As you might expect, it is possible to see associations between the mathematical meaning of the word *tree* and the images evoked by the everyday meaning of the word *tree*. A *tree* is basically a graph that can be drawn to look something like:

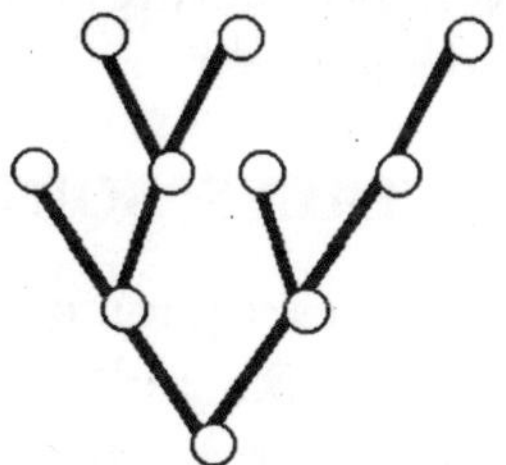

In my mind, the everyday meaning of the word *tree* evokes an image like:

Drawing vertices at the ends of the branches, and at the places where branches split gives a picture like:

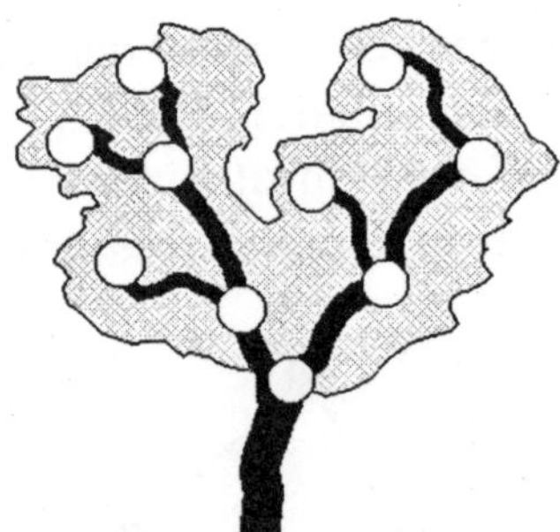

The branches of the (everyday meaning) *tree* can then be used for the edges of the (mathematical meaning) *tree* (the trunk of the tree doesn't connect two vertices, so it can't be used as an edge).

So, the mathematical meaning of the word *tree* can be related quite directly to the 'everyday' meaning of the word *tree*. However, this does not (at least in my mind) suggest what (mathematical) trees might be useful for.

1. Trees are connected graphs that contain no loops. What applications can you see for this kind of graph? Can you think of physical systems or networks that would be well represented by trees?
2. Trees are often used in information storage and retrieval systems. How could you use the structure of a tree to store or catalog information? What advantages would your system have over, say, storing information sequentially (that is, like a bunch of 35 cards in a card file box)?

AN APPLICATION FOR TREES: SORTING ALGORITHMS

A sorting algorithm is an outline or description of a procedure for arranging data in a particular order. For example, you might (like me!) have trouble remembering important information like people's birthdays, and keep a card file.

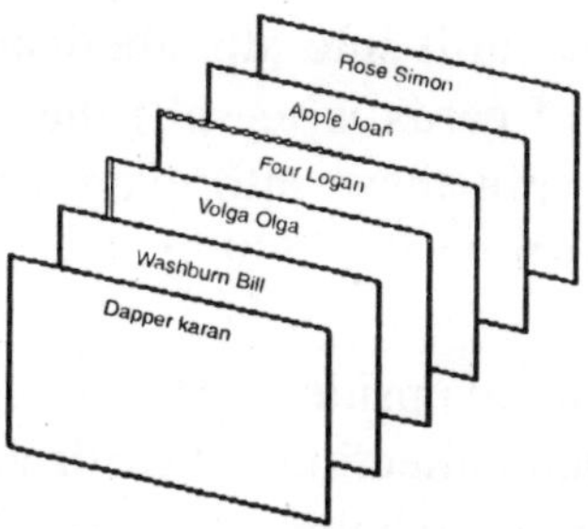

This file isn't very organized at the moment - you might end up having to go through every person's card to find the one you are looking for. To me, the most immediate way of organizing this file would be to arrange it in alphabetical order, by last name.

For a file of only six cards, you could just pick them all up, and put them back in the file box in alphabetical order. If your file had thousands of cards, then you would have to be more systematic. For example, you could go through the cards, comparing each card with the one immediately behind it. If the cards are out of alphabetical order, then you would switch them.

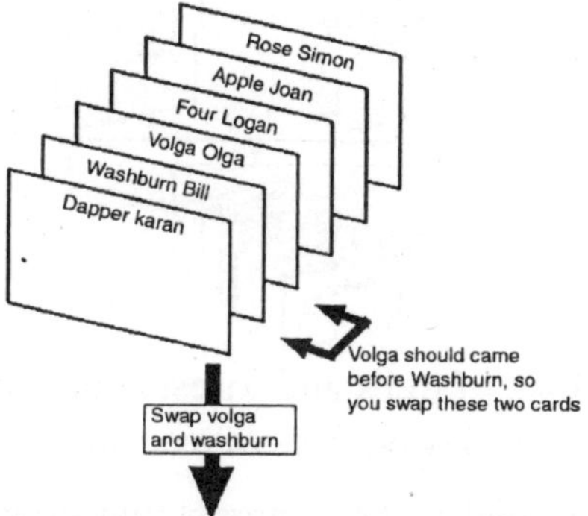

This kind of sorting algorithm, where you compare the cards pair by pair and swap them if the pair is not in alphabetical order is often called a *bubble sort*. As you can probably anticipate, bubble sorting is a tedious and time-consuming process (even for a computer).

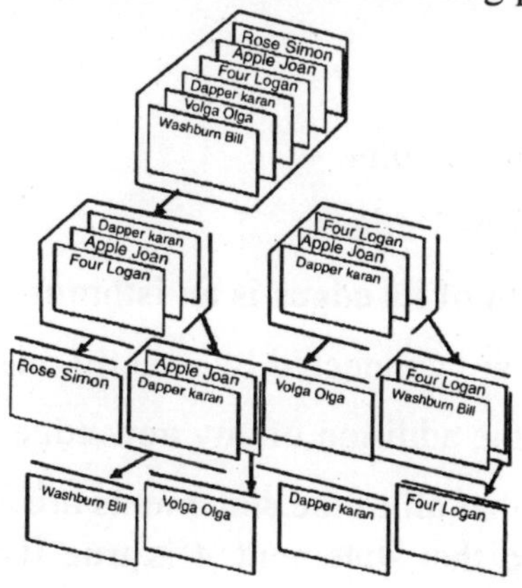

Trees can be used to sort the cards into alphabetical order in fewer steps. The idea behind using trees to sort a list of cards is to split the collection of cards into a lot of smaller collections, and then to split these into even smaller collections. For example, with the file of cards used to illustrate the bubble sort, the collection can be split up as shown below.

To sort the cards, the tree is turned upside down, and the cards recombined into steadily larger and larger collections. When collections of cards are put together, they are merged so that the resulting collection is in alphabetical order. When the process has run right down to the base or trunk of the tree, then you have the collection of cards that you started with, except that they are now in alphabetical order.

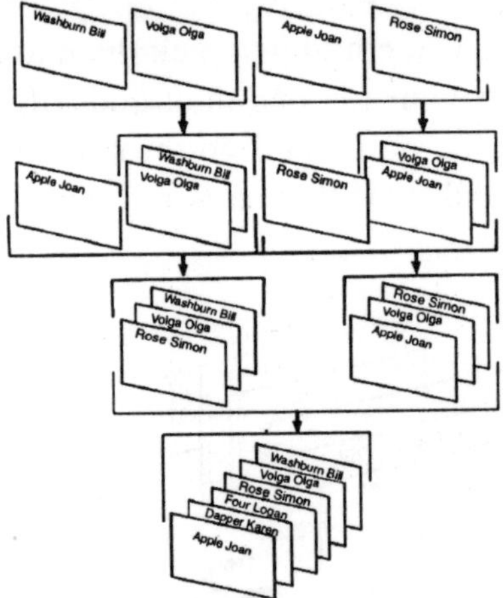

The mathematical definitions of trees and forests are given below. See if you can see how these definitions and the everyday notions of *tree* and *forest* correspond to each other.

Definition: A forest is defined to be a graph that contains no circuits. A tree is a connected forest.

Theorem: Suppose that T is a graph with n vertices. The following statements are equivalent:

1. T is a tree.
2. T has $n-1$ edges and no circuits.
3. T is connected and has $n-1$ edges.
4. T is connected and each of its edges is an isthmus.
5. Any two vertices of T are connected by exactly one simple path.
6. T has no circuits, but the addition of any new edge creates exactly one circuit.

Note: The theorem states that all of the statements are equivalent. Two statements, A and B, are equivalent provided that statement A is true if and only if B is true. This is another one of those "double barreled" statements, where you have to prove "both barrels" in order to prove the theorem. In this case, we would have to show that if statement 1 is true, then statement 2 is true.

Then, to do the other "barrel" of the proof, we'd have to show that if statement 2 is true, then statement 1 is true. This would have to be repeated for every pair of distinct statements in the theorem, and would amount to 30 little proofs that together constitute the proof of the theorem. This would be a pretty big job!

With a little organization, the proof of the theorem can be cut down to just proving six little proofs; that if statement 1 is true then statement 2 is true, if statement 2 is true then statement 3 is true, and so on, concluding with the proof that if statement 6 is true, then statement 1 is true. The proof ends up as a cycle of logical inferences.

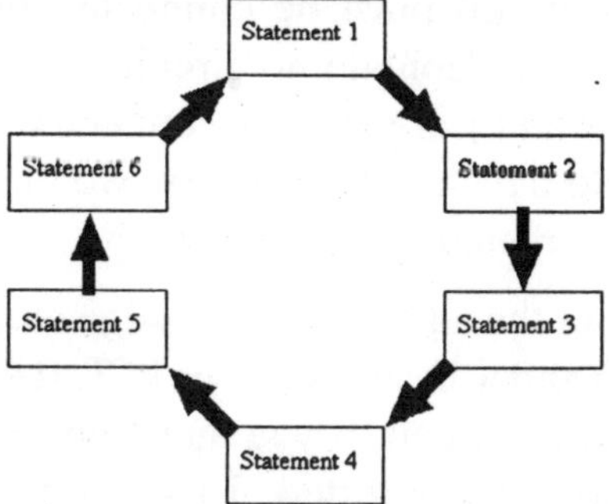

Proof: If $n = 1$, all six results are trivially true. So, assume $n >= 2$.

Use induction. Suppose the result is true for all graphs with fewer with n vertices.

Remove any edge, say from v to w. In the resulting graph, v and w lie in distinct connected components, A and B (each connected component is circled in the above diagram). Each of these connected components is a tree, say on r and s vertices respectively. A has r 1, and B has s *1* edges. Therefore, T has

$$(r - 1) + (s - 1) + 1 = r + s - 1 = n - 1 \text{ edges.}$$

$2 \Rightarrow 3$ T has $n - 1$ edges and no circuits. Assume that:

$$T = C_1 \cup L \cup C_3$$

where C_1 is a connected component of T with n, vertices. There, C, is a tree with n, – 1 edges. So, the number of edges in the graph T is:

$$\sum_{j=1}^{s}(n_j - 1) = \left(\sum_{j=1}^{s} n_j\right) - S = n - S.$$

But, T has $n - 1$ edges, so $s = 1$. Hence, T is connected.

$3 \Rightarrow 4$ T is connected and has $n - 1$ edges. Let e be any edge of T. Remove e and you get a graph with n vertices and $n-2$ edges. But such a graph cannot be connected (i.e. start with any vertex on its own, then the addition of each edge in turn adds one new vertex to the connected component containing v. As there are only $n - 2$ edges, the connected component containing v has at most $n - 1$ vertices. Therefore the graph is disconnected). As e was an arbitrary edge, then every edge of T disconnects T.

$4 \Rightarrow 5$ T is connected and every edge is an isthmus. Let v and w be any two vertices of T. T is connected, so there is a simple path from v to w. Assume that there are 2 or more such paths. Then the removal of any edge of the second path that is not part of the first path does not disconnect the graph, T. Therefore, that edges removed was not an isthmus, giving a contradiction. Therefore, there is only one simple path from v to w in T.

$5 \Rightarrow 6$ Every two vertices are connected by a simple path. If T contained a circuit, then any two vertices in this circuit must be connected by at least two simple paths, contradicting 5, implying that T can contain no circuit. Add a new edge, say from x to y. This creates a circuit because x and y are already connected in T. If more than one circuit was created, then x and y would have to be joined by two or more paths in T, which is not the case, so only one circuit is created.

$6 \Rightarrow 1$ T has no circuits, but the addition of any edge creates exactly one circuit. Must prove that T is connected. Let v and w be any vertices of T. If no edge exists from v to w then *add* such an edge and you obtain a circuit. If you then remove this edge, then you'll be left with a path from x to y in T, so T is connected. This completes the proof of the theorem.

SUBTREES AND SPANNING TREES

Definition: A subtree of a graph G is any subgraph of G that is a tree. A sub-forest is any subgraph of G that is a forest.

Definition: If G is a connected graph, then a spanning tree for G is any subgraph of G that is a tree, and includes *every* vertex of G.

Example: One pictorial representation of the cube graph is:

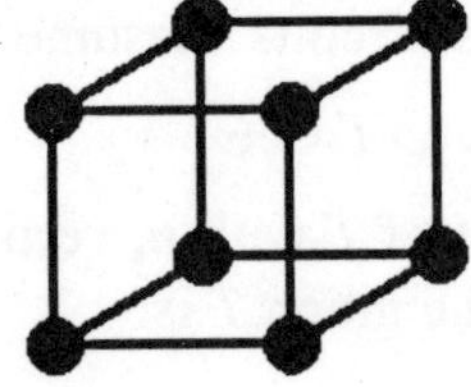

A spanning tree for the cube graph is:

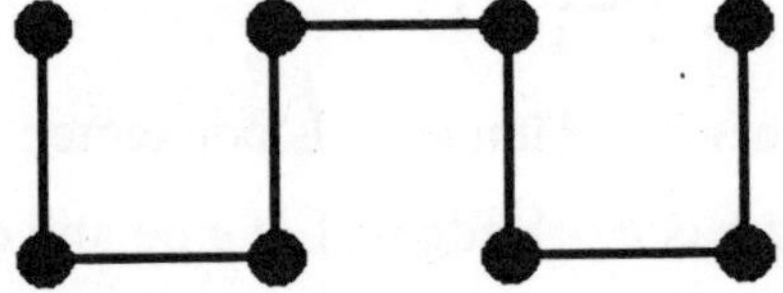

Coloring the vertices and edges shows how the spanning tree relates to the original cube graph.

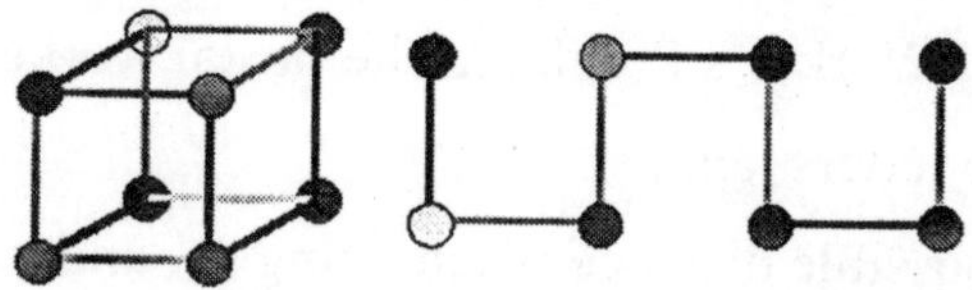

Proposition: For $n >= 2$, the complete graph on n vertices (i.e. K_n) has n^{n-2} different spanning trees.

Hint: You might approach this proof by examing the first few cases of the proposition ($n = 2$, $n = 3$, etc.) and then try to use the technique of proof by induction.

Definition: Any vertex, v, of a connected graph G with the property that the maximum distance from v to any other vertex is as small as possible is called a *central vertex* (or a center of the graph G). That maximum distance is called the *radius* of G.

Example: The graph,

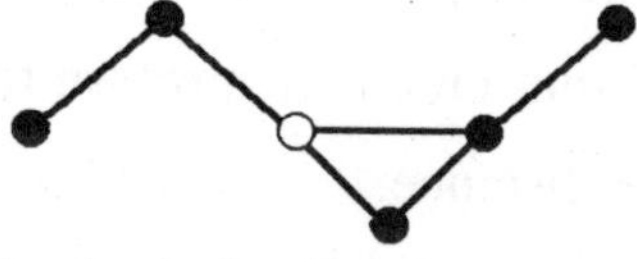

has radius 2 (the central vertex is shaded yellow).

THE MINIMAL CONNECTOR PROBLEM [H]

Suppose some real-life situation is modeled by a graph G with weights assigned to its edges. By 'weights' we mean that there is a weight function,

$$w : E(G) \to \{x \in \Re : x \geq 0\},$$

and for each edge, e, the weight, $w(e)$ usually represents some sort of cost (or profit) for using that edge.

Example: A possible pictorial representation of a graph with weights is shown below.

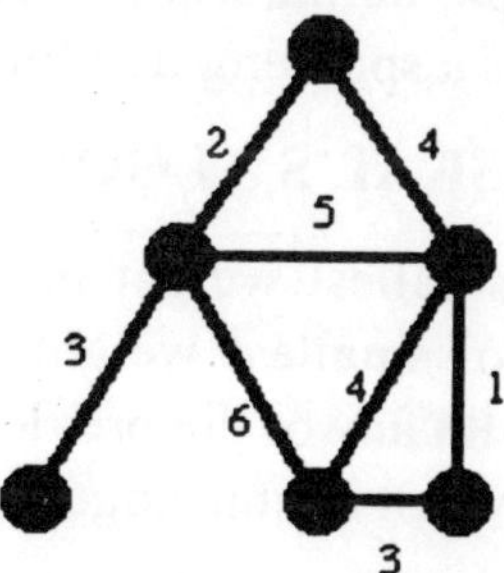

The weights could represent:

1. Costs,

2. Distances,
3. Waiting times (or just about anything else you care to imagine).

Example: The Postal Worker's Problem. The postal worker wants to:

1. Deliver all of the letters, and,
2. Cover the least possible distance when doing so, and,
3. Traverse each of the roads on the mail route at least once, and,
4. Return to the starting point (the post office).

This problem may be reformulated in terms of weighted graphs, where the graph corresponds to the network of roads, and the weight of each edge corresponds to the length of the road represented by that edge.

The requirement is to find a closed walk that includes every edge at least once, and has the least possible weight.

Example: The Traveling Sales Representative's Problem. The sales representative

1. Has to visit several different cities, then return to the starting point, and,
2. Travel the least possible distance.

This can also be reformulated in terms of weighted graphs. In this case, the requirement is to find a Hamiltonian cycle of least possible weight in a complete graph with weights.

Often, you want to find a connected subgraph of a graph G that uses every vertex of G (e.g. so you can go from any vertex to any other vertex in the subgraph) but so that the sum of the weights of all the edges in the subgraph is as small as possible. This is called the Minimal Connector Problem. The solutions to this problem is given by Kruskal's Algorithm.

This is an algorithm for finding a minimum weight spanning tree in a connected graph G that has weights assigned to its edges - in particular, it solves the minimal connector problem for G.

Definition: If T is any subgraph, define $w(T)$ = sum of the weights of the edges of T. A *minimum weight spanning tree* is a spanning tree for which $w(T)$ is as small as possible.

KRUSKAL'S ALGORITHM

First choose any edge e_1 of smallest weight in G. Then choose edges e_2, e_3,... by choosing at each stage an edge (of smallest weight not already chosen) subject to the condition that it forms no circuit with any of the previous edges. If there are no such edges left, then what you have already chosen are the edges of a minimum weight spanning tree.

Example: A cube graph with weights, and a minimum spanning tree. The weight of the minimum weight spanning tree is 34.

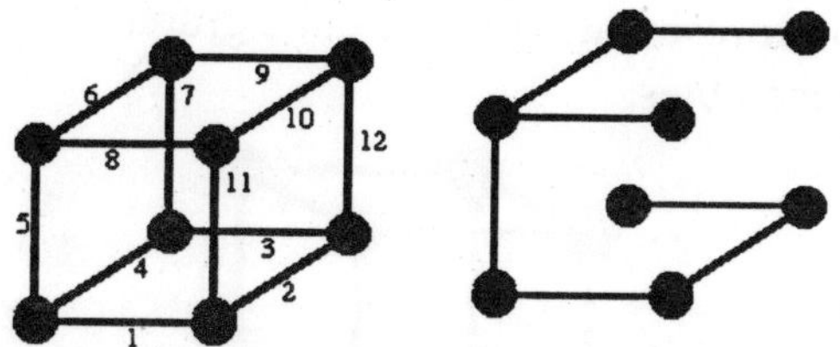

Proof: (That Kruskal's Algorithm always works for a connected graph G.)

First, note that if $e_1, e_2,..., e_m$ are the edges chosen by the time the algorithm finishes, then these edges must form a *connected* subgraph of G, that uses every vertex of G. (If not, then you would have a forest, not a tree, but G is connected so there will be some edges that could not be some edges that could be chosen as e_{m+1}, etc.)

It is necessary to prove that T is a minimum weight spanning tree. Let S be any spanning tree for G (of minimum weight). Consider e_1 to e_m in turn. Choose the first one of these that *isn't* an edge of S, say e_k. Add e_k to S. This creates a circuit in S that has an edge e that can't be in T (as T has no circuits). By the choice of e_k, $w(e_k) =< w(e)$. Remove e from S and rcplace it by e_k. This gives a new spanning tree S' with $w(S') =< w(S)$. Continue in this way, choosing the next e_l that isn't in S', etc. Eventually you get:

$$w(T) \leq \mathbb{K} \leq w(S''') \leq w(S'') \leq w(S') \leq w(S) \leq w(T).$$

So, $w(T) = w(S)$. Since S is a minimal spanning tree, T is a minimal spanning tree, too. This completes the proof.

PLANAR GRAPHS

Definition: A *planar graph* is one that can be drawn on a plane in such a way that there are no "edge crossings," i.e. edges intersect only at their common vertices.

Example: Gas, Water, Electricty Problem. Is there any way to connect each of the three houses to each of the three utilities in such a way that none of the supply lines cross?

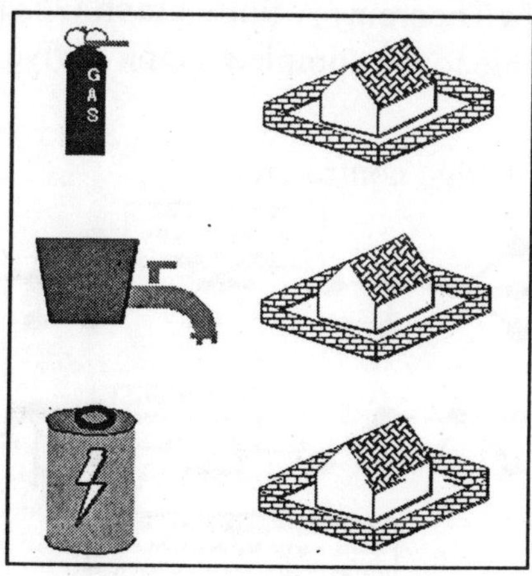

Example: A pictorial representation of the cube graph that makes it easy to see why it is called the cube graph resembles:

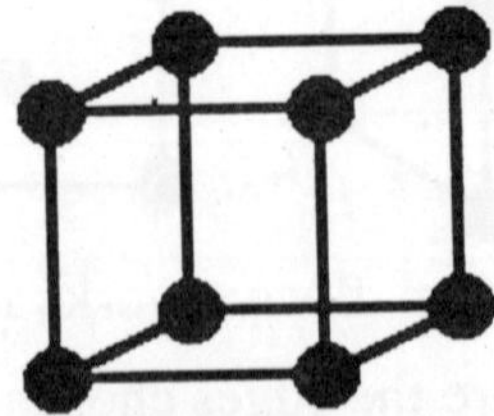

This representation includes many "edge crossing." It is possible to re- draw the cube graph so that no two edges cross. Coloring the vertices and edges shows how the graph is re-drawn more clearly.

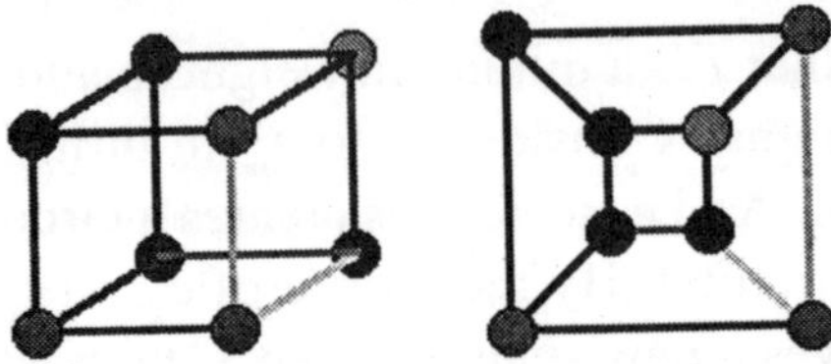

Definition: A *plane graph* is one that has been drawn in the plane in such a way that its edges intersect only at their common end-vertices.

Definition: Two graphs, G and H are defined to be homeomorphic if you can make one graph into the other by inserting vertices of degree 2

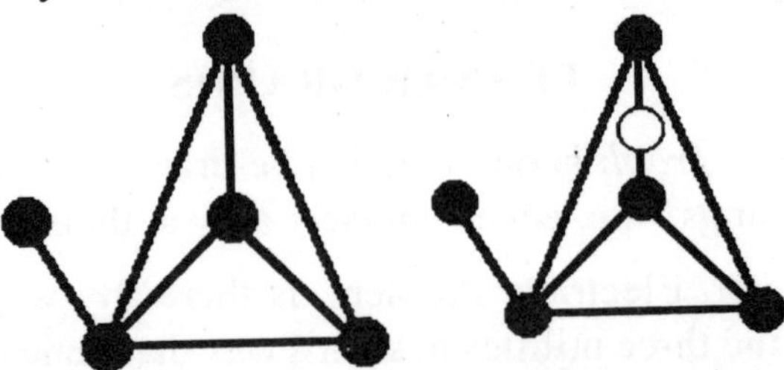

Theorem: (Kuratowski's Theorem) A finite graph G is planar if and only if it has no subgraph that is homeomorphic to the complete graph in five vertices (K_5) or the complete bipartite graph $K_{3,\,3}$.

Definition: Examples of edge contractions.

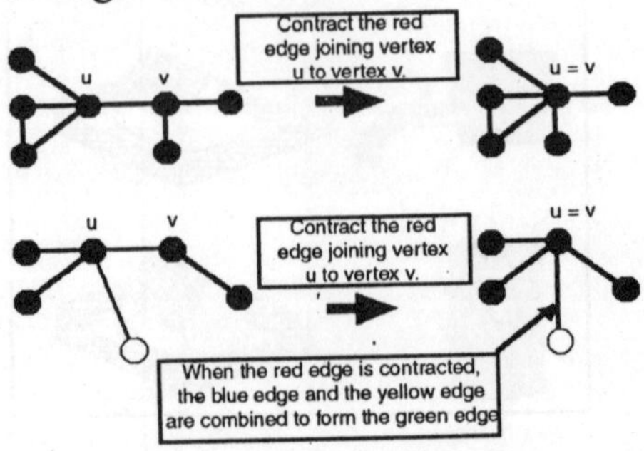

Theorem: A finite graph G is planar if and only if it contains no subgraph that is edge- contractible to K_5 or $K_{3,\,3}$.

Example: The Petersen graph may be represented pictorially as:

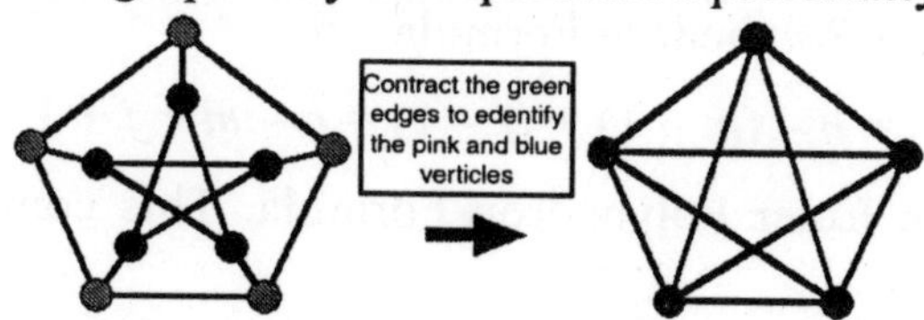

Since the Petersen graph is edge contractible to K_5, it is not a planar graph.

EULER'S POLYHEDRON FORMULA [H]

Definition Let G be a plane graph, and consider the regions bounded by the edges of G. These are called *faces*.

Example The following graph has 8 faces (one of them is colored - the outside of the graph is usually called the infinite face), 9 vertices and 15 edges.

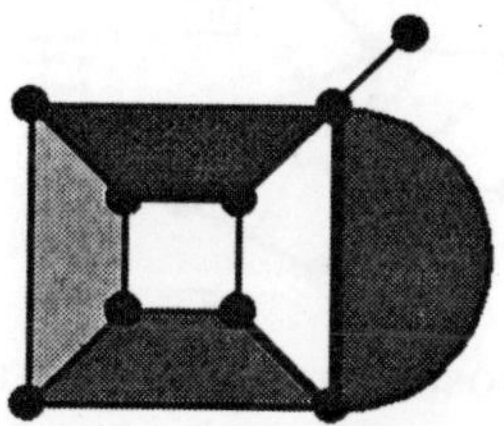

Theorem: If G is a connected plane graph with n vertices, m edges and f faces, then:

$$n - m + f = 2.$$

Proof: (By induction on m.) Suppose that $m = 0$. Then for G to be a connected graph, there may be only one vertex, and one face. So

$$n - m + f = 1 - 0 + 1 = 2.$$

and the formula works in the $m = 0$ case. For the $m = 1$ case, there are 2 possibilities.

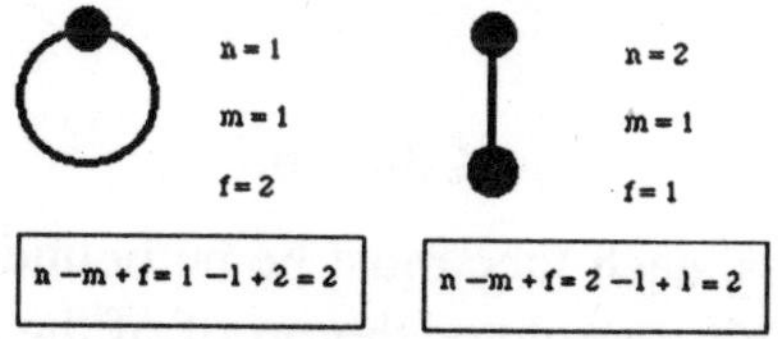

So, Euler's Polyhedron Formula works for m =< 0. For the inductive step, assume that $m > 1$ is given, and that the Euler Polyhedron Formula works for all graphs with m - 1 edges.

If G is a tree, then $f = 1$ and $m = n$- 1. So,

$$n - m + f = n - (n - 1) + f = n - n + 1 + 1 = 2.$$

provided that G is a tree. If G is not a tree, then remove an edge e belonging to some circuit. After removing the edge, e, you are left with a plane graph H on n vertices with $m - 1$ edges. Denote the number of vertices of H by n_H, the number of edges of H by m_H, and the number of faces of H by f_H. Then, $n_H = n$, $m_H = m - 1$, and $f_H = f - 1$. Since $m_H = m - 1$, H satisfies the Euler Polyhedron Formula, so:

$$2 = n_H - m_H + f_H = n - (m - 1) + (f - 1) = n - m + f + 1 - 1 = n - m + f.$$

Thus, G satisfies the Euler Polyhedron Formula. This completes the proof of the theorem.

Corollary: If G is a (connected) simple, finite planar graph with n vertices ($n >= 3$), then G has at most $3n - 6$ edges. Furthermore, if G contains no triangles, then G has at most $2n - 4$ edges.

Proof: Suppose that G has at most m edges, consider some plane drawing of G, with f faces. Consider the number of pairs (e, F) where e is one of the edges bounding the face F, viz:

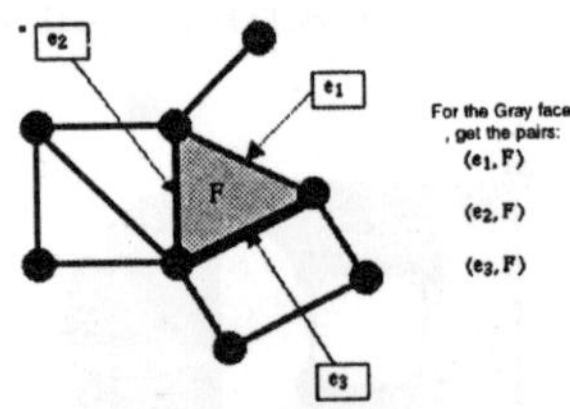

For each edge e, there are at most 2 faces that it bounds. So, the total number of these edge face pairs has to be less than $2m$. On the other hand, because G is a simple graph, each face is bounded by at least 3 edges. Therefore, the total number of edge-face pairs is greater than or equal to $3f$.

So, $3f =< 2m$.

By the Euler Polyhedron Formula, $n - m + f = 2$,, so, $3n - 3m + 3f = 6$. Since $3f =< 2m$,

$$3f = 6 - 3n + 3m \leq 2m$$

Re-arranging,

$$m \leq 3n - 6.$$

Next, if G has no triangles, each face must be be bounded by 4 or more edges. Then $2f =< m$, and $m =< 2n - 4$. This completes the proof of the corollary.

Note: This corollary is true even for disconnected, simple, finite planar graphs, provided that each connected component has at least three vertices.

Corollary: Every finite, simple, planar graph has a vertex of degree less than 6.

Proof: Suppose that G is such a graph. Without loss of generality, G is connected.

Suppose that G has n vertices, m edges, and that vertex j has degree d_j (for $1 =< j =< n$). Then,

$$\sum_{j=1}^{n} d_j = 2m \le 2(3n - 6) = 6n - 12$$

Then, the average degree of the vertices of the graph G, is given by,.

$$\textit{Average Degree} =< \frac{6n - 12}{n} = 6 \frac{12}{n} < 6..$$

So, at least one of the vertices has degree less than 6. This completes the proof of the corollary.

Definition: Suppose that G is a (finite) connected plane graph, then its dual G^* may be defined as follows:

Take as the vertices of G^* the faces of G, and for every pair of faces having an edge, e, in common, join the corresponding vertices in G^* by an edge that crosses the edge, e, only once.

Example: In the following, the graph is indicated in black, and its dual is drawn in Grey.

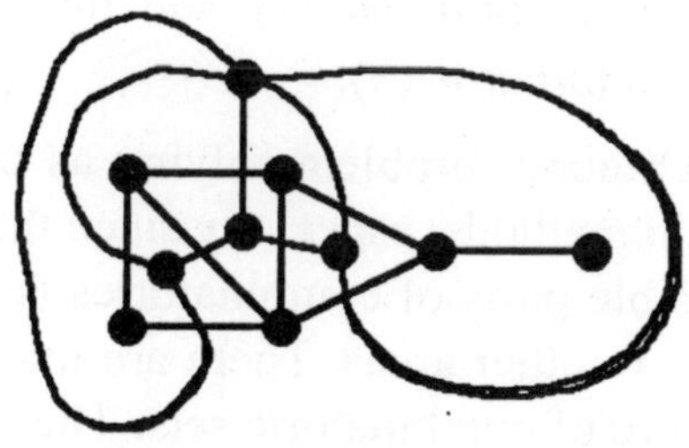

Chapter 7

Combinatorics

Combinatorics is a branch of pure mathematics concerning the study of discrete (and usually finite) objects. It is related to many other areas of mathematics, such as algebra, probability theory, ergodic theory and geometry, as well as to applied subjects in computer science and statistical physics. Aspects of combinatorics include "counting" the objects satisfying certain criteria (*enumerative combinatorics*), deciding when the criteria can be met, and constructing and analyzing objects meeting the criteria (as in *combinatorial designs and matroid theory*), finding "largest", "smallest", or "optimal" objects (*extremal combinatorics* and *combinatorial optimization*), and finding algebraic structures these objects may have (*algebraic combinatorics*).

Combinatorics is as much about problem solving as theory building, though it has developed powerful theoretical methods, especially since the later twentieth century. One of the oldest and most accessible parts of combinatorics is graph theory, which also has numerous natural connections to other areas. There are many combinatorial patterns and theorems related to the structure of combinatoric sets. These often focus on a partition or ordered partition of a set. See the List of partition topics for an expanded list of related topics or the List of combinatorics topics for a more general listing. Some of the more notable results are highlighted below.

An example of a simple combinatorial question is the following: What is the number of possible orderings of a deck of 52 distinct playing cards? The answer is 52! (52 factorial), which is equal to about 8.0658×10^{67}.

Another example of a more difficult problem: Given a certain number n of people, is it possible to assign them to sets so that each person is in at least one set, each pair of people is in exactly one set together, every two sets have exactly one person in common, and no set contains everyone, all but one person, or exactly one person? Combinatorics is used frequently in computer science to obtain estimates on the number of elements of certain sets. A mathematician who studies combinatorics is often referred to as a *combinatorialist* or *combinatorist*.

HISTORY OF COMBINATORICS

The appearance of the Fibonacci number five in prosody. There is one way to arrange one beat, two for two, three for three, and five for four beats.

The earliest books about combinatorics are from India. A Jainist text, the Bhagabati Sutra, had the first mention of a combinatorics problem; it asked how many ways one could take six tastes one, two, or three tastes at a time. The Bhagabati Sutra was written around 300 BC, and thus was the first book to mention the choice function. The next ideas of Combinatorics came from Pingala, who was interested in prosody. Specifically, he wanted to know how many ways a six syllable meter could be made from short and long notes. He wrote this problem in the Chanda sutra (also Chandahsutra) in the second century BC. In addition, he also found the number of meters that had *n* long notes and *k* short notes, which is equivalent to finding the binomial coefficients.

The ideas of the Bhagabati were generalized by the Indian mathematician Mahariva in 850 AD, and Pingala's work on prosody was expanded by Bhaskara and Hemacandra in 1100 AD. Bhaskara was the first known person to find the generalized choice function, although Brahmagupta may have known earlier. Hemacandra asked how many meters existed of a certain length if a long note was considered to be twice as long as a short note, which is equivalent to finding the Fibonacci numbers.

While India was the first nation to publish results on Combinatorics, there were discoveries by other nations on similar topics. The earliest known connection to Combinatorics comes from the Rhind papyrus, for the implementation of a geometric series. The next milestone is held by the I Ching. The book is about what different hexagrams mean, and to do this they needed to know how many possible hexagrams there were. Since each hexagram is a permutation with repetitions of six lines, where each line can be one of two states, solid or dashed, combinatorics yields the result that there are $2^6 = 64$ hexagrams. A monk also may have counted the number of configurations to a game similar to Go around 700 AD. Although China had relatively few advancements in enumerative combinatorics, they solved a combinatorial design problem, the magic square, around 100 AD.

In Greece, Plutarch wrote that the Xenocrates discovered the number of different syllables possible in the Greek language. This, however, is unlikely because this is one of the few mentions of Combinatorics in Greece. The number they found, $1.002.10^{12}$ also seems too round to be more than a guess.

Magic squares remained an interest of China, and they began to generalize their original 3×3 square between 900 and 1300 AD. China corresponded with the Middle East about this problem in the 13th century. The Middle East also learned about binomial coefficients from Indian work, and found the connection to polynomial expansion.

COMBINATORICS IN THE WEST

Combinatorics came to Europe in the 13th century through two mathematicians, Leonardo Fibonacci and Jordanus de Nemore. Fibonacci's Liber Abaci introduced many of the Arabian and Indian ideas to Europe, including that of the Fibonacci numbers. Jordanus was the first person to arrange the binomial coefficients in a triangle, as he did in proposition of *De Arithmetica*. This was also done in the Middle East in 1265, and China around 1300. Today, this triangle is known as Pascal's triangle. Pascal's contribution to the triangle that bears his name comes from his work on formal proofs about it, in addition to his connection between it and probability. Together with Leibniz and his ideas about partitions in the 17th century, they are considered the founders of modern combinatorics.

Both Pascal and Leibniz understood that algebra and combinatorics corresponded (aka, binomial expansion was equivalent to the choice function). This was expanded by De Moivre, who found the expansion of a multinomial. De Moivre also found the formula for derangements using the principle of inclusion-exclusion, a method different from Nikolaus Bernouli, who had found them previously. He managed to approximate the binomial coefficients and factorial. Finally, he found a closed form for the Fibonacci numbers by inventing generating functions. In the 18th century, Euler worked on problems of combinatorics. In addition to working on several problems of probability which link to combinatorics, he worked on the knights tour, Graeco-Latin square, Eulerian numbers, and others. He also invented graph theory by solving the Seven Bridges of Kφnigsberg problem, which also led to the formation of topology. Finally, he broke ground with partitions by the use of generating functions.

ENUMERATIVE COMBINATORICS

Counting the number of ways that certain patterns can be formed is the central problem of enumerative combinatorics. Two examples of this type of problem are counting combinations and counting permutations. More generally, given an infinite collection of finite sets $\{S_i\}$ indexed by the natural numbers, enumerative combinatorics seeks to describe a *counting function* which counts the number of objects in S_n for each n. Although counting the number of elements in a set is a rather broad mathematical problem, many of the problems that arise in applications have a relatively simple combinatorial description.

The simplest such functions are *closed formulas*, which can be expressed as a composition of elementary functions such as factorials, powers, and so on. For instance, as shown below, the number of different possible orderings of a deck of n cards is $f(n) = n!$. Often, no closed form is initially available. In these cases, we frequently first derive a recurrence relation, then solve the recurrence to arrive at the desired closed form.

Finally, $f(n)$ may be expressed by a formal power series, called its *generating function*, which is most commonly either the ordinary generating function

$$\sum_{n\geq 1} f(n)x^n$$

or the exponential generating function

$$\sum_{n\geq 1} f(n)\frac{x^n}{n!}.$$

Often, a complicated closed formula yields little insight into the behavior of the counting function as the number of counted objects grows. In these cases, a simple asymptotic approximation may be preferable. A function $g(n)$ is an asymptotic approximation to $f(n)$ if $f(n)/g(n) \rightarrow 1$ as $n \rightarrow$ infinity. In this case, we write $f(n) \sim g(n)$ Once determined, the generating function may allow one to extract all the information given by the previous approaches. In addition, the various natural operations on generating functions such as addition, multiplication, differentiation, etc., have a combinatorial significance; this allows one to extend results from one combinatorial problem in order to solve others.

Permutations with Repetitions

When the order matters, and an object can be chosen more than once, the number of permutations is n^r where n is the number of objects from which you can choose and r is the number to be chosen.

For example, if you have the letters A, B, C, and D and you wish to discover the number of ways to arrange them in three letter patterns (trigrams)

1. Order matters (e.g., A-B is different from B-A, both are included as possibilities)
2. An object can be chosen more than once

you find that there are 4^3 or 64 ways. This is because for the first slot you can choose any of the four values, for the second slot you can choose any of the four, and for the final slot you can choose any of the four letters. Multiplying them together gives the total.

Permutations without Repetitions

When the order matters and each object can be chosen only once, then the number of permutations is

$$(n)_r = \frac{n!}{(n-r)!}$$

where n is the number of objects from which you can choose, r is the number to be chosen and "!" is the standard symbol meaning factorial.

For example, if you have five people and are going to choose three out of these, you will have 5!/(5"3)! = 60 permutations.

Note that if $n = r$ (meaning the number of chosen elements is equal to the number of elements to choose from; five people and pick all five) then the formula becomes

$$\frac{n!}{(n-n)!} = \frac{n!}{0!} = n!$$

where $0! = 1$.

For example, if you have the same five people and you want to find out how many ways you may arrange them, it would be 5! or 5×4×3×2×1=120 ways. The reason for this is that you can choose from 5 for the initial slot, then you are left with only 4 to choose from for the second slot etc. Multiplying them together gives the total of 120.

Combinations without Repetitions

When the order does not matter and each object can be chosen only once, the number of combinations is the binomial coefficient:

$$\left(\frac{n}{k}\right) = \frac{n!}{k!(n-n)!}$$

where n is the number of objects from which you can choose and k is the number to be chosen.

For example, if you have ten numbers and wish to choose 5 you would have 10!/(5!(10"5)!) = 252 ways to choose. The binomial coefficient is also used to calculate the number of permutations in a lottery.

Combinations with repetitions

When the order does not matter and an object can be chosen more than once, then the number of combinations is

$$\frac{(n+k-1)!}{k!(n-1)!} = \binom{n+k-1}{k} = \binom{n+k-1}{n-1}$$

where n is the number of objects from which you can choose and k is the number to be chosen.

For example, if you have ten types of donuts (n) on a menu to choose from and you want three donuts (k) there are (10 + 3 " 1)! / 3!(10 " 1)! = 220 ways to choose.

Fibonacci Numbers

Let $f(n)$ be the number of distinct subsets of the set $S(n) = \{1, 2, 3,..,n\}$ that do not contain two consecutive integers. When $n = 4$, we have the sets {}, {1}, {2}, {3}, {4}, {1,3}, {1,4}, {2,4}, so $f(4) = 8$. We count the desired subsets of $S(n)$ by separately counting those subsets that contain element n and those that do not. If a subset contains n, then it does not contain element n " 1. So there are exactly $f(n$ " 2) of the desired subsets that contain element n. The number of subsets that do not contain n is simply $f(n$ " 1). Adding these numbers together, we get the recurrence relation:

$$f(n) = f(n-1) + f(n-2),$$

where $f(1) = 2$ and $f(2) = 3$.

As early as 1202, Leonardo Fibonacci studied these numbers. They are now called Fibonacci numbers; in particular, $f(n)$ is known as the $n + 2$nd Fibonacci number. Although the recurrence relation allows us to compute every Fibonacci number, the computation is inefficient. However, by using standard techniques to solve recurrence relations, we can reach the closed form solution:

$$f(n) = \frac{\phi^{n+2} - (1-\phi)^{n+2}}{\sqrt{5}}$$

where $\phi = (1 = \sqrt{5})/2$, the golden ratio.

In the above example, an asymptotic approximation to $f(n)$ is:

$$f(n) \sim \frac{\phi^{n+2}}{\sqrt{5}}$$

as n becomes large.

STRUCTURAL COMBINATORICS

GRAPH THEORY

Graphs are basic objects in combinatorics. The questions range from counting (e.g. the number of graphs on n vertices with k edges) to structural (e.g. which graphs contain Hamiltonian cycles) to algebraic questions (e.g. given a graph G and two numbers x and y, does the Tutte polynomial $T_G(x, y)$ have a combinatorial interpretation?). It should be noted that while there are very strong connections between graph theory and combinatorics, these two are sometimes thought of as separate subjects.

DESIGN THEORY

A simple result in the block design area of combinatorics is that the problem of forming sets, described in the introduction, has a solution only if n has the form $q^2 + q + 1$. It is less simple to prove that a solution exists if q is a prime power. It is conjectured that these are the *only* solutions. It has been further shown that if a solution exists for q congruent to 1 or 2 mod 4, then q is a sum of two square numbers.

This last result, the Bruck-Ryser theorem, is proved by a combination of constructive methods based on finite fields and an application of quadratic forms. When such a structure does exist, it is called a finite projective plane; thus showing how finite geometry and combinatorics intersect.

MATROID THEORY

Matroid theory abstracts part of geometry. It studies the properties of sets (usually, finite sets) of vectors in a vector space that do not depend on the particular coefficients in a linear dependence relation. Not only the structure but also enumerative properties belong to matroid theory.

For instance, given a set of n vectors in Euclidean space, what is the largest number of planes they can generate? Answer: the binomial coefficient

$$\binom{n}{2}.$$

Is there a set that generates exactly one less plane? (No, in almost all cases.) These are extremal questions in geometry, as discussed below.

EXTREMAL AND PROBABILISTIC COMBINATORICS

Many extremal questions deal with set systems. A simple example is the following: what is the largest number of subsets of an n-element set one can have, if no two of the subsets are disjoint? Answer: half the total number of subsets. Proof: Call the n-element set S. Between any subset T and its complement S " T, at most one can be chosen. This proves the maximum number of chosen subsets is not greater than half the number of subsets. To show one can attain half the number, pick one element x of S and choose all the subsets that contain x.

A more difficult problem is to characterize the extremal solutions; in this case, to show that no other choice of subsets can attain the maximum number while satisfying the requirement.

Often it is too hard even to find the extremal answer $f(n)$ exactly and one can only give an asymptotic estimate.

RAMSEY THEORY

Ramsey theory is a celebrated part of extremal combinatorics. It states that any sufficiently large random configuration will contain some sort of order.

Frank P. Ramsey proved that for every integer k there is an integer n, such that every graph on n vertices either contains a clique or an independent set of size k. This is a special case of Ramsey's theorem. For example, given any group of six people, it is always the case that one can find three people out of this group that either all know each other or all do not know each other. The key to the proof in this case is the Pigeonhole Principle: *either A knows three of the remaining people, or A does not know three of the remaining people.*

Here is a simple proof: Take any one of the six people, call him A. Either A knows three of the remaining people, or A does not know three of the remaining people. Assume the former (the proof is identical if we assume the latter). Let the three people that A knows be B, C, and D. Now either two people from {B,C,D} know each other (in which case we have a group of three people who know each other–these two plus A) or none of B,C,D know each other (in which case we have a group of three people who do not know each other--B,C,D). QED.

Extremal Combinatorics

The types of questions addressed in this case are about the largest possible graph which satisfies certain properties. For example, thc largest triangle-free graph on *2n* vertices is a complete bipartite graph $K_{n,n}$.

Probabilistic Combinatorics

Here the questions are of the following type: what is the probability of a certain graph property for a random graph (within a certain class) E.g. what is the average number of triangles in a random graph? Probabilistic methods are also used to determine the existence of combinatorial objects with certain prescribed properties (for which explicit examples might be difficult to find), simply by observing that the probability of randomly selecting an object with those properties is greater than 0.

GEOMETRIC COMBINATORICS

Geometric combinatorics is related to convex and discrete geometry. It asks, e.g. how many faces of each dimension can a convex polytope have. Metric properties of polytopes play an important role as well, e.g. the Cauchy theorem on rigidity of convex polytopes. Special polytopes are also considered, such as permutohedron, associahedron and Birkhoff polytope.

PARTITIONS INTO DISTINCT PARTS

For any positive integers *n* and k, let $p_k(n)$ denote the number of ways in which the integer *n* can be expressed as a sum of exactly *k* distinct positive integers, without regard to order. For example, the integer $n = 12$ can be expressed as a sum of three distinct positive integers in the following seven ways:

1+2+9 = 1+3+8 = 1+4+7 = 1+5+6 = 2+3+7 = 2+3+7 = 2+4+6 =3+4+5

Therefore we have $p_3(12) = 7$. Obviously $p_1(n) = 1$ for all n, so the generating function for $p_1(n)$ is just the geometric series

$$\frac{x}{1-x} 1x^1 + 1x^2 + 1x^3 + \ldots$$

The values of $p_2(n)$ for $n = 1, 2, 3, \ldots$ are easily seen to be 0, 0, 1, 1, 2, 2, 3, 3, 4, 4, … and so on. In other words, $p_2(1+2j) = p_2(2+2j) = j$. Factoring x^3 and $1+x$ from the power series with these coefficients, we get

$$1x^3 + 1x^4 + 2x^5 + 2x^6 \ldots = x^3(1 + x)(1 + 2x^2 + 3x^4 + 4x^6 \ldots)$$

The infinite series on the right is just the derivative of the previous geometric series, except with x^2 in place of x, so we can differentiate the closed form of the geometric series and insert into this expression to give the generating function for $p_2(n)$

$$\frac{x^3(1+x)}{(1-x^2)^2} = \frac{x^3}{(1-x)(1-x^2)}$$

The first several values of $p_3(n)$ are listed (by rows) below.

0	0	0	0	0	1
1	2	3	4	5	7
8	10	12	14	16	19
21	24	27	30	33	37
40	44	48	52	56	61

Just as every second term of the $p_2(n)$ series gives an arithmetic sequence, we expect that the values of $p_3(n)$ should give sequences of degree 2, and this is indeed the case, as can be seen by taking every sixth term. The values in the columns are given by second-degree polynomials, i.e., we have

$$p_3\,(1+6j) = 3j^2 - 2j$$
$$p_3\,(2+6j) = 3j^2 - 1j$$
$$p_3\,(3+6j) = 3j^2$$
$$p_3\,(4+6j) = 3j^2 + 1j$$
$$p_3\,(5+6j) = 3j^2 + 2j$$
$$p_3\,(6+6j) = 3j^2 + 3j + 1$$

The generating functions for these individual sub-sequences are

$$x^2\frac{1+5x^6}{(1-x^6)^3} = 0x^1 + 1x^7 + 8x^{13} + 21x^{19} + \ldots$$

$$x^8\frac{2+4x^6}{(1-x^6)^3} = 0x^2 + 2x^8 + 10x^{14} + 24x^{20} + \ldots$$

$$x^9\frac{3+3x^6}{(1-x^6)^3} = 0x^3 + 3x^9 + 12x^{15} + 27x^{21} + \ldots$$

$$x^{10}\frac{4+2x^6}{(1-x^6)^3} = 0x^4 + 4x^{10} + 14x^{16} + 30x^{22} + \ldots$$

$$x^{11}\frac{5+1x^6}{(1-x^6)^3} = 0x^5 + 5x^{12} + 16x^{17} + 33x^{23} + \ldots$$

$$x^6\frac{1+4x^6+x^{12}}{(1-x^6)^3} = 1x^1 + 7x^{12} + 19x^{18} + 37x^{24} + \ldots$$

The sum of these sub-sequences gives the generating function for the entire $p_3(n)$ sequence. The numerator of the sum is a (shifted) palindromic polynomial

$$x^6 + x^7 + 2x^8 + 2x^9 + 4x^{10} + 5x^{11} + 4x^{12} + 5x^{13} + 4x^{14} + 3x^{15}$$
$$+2x^{16} + x^{17} + x^{18} = x^6(x+1)^2(x^2+x+1)^2(x^2-x+1)^3$$

Dividing this by the denominator and simplifying, we get the complete generating function for the number of partitions of n into 3 distinct parts

$$\frac{x^6}{(1-x)(1-x^2)(1-x^3)} = 1x^6 + 1x^7 + 2x^8 + 3x^9 + 4x^{10}$$
$$+5x^{11} + 7x^{12} + \ldots$$

Notice that if x^6 is replaced with 1 in the numerator the resulting expression would be the generating function for the number of partitions of n into positive integers (not necessarily distinct) no greater than 3. We can see this immediately because each of the factors in the denominator contributes a geometric series in x^k where k is 1, 2, and 3 respectively, and hence the coefficient of x^n for n greater than 0 is the number of ways in which n is the sum of those three elements. Therefore, letting $P_j(n)$ denote the partitions of n into positive integers no greater than j (and defining $P_3(0) = 1$), we have shown that $P_3(n) = p_3(n+6)$ for all n. This correspondence can be represented pictorially as shown below for the case $n = 5$.

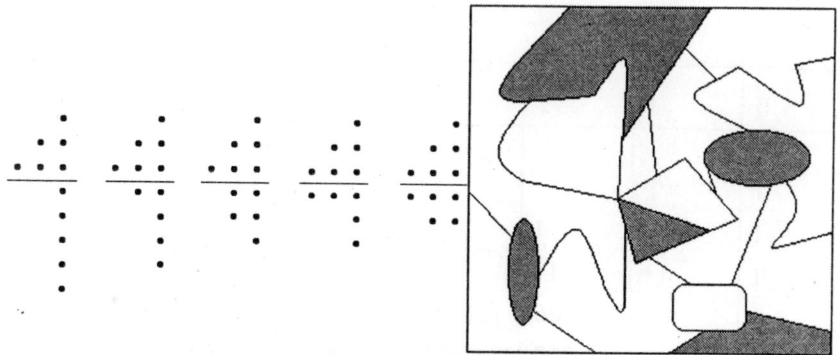

Each partition of 11 into distinct positive integers must have a smallest part of at least 1, a second smallest part of at least 2, and a third smallest part of at least 3, so we can begin with this triad of 6 dots, and then to each column we add zero or more dots, with the requirement that the number of added dots can never decrease as we go from the left-most to the right-most column. The total number of additional dots to be added is 5, and with the stated restriction these rows can be interpreted as the partitions of 5 into positive integers no greater than 3. By the same reasoning it follows that, for any positive integer k, we have $P_k(n) = p_k(n + k(k+1)/2)$. In other words, the number of partitions of n into positive integers no greater than k is equal to the number of partitions of $n + k(k + 1)/2$ into k distinct positive integers. Thus the generating function for the latter is

$$\frac{x^{k(k+1)2}}{(1-x)(1-x^2)\cdots(1-x^k)}$$

EULERIAN NUMBERS

Beginning with the geometric series

$$\sum_{k-1}^{\infty} x^k = \frac{x}{1-x}$$

which converges for all $|x| < 1$, we can repeatedly differentiate and multiply by x to give the sequence of identities

$$\sum_{k-1}^{\infty} kx^k = \frac{x}{(1-x)^2}(1)$$

$$\sum_{k-1}^{\infty} k^2 x^k = \frac{x}{(1-x)^3}(1+x)$$

$$\sum_{k-1}^{\infty} k^3 x^k = \frac{x}{(1-x)^4}(1+4x+x^2)$$

and so on. The coefficients of the numerator polynomials on the right side are as shown below.

m	0	1	2	3	n 4	5	6	7	8
1	1								
2	1	1							
3	1	4	1						
4	1	11	11	1					
5	1	26	66	26	1				
6	1	57	302	302	57	1			
7	1	120	1191	2416	1191	120	1		
8	1	247	4293	15619	15619	4293	247	1	
9	1	502	14608	88234	156190	88234	14608	502	1

The sum of the mth row equals the factorial of m. Also, the mth row interpreted as a histogram approaches a normal distribution.

Letting $E(m, n)$ denote the value from t! ꞉ mth row and nth column, we have the general identity

$$\sum_{k-1}^{\infty} k^m x^k = \frac{x}{(1-x)^{m-1}} \sum_{n-0}^{m-1} E(m, n) x^n$$

Not surprisingly, the Eulerian numbers have many combinatorial applications. The most famous is the fact that the nth number in the mth row represents the number of permutations of *m* objects that have exactly *n* "rises".

For example, the six possible permutations of three objects, along with the number of "rises" (i.e., the number of times it goes from a lower to a higher number, reading left to right) are shown below.

Premutation	*Rises*
123	2
132	1
213	1
231	1
312	1
321	0

Thus the numbers of permutations having exactly 0, 1, and 2 rises are 1, 4, and 1 respectively, and these numbers comprise the 3rd row of Eulerian numbers. If we arrange the Eulerian Numbers with symmetrical indices as shown below:

1
m 1 1 *n*
1 4 1
1 11 11 1
1 26 66 26 1
1 57 302 302 57 1

then they can be generated by the well-known recurrence

$$E[m, n] = nE\,[m-1, n] + mE\,[m, n-1]\,.$$

In addition, they have the interesting generating function

$$f(x, y, t) = \frac{1}{xyt} \ln\left(\frac{xe^{-yt} - ye^{xt}}{x-y}\right).$$

The coefficient of $x^{m-1}\, y^{n-1}\, (-t)^{m+n-1} / (m+n)!$is the Eulerian number $E[m, n]$.Thus we have

$$f(x,y,t)=\frac{1}{2}t+\frac{x+y}{6}t^2-\frac{x^2+4xy+y^2}{24}t^3$$
$$+\frac{x^3+1x^2y+11xy^2+y^3}{120}t^3-\ldots$$

I found this by a very indirect approach.Is there a simple procedure for deducing the generating function–if one exists–for a given array of numbers?

There's also a nice relation between Eulerian numbers and the generalized Bernoulli numbers.

Let $W(m; n)$ be a weighted average of the nth powers of the first m natural numbers, using the Eulerians as the weights. For example

$$W(6.5)=\frac{(1)1^5+(26)2^5+(66)3^5+(26)4^5+(1)5^5}{120}=\frac{777}{2}$$

For $m>n$ it turns out that $W(m,n)$ equals $B(-m,n)$, the generalized Bernoulli number, which can be defined as the coefficients in the expansion

$$\left(\frac{e^x-1}{x}\right)^m=\sum_{n-0}^{\infty}\frac{B(-m,n)}{n!}x^n$$

Beginning at the "diagonal" where $m = n$, the values of W and B begin to differ, although they differ only half the time on the diagonal. Specifically, the differences $B(-m, m)-W(m, m)$ for $m = 0,1,2,3...$ along the diagonal are

$$1,-\frac{1}{2},\frac{1}{6},\ 0,\ -\frac{1}{30},\ 0,\ \frac{1}{42},\ 0\ -\frac{1}{30},\ 0,\ \frac{5}{66},\ 0,\ \frac{-691}{2730},\ldots.$$

which of course are just the ordinary Bernoulli numbers.

There is also an interesting relation between the Eulerian Numbers and the binomial coefficients. The most obvious recurrence relation satisfied by the binomial coefficients is

$$C(m,n)=C(m+1,n)+C(m,n-1)$$

but they also satisfy the recurrence

$$C(m,n)=(n+1,n)C(m+1,n)-(m+2)\,C(m,n-1)$$

Interestingly, this formula also generates the Eulerian numbers, so both sets of numbers can be regarded as simply different regions of a single array. Here's a brief table of the combined binomial and Eulerian numbers.

m	n: −1	0	1	2	3	4	5	6	7
−8	0	1	120	4293	88234	1310354	1310354		
−7	0	1	57	1191	15619	156190	1310354		
−6	0	1	26	302	2416	15619	88234	455192	
−5	0	1	11	66	302	1191	4293	14608	47840
−4	0	1	4	11	26	57	120	247	502
−3	1	1	1	1	1	1	1	1	1
−2	0	0	0	0	0	0	0	0	0
−1	1	0	0	0	0	0	0	0	0
0	0	1	0	0	0	0	0	0	0
1	0	1	1	0	0	0	0	0	0
2	0	1	2	1	0	0	0	0	0
3	0	1	3	3	0	0	0	0	0
4	0	1	4	6	4	1	0	0	0
5	0	1	5	10	10	5	1	0	0
6	0	1	6	15	20	15	6	1	0
7	0	1	7	21	35	35	21	1	1

On the other hand, it is sometimes convenient to write the array of Eulerian numbers in "rectangular form" as follows:

m	n: 1	2	3	4	5	6	7	8
0	1	0	0	0	0	0	0	0
−1	1	1	1	1	1	1	1	1
−2	1	4	11	26	57	120	247	502
−3	1	11	66	302	1191	4293	14608	
−4	1	26	302	2416	15619	88234		
−5	1	57	1191	15619	156190			
−6	1	120	4293	88234				
−7	1	247	14608					
−8	1	502						
−9	1							

With these indicies the Eulerians satisfy the recurrence relation

$$E[m, n] = -mE[m, n-1] + nE[m+1, n]$$

The fact that I've negated the "m" index may seem strange, but it leads to a natural generalization. Suppose that instead of dealing with infinite series, we want an expression for the *finite* sum

$$\left(1^N\right)x^1 + (2^N)x^2 + (3^N)x^3 + \ldots + ((k-1)^N)x^{k-1}$$

Obviously if $N = 0$ this sum is given by the simple Euclidean expression

$$\frac{x^k - 1}{x - 1}$$

but what about higher values of N? Clearly if we had an expression for the infinite sum

$$\left(k^N\right)x^k + \left((k+1)^N\right)x^{k+1} + ((k+2)^N)x^{k+2} + \ldots$$

we could subtract it from our original infinite sum (1) to give the desired finite sum.So, our objective is to find a closed form expression for (2).Notice that if we multiply the basic geometric series by x^k, we can proceed as before, alternately differentiating and multiplying x to give

$$\frac{x^k}{(1-x)^{N+1}} \sum_{j-0}^{N} E_k\{N, j\}x^j = k^N x^k + (k+1)^N x^{k+1} + \ldots$$

where the coefficients $E_k\{N,j\}$ are taken from an array of number somewhat similar to the Eulerian numbers.Of course, if $k = 0$ these coefficients *are* the Eulerian numbers. For other values of k we can generate the coefficients $E_k\{m,n\}$ (using "rectangular indicies") with the same recurrence as for the standard Eulerians, except that we "slide the indicies around the corner". This is best illustrated with an example. For the case $k = 5$ we are seeking a closed-form expression for the infinite sum

$$5^N x^5 + 6^N x^6 + 7^N x^7 + \ldots$$

This is given by (3) with the values of $E_k\{N, j\}$ taken from the following table for $E_5\{m,n\}$ (and appropriately converting their indicies from the rectangular to the triangular array format):

			n			
m	5	6	7	8	9	10
4	1	−4	16	−64	256	1024
3	5	−39	229	−1199	5901	
2	25	−284	2171	−13934		
1	125	−1829	17026			
0	625	−10974				
−1	3125					

So, with $k = 5$ and $N = 3$, equation (3) gives the identity

$$\frac{x^5}{(1-x)^4}(125 - 284x + 229x^2 - 64x^3) = 125x^5 + 216x^6 + ...$$

Subtracting this from (1) with N=3 gives, as expected

$$1x + 8x^2 + 27x^3 + 64x^4$$

For a less trivial example, suppose we want an abbreviated expression for the finite sum

$$f(x) = 1^3 x + 2^3 x^2 + 3^3 x^3 + ... + 99^3 x99^3 + 100^3 x^{100}$$

The procedure is just the same, except that we have $k = 101$ instead of $k = 3$, so we need to slide the indicies further around the corner. In other words, the coefficient array with $k = 101$ is

m	101	n 102	103	104
100	1	−100	10000	−100000
99	101	−20199	3029701	
98	10201	−3059996		
97	1030301			

By the way, a convenient check on the accuracy of these coefficients is to notice that. just as in the case of the basic Eulerian numbers, the sum of the jth diagonal is j!. This table gives us the closed-form expression for the infinite sum

$$101^3 x^{101} + 102^3 x^{102} + 103^3 x^{103} + ...$$
$$= \frac{x^{101}}{(1-x)^4}(1030301 - 3059996x + 3029701x^2 - 100000x^3)$$

so again we can subtract this from the simple closed-form expression for the infinite sum given by (1) with N=3 to represent the polynomial $f(x)$ of degree 100.

Incidentally, the process of differentiating and multiplying by x can be reversed to give expressions for infinite sums whose coefficients are the inverses of powers of integers. To reverse the process we integrate both sides of the basic geometric series, which gives the well-known expansion for the natural log

$$\sum_{n-1}^{\infty} \frac{x^n}{n} = -\ln(1-x)$$

Diving by x and integrating then gives the relation

$$\sum_{n-1}^{\infty} \frac{x^n}{n} = -\int \frac{\ln(1-x)}{x} dx$$

This is the same formula that appears in the note Average of s(n)/n describing the relation between the sum-of-divisor function and the sum of the inverse squares.

PERMUTATION LOOPS

Suppose *a, b, c, d, e* are elements of a group. We'll call a circular arrangement of such elements a "loop". For example, we might choose to arrange those elements in alphabetical order, giving the loop {*abcde*}, where the last element is understood to wrap around to the first. Note that {*abcde*} and {*eabcd*} are both valid representations of the same loop, since they are just rotations of each other.

For any given loop, a "cycle" consists of the sequence of elements going once around the loop from any specific starting point. Thus, a loop of length *N* has *N* cycles. For example, the five cycles of the loop {*abcde*} are [*abcde*], [*eabcd*], [*deabc*], [*cdeab*], and [*bcdea*].

Using the group operation we can evaluate each cycle of a given loop. (A group operation is associative, so there's no ambiguity about how to evaluate the composition of each cycle.) The resulting sequence of elements constitutes another loop, which we will call the "successor" of the original loop. Of course, if the group operation is commutative, each cycle of a given loop will evaluate to the same element, so a successor loop is not very interesting. However, if the group operation is NOT commutative, then successor loops will typically be non-trivial.

Even if the group operation is not commutative, it's clear that the elements of a successor loop must all belong to a single conjugacy class. To see why, suppose the cycle [*abcde*] evaluates to the group element *x*. The next cycle is [*eabcd*], which can be written as $e\,x\,e^{-1}$, so it is conjugate to x. If we call this new element y, then the next element is [deabc], which can be written as d $y\,d^{-1}$, so it too is in the conjugacy class of *x* and *y*. And so on.

It's interesting to see how the successor operation affects loops consisting of elements of the symmetric group S_n. I'll represent the elements of S_n by the permutations of *n* objects, and accordingly I'll refer to these loops as "permutation loops". Also, I'll focus on the effects of the successor operation on *complete* permutation loops, consisting of some arrangement of all n! elements in a loop of length n!.

The elements of S_2 are just the two permutations represented by I=[12] and A=[21]. There's only one complete permutation loop of order 2, namely, {IA}, which is the same loop as {AI}. The first successor of this loop is {AA}, and the next successor is {II}. All subsequent successors are simply {II}.

For a slightly less trivial example, consider the symmetric group S_3, whose sixe elements are

I = 123	C = 231
A = 132	D = 312
B = 213	E = 321

One possible loop of these elements is {IDEACB}. The six cycles of this loop are evaluated as

IDEACB = A

DEACBI = A

EACBID = B

ACBIDE = A

CBIDEA – A

BIDEAC = E

Therefore, the successor of {IDEACB} is {AABAAE}. Continuing in this way we can generate the sequence of successors as shown below:

{IDEACB}, {AABAAE}, {DCDCDC}, {IIIIII}, {IIIIII},...

After three applications of the successor operation, we arrive at the "null loop", i.e., the loop consisting entirely of identity elements. (All subsequent successors are obviously null loops.) Therefore, we say the original loop {IDEACB} has a "persistence" of 3. Of the 120 complete cycles that can be formed from S_3, 72 have a persistence of 3, and 48 have a persistence of just 2.

If we go on to consider the group S_4, the total number of complete permutation loops is very large (23!), but evidently every such loop has persistence of 1, 2, 3, or 4. (It's interesting to trace the sequence of conjugacy classes followed in each of these cases.) Thus, every complete loop of the symmetric group S_n for n less than 5 has persistence less than or equal to n.

However, for n=5 and above the behavior is suddenly quite different. In these cases the persistence T of a randomly chosen complete permutation loop of order n has essentially an exponential distribution

$$\text{dens}(T) = L \exp(-LT)$$

where $L = 2/(n!)$. It's clear that since the total number of transpositions in all the elements of S_n (n>4) is even, each successor must represent an "even" conjugacy class, of which there are n!/2 possible representatives. Exactly one of those is the null class, so on each iteration the probability of reaching null is $2/n!$. This is a Poisson process leading to (approximately) the exponential distribution with rate parameter $2/(n!)$. (It's interesting that random permutation loops under the successor operation exhibit a "half-life" identical to radio-active decay.) Of course, the distribution cannot really be perfectly exponential, because there is only a finite (although huge) number of distinct complete loops, so only a finite number of persistences can occur. Still, it's remarkable how accurately the distribution of persistences for S_5 matches the exponential density, even out to extremes. The average persistence for a loop of S_5 is $2/(5!) = 60$, but rare persistences in the range of several hundreds occur at just the rates predicted by the exponential density.

After much computer searching I've found two permutation loops of order 5 with persistences greater than 1000. These are extremely rare, considering that the mean persistence is only 60. Here is a loop with persistence 1009 (read by rows):

25134 13254 12354 13245 42153 53124 45123 45213 24153 24135
23541 21543 13524 25431 52134 14352 14253 21453 53412 21534
15243 24351 31452 32415 31542 45321 42135 51432 21435 34512
54321 31425 12543 45312 41325 34215 43215 53421 35421 34125
23451 41532 52314 43521 31245 41352 31524 25143 53241 12453
43152 31254 42513 51234 43125 54123 32514 54132 54312 23514
52341 34521 23154 43512 32541 34152 42531 24315 45132 15342
32451 41235 51243 52431 24531 15324 25413 15234 24513 54231
14523 14532 13425 14235 42315 52143 15432 25314 51324 51423
35124 23145 35241 12345 21354 43251 13542 32154 12435 41523
34251 42351 52413 53142 13452 23415 45231 32145 35142 54213
14325 21345 12534 53214 51342 15423 35412 35214 41253 25341

Here is the Methuselah of permutation loops of order 5, having a persistence of 1203 (read by rows):

12453 45231 52341 31254 53421 32514 12435 52413 23154 15243
13254 35214 24135 45123 45132 25143 12543 42513 54321 51423
32541 42153 14532 32154 32451 41325 25431 24513 51324 13245
51243 21453 15234 32145 24531 14253 43215 12534 15342 23541
52431 24351 34152 31524 12354 25413 54132 52314 13425 43125
54213 14523 14325 31542 43251 52134 42135 45321 41235 34512
35142 23415 51342 25314 54123 45312 54312 41352 42315 35124
13542 41532 53142 52143 21543 23145 43152 21534 35421 15432
34521 25341 53241 51432 31425 34251 21345 23451 34215 45213
35241 53214 23514 51234 31452 43521 53412 24153 14352 13524
21354 15423 41523 42531 35412 32415 31245 42351 53124 24315
43512 21435 14235 54231 12345 34125 25134 41253 13452 15324

Finding this 1203 loop seems to have been a remarkable stroke of luck, because a loop of order 5 with a persistence of 1203 is expected to be about a "1-in-500 million" loop, whereas I had only generated less than 20 million "random" loops. I've never found another one close to this.

It's interesting that for any given initial loop there is normally a (small) set of transpositions that leave the persistence unchanged, whereas any other single transposition generally results in a completely uncorrelated persistence.

It's also worth noting that if an odd number of components is removed from a complete permutation loop, it appears that the successors never degenerate to the null string. (By the way, the greatest persistence for a complete permutation loop of order 6 that I've found is 3952, but I haven't searched very much.)

Regarding S_5 loops, one of two things must be true: either there is an S_5 loop with the absolute maxiumum persistence, or there is an S_5 loop that NEVER reaches null, and instead goes periodic. I suspect the former is true. I believe every complete loop of a symmetric group eventually decomposes to a null loop, but I don't know how to prove it.

If true, this is an interesting higher-order analog to "solveability" of groups in Galois theory. In fact, the successor sequences for S_3 and S_4 leading to the null loop parallel the normal sub-group chains of solvability for the groups related to cubics and quartics.

I find it very interesting that, although the behavior is very different for S_5 and above, it is apparently always possible to reduce any group to the null loop by repeated applications of the successor operation.

By the way, two elements of S_n are in the same conjugacy class if and only if they have the same cyclic structure, so the number of conjugacy classes of S_n is p(n), the number of partitions of n. Also, the number of members of a given conjugacy class can be computed from the corresponding partition.

If we let a_k denote the number of k's in a given partition of n, then the number of members of the conjugacy class corresponding to that partition is given by the multinomial coefficient

$$\frac{n!}{(1a_1)(1a_1)!(2a_2)(a_2)!\dots(na_n)(a_n)!}$$

These coefficients are called M2 in Abramowitz and Stegun's "Handbook of Mathematical Functions", tabulated for all the permutations of the numbers $n = 1$ to 10.

For example, with n=5 we have $p(5) = 7$ partitions, so there are 7 conjugacy classes of S_5. The numbers of elements in each of these classes are given by the multinomial coefficients as shown below

Cyclic Structure	*Number of Conjugacy Classes (M2)*	*Number of Transpositions*	*Order*
5	24	4	5
4+1	30	3	4
3+2	20	3	6
3+1+1	20	2	3
2+2+1	15	2	2
2+1+1+1	10	1	2
1+1+1+1+1	1	0	1
Total 120			

As mentioned previously, a loop cycle of order 5 or above must always give an "even" conjugacy class, i.e., a class whose cyclic structure has an even number of transpositions. This is confirmed by examining the elements of the successor loops to various complete loops of S_5, which always consist of one of the following conjugacy classes:

Cyclic Structure	*Number of Conjugacy Classes (M2)*	*Number of Transpositions*	*Order*
5	24	4	5 3+1+1 20 2
3			
2+2+1	15	2	2
1+1+1+1+1	1	0	1
Total 60			

Evidently a cycle of random loop of S_5 has an equal chance of evaluating to any of these 60 even permutations (which fixes the conjugacy class for that generation), so the sequence of successors typically jumps between the conjugacy classes {5}, {3+1+1}, and {2+2+1}, with only a 1/60 chance of hitting the identity class {1+1+1+1+1} on each generation.

This makes the successor operation a Poisson process, and explains why the persistences have essentially an exponential distribution with a "rate" of 2/n!. Figure shows the sequence of conjugacy classes for the Methuselah sequence of persistence 1203. Each small vertical stripe represents a loop, and the color denotes the conjugacy class, with blue={5}, red={3+1+1}, and green= {2+2+1}.

The initial white loop on the top row represents the initial Methuselah arrangement, and thereafter each row shows 200 loops, except for the last row, which shows 203 loops to give the total 1203. All the loops subsequent to the 1203rd are the null loop {1+1+1+1+1}.

THE FOUR COLOR THEOREM

How many different colors are sufficient to color the countries on a map in such a way that no two adjacent countries have the same color? The figure below shows a typical arrangement of colored regions.

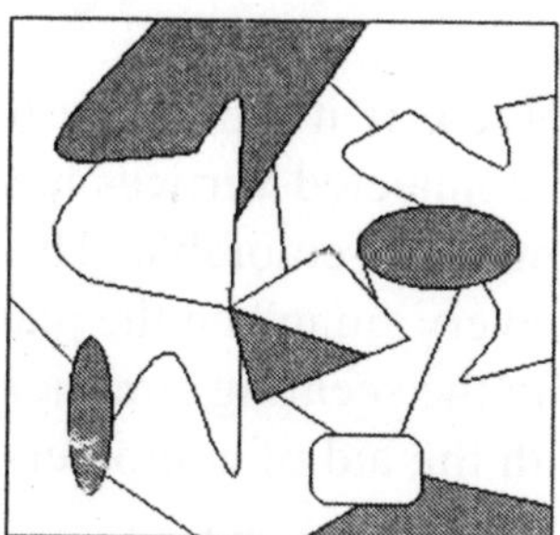

Notice that we define adjacent regions as those that share a common boundary of non-zero length. Regions which meet at a single point are not considered to be "adjacent".

After examining a wide variety of different planar graphs, one discovers the apparent fact that every graph, regardless of size or complexity, can be colored with just four distinct colors. This "four-color conjecture" was first noted by August Ferdinand Mobius in 1840. In 1852 a young man named Francis Guthrie wrote about the problem in a letter to his brother Frederick, then a student at University College in London. Neither of the brothers was able to prove the conjecture, so Frederick asked one of his professors, Augustus DeMorgan.

DeMorgan too was unable to prove the conjecture, and after recognizing the difficulty of the problem, he wrote to Sir William Rowan Hamilton to ask for help.

It might seem that this problem would be irresistible to Hamilton, since anything having to do with the number "four" immediately suggests a connection with his beloved quaternions. Also, Hamilton made contributions to graph theory (such as the idea of a Hamiltonian circuit, i.e., a path along the edges of a graph that visits each vertex exactly once), a subject that was developed largely through efforts to prove the four color conjecture. Nevertheless, Hamilton immediately wrote back that he did not believe he would solve DeMorgan's "quaternion of colors" any time soon.

The coloring of geographical maps is essentially a topological problem, in the sense that it depends only on the connectivities between the countries, not on their specific shapes, sizes, or positions. We can just as well represent each country by a single point (vertex), and the adjacency between two bordering countries can be represented by a line (edge) connecting those two points. It's understood that connecting lines cannot cross each other. A drawing of this kind is called a planar graph. A simple map (with just five "countries") and the corresponding graph are shown below.

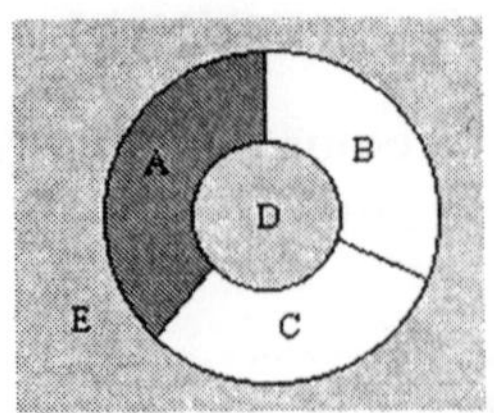

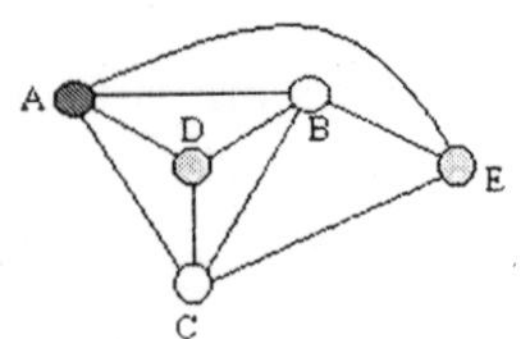

A graph is said to be n-colorable if it's possible to assign one of *n* colors to each vertex in such a way that no two connected vertices have the same color. Obviously the above graph is not 3-colorable, but it is 4-colorable. The Four Color Theorem asserts that every planar graph–and therefore every "map" on the plane or sphere–no matter how large or complex, is 4-colorable. Despite the seeming simplicity of this proposition, it was only proven in 1976, and then only with the aid of computers.

Notice that the above graph is "complete" in the sense that no more connections can be added (without crossing lines). The edges of a complete graph partition the graph plane into three-sided regions, i.e., every region (including the infinite exterior) is bounded by three edges of the graph. Every graph can be constructed by first constructing a complete graph and then deleting some connections (edges). Clearly the deletion of connections cannot cause an n-colorable graph to require any additional colors, so in order to prove the Four Color Theorem it would be sufficient to consider only complete graphs.

Although DeMorgan was unable to prove the four-color conjecture, he did observe that no more than four regions on the plane can all be in mutual contact with each other. The graph of a set of three mutually adjoining regions is simply a topological triangle, and if we add a fourth region, it is represented by a fourth vertex in the graph, which must be located either inside or outside the triangle formed by the graph of the original three vertices. In either case, we must then connect this fourth vertex with each of the three original vertices so that all four of the regions are mutually adjoining. Having done this, we can slide the vertices around (if necessary) to bring the graph into the form shown below.

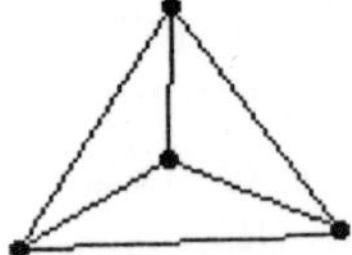

This is the unique graph of four mutually adjoining plane regions (V = 4, E = 6), and also of the tetrahedron. It clearly divides the graph plane into four isolated regions, one region exterior to the big triangle, and the three interior regions. A fifth vertex added to this graph will be able to have edges that reach only three of the four existing vertices, because each of the four regions is completely bounded by three edges that block access to one of the existing vertices. Hence there does not exist a plane graph of five mutually connected vertices, so there does not exist a set of five mutually adjoining regions on the plane.

Of course, the non-existence of a graph with five mutually adjoining vertices does not automatically imply the non-existence of a graph requiring five distinct colors. For example, there exist graphs in which the largest subset of mutually adjoining vertices is two, and yet for which three colors are required. This is the case with any simple loop with an odd number of vertices, as illustrated in the left-hand figure below. Similarly there are graphs containing no subset of four mutually adjoining vertices that nevertheless require four colors, such as the graph shown on the right.

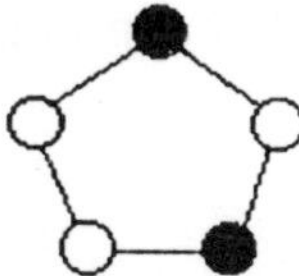

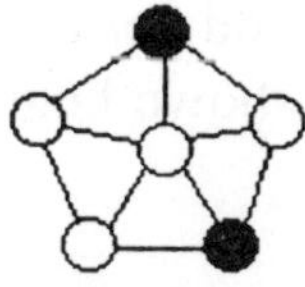

Therefore, we must consider the possibility that five colors might be required for some graph, even though it's not possible for a graph to contain any set of five mutually connected vertices.

On the other hand, it could be argued that the two example shown above can each be reduced, clarifying the causal links between different parts of the graphs. For example, in the left figure we have an alternating sequence of red, green, red, green, and if these were the only two available colors, it's clear that the chain must continue to alternate, no matter how long–or how short. Thus we could just as well replace any chain of the form rgrg...rg with the simple pair rg, in which case the left-hand figure above consists of three mutually connected vertices. Likewise if we replace the circumferential chain rgrg around the central yellow vertex in the right hand figure with just the pair rg, we get a set of four mutually connected vertices.

In view of this, it's tempting to think that DeMorgan's observation somehow, perhaps indirectly, does ultimately account for the truth of the four-color theorem. Indeed if we construct a graph by adding vertices, one at a time, and make the maximum number of connections at each stage, we will always find the graph plane divided into "triangular" regions, each of which has access to only three vertices. Thus it follows that four colors suffice for any graph that can be constructed in this way. For example, consider the two graphs shown below.

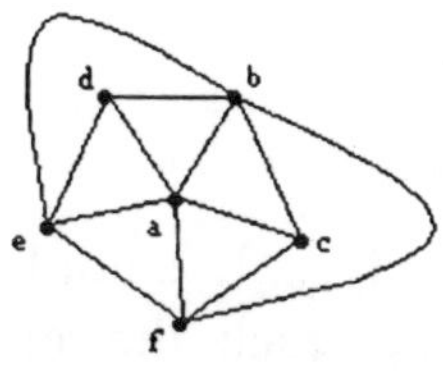

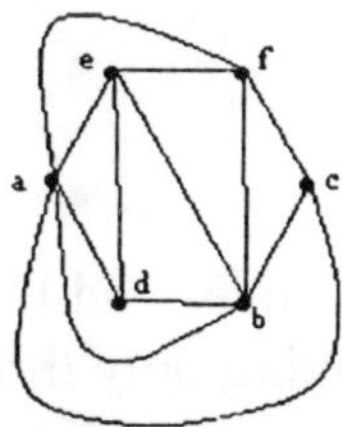

These two graphs each have V = 6 vertices and E = 12 edges, and in fact the two graphs are topologically identical. The plane regions represented by the vertices "*a*" and "*b*" each have five adjoining neighbors, whereas the vertices "e" and "f" each have four, and the vertices "*c*" and "*d*" each have three. These are complete graphs and, as noted above, a complete graph divides the entire graph plane into "triangular" regions, i.e., regions bounded by three edges connecting three vertices. Nevertheless, depending on how we add the points, it is possible that when a vertex is added to a graph it has more than three neighbors, so we cannot say automatically that four colors would suffice. For example, if vertex "a" is the last to be added, it would have five pre-existing neighbors, and if four colors have already been used in those five vertices, the vertex "a" would require a fifth color.

However, we need not add vertex "*a*" last. Another topologically equivalent way of drawing the above graph is shown below

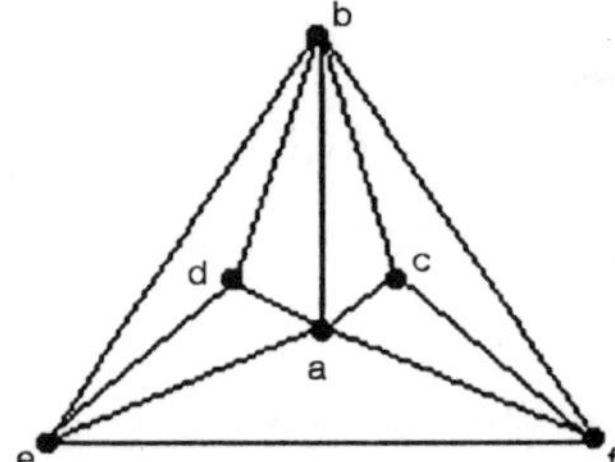

This shows that we could first assign three distinct colors to the vertices e,b,f, and then place the vertex "*a*" in this triangle, connect it to each of the three surrounding vertices, and give it a fourth color. Then we can place vertex d inside the triangle abe and give it the same color as *f*. Then we can place vertex c inside the triangle abf and give it the color of e. Hence the graph is 4-colorable. Moreover, any graph, or portion of a graph bounded by a triangle such as ebf, and having this hierarchical pattern of nested triangles, is 4-colorable. This is the case when the graph contains some vertices with only three edges, and when those vertices and edges are removed, the remaining graph has some vertices with only three edges, which can be removed, and so on, until finally all that remains is a single triangle.

Unfortunately (for the prospect of a simple proof), not every graph is of this hierarchical form. For example, consider the complete graph shown below.

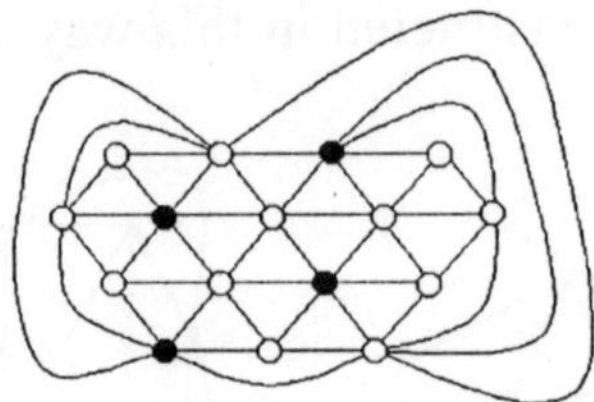

This graph has V = 16 vertices and E = 42 edges. It is not hierarchical, because after removing the two vertices having only three edges each, the remaining graph has four or more edges attached to each vertex. (We can also infer that this graph is not hierarchical from the fact that no three mutually connected vertices are directly connected to a fourth vertex.) Nevertheless, this graph is 4-colorable, as shown in the figure.

The above graph contains a "flat" patch consisting of a uniform hexagonal grid. An *infinite* hexagonal grid is obviously 4-colorable by means of the alternating layered pattern shown below.

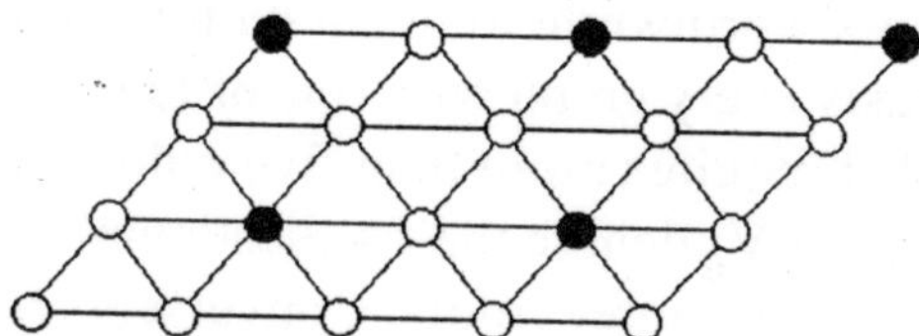

In a sense, two extreme types of graph structures are the purely hierarchical graphs and the perfectly "flat" graphs, both of which are easily seen to be 4-colorable.

Notice that we need not consider graphs containing any configurations of four mutually connected vertices, because, as discussed previously, the edges of such a configuration necessarily split the graph plane into four separate regions, each of which has access to only three of the four vertices.

Hence if we can 4-color the interior of such a triangle, we can 4-color the tetrahedral "frame" as well. If we could reduce all the causal links in a graph to their bare minimums, it might be possible to reduce every n-colorable graph to a graph with just *n* mutually connected vertices, and hence *n* could never exceed four. However, the reduction of graphs with more than two colors is not a trivial task.

If we denote the number of vertices, edges, and faces (i.e., the bounded regions) of a planar graph by V, E, and F respectively, then Euler's formula for a plane (or a sphere) is $V–E + F = 2$. Furthermore, each face of a complete graph is bounded by three edges, and each edge is on the boundary of two faces, so we have $F = 2E/3$, and Euler's formula for a complete planar graph is simply $E = 3V–6$. Now, each edge is connected to two vertices, so the total number of attachments (in a complete graph) is $2E = 6V–12$, and hence the average number of attachments per vertex is $6–12/V$. For any incomplete graph, the total number of attachments is less. Consequently the average number of attachments per vertex for any graph (with a finite number of vertices) is less than 6, which implies that at least one vertex has only five or fewer attachments.

If we have six available colors, a vertex with only five neighbors obviously imposes no constraint on the coloring of the other vertices, because, regardless of the colors of its five (or fewer) neighbors, we can assign it a color without exceeding the six available colors. Therefore if we delete this vertex and all its connections from the graph, creating a graph with one fewer vertices, it's clear that if the resulting graph is 6-colorable, then so was the original graph. Moreover, Euler's formula assures us that this reduced graph also contains at least one vertex with five or fewer neighbors, so we can apply this procedure repeatedly, reducing the graph eventually to one with just 6 vertices, which is obviously 6-colorable. Hence the original graph is 6-colorable.

So, we've seen that Euler's formula immediately implies that no graph can require more than six colors. Furthermore, with just a little more work, we can also show that no graph can require more than *five* colors. (Ultimately we will see that no graph can require more than four colors, but it's worthwhile to begin with the proof of the 5-colorability of every planar graph.) Obviously every graph with five or fewer vertices is 5-colorable, so if there exists a finite graph that requires more than five colors, it must have more than five vertices. Let the positive integer V_6 denote the smallest number of vertices on which there exists a graph that requires six (or more) colors. Conversely, every graph with fewer than V_6 vertices is 5-colorable.

Now, assuming the existence of a graph that requires more than five colors, we can consider one that has exactly V_6 vertices. By Euler's formula, this graph must contain at least one vertex with five or fewer connections. However, it cannot contain any vertex with just four (or fewer) connections, because if it did, we could delete such a vertex and leave a graph with just V_6–1 vertices, which is 5-colorable by definition. Re-inserting the deleted vertex would clearly have no effect on the 5-colorability of the graph, because the vertex has only four (or fewer) neighbors, so the original graph must be 5-colorable, contradicting our assumption. Therefore, a graph with V_6 vertices that requires more than five colors cannot contain any vertex with just four or fewer neighbors.

Since Euler's formula implies that the graph contains at least one vertex with five or fewer connections, the only remaining possibility is that the graph contains a vertex with exactly five connections. However, this too leads to a contradiction. To show this, it's helpful to introduce the notion of a *k-cluster*, which is specified by a set of *k* distinct colors and one particular vertex that has one of those colors. The original vertex is included in the cluster, and, in addition, every vertex with one of the *k* specified colors that neighbors a vertex in the cluster is also in the cluster. By definition the only vertices outside a cluster that are directly connected to vertices inside the cluster have colors that are not in the specified set of *k* colors. Therefore, we can apply any permutation of the *k* colors to the vertices in a cluster without invalidating the coloration. In particular, a 2-cluster is a contiguous set of vertices, each with one of two specified colors, and we can transpose these two colors without upsetting the coloration of a graph.

Now consider a graph containing a vertex with exactly five immediate neighbors, of five distinct colors, as illustrated below.

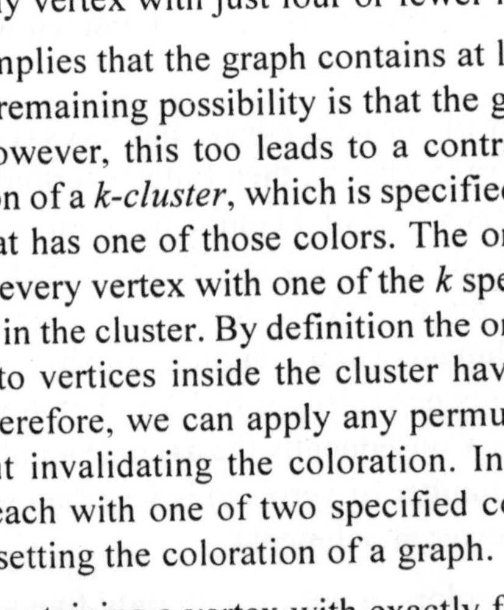

Since the uncolored vertex in the center has neighbors of five distinct colors, it might seem that a sixth color is required. However, notice that we can transpose the blue and green colors in the blue/green 2-cluster attached to the upper left vertex of the central pentagon, so that the upper left vertex is green instead of blue. Once we have done this, the uncolored vertex in the center has neighbors of only four distinct colors. If we delete the central vertex, the overall graph has V_6-1 vertices, so it is 5-colorable, but re-inserting this vertex requires no sixth color (once we have transposed the blue and green in the small 2-cluster as described), so the original graph is 5-colorable.

The only possible objection to the transposing of the colors in a cluster as a means of reducing the number of distinct colors around the perimeter of the pentagon is that we cannot necessarily change the color of any vertex to the color of one of the opposite vertices of the pentagon, because two opposite vertices of a pentagon might be in the same 2-cluster. This is the case, for instance, with the yellow and red vertices of the pentagon in the figure above. If we transpose the red and yellow colors in this 2-cluster, there would still be five distinct colors around the perimeter of the pentagon. However, if such a cluster exists, connecting two opposite vertices of the pentagon, we are guaranteed that at least one other pair of vertices are cut off from each other, i.e., they cannot be in the same 2-cluster. For example, the blue/green 2-cluster outlined in red in the above figure cannot be part of the blue/green 2-cluster attached to the upper right of the pentagon. In general, for any vertex connected to exactly five other vertices, we can always (without invalidating the coloration) change the color of at least one of the five neighbors so that there are only four distinct colors. It follows, using the reduction argument described in the preceding paragraph, that no graph can require more than five colors.

This still leaves open the question of whether five colors are ever actually required, or whether four colors will always suffice. If there exist planar graphs requiring five distinct colors, there must be a positive integer V_5 that is the smallest number of vertices on which such a graph can be constructed. Let's call a graph with V_5 vertices that requires five colors a minimal 5-color graph. By the reduction argument described previously, such a graph obviously cannot contain any vertex with three or fewer connections. Moreover, by considering 2-clusters again, we can prove that such a graph cannot contain any vertex with just four connections. To see why, consider the portion of a graph illustrated below.

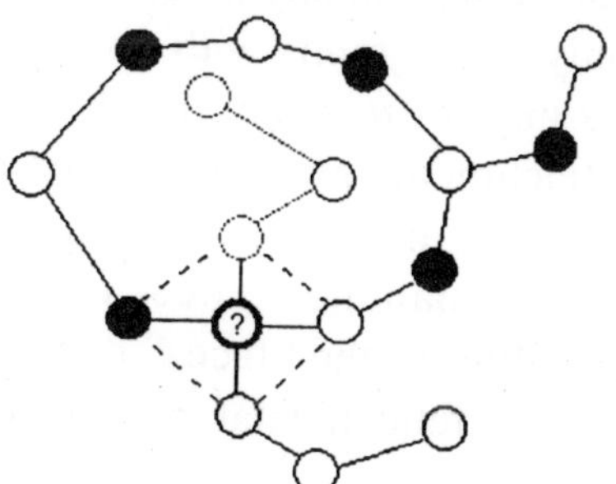

The uncolored vertex in the center has just four neighbors, which we've colored red, yellow, green, and blue. This might seem to require a fifth color for the central vertex, but in fact by permuting the yellow and blue colors in the 2-cluster (outlined in red) above the central vertex we can transform this coloration so that the central uncolored vertex has neighbors of only three distinct colors (red, green, and blue).

Since the overall graph from which this configuration was taken is assumed to have exactly V_5 vertices, it follows that if we delete the central vertex, the resulting graph has only V_5-1 vertices, so it is 4-colorable, but we can clearly add the central vertex back without requiring a fifth color, so the whole graph must be 4-colorable, contradicting our original assumption. This shows that it's always possible to modify a graph containing a vertex with only four connections in such a way that the vertex is connected to only three distinct colors. At most, only one of the two pairs of opposite vertices can be linked by a common 2-cluster, because if one pair is so linked, the other cannot be linked.

To summarize the argument up to this point, we've shown that no graph can require more than five colors, and we've also shown that if there exists a graph requiring five colors, there must exist a minimal 5-color graph, and such a graph cannot contain any vertex with fewer than five connections. Also, by adding connections, we can "complete" any graph on V vertices so that all the faces are triangular and there are a total of exactly 6V-12 attachments. This graph must still be a minimal 5-color graph because we haven't increased the number of vertices, and adding connections cannot reduce the number of colors required, whereas no graph requires more than five colors. Therefore, letting a_5, a_6, a_7,... denote the number of vertices with precisely 5, 6, 7,... attachments respectively, we must have a minimal 5-color graph such that

$$6V-12 = 5a_5 + 6a_6 + 6a_6 + 7a_7 + 8a_8 + 9a_9....$$

where $V = a_5 + a_6 + a_7 + a_8 + a_9 +...$

Substituting this expression for V into the above formula and re-arranging, we have the condition

$$12 = a_5 - a_7 - 2a_8 - 3a_9 -....$$

This places fairly severe constraints on any putative complete minimal 5-color graph. For example, if we consider only such graphs with no more than six attachments at any single vertex, then this formula implies $a_5 = 12$. In other words, a complete minimal 5-graph with no more than six attachments per vertex must have exactly 12 vertices with five attachments each. This suggests that these 12 vertices are arranged globally in the pattern of an icosahedron, and the remaining vertices with six attachments per vertex are in a regular hexagonal pattern, filling in the "faces" of the icosahedron. (This pattern is the basis for "geodesic domes"). The fundamental graph of this type, with $a_6 = 0$, is shown below.

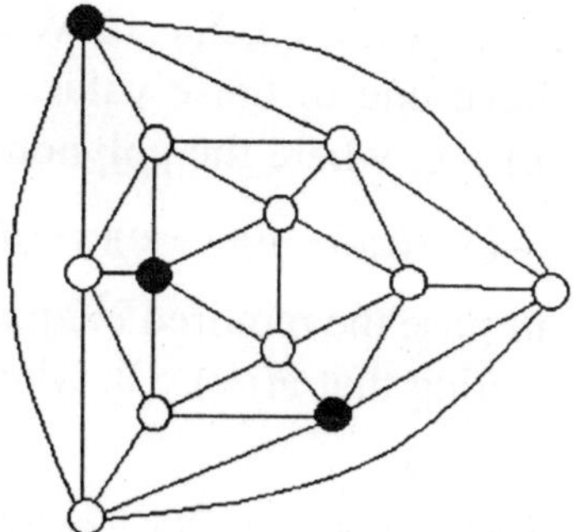

We can give explicit 4-colorings of every such pattern, so such graphs can be shown explicitly to require no more than four colors. Therefore, if a complete minimal 5-color graph exists, it must contain at least one vertex with seven or more attachments.

It was not until 1976 that, with the help of modern computers, the 4-color conjecture was finally proven to be true. The proof, developed by Appel and Haken based on ideas of several earlier people (Kempe, Heawood, Birkhoff, etc) consisted of finding a set of subgraphs, at least one of which must be contained in any normal planar graph (i.e., the set is unavoidable), and such that if a graph containing one of those sub-graphs is not 4-colorable, then neither is a graph with one fewer vertices. This leads immediately to a contradiction, because it implies that, if we hypothesize the existence of a graph that is not 4-colorable, we can always construct another graph, with fewer vertices, that is also not 4-colorable, eventually arriving at a graph with only four vertices, but this graph obviously *is* 4-colorable. Hence we have a contradiction, so we can conclude that the original hypothesis was false, i.e., there does *not* exist a graph that is not 4-colorable. The unavoidable set found by Appel and Haken consisted of nearly 1500 subgraphs, and many of these required considerable analysis to prove that they were "reducible". The development of this unavoidable set, and the proofs of reducibility for its members, was carried out largely on computers, so the proof is notable (and slightly controversial) as an early example of a mathematical proposition whose proof can only be carried out on a computer, and is, in some sense, beyond human mental capabilities to verify.

It's interesting to consider the problem of determining a 4-coloring for any given graph from a purely algebraic standpoint (harking back to Hamilton's quaternions). We might stipulate that each vertex is to be assigned one of four possible values, and the conditions imposed by the connections would be represented by algebraic equations, one equation for each connection. Part of the difficulty of this approach is that the condition of adjacency is not an equality, it is an *in*equality, i.e., we require that two vertices have unequal values (from among the four possible values). Also, there is necessarily a great deal of ambiguity in the solution, because any permutation of the colors leaves a solution intact. In addition, there is even more ambiguity, since in general there can be more than one distinct 4-coloration of a graph.

Given the four allowable (distinct) values a, b, c, d, we could algebraically impose the requirement for every vertex to have one of these values by requiring that the value u assigned to any vertex satisfies $f(u) = 0$, where the polynomial f is defined as

$$f(x) = (x-a)(x-b)(x-c)(x-d)$$

Then we could algebraically impose the required inequality on the values u,v assigned to two connected vertices by stipulating that $g(uv) = 0$, where the polynomial g is defined as

$$g(x) = (x-ab)(x-ac)(x-ad)(x-bc)(x-bd)(x-cd)$$

The equation $g(uv) = 0$ is satisfied if and only if u and v have distinct values from the set {a,b,c,d}. For example, the set {-2,-1,1,2} gives

$$f(x) = (x^2-1)(x^2-4) \qquad g(x) = (x+1)(x+4)(x^2-4)$$

To illustrate, consider again the octahedral graph (which is 3-colorable):

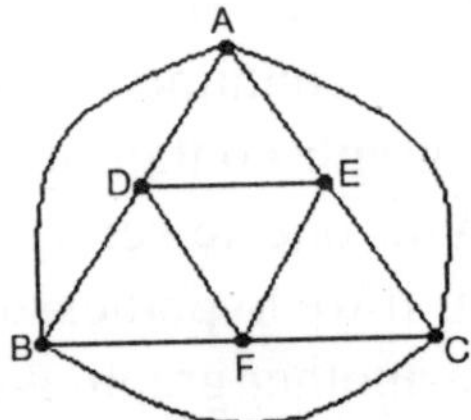

The algebraic conditions for the individual "color" values assigned to the six variables A,B,C,D,E,F are

$$f(\mathrm{A}) = f(\mathrm{B}) = f(\mathrm{C}) = f(\mathrm{D}) = f(\mathrm{E}) = f(\mathrm{F}) = 0$$

and the algebraic conditions imposed by the twelve connections are

$$\begin{aligned} g(\mathrm{AB}) &= g(\mathrm{AC}) = g(\mathrm{AD}) = g(\mathrm{AE}) \\ &= g(\mathrm{BC}) = g(\mathrm{BD}) = g(\mathrm{BF}) = g(\mathrm{CE}) \\ &= g(\mathrm{FC}) = g(\mathrm{DE}) = g(\mathrm{DF}) = g(\mathrm{EF}) = 0 \end{aligned}$$

This amounts to 18 equations in the 6 variables. We can, however, without loss of generality stipulate the "colors" of three of the variables, say, A = 1, B = -1, C = 2, since these three must have mutually distinct values. This leaves us with the following twelve equations in the three unknowns

$$f(D) = f(\mathrm{E}) = f(\mathrm{F}) = 0$$

and

$$\begin{aligned} g(\mathrm{D}) &= g(\mathrm{E}) = g(-\mathrm{D}) = g(-\mathrm{F}) = g(2\mathrm{E}) \\ &= g(2\mathrm{F}) = g(\mathrm{DE}) = g(\mathrm{DF}) = g(\mathrm{EF}) = 0 \end{aligned}$$

The three equations $f(\mathrm{D}) = 0$, $g(\mathrm{D}) = 0$, and $g(-\mathrm{D}) = 0$ jointly have only two solutions, namely, D = 2 and D = -2. Therefore those three equations can be replaced with the single equation $D^2–4 = 0$. Likewise the three equations involving only E can be reduced to $E^2 + 3E + 2 = 0$, and the three equations involving only F can be reduced to $F^2 + F–2 = 0$. Thus we have six equations in the three unknowns.

A different approach would be to assign a four-dimensional vector to each vertex, and then the connection between two given vertices would be represented by requiring the dot product of those two vectors to vanish, i.e., requiring that the two vectors be mutually orthogonal. The three vectors assigned to the vertices of any triangle would then be an orthogonal triad with some orientation in 4-dimensional space. Likewise the four vectors assigned to the vertices of a tetrahedral graph would be a complete set of four mutually orthogonal vectors (a tetrad), but still with arbitrary orientation. For example, the conditions on the six vectors A,B,C,D,E,F assigned to the vertices of the octahedral graph shown above would be simply

$$\begin{aligned} A \cdot B &= A \cdot C = A \cdot D = A \cdot E \\ &= B \cdot C = B \cdot D = B \cdot F = C \cdot E \\ &= C \cdot F = D \cdot E = D \cdot F = E \cdot F = 0 \end{aligned}$$

In this context the four color theorem tells us that a space of four dimensions is sufficient to enable us to assign one of the four basis vectors to each vertex of a planar graph in such a way that the vectors of every pair of adjacent vertices are orthogonal. We can assume each vector is of unit length, so it has three independent components. Therefore, we have 18 independent components in all, constrained by 12 equations. Naturally the system is underspecified, because we can apply an arbitrary permutation to the vectors.

Since the edges of a complete graph partition the graph plane into three-sided regions whose vertices are three mutually connected points, we can uniquely assign the fourth color to each of these regions. The result is useful for visualizing the symmetries of the coloring. For example, the 4-colored icosahedral graph discussed above looks like this:

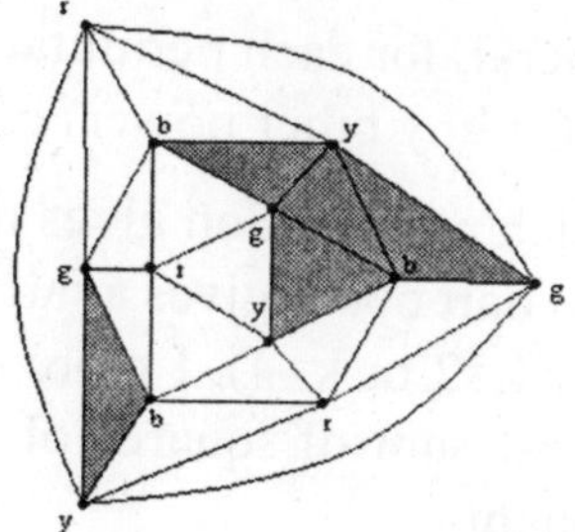

THE DARTBOARD SEQUENCE

The arrangement of the numbers around the circumference of a standard dart board is as shown below

20 1 18 4 13 6 10 15 2 17 3 19 7 16 8 11 14 9 12 5

Oddly enough, no one seems to know how this particular arrangement was selected. It evidently dates back over 100 years, but I've been unable to find any simple criterion that uniquely singles out this arrangement as the best possible in any quantitative sense. It may be just an accident of history that this particular arrangement has been adopted as the standard dart board format.

However, it got me thinking about possible criteria for choosing a circular arrangement of the first n positive integers. I think the most natural idea, in order to get as "flat" a distribution as possible, is to minimize the sum of the squares of each k consecutive terms. For example, setting $k = 3$ the standard dard board sequence gives

$$(20+1+18)^2 + (1+18+4)^2 + (18+4+13)^2 + ... + (5+20+1)^2 = 20478$$

In contrast, if you arrange the numbers by just inter-weaving the largest and smallest numbers like this

20 1 19 2 18 3 17 4 16 5 15 6 14 7 13 8 12 9 11 10

the resulting sum of squares of each 3 consecutive elements is 20510, so the standard dart board is, in this sense, a more flat distribution. Both of these arrangements are much more flat than the natural sequence

1 2 3 4 5 6 7 8 9 10 11 12 13 14 15 16 17 18 19 20

which has a sum of 72000. In general, for each permutation of the integers from 1 to n we can compute the sum of squares (or any other power) of every k consecutive terms.

This raises the question: Which permutation gives the smallest sum, and which gives the largest? Although the standard dart board gives a fairly low sum, it is not the lowest for n=20,k=3 (nor is it the lowest for k=2 or k=4). I more or less randomly checked several other arrangements, and the lowest sum of squares of 3 consecutive terms I found was 20050 for the following arrangement

7 18 4 9 20 2 13 17 1 15 12 3 16 6 5 19 10 8 11 14

Subsequently, Ben Kibel found a better sequence

20 1 12 17 3 13 14 4 15 11 6 16 8 7 18 5 9 19 2 10

which yields a total of 19874. Does anyone know of a better one? I.e., a circular arrangement of the integers 1 through 20 such that the sum of squares of every 3 consecutive terms is less than 19874?

In general if H(n,k) and L(n,k) are the highest and lowest sums of squares of every *k* consecutive elements in a circular arrangement of the first *n* positive integers, are the values of H(*n*, *k*) and L(*n*, *k*) well known and/or easily computed?

When considering the possible ways of "optimizing" the arrangement of the numbers 1 through 20 on a dart board, one proposal was to minimize the sum of the squares of every sum of two consecutive numbers. In general, I think the minimum sum of squares of every sum of two consecutive numbers in a cyclical arrangement of the integers 1 through *N* is

$$S_{min}(N) = N^3 + 2N^2 + 2N - j$$

where j is 1 if *N* is odd, and j is 2 if *N* is even. For the particular case N=20 this formula gives a minimum sum of 8838.

For even *N* the minimum arrangement has the odd and even numbers restricted to separate halfs of the cycle, as illustrated below for N=20

1 3 5 7 9 10 8 6 4 2

19 17 15 13 11 12 14 16 18 20

For odd *N* the minimum arrangement is very simple, as shown below for N=19.

1 3 5 7 9 11 13 15 17 19

18 16 14 12 10 8 6 4 2

This raises some interesting questions. Given any circular arrangement of the integers 0 through n-1, let S denote the sum of squares of every sum of two contiguous numbers, and let v(n) denote the number of distinct values of S for all n! possible arrangements. Following is a table of the number of distinct values of *v*(*n*)

n	*v*(*n*)
1	1
2	1
3	1
4	3
5	8
6	21
7	43
8	69
9	102
10	145

The sequence of integers v(n) does not appear to have been considered before. (It doesn't appear in Sloane's Encyclopedia of Integer sequences.) Hugh Montgomery, A. M. Odlyzko, and Bjorn Poonen developed a very nice approach to this problem, showing that the general term with n>6 is given by

$$v(n) = \begin{cases} (n^3 - 16n + 27)/6 & \text{if } n \text{ is odd} \\ (n^3 - 16n + 30)/6 & \text{if } n \text{ is even} \end{cases}$$

A whole family of interesting sequences can be produced by generalizing the definition as follows: Given any circular arrangement of the integers 0 through n-1, let S denote the sum of the qth powers of every sum of k contiguous numbers. Then let $v(q, k, n)$ denote the number of distinct values of S for all possible arrangements.

With this nomenclature, the previous sequence is denoted as $v(2,2,n)$. Of course, we have $v(1, k, n) = 1$ for all k and n, because the sum of the 1st powers is independent of the arrangement. We also have $v(q,1, n) = 1$ because the sum of any fixed power of the individual numbers is also independent of the arrangement. Also, for fixed values of q and n, the function v is PERIODIC in k.

Another generalization is to add some constant integer j to each of the numbers 0 to n–1. Thus, the general function has four indices, $v(q, k, j, n)$. Notice that v is independent of j for $q<3$, but for larger values of q, j becomes significant. Can $v(q, k, j, n)$ be expressed in closed form as a function of the indices? Which other integer sequences are contained in this family? Which continuous functions (e.g., sin(x), cos(x), exp(x), etc) can be approximated by sequences of this form?

MEETING COMBINATIONS AND INTERVAL GRAPHS

Let n continuous segments with freely defined starting and ending points be placed along the x axis, and consider the intersections between these segments. For any given arrangement there may be some regions where no segment is present, some where just one segment is present, some where exactly two segments overlap, some where exactly three segments overlap, and so on. Any such arrangement establishes a set of "meetings" between two or more segments. To illustrate, the figure below shows the placements of six segments, and identifies the meetings between them.

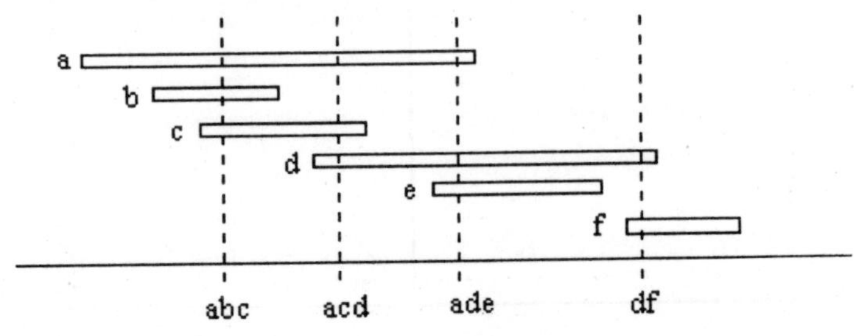

The meetings for this arrangement are abc, acd, ade, and df. In general, a meeting is a local maximum point on the plot of the number of segments present as a function of position. The plot for the above arrangement of six segments is shown in the figure below.

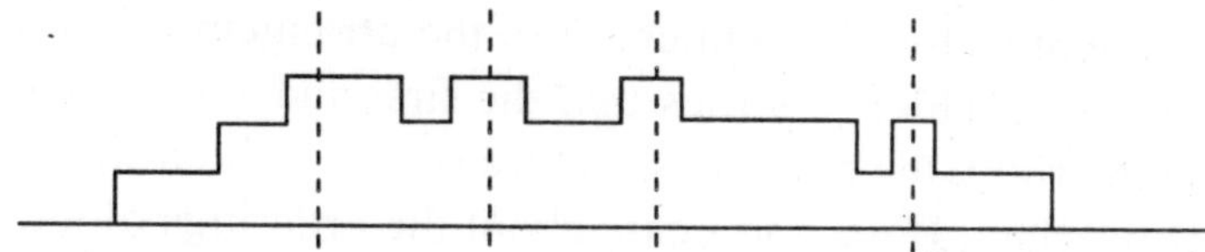

The number of different arrangements of the starting and stopping points for *n* segments (with distinct identities) is easy to determine. For the first segment there is obviously only one arrangement. This segment divides the axis into three regions, namely, the region to the left of the starting point, the region between the starting and ending points (i.e., the region covered by the segment), and the region to the right of the ending point. Therefore, the starting point of a second segment can be placed in any one of those three regions, and the ending point can be placed in the same region or any other region to the right of the region containing the starting point. Thus there are six distinct ways of placing the second segment.

In general, as we prepare to add the kth segment, there will already be k–1 segments in place, and the 2(k–1) terminal points of these segments divide the axis into 2k–1 regions. The starting point for the kth segment can be placed in any of these, and then the ending point is placed in the same or more rightward region. Thus the number of possibilities for placing both these points is (2k–1) + (2k–2) +... + 2 + 1, which equals k(2k–1).It follows that if we let N_j denote the number of distinct arrangements of j segments, then we have

$$N_k = k(2k-1)N_{k-1} = \frac{(2n)!}{2^n}$$

This applies only if the n segments have specified identities. If we relax this condition, allowing the labels of the segments to be interchangeable, then the number of distinct arrangements is much less. In general we will have n starting points, so it only remains to place the n ending points. The only restriction is that each ending point must follow its starting point. Thus the ending point for the first segment (i.e., the segment with the left-most starting point) can be placed in any of n regions, delimited by the n starting points. The ending point of the second segment can be placed in any of the n–1 regions to the right of the second starting point. And so on. (The arrangement of the ending points in between consecutive starting points has no effect on the presence of intersections.) Consequently the number of arrangements for indistinguishable segments is simply (n!).

However, this still distinguishes between the 1st segment, the 2nd segment, and so on. It is possible for two or more distinct arrangements to yield the same "meeting structure" up to permutation of the segments. For a trivial example, consider the two arrangements of three segments shown below.

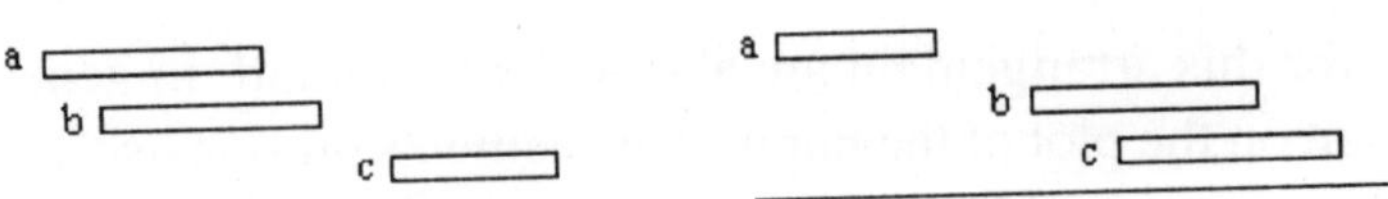

(The segment labels here merely indicate the order of the starting points, not implying any distinguishable identity for the segments.) In the arrangement shown on the left, the meeting structure is "ab", which signifies that the first and second segments overlap. In the arrangement on the right, the meeting structure is "*bc*", which signifies that the second and third segments overlap. If we don't care about the orderings of the start times for the segments in the meeting structure, then these two arrangements induce the same meeting structure. On this basis, what is the number of distinct meeting structures for *n* segments?

Let *M*(*n*) denote the number of distinct meeting structures for *n* segments. Obviously for $n = 0$ and $n = 1$ segments, no meetings are possible, so there is only one meeting structure in each of these cases, namely the null structure, denoted by {Z}. Thus we have M(0) = M(1) = 1. For $n = 2$ segments there are two possible meeting structures, corresponding to the fact that either these segments overlap or they do not. We will let {Z, ab} denote these two possibilities, and we have M(2) = 2.

For $n = 3$ segments, there are 3! = 6 distinct arrangements, which we can express symbolically as

$$\varnothing \quad ab \quad bc \quad ab+bc \quad ab+ac \quad abc$$

However, as noted above, the arrangements *ab* and *bc* represent the same meeting structure. Likewise the arrangement $ab + bc$ is the same structure as $ab + ac$. In both cases, one segment meets with each of the other two, but the other two don't overlap with each other. Therefore, the number of distinct meeting structures for $n = 3$ segments is just four, consisting of the set {Z, *ab*, $ab + bc$, *abc*}, and so we have M(3) = 4. At this point we should make it clear that although our notation for meeting structures looks like Boolean logical notation, it is not. For example, the meeting "*abc*" can also be written in the form "$ab + ac + bc$", because these both signify a triple intersection between the segments a, b, and c. In this notation, conjoined characters signify a mutual meeting of those segments, and the "+" symbol represents logical "AND".

Thus both of these operations are forms of "intersection". It's important to note that if *a,b,c* represented arbitrary sets, then $ab + ac + bc$ would *not* be equivalent to abc, because it would be possible to have those three pair-wise intersections with no three-way intersection. However, since our basic sets are continuous segments on the one-dimensional *x* axis, the three pairwise meetings of three segments imply that the first ending point occurs after the last starting point, so there is a region of three-way intersection. (It's interesting that the topology of continuous regions in one-dimensional space is induced by imposing the equality $ab + ac + bc = abc$, raising the question of whether other geometrical contexts can be specified in terms of some other Boolean equalities.) On the other hand, the "meeting" relation between segments is not transitive, because $ab + bc$ does not imply *ac*.

For $n = 4$ segments, we examine all $n! = 24$ distinct arrangements and find the following ten distinct meeting structures

(1)	$\varnothing$	(1)	$ab(c+d)$
(3)	ab	(3)	$a(bc+d)$
(4)	$a(b+c)$	(4)	$ab+bc+cd$
(1)	$ab+cd$	(1)	abc
(2)	$a(b+c+d)$	(2)	$abcd$

(The number to the left of each structure is the number of occurrences of that structure in the n! arrangements.) Hence we have M(4) = 10.

To determine the value of M(5), we could examine each of the 5! = 120 arrangements by hand, and check for the unique structures, but this would be very laborious. To automate the process, we can step through each of the n! placements of the n ending points, and for each arrangement we can easily determine the pairwise meetings. We can then count how many pairs contain the first segment, how many the second, and so on. We can then count how many segments appear in exactly one pair, how many in exactly two, and so on. This gives a set of numbers $d_1, d_2,...$ that characterize the meeting structure, independent of the orderings of the segments. If two structures are characterized by different values of d_j, then they are certainly distinct structures, so we can determine a lower bound on the value of M(n) by determining the number of unique characteristics $\{d_1, d_2,...\}$.

For the case $n = 5$, this criterion turns out to be nearly sufficient to distinguish all the structures. There are 26 distinct sets of $\{d_j\}$ indices, so we know M(5) is at least 26. However, it is possible for distinct meeting structures to have the same set of $\{d_j\}$ values. This occurs for the structures $ab + ac + bc + de$ and $ab + bc + cd + de$. In each case, two letters appear exactly once, and the other three letters appear twice each. However, they are obviously different structures, as seen by the fact that the first one reduces to abc+de whereas the second is irreducible. So, there are actually M(5) = 27 distinct meeting structures with $n = 5$. These are shown below.

$\varnothing$	$a(b+c+d)+de$	$a(ac+de)$
ab	$a(bc+d)$	$a(b(c+d)+e)$
$ab+cd$	$abc+d(a+e)$	$a(b(c+d)+de)$
$a(b+d)$	$a(b+c+d)+c(d+e)$	$a(bcd+e)$
$a(b+c)+de$	$ab(c+d)$	$ab(c+d+e)$
$ab+ac+cd$	$ab(c+d)+de$	$ab(cd+e)$
abc	$abcd$	$abc(d+e)$
$abc+de$	$a(b+c+d+e)$	$ab+bc+cd+de$
$a(b+c+d)$	$a(bc+d+e)$	$abcde$

To help distinguish structures that have the same $\{d_j\}$ sets, consider again the two structures ab+ac+bc+de and ab+bc+cd+de. If we replace each letter by the number of pairs in which it appears, these become $2{\cdot}2 + 2{\cdot}2 + 2{\cdot}2 + 1{\cdot}1$ and $1{\cdot}2 + 2{\cdot}2 + 2{\cdot}2 + 2{\cdot}1$ respectively, clearly distinct. Carrying out the multiplications and additions numerically, these give the values 13 and 12 respectively. These sums are an additional discriminator that can be used to help count the distinct meeting structures. On this basis, we can distinguish 91 meeting structures for $n = 6$. However, there are actually M(6) = 92 distinct meeting structures. The two that are indistinguishable based on the criteria we've described so far are ab + bc + de + df + ef and ab + bc + cd + de + ef. These have the same set of valences, and the same pairings of valences, but they are nevertheless distinct structures, as shown by the fact that the first can be reduced to ab + bc + def, whereas the second is irreducible.

We can define another discriminator by replacing each letter not with the valence of that letter, but with the sum of the valences of all the letters with which it is paired. For example, in the first structure e is paired with d and with f, and these two letters appear twice each, for a total of 4, so we replace the letter e with 4. Doing the same for all the other letters in these structures, and carrying out the multiplications and additions numerically, we get

$$2.2 + 2.2 + 4.4 + 4.4 + 4.4 = 56$$

$$2.3 + 3.4 + 4.4 + 4.3 + 3.2 = 52$$

So this criterion enables us to distinguish between these two structures. On the same basis we can determine that M(7) = 369, meaning that there are 369 distinct meeting structures for $n = 7$ segments. However, for $n = 8$ segments the criteria we have described so far are not sufficient to distinguish between all the different structures. We find that there are at least 1800 such structures by these criteria, but in fact there are 1807. It's interesting to try to identify the most efficient discriminators between these structures. Obviously if we are given two sets of pairing expressions we can simply try all possible permutations of the letters and terms, to see if they can be brought into agreement, but this in not feasible for large values of n.

It's worth noting that if $a_j a_k$ is a connected pair with $j < k$, then $a_j a_i$ is also a connected pair for all indices i between j and k. For example, if bf is a connected pair, then bc, bd, and be are also connected pairs. It is this requirement that reduces the number of possible structures from $2^{n(n-1)/2}$ down to n!.

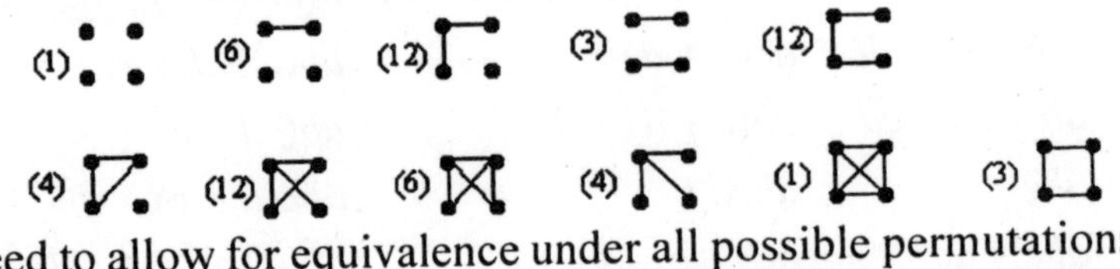

(1) (6) (12) (3) (12)

(4) (12) (6) (4) (1) (3)

We then need to allow for equivalence under all possible permutations of the letters to reduce the number of distinct structures from n! down to M(n). These structures are often called "interval graphs", because they can be represented pictorially as a set of connections between the n nodes of a plane graph. Consider, for example, all $2^{4(4-1)/2} = 64$ plane graphs involving four nodes, shown below.

The numbers in parentheses indicate how many of each of these structures there are if we regard the nodes as having distinct identities. (A convenient check is given by confirming that the total numbers of graphs with k edges is C(6, k).) If, on the other hand, the nodes are regarded as indistinguishable (as is the convention for graphs), then the eleven arrangements illustrated above represent all the distinct graphs on four nodes. Ten of these graphs are interval graphs, corresponding to the ten "meeting structures" described previously. However, the "square cycle" graph shown on the right (in red) is *not* an interval graph, i.e., it does not correspond to any meeting structure of overlapping continuous intervals on a line. This stands to reason, because an interval graph cannot contain any minimal closed cycle of length greater than three.

One way of characterizing the unique meeting structures for n segments is in terms of the determinant of a certain n′n matrix. Let a_k with $k = 1, 2, .., n$ designate the n segments, and define an n′n matrix A with the non-zero components $A_{j,k} = 1$ for each pair of indices j, k such that $a_j a_k$ (or $a_k a_j$) is a connected pair. In addition, we set the diagonal values of A to the parameter x, so we have $A_{j,j} = x$ for $j = 1$ to n. (Note that a meeting structure is specified as a union of m pair-wise connections. If no two segments meet, then $m = 0$. If there is a connection between every pair of segments, then $m = n(n-1)/2$.)

Given the matrix A for a particular structure, the determinant of A is a polynomial in the parameter x. Since the determinant of a matrix is invariant under permutations of the row/columns, it follows that this polynomial characterizes a full equivalence class of meeting structures. Thus we can use these polynomials to discriminate between distinct structures. To illustrate, consider the following two meeting structures for $n = 7$ segments:

$$a_2a_3 + a_3a_4 + a_3a_5 + a_4a_5 + a_4a_6 + a_4a_7 + a_5a_6 + a_5a_7$$

$$a_2a_3 + a_2a_4 + a_2a_5 + a_2a_6 + a_3a_4 + a_3a_5 + a_5a_6 + a_5a_7$$

These structures have the same valence sets (i.e., they each have two characters that appear in four pairs, one that appears in three pairs, two that appear in two pairs, and one that appears in one pair), so in order to distinguish between them using the previous method we need to go to higher-order valences. However, these structures correspond to the matrices

$$\begin{bmatrix} x & 0 & 0 & 0 & 0 & 0 & 0 \\ 0 & x & 1 & 0 & 0 & 0 & 0 \\ 0 & 1 & x & 1 & 1 & 0 & 0 \\ 0 & 0 & 1 & x & 1 & 1 & 1 \\ 0 & 0 & 1 & 1 & x & 1 & 1 \\ 0 & 0 & 0 & 1 & 1 & x & 0 \\ 0 & 0 & 0 & 1 & 1 & 0 & x \end{bmatrix} \qquad \begin{bmatrix} x & 0 & 0 & 0 & 0 & 0 & 0 \\ 0 & x & 1 & 1 & 1 & 1 & 0 \\ 0 & 1 & x & 1 & 1 & 0 & 0 \\ 0 & 1 & 1 & x & 0 & 0 & 0 \\ 0 & 1 & 1 & 0 & x & 1 & 1 \\ 0 & 1 & 0 & 0 & 1 & x & 0 \\ 0 & 0 & 0 & 0 & 1 & 0 & x \end{bmatrix}$$

and the determinants of these matrices are

$$x^2(x-1)(x^4+x^3-7x^2-x+4)$$
$$x(x^6-8x^4+6x^3+7x^2-4x-1).$$

This shows how we can immediately distinguish between these two structures. (For automated numerical purposed, we can simply evaluate these polynomials for a few arbitrary values of x to discriminate between them.)

The matrices and characteristic polynomials for the four distinct meeting structures for $n = 3$ segments are shown below.

$$\det\begin{bmatrix} x & 0 & 0 \\ 0 & x & 0 \\ 0 & 0 & x \end{bmatrix} = x^3 \qquad \det\begin{bmatrix} x & 1 & 0 \\ 1 & x & 0 \\ 0 & 0 & x \end{bmatrix} = x(x^2-1)$$

$$\det\begin{bmatrix} x & 1 & 0 \\ 1 & x & 1 \\ 0 & 1 & x \end{bmatrix} = x(x^2-2) \qquad \det\begin{bmatrix} x & 1 & 1 \\ 1 & x & 1 \\ 1 & 1 & x \end{bmatrix} = x(x-1)^2(x+2)$$

Likewise we can determine the characteristic polynomials for each of the 10 interval graphs on $n = 4$ nodes, shown below.

∅	x^4
ab	$x^2(x-1)(x+1)$
ab+bc	$x(x-1)^2$
ab+cd	$(x-1)^2(x+1)^2$
abc	$x(x+2)(x-1)^2$
ab+bc+cd	$(x^2+x-1)(x^2-x-1)$
abc+cd	$(x-1)(x^3+x^2-3x-1)$
abc+cd	$x(x-1)(x^2+x-4)$
abc+abd	$x^2(x^2-3)$
ab+ac+ad	$(x+3)(x-1)^3$

As noted previously, the only other plane graph on four nodes is the four-loop

ab + bc + cd + da $x^2(x-2)(x+2)$

BINARY GAMES

Here's an interesting game: I announce a sequence of binary digits, such as "1 1 0 1 0 0 1...etc". These digits represent a sequence of numbers, with the first announced digit being the most significant and the latest being the least significant. Thus, the string of digits just mentioned corresponds to the sequence of numbers 1, 3, 6, 13, 26, 52, 105,...

Your object is to interrupt me at some stage and supply one more digit such that the resulting number is divisible by (at least) one of a specified set of primes, say {7,11,13,17}. My object is to avoid that outcome.

What interests me about this game is that, depending on the specified set of primes, it may or may not be possible for me to avoid eventually giving you a winning number. For example, with the set {7,11,13,23} I can go on indefinitely without giving you a winning opportunity, whereas with the set {7,11,13,17} you are sure to have a winning opportunity no later than the 16th digit. (This is because any number larger than 9705 can be truncated at some point in its binary sequence and made divisible by one of {7,11,13,17} by flipping the least significant digit if necessary.)

Let [*p*,...,*q*] denote the largest number such that no sequential substring of its binary digits (including the most significant) can be made divisible by one of {p,..,q} by flipping the least significant digit. Then we have the following miscellaneous results

$$
\begin{aligned}
[3] &= 1\\
[5] &= \text{oo?}\\
[5,7] &= 3\\
[5,11] &= \text{oo?}\\
[5,13] &= 7\\
[7,11,13,17] &= 9705\\
[7,11,13,19] &= 17\\
[7,11,13,23] &= \text{oo?}\\
[11,13,17,19,31] &= 59\\
[11,13,17,19,29,31] &= 15\\
[13,19,23,29,31,41,47,67] &= 85
\end{aligned}
$$

I think each of these is "minimal" in the sense that removing any one of the primes would change the result. QUESTION: Is there a simple way of characterizing sets of primes that "cover" all the integers above a certain number (in the sense described above)?

Analagous problems arise in more complicated games where two or more players each select a digit in turns, and each player trys to cause the constructed number to be divisible by a certain set of primes. Also, there are interesting variations using base-3 or base-4 numbers, etc.

Obviously if the set T of "target" primes includes 2, you can win at any stage by simply placing a "0" at the end of the present sequence. If T includes 3 you can win after I supply the first (non-zero) digit by supplying a "1". On the other hand, if T = {5}, you are not guaranteed an opportunity of winning. However, with T = {5,7} you are guaranteed a winning opportunity.

Here is a short list of target sets for which you are guaranteed a winning opportunity:

{2}

{3}

{5,7}

{7,11,13,17}

{11,13,17,19,31}

{13,17,23,29,31,37,41}

{17,23,29,31,37,41,53,59,61,67,73,79,97,251,1049,1571}

{19,23,29,31,37,41,43,53,61,67,73,79,97,173,283,643,701,1049, 1123,1571,3137,4201}

For any given prime p, does there exist a set T with smallest member p such that you are guaranteed a winning opportunity? What is the minimum number of elements for such a set?

MEETING PROBABILITIES

Given n independent random variables, each evenly distributed over the interval 0 to 1, the probability that all n are within q of each other (for any $q < 1$) is

$$p_n = nq^{n-1} - (n-1)q^n$$

This can be derived in several different ways. For example, we can divide the unit interval into k equal segments, and note that the probability of n randomly selected points all falling within j consecutive segments corresponds approximately to the probability that all n points fall within $q = (j/k)$ of each other. This correspondence becomes exact in the limit as k goes to infinity (holding q constant). Equation (1) can also be derived from a geometrical point of view. Given a unit "cube" in n dimensions, equation (1) represents the fraction of the cube's content ("volume") consisting of points with orthogonal coordinates $[x_1, x_2,..., x_n]$ such that $|x_i - x_j| < q$ for all i, j.

One possible generalization of this is to allow different tolerances on the different events. For example, suppose each of n people are to arrive at a certain location at some randomly chosen time between 1:00 PM and 2:00 PM, and each person will wait a certain amount of time before leaving. Say, for example, with $n = 2$ people, one can wait for w_1 and the other can wait for w_2 (both expressed as fractions of the total time interval). What is the probability that they will meet? Geometrically it's easy to see that this is just given by the are of the shaded region in the unit square shown below:

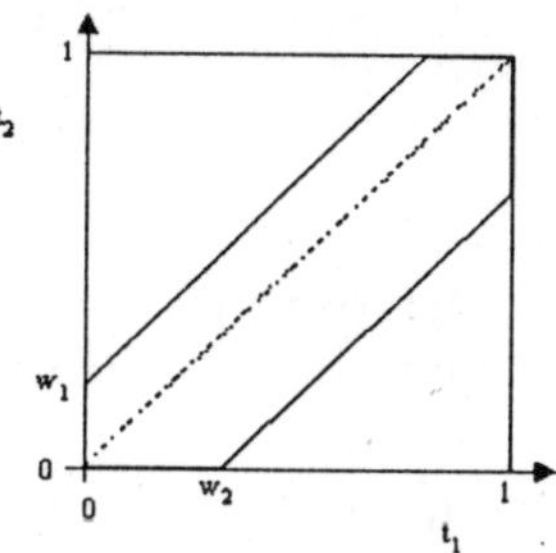

The shaded area equals to total square area minus the two excluded triangles, so we have

$$P_2 = 1 - \frac{(1-w_1)^2}{2} - \frac{(1-w_2)^2}{2} = w_1 + w_2 - \frac{1}{2}w_1^2 - \frac{1}{2}w_2^2$$

More generally, for n people with waiting times $w_1, w_2, .., w_n$, we just need to evaluate the integral of 1 over the following ranges of the arrival times $t_1, t_2, .., t_n$:

variable	lower limit	*upper* limit
t_1	0	1
t_2	$\max\ (0, t_1 - w_2)$	$\max\ (1, t_1 + w_1)$
t_3	$\max\ (0, t_1 - w_3, t_2 - w_3)$	$\max\ (1, t_1 + w_1 - t_2 + w_2)$
t_4	$\max\ (0, t_1 - w_4, t_2 - w_3 - w_4)$	$\max\ (1, t_1 + w_1, t_2 + w_2, t_3 + w_3)$

etc

Obviously this becomes extremely complicated as n increases. To illustrate, consider the case $n = 3$. We need to divide the problem into 12 separate double integrals over t_1 and t_2 in order to specify fixed integration limits for each one. Letting w_1 denote the smallest waiting time, and w_3 the largest, these 12 regions are illustrated below as the regions denoted by A through L.

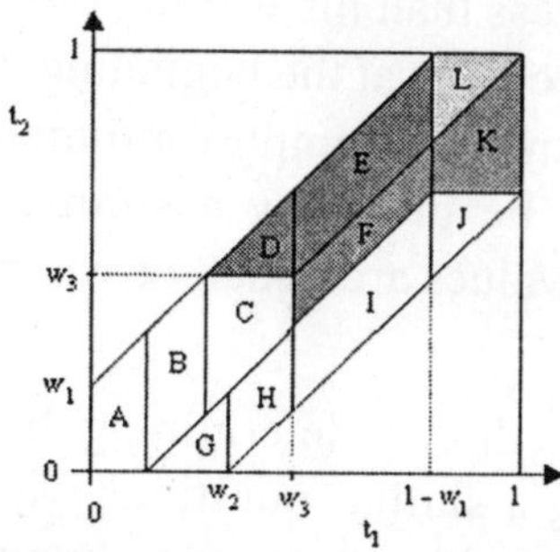

Within each of these regions we need to integrate the difference between the high and low integration limits for t_3. The integration limits and integrands are listed below

Region	t_1 Limits		t_2 Limits		Integrand
A	0	$w_2 - w_1$	0	$t_1 + w_1$	$t_1 + w_1$
B	$w_2 - w_1$	$w_3 - w_1$	$t_1 - w_2 + w_1$	$t_1 + w_1$	$t_1 + w_1$
C	$w_3 - w_1$	w_3	$t_1 - w_2 + w_1$	w_3	$t_1 + w_1$
D	$w_3 - w_1$	w_3	w_3	$t_1 + w_1$	$(t_1 + w_1) - (t_2 - w_3)$
E	w_3	$1 - w_1$	t_1	$t_1 + w_1$	$(t_1 + w_1) - (t_2 - w_3)$
F	w_3	$1 - w_1$	$t_1 - w_2 + w_1$	t_1	$(t_1 + w_1) - (t_1 - w_3)$
G	$w_2 - w_1$	w_2	0	$t_1 - w_2 + w_1$	$t_2 + w_2$
H	w_2	w_3	$t_1 - w_2$	$t_1 - w_2 + w_1$	$t_2 + w_2$
I	w_2	$1 - w_1$	$t_1 - w_2$	$t_1 - w_2 + w_1$	$(t_2 + w_2) - (t_1 + w_3)$
J	$1 - w_1$	1	$t_1 - w_2$	$t_1 - w_2$	$(t_2 + w_2) - (t_1 + w_3)$
K	$1 - w_1$	1	$1 - w_2$	t_1	$1 - (t_1 + w_3)$
L	$1 - w_1$	1	t_1	1	$1 - (t_1 + w_3)$

Evaluating all twelve of these integrals and simplifying gives the result

$$P_3 = w_1 w_2 + w_1 w_3 + w_2 w_3 - \frac{1}{2}\left(w_1 w_2^2 + w_1 w_3^2 + w_2 w_3^2\right)$$

$$-\frac{1}{3} w_1^2 - \frac{1}{6} w_2^3$$

For example, with 3 people who can wait $w_1 = 1/4$, $w_2 = 1/3$, and $w_3 = 1/2$ (15 minutes, 20 minutes, and 30 minutes, respectively) the probability of all three of them being at the location simultaneously during the hour is exactly 1435/5184, which is about 0.27681.

Naturally both the formulas for $n = 2$ and $n = 3$ reduce to equation when all the waiting times are equal. It's also worth noting that the formula for $n = 3$ is not symmetrical in the three waiting times, because the boundaries of the 3-dimensional solid representing the meeting region in phase space depend on a set of min and max selections, so the result depends on the ordering of the parameters, with w_1 the smallest and w_3 the longest waiting time. As n increases the direct integration method for arbitrary and unequal waiting times rapidly becomes much more complicated, because the waiting times must be truncated whenever someone arrives with less than his waiting time remaining before the end of the interval, and a similar truncation occurs at the beginning of the hour. Geometrically, we're dealing with the volume of an extremely complex n-dimensional asymmetrical polyhedron whose volume can't generally be given by a simple formula, except in the highly symmetrical case when all the d values are equal, or when n is very small (like the cases of $n = 2$ and $n = 3$ given above).

To derive a more efficient method of determining the probabilities for larger values of n, we may begin by considering a similar questions whose answers can immediately be given by fairly simple formulas with arbitrary n. Suppose we eliminate the boundary truncation in the original problem by wrapping the 1-hour interval around in a circle of unit circumference, and stipulating that the jth person is present for a continuous span of exactly w_j on that interval, and the location of each w_j is uniformly and independently distributed around the circle. Now we can ask for the probability that n spans of length w_1, $w_2,\ldots,w_n$ will have some common overlap. Let's also assume that the d values are all fairly small relative to the entire interval, so that $w_1 + w_2 + \ldots + w_n$ is less than 1. (This allows us to avoid complications due to wrap-around.)

Consider first the discrete case, i.e., suppose the circumference of the circle is divided into N equal increments, and the n randomly placed spans have discrete lengths of m_1, $m_2,\ldots, m_n$ increments respectively. The total number of possible arrangements of the spans is N^n, so we need to determine how many of those contain at least one increment of overlap between all the spans.

Obviously each configuration with overlap can be translated into N different positions, so we really just need to find the number of distinct intrinsic overlap configurations of these n spans.

If we fix one particular increment on the circle and require that all the spans contain that increment, then the number of arrangements that satisfy this requirement is obviously the product $m_1 m_2 \ldots m_n$. However, this counts the intrinsic arrangements with two increments of overlap twice, because of translation. Likewise it counts the arrangements with three increments of overlap three times, and so on. Thus the product of lengths represents an overestimation of the number of intrinsic arrangements with overlap.

But suppose we fix *two* particular consecutive increments on the circle and require that all the spans contain both of them. The number of arrangements that satisfy this requirement is obviously the product $(m_1-1)(m_2-1)\ldots(m_n-1)$. This counts the arrangements with two increments of overlap only once, and it counts the arrangements with three increments of overlap twice, and so on. Therefore, if we subtract this product from the previous product, the result counts the number of arrangements with any number of increments of overlap exactly once. Hence the number of such arrangements is

$$m_1 m_2 \ldots m_n - (m_1 - 1)(m_2 - 1)\ldots.(m_n - 1)$$

It follows that the probability of overlap for these n spans placed randomly around the unit circle is

$$P_n = \frac{N\left[m_1 m_2 \ldots m_n - (m_1 - 1)(m_2 - 1)(m_n - 1)\right]}{N^n}$$

For example, with $n = 3$ the result is

$$P_3 = \frac{1 - m_1 - m_2 - m_3 + m_1 m_2 + m_1 m_3 + m_2 m_3}{N^2}$$

Obviously if we hold the ratios $w_1 = m_1/N$, $w_2 = m_2/N$, and $w_3 = m_3/N$ constant and increase N, all the terms in the numerator of degree less than two drop out, and we're left with the result for the continuous case

$$P_3 = w_1w_2 + w_1w_3 + w_2w_3 = w_1w_2w_3\left(\frac{1}{w_1}+\frac{1}{w_2}+\frac{1}{w_3}\right)$$

In general for n spans of lengths $w_1, w_2,..,w_n$ randomly distributed uniformly around a circle of unit circumference, with the condition that the sum $w_1 + w_2 +...+ w_n$ is less than 1, the probability of some common overlap among all n of the spans is

$$P_n = w_1w_2....w_n\left(\frac{1}{w_1}+\frac{1}{w_2}+....+\frac{1}{w_n}\right)$$

Now let's return to our original problem, to find the probability of meeting between n people with distinct waiting times in a fixed interval. Again we can start with the discrete case, and then go over to the continuous case at the end. We divide the unit interval into N segments, and let the integer m_j for $j = 1,2,..,n$ denote the waiting times (corresponding to Nw_j), arranged so that $1 \le m_1 \le m_2 \le ... \le m_n \le N$. We apply the same reasoning as in the case of the circle, except that now each fixed segment of overlap must be treated separately, because the segments near the starting time are truncated by the boundary. For example, the first waiting interval m_1 has only one position that includes the first segment, and it has only two positions that overlap with the second segment, and so on.

Only when we reach segments more than m_j from the boundary are there m_j intersecting positions. In general, the number of positions of the kth waiting interval that intersect with the jth segment is $\min(j, m_k)$. Hence, instead of having N times the product of the m_k values, we must evaluate the sum over j from 1 to N of the products of the $\min(j, m_k)$ values. Of course, as in the case of the circle, this represents an over-estimate, because if counts each configuration multiple times.As before, this is corrected by subtracting the products of the quantities $[\min(j, m_k)-1]$.

Consequently, the probability of meeting for n people with waiting times $m_1, m_2,...,$ m_n is given by

$$P_n = \frac{1}{N^n}\sum_{j=1}^{N}\left\{\left(\prod_{i=1}^{n}\min(j,m_i)\right)-\left(\prod_{i=1}^{n}\min(j,m_i)-1\right)\right\}$$

After expanding the products, the terms of degree n cancel, leaving only terms of degree n–1 and lower. Bringing $1/N^{n-1}$ in from the leading factor, and letting w_j and t denote the ratios m_j/N and j/N respectively, the resulting expression is purely a function of the w_j and the variable t, plus terms divided by some power of N. The latter vanish in the limit as N goes to infinity, so only the terms of degree n–1 remain. (These are comprised of the n products of n–1 elements of the set $\min(j, m_j), j = 1,2,..,n$.) In the continuous limit, the summation over the index j can be evaluated as an integration over the variable t. Noting that $dj = Ndt$, the remaining power of 1/N in the leading factor is cancelled, and we have in the continuous limit

$$P_n = \int_{t=0}^{1} s_{n-1}(\min\ (t, w_1), \min\ (t, w_1), \ldots, \min(t, w_n))\,\mathrm{dt}$$

where $s_{n-1}(x_1, x_2,..,x_n)$ denotes the sum of all products of n–1 arguments. Splitting up this integral into the ranges from 0 to w_1, from w_1 to w_2, and so on, we can eliminate the min functions and write this as a sum of elementary integrals. To simplify the expressions, it's convenient to define the following symbols for the kth and (k–1)th symmetric functions of the k smallest waiting times:

$$\pi_k = w_1 w_2 \ldots w_1 \quad \sigma_k = \left(\frac{1}{w_1} + \frac{1}{w_2} + \ldots \frac{1}{w_k}\right)\pi_k$$

with $p_0 = 1$ and $s_0 = 0$. Also, let $w_0 = 0$ and $w_{n+1} = 1$. In terms of these parameters we can write

$$P_n = \sum_{j=1}^{n}\left[\int_{w_j}^{w_{j+1}} \left\{(n-j)p_j t^{n-1-j} + \sigma_j t^{n-j}\right\} dt\right]$$

Evaluating the integrals, this becomes

$$P_n = \sum_{j=1}^{n-1}\left[\left(w_{j+1}{}^{n-j} - w_j{}^{n-j}\right)\pi_j\right]$$

$$+\sum_{j=1}^{n}\left[\left(w_{j+1}{}^{n-j+1} - w_j{}^{n-j+1}\right)\frac{\sigma_j}{n-j+1}\right]$$

In view of the identity $p_{j+1} = p_j\ w_{j+1}$, the argument of the first summation can be written as

$$\left(w_{j+1}{}^{n-j} - w_j{}^{n-j}\right)\pi_j = \pi_{j+1} w_{j+1}{}^{n-[j+1]} - \pi_j w_j{}^{n-j}$$

Hence these terms constitute a telescoping sequence, with the net sum π_n. Therefore, we have the result

$$p_n = \pi_n + \sum_{j=1}^{n}\left[\left(w_{j+1}{}^{n-j+1} - w_j{}^{n-j+1}\right)\frac{\sigma_j}{n-j+1}\right]$$

Notice that this formula reduces to equation (1) if all the waiting times are equal. Also, it's easily verified that this formula gives the results previously derived for $n = 2$ and $n = 3$. For another illustration, consider the case $n = 4$. Using this expression, the probability of all parties meeting is

$$P_4 = \pi_4 + \frac{w_2{}^4 - w_1{}^4}{4}\sigma_1 + \frac{w_3{}^3 - w_2{}^3}{3}\sigma_2 + \frac{w_4{}^4 - w_3{}^4}{2}\sigma_3 + \frac{1 - w_4}{4}\sigma_4$$

Inserting the explicit expressions for the symmetric functions, and simplifying, this gives the formula

$$P_4 = w_1 w_2 w_3 + w_1 w_2 w_4 + w_1 w_3 w_4 + w_2 w_3 w_4$$
$$-\frac{w_4^{\,2}}{2}(w_1 w_2 + w_1 w_3 + w_2 w_3) - \frac{w_3^{\,3}}{6}(w_1 + w_2)$$
$$-\frac{w_1 w_2 w_3^{\,2}}{2} - \frac{w_1 w_2^{\,2}}{3} - \frac{w_1^{\,4}}{4} - \frac{w_2^{\,4}}{12}$$

To give the explicit formula more directly, the summation in equation (2) can be made more explicit if we collect terms by w_j. This gives

$$P_n = \pi_n - \frac{w_1^{\,n}}{n}\sigma_1 + \sum_{j=1}^{n-1}\left[\frac{w_{j+1}^{\,n-j+1}}{n-j+1}\sigma_j - \frac{w_{j+1}^{\,n-j}}{n-j}\sigma_{j+1}\right] + \sigma_n$$

Making use of the identity $s_{j+1} = w_{j+1}\, s_j + p_j$, the summation splits into a sum over the p_j and a sum over the s_j as follows

$$P_n = \pi_n - \frac{w_1^{\,n}}{n}\sigma_1 + \sum_{j=1}^{n-1}\left[\frac{w_{j+1}^{\,n-j}}{n-j}\pi_j\right] - \sum_{j=1}^{n-1}\left[\frac{w_{j+1}^{\,n-j+1}}{(n-j)(n-j+1)}\sigma_j\right] + \sigma_n$$

The argument of the first summation when $j = n{-}1$ is simply $w_n \pi_{n-1} = \pi_n$, so this cancels with the leading p_n term. Also, since $\sigma_1 = \pi_0 = 1$, we can bring the term involving w_1 into the first summation, shift the indices, and combine one power of w_j with π_{j-1}, to give the result

$$P_n = \sigma_n - \sum_{j=1}^{n-1}\left[\frac{w_j^{\,n-j}}{n-j+1}\pi_j\right] - \sum_{j=1}^{n-1}\left[\frac{w_{j+1}^{\,n-j+1}}{(n-j)(n-j+1)}\sigma_j\right]$$

(For comparison, recall that the circular case gave simply $P_n = \sigma_n$.) Each s_j represents j terms, so the total number of terms (i.e., individual products) in P_n is

$$n + (n+1) + \frac{(n-1)n}{2} = \frac{n^2 + 3n - 2}{2}$$

However, even though the number of terms increases as the square of n, the computation of P_n can be completed in O(n) steps, because the σ_j quantities can be computed recursively in O(n) steps.

The other patterns of intersection that can arise when n intervals of various lengths are placed with their starting points on the unit interval are discussed in Meeting Combinations and Interval Graphs. The probabilities of these combinations in the case of equal intervals are derived in Probability of Intersecting Intervals.

Chapter 8

Symmetric Pseudoprimes

Fermat's Little Theorem states that for any integer c, if p is a prime, then $c^p - c$ is divisible by p. This is a necessary but not quite sufficient condition for primality, because there are (rare) composites that satisfy this type of test for certain values of c. For example, $2^{341} - 2$ is divisible by 341, even though 341 is composite. This is why 341 is called a pseudoprime relative to the base 2. There are even composites, called Carmichael numbers, that are pseudoprimes relative to *every* integer base. The smallest Carmichael number is 561.

Another way of stating Fermat's Little Theorem is that, given a polynomial of the form $f(x) = x - c$, if p is a prime, then the sum of the pth powers of the roots of f is congruent to the sum of the first powers (mod p). Since f has just the single root c, this may seem like a trivial restatement of the previous one, but it can be immediately generalized to integers of any degree, yielding much stronger primality tests. For example, consider the Fibonacci polynomial $x^2 - x - 1$, and let a,b denote the roots. The nth term of the Lucas sequence is just $L_n = a^n + b^n$, and so, since $a + b = 1$, it follows from Fermat's Little Theorem that $L_p = 1$ (mod p) for every prime p. A composite number c such that $L_c = 1$ (mod c) is called a pseudoprime relative to the polynomial $x^2 - x - 1$. The smallest such number is 705.

Most discussions of generalized psuedoprime tests have imposed just the congruence condition on the sum of the nth powers of the roots, i.e., on the first symmetric function of the roots. A better generalization is to impose the appropriate congruence condition on *all* symmetric functions of the roots. This is based on the following generalization of Fermat's Little Theorem:

For any monic polynomial f with integer coefficients, if θ is a (complex) root of f and p is a prime, then $f(\theta^{p}) = 0$ (mod p).

Lucas considered this generalization, observing that if s(k) denotes the sum of the kth powers of the roots of f then we have s(np)´I s(n) (mod p) for any integer n and prime p. In particular he gave the (insufficient) primality criterion s(N)´I 0 (mod N) for the polynomial f(x) = x³ □ x □ 1. This criterion was also discussed by Perrin, and again by Shanks and Adams, who defined an "acceptable composite" as a composite integer N such that the six quantities s(±N), s(±(N+1)), and s(±(N−1)) satisfy certain congruence conditions (mod N). They reported that, at least for integers up to $50(10)^9$, the congruences involving s(±(N+1)) and s(±(N−1)) were superfluous, so the test essentially consisted of the requirement s(±N)′I s(±1) (mod N). Although this "two-sided" test can easily be applied to polynomials of higher degree, it does not fully embody the primality criterion implicit in a given polynomial of degree greater than 3 (or of cubics for which the constant coefficient is not equal to ±1). In general, we expect a polynomial with d distinct roots to represent d non-redundant primality conditions, one for each root.

Accordingly, a *symmetric pseudoprime* relative to f is defined as any composite integer c such that $f(\theta^c) = 0$ (mod c) for every root θ of f. In other words, given a monic polynomial f with integer coefficients, a symmetric pseudoprime relative to f is defined as a composite integer N such that every elementary symmetric function of the Nth powers of the roots of f is congruent (mod N) to the same function of the first powers.

Equivalently, we could define symmetric pseudoprimes in terms of Newton's sums for the roots of a polynomial. Let f denote a monic polynomial of degree d with integer coefficients, and let s(k) denote the sum of the kth powers of the roots of f. Lucas observed that if p is a prime then $s(p^k) = s(k)$ (mod p) for every integer k. A composite integer c such that $s(c^k) = s(k)$ (mod c) for every positive integer k is a symmetric pseudoprime relative to f. For example, the smallest symmetric pseudoprime relative to f(x) = x–2 is 341 = (11)(31). The relative rarity of these pseudoprimes, along with the fact that s(N) (mod N) can be evaluated in O[log(N)] multiplications modulo N, makes them useful for primality testing. Symmetric pseudoprimes tend to be more rare relative to polynomials with larger Galois groups. The Galois group of the quintic polynomial

$$f(x) = x^5 - x^3 - 2x^2 + 1$$

is the fully symmetric group S_5. The smallest symmetric pseudoprime relative to this polynomial that is the product of two splitting primes is

$$N = 2258745004684033 = (27439297)(82317889)$$

The basic propositions and computational techniques associated with general linear recurrences and symmetric pseudoprimes are presented, along with specific examples relative to selected polynomials of degrees 1 to 5. The final section describes complete congruence conditions on the terms of arbitrary linear recurring sequences.

BASIC PROPOSITIONS ON SYMMETRIC PSEUDOPRIMES

NOMENCLATURE AND DEFINITIONS

Let Z[x] denote the set of polynomials (of finite degree) with integer coefficients. For any given monic polynomial f(x) ∈ Z[x] of degree d, let $Z[x]_f$ denote the elements of Z[x] modulo f. This signifies that two elements u(x), v(x) of Z[x] are equivalent in $Z[x]_f$ if and only if u(x) – v(x) is divisible by f(x).

Proposition: Each element of $Z[\theta]_f$ has a unique expression of degree d–1.

Proof: Letting the modulus polynomial f(x) be

$$x^d = -c_1 x^{d-1} - c_2 x^{d-2} - \ldots - c_d$$

we see that any power of x greater than or equal to d can be expressed in terms of lower powers of x while preserving equivalence in $Z[x]_f$ by making the substitution Thus, every element of $Z[x]_f$ can be reduced to a polyomial of degree ≤ d–1. The reduced polynomial is unique because if any two (distinct) such polynomials were equivalent, it would imply that their difference, a non-vanishing polynomial of degree d-1, is divisible by f(x), which is impossible. ◊

Thus each element of $Z[x]_f$ can be expressed in the form

$$a_0 + a_1 x + a_2 x^2 + \ldots + a_{d-1} x^{d-1}$$

where $a_i \in Z$. If the coefficients a_k are considered as elements of Z_n (that is, the integers modulo n), then the set of polynomials of this form is denoted as $Z_n(x)_f$. For any u,v ∈ Z[x] and n ∈ Z, the notation u = v (mod f,n) signifies that u(x) and v(x) are equivalent elements of $Z_n[x]_f$.

The following generalization of Fermat's Little Theorem is the basis for the definition of a symmetric pseudoprime:

Proposition: If p is a prime, then $f(x^p) = 0$ (mod f,p).

Proof: Raising both sides of the congruence f(x) = 0 (mod f,p) to the power p, noting that all cross-product coefficients in the multinomial expansion are divisible by p, and that $c_j^p = cj$ (mod p), we have the result. ΰ

Definition: A positive composite integer N such that $f(x^N) = 0$ (mod f;N) is called a *symmetric pseudoprime* relative to f.

BASIS SEQUENCE COMPUTATIONS

For any non-negative integer n there is a unique set of integers $b_k(n)$ such that

$$x_n = \sum_{k=0}^{d-1} b_k(n) x^k$$

For each k the sequence of integers $b_k(n)$ (n = 0,1,2,...) is called a basis sequence of f. Each basis sequence satisfies the fundamental recurrence relation of f

$$\sum_{j=0}^{d} c_j b_k(n-j) = 0$$

with the initial values $b_k(j) = \begin{cases} 0 \text{ if } k \neq j \\ 1 \text{ if } k \neq j \end{cases}$

for j = 0, 1,.., d-1.

Proposition: For any non-negative integers $n_1, n_2,..., n_j$ and r, and for k = 0, 1,.., d-1, we have the identity

$$b_k\left(n_1+\ldots+n_j+r\right)=b_k\left(\alpha_1+\ldots+\alpha_j+r\right)b\alpha_1\left(n_1\right)\ldots b\alpha_j\left(n_j\right)$$

where summation from 0 through d–1 is implied over each α_i.

Proof: By direct multiplication of the basis representations of each power of x we have

$$x^{n_1}x^{n_2}\cdots x^{n_j}=B_0x^0+B_1x^1+B_2x^2\cdots+B_{j(d-1)}x^{j(d-1)}$$

where B_g equals the sum of all products of the form

$$b_{\alpha 1}\left(n_1\right)b_{\alpha 2}\left(n_2\right)\cdots b_{\alpha j}\left(n_j\right)$$

with $a_1 + a_2 +... + a_j = g$. Substituting the representation for each power of x on the right side of the previous expression, the coefficient of x^k in the representation of $x^{n_1}x^{n_2}\ldots x^{n_j}$ is given by

$$b_k\left(n_1+n_2+\ldots+n_j\right)=\sum_{g=0}^{j(d-1)}B_g b_k(g)$$

$$=\sum_{\alpha_1=0}^{d-1}\sum_{\alpha_2=0}^{d-1}\cdots\sum_{\alpha_j=0}^{d-1}b_k\left(\alpha_1+\alpha_2+\ldots+\alpha_j\right)b_{\alpha_1}\left(n_1\right)b_{\alpha_2}\left(n_2\right)\cdots b_{\alpha j}\left(n_j\right)$$

Clearly, this equality still holds if the argument of the left hand side and of the first term in the summation are both increased by any integer r. ◊

Notation: Throughout this paper, whenever a Greek symbol appears both as a sequence index and in a sequence argument in a single product, summation over that symbol from 0 through d–1 is implied. Similarly, whenever a Latin symbol appears both as an index and an argument in a given product, summation over that symbol from 0 through d is implied (except when other summation limits are explicitly specified). For computational purposes, a useful special case of (2) can be expressed using the index-argument convention as

$$b_j(2n+r)=b_j(\alpha+\beta+r)b_\alpha(n)b_\beta(n)$$

By applying this formula $\log_2 N$ times, with r equal to either 0 or 1 at each step according to the binary bits of N, we can efficiently compute the basis elements $b_j(N)$ (j = 0, 1,.., d–1). The quantities $b_j(\alpha+\beta+r)$ are fixed constants for any given recurrence, so the only full multiplications required in each evaluation of (3) are the d(d+1)/2 distinct products of the form $b_\alpha(n)b_\beta(n)$.

Once the basis elements $b_j(N)$ have been determined, the basis representations of x^{2N}, x^{3N},.., x^{dN} can be found by applying the formula

$$b_j(kN) = b_j(\alpha+\beta)b_\alpha(k-1)N)b_\beta(N)$$

with k = 2, 3,.., d. With these we can determine whether N satisfies the congruence $f(x^N) = 0$ (mod f,N), which is equivalent to the set of congruences

$$c_k b_j((d-k)N) \equiv 0(\text{mod } N)\, j = 0, 1, \ldots, d-1$$

If N satisfies, then either N is a prime or N is a symmetric pseudoprime relative to f.

ELEMENTARY SYMMETRIC FUNCTIONS

Let $\theta_1, \theta_2, \ldots \theta_d$ denote the complex roots of f, let f_n denote the monic polynomial whose roots are the nth powers of the roots of f, and let $c_j(n)$ denote the coefficient of x^{d-j} in $f_n(x)$. The jth elementary symmetric function of the roots of f_n is defined as the sum of all products of the roots taken j at a time, and is denoted by $s_j(n)$. (For convenience we define $s_0(n) = 1$ for all n.) It follows that

$$s_j(n) = (-1)^j c_j(n) \qquad (j = 0, 1, \ldots d)$$

Hereafter, whenever s(n) is written without subscript, it is understood to signify the first symmetric function, i.e., the sum of the kth powers of the roots of f_n.

The index-argument summation convention previously defined for basis sequences also applies to both of the sequences $s_j(n)$ and $c_j(n)$, and to combinations of these sequences. For example, $s(\alpha)b_\alpha(n)$ signifies the sum

$$s(0)b_0(n) + s(1)b_1(n) + \ldots + s(d-1)b_{d-1}(n)$$

Proposition: The integer N is a prime or a symmetric pseudoprime relative to f if and only if $s_j(N) = s_j(1)$ (mod N) for j = 0, 1,.., d.

Proof: The condition $s_j(N) = s_j(1)$ (mod N) for j = 0, 1,.., d implies that the coefficients of f and f_N are congruent (mod N). Therefore, since each x^N is a root of f_N, we also have $f(x^N) = 0$ (mod f,N), proving that N is a symmetric pseudoprime relative to f. Conversely, if N is a symmetric pseudoprime relative to f, we have $f(\theta_i^N) = 0$ (mod f,N) for each root θ_i^N of f_N, which implies that the coefficients of f are congruent to the coefficients of f_N (mod N). ◊

In view of Proposition we have the following alternative definition of symmetric pseudoprimes:

Definition: Given a monic polynomial f of degree d with integer coefficients, a symmetric pseudoprime relative to f is defined as a positive composite integer N such that every elementary symmetric function of the Nth powers of the roots of f is congruent (mod N) to the same function of the first powers.

We will occassionally compare symmetric pseudoprimes with "ordinary" pseudoprimes, defined as follows:

Definition: Given a monic polynomial f of degree d with integer coefficients, an *ordinary pseudoprime* relative to f is defined as a positive composite integer N such that the sum of the Nth powers of the roots of f is congruent (mod N) to sum of the first powers.

For example, if $\theta_1, \theta_2, \theta_3$ are the roots of a cubic polynomial f, then an ordinary pseudoprime relative to f is a composite integer N co-prime to 6 and such that

$$\theta_1{}^N + \theta_2{}^N + \theta_3{}^N \equiv \theta_1 + \theta_2 + \theta_3 \pmod N$$

whereas a symmetric pseudoprime relative to the same cubic f is a composite integer N such that

$$\left.\begin{aligned} \theta_1{}^N + \theta_2{}^N + \theta_3{}^N &\equiv \theta_1 + \theta_2 + \theta_3 \\ \theta_1{}^N\theta_2{}^N + \theta_1{}^N\theta_3{}^N + \theta_2{}^N\theta_3{}^N &\equiv \theta_1\theta_2 + \theta_1\theta_3 + \theta_2\theta_3 \\ \theta_1{}^N + \theta_2{}^N + \theta_3{}^N &\equiv \theta_1\theta_2\theta_3 \end{aligned}\right\} \pmod N$$

Proposition: If N is a prime or a symmetric pseudoprime relative to f, we have $s_j(kN) = s_j(k) \pmod N$ $(j = 0, 1,.., d)$ for *every* positive integer k.

Proof: The elementary symmetric functions of θ_i^k (i = 1, 2,.., d) can be expressed as polynomials in the elementary symmetric functions of the values θ_i. Similarly, the elementary symmetric functions of θ_i^{Nk} (k = 1, 2,.., d) can be expressed by the same polynomials in the elementary symmetric functions of the values θ_i^N. By Proposition each elementary symmetric function of the θ_i is congruent (mod N) to the same function of the θ_i^N, so the elementary symmetric functions of the values θ_i^k are all congruent (mod N) to the same functions of the values θ_i^{Nk}. ◊

Proposition: A positive composite integer N co-prime to d! is a symmetric pseudoprime relative to f if and only if $s(kN) = s(k) \pmod N$ for k = 1, 2,.., d.

Proof: The necessity of the condition follows immediately from Proposition. To prove sufficiency, consider "Newton's relations"

$$c_j(n) = -\frac{1}{j}\sum_{k=0}^{j-1} c_k s(n(j-k)) \qquad j = 1, 2, ..., d$$

for any integer n. Given the set of values s(n), s(2n),... s(dn) modulo some positive integer M, these relations uniquely determine the coefficients of f_n (mod M), provided that M is co-prime to d! (to ensure that all the divisions by j are unique modulo M). Therefore, if $s(kN) = s(k)$ (mod N) for k = 1, 2,.., d, it follows that the coefficients of f are congruent (mod N) to the coefficients of f_N, which proves that for any $u \in Z[x]$ we have $f(u) = f_N(u)$ (mod f,N). In particular, $f(\theta^N) = f_N(\theta^N) = 0$ (mod f,N), so N is a symmetric pseudoprime relative to f. ◊

REVERSIBLE SEQUENCES

If the constant coefficient c_d of f equals +1 or -1, then the values of $c_1(k)$ are integers for all integers k, positive and negative. In this case, the sequence $c_1(k)$ and the polynomial f are both called reversible.

Proposition: If N is a prime or a symmetric pseudoprime relative to f, and the coefficient c_d of f is co-prime to N, then $s_j(kN) = s_j(k)$ (mod N) (j = 0, 1,.., d) for every integer k, positive and negative.

Proof: Since c_d is co-prime to N, the fundamental recurrence relations can be solved uniquely for s(n-d) and s(N(n-d)) respectively, and the corresponding coefficients remain congruent (mod N). Therefore, since the two sequences have congruent initial values and congruent recurrence coefficients (mod N), the sequences are entirely congruent (mod N) in both directions, positive and negative. The coefficients of the polynomials f_n and f_{Nn} are congruent for all n, positive and negative. ◊

In particular, Proposition applies to every reversible sequence for any prime or symmetric pseudoprime N, because $c_d = \pm 1$ is obviously co-prime to N.

Proposition: If f is reversible, then an odd positive composite integer N co-prime to (d-1)! is a symmetric pseudoprime relative to f if and only if s(kN) = s(k) (mod N) for k = 1, 2,.., d-1.

Proof: The necessity follows from Proposition. To prove sufficiency, note that the d-1 specified congruences, together with the trivial congruence s(0N) = s(0), provide a complete set of congruent "initial values" for the two sequences s(kN) and s(k). Also, by the same argument used to prove Proposition, "Newton's relations" uniquely determine the coefficients $c_1()$ through $c_{d-1}()$ of the two recurrence relations. In addition, since $c_d = \pm 1$, we have $c_d(k) = c_d(Nk)$ for every odd integer N. Thus, a complete set of initial values and all the recurrence coefficients of the two sequences are congruent (mod N), so we have s(k) = s(kN) (mod N) for every integer k. By Proposition, it follows that N is a symmetric pseudoprime relative to f. In the following proposition and proof we use the symbol {x} signify the greatest integer less than or equal to x.

Proposition: If f is reversible, then an odd positive composite integer N co-prime to {d/2}! is a symmetric pseudoprime relative to f if and only if s(kN) = s(k) (mod N) for k = ±1, ±2,..,ε{d/2}, where ε = ±1 if d is odd and +1 if d is even.

Proof: The coefficient $c_d(n)$ of f_n is related to the roots of f by

$$c_d(n) = (-1)^d (\theta_1\theta_2 \ldots \theta_d)^n$$

Substituting $(-1)^d c_d$ for $(\theta_1\theta_2 \ldots \theta_d)$n gives

$$c_d(n) = (-1)^d \left((-1)^d c_d\right)^n$$

Now, observing that the roots of f_{-n} are just the inverses of the roots of f_n, we have

$$c_j(n) = c_d(n) c_{d-j}(-n) = (-1)^d \left((-1)^d c_d\right)^n c_{d-j}(-n)$$

Given the values of s(kN) (mod N) for k=±1, ±2,.., ε{d/2}, equation uniquely determines the values (mod N) of c_j(N) and c_j(-N) for j = 1, 2,.., {d/2}, provided that N is co-prime to {d/2}! (to make the divisions by j unique (mod N)). Then by equation we can use the c_j(-kN) to compute the values of c_j(kN) (mod N) for j = {d/2} + 1,.., d. Thus, the two sequences s(k) and s(kN) have a set of d consecutive values congruent (mod N) (including the equality with k = 0), and their recurrence coefficients are all congruent, so the entire sequences are congruent (mod N). ◊

COEFFICIENT SEQUENCES

The basis sequence relations provide an efficient and convenient means of evaluating the elements of the c_1(k) sequence for any given polynomial f, it is sometimes useful to have formulas expressed entirely in terms of the c_1(k) (or s(k)) sequence elements themselves. The fundamental recurrence for f_n can be written as

$$c_j(n)c_1(n(k-j))=0$$

Since the value of k is arbitrary, we can set k = m + (r/n) for any integers m and r to give

$$c_j(n)c_1(n(m-j)+r)=0$$

Thus, the recurrence relation of f_n applies not only to the sequence c_1(nk) (k = 0, 1, 2,...), but also to the sequences c_1(nk+r) for any integer r. Furthermore, the coefficients of this recurrence relation can be expressed in terms of the c_1(nk) sequence elements by solving equation explicitly for the c_j(n), which gives

$$c_0(n)=1$$

$$c_1(n)=c_1(n)$$

$$c_2(n)=\frac{c_1(n)^2+c_1(2n)}{2}$$

$$c_3(n)=\frac{c_1(n)^3+3c_1(n)c_1(2n)+2c_1(3n)}{6}$$

In general, we have (Waring's formula)

$$c_m(n)=\sum\frac{\prod_{j=1}^{m}c_1(jn)a_j}{\prod_{j=1}^{m}j^{aj}j!}$$

where the summation is evaluated over all sets of m non-negative integers a_j such that

$$\prod_{j=1}^{m}j^{aj}j=m$$

These equations can be used to implement a $\log_2 N$ algorithm for computing $c_1(N)$. For example, we can compute the recurrence coefficients $c_k(n)$ for n = 1, 2, 4, 8,.., 2s, where s = $[\log_2 n]$, in a "bootstrap" procedure, and then use these coefficients to compute $c_1(N)$. Algorithms such as this can be applied to polynomials of any degree, but they generally involve more full multiplications per step, and more elaborate bookkeeping, than the basis sequence algorithm.

However, in some special cases it's possible to devise a coefficient sequence algorithm that is as efficient as a basis sequence algorithm. If the sequence is reversible, then the terms of $c_j(n)$ can be replaced by $\pm c_{d-j}(-n)$ as given by equation. Thus, can be expressed entirely in terms of the coefficient sequences $c_j(\pm n)$ (or symmetric functions $s_j(\pm n)$) with $j \le (d/2)$. This is particularly convenient for the case of a reversible cubic.

RELATIONS BETWEEN BASIS SEQUENCES AND SYMMETRIC FUNCTIONS

Given any sequence of values A(k) (k = 0, 1, 2,...) that satisfies the fundamental recurrence relation of f, we have

$$A(n) = A(\alpha) b_\alpha(n)$$

In particular, $s(n) = s(\alpha) b_\alpha(n)$

Another expression for the first symmetric function s(n) is given by a "contraction" of the basis sequences

$$s(n) = b_\alpha(n+\alpha)$$

If we define the sequence of square matrices B_n (n = 0,1,2,..) with components given by $B_n(i,j) = b_j(n+i)$, then s(n) can also be regarded as the *trace* of B_n.

It follows directly from that

$$b_\alpha(n+\alpha) b_\beta(m+\beta) = s(n)s(m)$$

If the symbols α and β are permuted in the arguments, we have the slightly less obvious identity

$$b_\alpha(n+\beta) b_\beta(m+\beta) = s(n+m)$$

Equations are just two special cases of the following proposition.

Proposition: Let {i,j,..,k} be a permutation of {1,2,..,q} composed of the cyclic components $C_1, C_2,.., C_t$. Then

$$b_{\alpha_1}(n_1+\alpha_i) b\alpha_2(n_2+\alpha_j) \cdots b\alpha_q(n_q+\alpha_k) = \prod_{u=1}^{t} s\left(\sum_u n\right)$$

where $\sum_u n$ signifies the sum of all the n_r such that r is in the cycle C_u.

Proof: Proposition gives both

$$b_\alpha(k+\gamma) = b_\alpha(\sigma+\gamma)b_\sigma(k)$$

and

$$b_\alpha(n+m+\beta) = b_\alpha(\sigma+\gamma)b_\sigma(n)b_\gamma(m+\beta)$$
$$= b_\alpha(\sigma+\gamma)b_\gamma(n+\beta)b_\sigma(m)$$

Substituting Equations (with k = n or m) gives

$$b_\alpha(n+m+\beta) = b_\alpha(n+\gamma)b_\gamma(m+\beta)$$
$$= b_\gamma(n+\beta)b_\alpha(m+\gamma)$$

which is just the rule of multiplication for the components of matrices. Thus, equation is equivalent to the matrix product $B_{n+m} = B_n B_m = B_m B_n$. Setting $\alpha = \beta$ in (12) gives, by contraction, equation. Repeated substitutions of either leads to the general result. ◊

For example, the quantity

$$b_\alpha(n_1+\mu)b_\beta(n_2+\beta)b_\mu(n_3+\alpha)$$

corresponds to the permutation $[\alpha,\beta,\mu] \to [\mu,\beta,\alpha]$, the cyclic components of which are $[\alpha,\mu] \to [\mu,\alpha]$ and $[\beta] \to [\beta]$. Thus, Proition 10 gives the identity

$$b_\alpha(n_1+\mu)b_\beta(n_2+\beta)b_\mu(n_3+\alpha) = s(n_2)s(n_1+n_3)$$

PRIME TYPES RELATIVE TO F(X)

We will classify each prime p relative to the polynomial f according to how f factors in $Z_p[x]$. The type of each prime p will be denoted by a symbol of the form $\{1_{k_1}\ 2_{k_2} \cdots d_{k_4}\}$, signifying that in the ring Z_p the polynomial f is the product of k_1 irreducible factors of degree 1, and k_2 irreducible factors of degree 2, and so on. If a subscript is zero, the term will normally be omited from the symbol. For example, if f is the product of one irreducible cubic factor and two linear factors in $Z_p[x]$ we will say that p is a prime of type $\{1_2 3_1\}$ relative to f.

The type of the prime p relative to f is also related to the general period of the characteristic recurrence of f (mod p), defined as the least positive integer n such that $b_j(n) = b_j(0)$ (mod p) (j = 0, 1,.., d-1). We will use the symbol τ_p to signify the general period of f (mod p).

Recall that Proposition gives $f(x^p) = 0$ (mod f,p). By iterating the method of proof, this result can be extended to

$$f\left(x^{p^k}\right) \equiv 0 \pmod{f, p}$$

for any non-negative integer k. Thus, each number θ^{p^k} is a root of f in $Z_p[x]_f$.

Proposition: If f is irreducible in $Z_p[x]$, then the sequence of values x^{p^k} (mod f,p) has a period of d, i.e.,

$$x^{(p^d)} \equiv x \pmod{f, p}$$

Proof: If p is a prime, then $Z_p[x]_f$ is isomorphic to the finite field of order p^d. It is well known that if f is irreducible in Z_p, then the proposition follows.

Since f is irreducible in $Z_p[x]$, it follows that the ring of polynomials $Z_p[x]_f$ possesses unique factorization, and therefore f has exactly d distinct factors in $Z_p(\theta)[x]$. Dividing by x gives

$$x^{(p^d - 1)} \equiv 1 \pmod{f, p}$$

This shows that τ_p must divide p^{d-1}. For some common special cases this can be further refined by noting that the values of x^{p^k} (k = 0, 1,.., d-1) are the d roots of f in $Z_p[x]_f$. For example, since the product of the roots equals c_d, it follows that if $c_d = (-1)^d$, then τ_p divides $1 + p + .. + p^{d-1} = (p^d - 1)/(p - 1)$. On the other hand, if $c_d = -(-1)^d$, then $x^{(1+p+..+p^{d-1})} \equiv -1$, so τ_p divides $2(p^d - 1)/(p - 1)$.

If f is reducible in $Z_p[x]$, then the general period of its recurrence divides the least common multiple of the periods of the irreducible factors. For example, if p is a prime of type {2,3} relative to f, then t_p divides $\text{lcm}[(p^2 - 1), (p^3 - 1)]$. Since $p^2 - 1$ and $p^3 - 1$ have the common factor $p - 1$, it follows that τ_p divides

$$\frac{(p^2 - 1)(p^3 - 1)}{(p - 1)} = (p + 1)(p^3 - 1)$$

In general, the sequence $x, x^p, x^{p^2}, x^{p^3}, \ldots$ has a period that divides the least common multiple of the degrees of the irreducible factors of f in $Z_p[x]$. Thus, we can write

$$\tau_p = \frac{\text{lcm}\left[\left(p_1^{d_1} - 1\right), \ldots, \left(p_1^{d_\tau} - 1\right)\right]}{k_p}$$

where k_p is a divisor associated with the prime p.

The number of distinct roots of f in $Z_p[x]_f$ can exceed the degree of f if f is reducible, because then unique factorization does not hold. For example, if p is of the type $\{2_1, 3_1\}$ relative to f, then each of the six values $x, x^p, \ldots, x^{p^5}$ is a distinct root of f in $Z_p[x]_f$. However, in related rings these six roots are not distinct. We have monic irreducible polynomials g and h of degree 2 and 3 respectively such that f = gh (mod p), and we find that the sequence $x, x^p, x^{p^2}, x^{p^3}, \ldots,$ has period 2 in the ring $Z_p[x]_g$ and period 3 in the ring $Z_p[x]_h$. Specifically we have

$$\left.\begin{array}{l} x \equiv x^{\left(p^2\right)} \equiv x^{\left(p^4\right)} \\ x^p \equiv x^{\left(p^1\right)} \equiv x^{\left(p^3\right)} \end{array}\right\} \pmod{g.p}$$

and

$$\left.\begin{array}{l} x \equiv x^{\left(p^1\right)} \\ x^p \equiv x^{\left(p^4\right)} \\ x^{\left(p^2\right)} \equiv x^{\left(p^3\right)} \end{array}\right\} \pmod{h.p}$$

Letting a_1,a_2 denote the roots of g, and β_1,β_2,β_3 denote the roots of h, we can say the prime p induces the permutation $[\alpha_1,\alpha_2,\beta_1,\beta_2,\beta_3] \rightarrow [\alpha_2,\alpha_1,\beta_2,\beta_3,\beta_1]$ on the roots of f. This permutation has two cyclic components, α_i and β_i, with periods 2 and 3 respectively. The periods of the cyclic components of the permutation induced by a prime p determines the "type" of p (relative to f). Furthermore, the cyclic structure corresponding to each type determines the asymptotic proportion of primes of that type.

If each of the d! possible permutations of the d roots of f is equally likely to be induced by any given prime, we can see immediately the expected asymptotic proportion of primes of each type, based on the proportion of the permutations that possess each of the cyclic structures, corresponding to the partitions of d. For example, with d = 5 there are seven possible cyclic structures: 5, 4+1, 3+2, 3+1+1, 2+2+1, 2+1+1+1, and 1+1+1+1+1. The cyclic structure corresponding to "5" is just a single cycle of length 5, so the first element can go to one of four places, then one of three, then one of two, and then the one remaining place before it repeats. Hence there are 4! permutations with this cyclic structure, which implies that 4!/5! = 1/5 of all primes will be of type $\{5_1\}$. At the other extreme, the cyclic structure "1+1+1+1+1" is given by only one permutation, so only 1/5! = 1/120 of all primes would be splitting primes. In general, assuming each of the d! permutations is equally likely, the number of permutations for the prime type $\left\{a_{a_1}\, 2_{a_2} \cdots d_{a_4}\right\}$ is given by the multinomial coefficient

$$\frac{d!}{1^{a_1}a_1!2^{a_2}a_2!\cdots d^{a_4}a_d!}$$

so the asymptotic density of each type is just 1 over the denominator.

However, for many polynomials of degree d the set of possible permutations of the roots is more restricted. The most general polynomial of degree d allows all d! permutations, in which case we say the (Galois) group of the polynomial is the fully symmetric group S_d, but there are also polynomials of degree d whose Galois group is a subgroup of S_d. For example, a quintic that is the product of 5 linear factors in Z[x] will obviously split completely in every ring $Z_p[x]$. In this case the basic Galois group is just the trivial identity group, I, with only one element, so every prime will be of the type $\{1_5\}$. On the other hand, some polynomials have Galois groups G that are intermediate between the identity group I and the fully symmetric group S_d. In those cases we can compute the asymptotic density of each "type" simply by computing the fraction of the permutations contained in G that have the cyclic structure of the respective type.

RECURRENCES MODULO COMPOSITES

Given the general period of the recurrence of f for each prime in the factorization of $n = p_1^{a_1} p_2^{a_2} \cdots p_k^{a_k}$, the general period of the recurrence (mod n) is given by

$$\tau_n = \text{lcm}\left[p_1^{a_1-1}\tau_{p_1}, p_2^{a_2-1}\tau_{p_2}, \dots, p_k^{a_k-1}\tau_{pk} \right]$$

where "lcm" signifies the least common multiple. For example, given the prime periods

$$\tau_2 = 7 \qquad \tau_3 = 13 \qquad \tau_5 = 24$$

relative to the polynomial $f(x) = x^3 - x - 1$, the general period of the composite integer 360 = $(2)^3(3)^2(5)$ equals the least common multiple of $2^2(7)$, $3^1(13)$, and $5^0(24)$. Therefore, $\tau_{360} = 2184$.

Using the equation above, we can determine highly "resonant" composite moduli relative to a given polynomial. For example, relative to $x^3 - x - 1$ we have $\tau_{449} = 224 = 2^5(7)$, so the general period of the recurrence modulo $2^6(449)$ equals the least common multiple of $2^5(7)$ and $2^5(7)$, which is just 224. This shows that all the elements of any given sequence satisfying the recurrence $A_n = A_{n-2} + A_{n-3}$ fall into no more than 224 of the residue classes modulo 28736, which is less than 1% of all the residue classes.

To construct another example, we can begin with the fact that

$$\tau_{769} = 591360 = (2^9)(3)(5)(7)(11)$$

and search for other primes whose periods divide this number. We find the following values

$$\tau_2 = 7 \qquad \tau_5 = 2^3 \cdot 3 \qquad \tau_7 = 2^4 \cdot 3 \qquad \tau_{11} = 2^3 \cdot 3 \cdot 5$$

$$\tau_{43} = 3 \cdot 7 \cdot 11 \qquad \tau_{211} = 2 \cdot 3 \cdot 5 \cdot 7 \qquad \tau_{449} = 2^5 \cdot 7 \qquad \tau_{463} = 2 \cdot 7 \cdot 11$$

$$\tau_{3361} = 2^5 \cdot 3.57 \qquad \tau_{6337} = 2^6 \cdot 3 \cdot 11 \qquad \tau_{9241} = 2^2 \cdot 3 \cdot 5 \cdot 711 \qquad \tau_{9857} = 2^7 \cdot 7 \cdot 11$$

Therefore, all the elements of any given sequence satisfying the recurrence $A_n = A_{n-2} + A_{n-3}$ fall into no more than 591360 distinct residue classes when evaluated modulo the number

$$2^{10} \cdot 5^2 \cdot 7^2 \cdot 11^2 \cdot 43 \cdot 211 \cdot 449 \cdot 463 \cdot 769 \cdot 3361 \cdot 6337 \cdot 9241 \cdot 9857$$

which equals

$$42711216852139799644112540232087040 0$$

In other words, the values of the sequence can fall into only 1 in $(7.223)10^{29}$ of all the residue classes.

SYMMETRIC PSEUDOPRIMES RELATIVE TO SELECTED POLYNOMIALS

This section presents examples of symmetric pseudoprimes relative to selected polynomials of degrees 1 through 5. For degrees 1, 2, and 3 the discussion focuses on familiar polynomials (such as the Fibonacci polynomial of degree 2 and Perrin's polynomial of degree 3) to show how these well known cases fit within the context of symmetric pseudoprimes. For higher degrees there do not appear to be such well established examples, so we have focused on irreducible polynomials with small integer coefficients and a variety of splitting field structures. In particular, we exhibit a symmetric pseudoprimes relative to a fully symmetric quintic.

POLYNOMIALS OF DEGREE 1

By Proposition, the symmetric pseudoprimes relative to any given first degree polynomial $f(x) = x - c$ ($c \in Z$) are the integers N such that $s(N) = s(1)$ (mod N), which is to say, $c^N = c$ (mod N). Therefore, the symmetric pseudoprimes relative to f are identical to the ordinary pseudoprimes relative to f.

All the symmetric pseudoprimes less than 10^4 relative to the first degree polynomials $f(x) = x - 2$, $f(x) = x - 3$, and $f(x) = x - 5$ are listed in Table. The entries listed in bold are Carmichael numbers, which are symmetric pseudoprimes relative to every first degree polynomial. The entries listed in parentheses are those divisible by c.

The importance of imposing congruence conditions on all the symmetric functions (not just the sum) of the roots can be shown by considering products of first degree polynomials. For example, the ordinary pseudoprimes relative to the polynomial

$$f(x) = (x-2)(x-3) = x^2 - 5x + 6$$

include not only the integers that are pseudoprimes relative to both $x - 2$ and $x - 3$, but also the integers n for which $2^n + 3^n = 2 + 3$ even though $2^n \neq 2$ and 3^n′H 3 (mod n). (The values listed in parentheses are those divisible by 2 or 3). Also shown are the values of $2^n 3^n$ (mod n), demonstrating that of all these ordinary pseudoprimes only 15 is a symmetric pseudoprime; the others each satisfy the congruence condition on the first symmetric function (the sum), but not on the second summetric function (the product). Table 3 gives the ordinary pseudoprimes less than 10^6 relative to $f(x) = (x - 2)(x - 3)(x - 5)$, excluding those that are pseudoprimes relative to all three of the individual polynomials x–2, x–3, and x–5. Notice that 50721 and 811921 satisfy the required congruence conditions on the third as well as the first symmetric function. Similarly, the integer 49 (not shown) satisfies the second and third congruence but not the first.

To complete the demonstration that no two of the congruences, for a general cubic, is sufficient to ensure that all three are satisfied, we need to show a case that satisfies the first two but not the third. We have not found such a case relative to $f(x) = (x - 2)(x - 3)(x - 5)$. However, relative to the polynomial

$$f(x)=(x+17)(x-23)(x-31)$$
$$=x^3-37x^2-250x+12121$$

the composite integer N = 703 = (19)(37) satisfies the congruence s(kN) = s(k) (mod N) for k = 1 and 2, but not for k = 3. To prove this, we first form the Nth powers of the roots (mod N) as follows:

$$(-17)^N=572 \qquad 23^N=384 \qquad 31^N=487 \quad (\text{mod } N)$$

The elementary symmetric functions of these powers are

$$s_1(N)=37 \qquad s_2(N)=498 \qquad s_3(N)=96 \quad (\text{mod } N)$$

Since 498 = -205 (mod N), we see that $s_1(N) = s_1(1)$ and $s_2(N) = s_2(1)$ (mod N). However, 96 is not congruent to -12121 (mod N), so N is not a symmetric pseudoprime relative to f. This fact can also be shown in terms of the sequence of $s_1(k)$ values (mod N) as follows

$$s(1)=37 \qquad s(2)=373 \qquad s(3)=489$$
$$s(1N)=37 \qquad s(2N)=373 \qquad s(3N)=584$$

Thus, the three congrunces together are a stronger test than just the first two.

POLYNOMIALS OF DEGREE 2

The odd symmetric pseudoprimes relative to any given second degree polynomial $f(x) = x^2 + c_1x + c_2$ ($c_i \in Z$) are the composite integers N for which the congruences s(N) = s(1) and s(2N) = s(2) are both satisfied (mod N). Noting that s(0) = 2, s(1) = $-c_1$, and s(2) = $c_1{}^2 - 2c_2$ allows us to express the conditions s(N) = s(1) and s(2N) = s(2) as follows

$$2b_0(N)-c_1\,b_1(N)\equiv -c_1$$
$$2b_0(2N)-c_1\,b_1(2N)\equiv -c_1{}^2-2c_2$$

These and all other congruences in this section are understood to be (mod N) unless some other modulus is specified. Notice that if N is co-prime to $2c_1$, then equation implies

$$\gcd(b_0(N),N)=\gcd(b_1(N)-1,N)$$

where gcd(u,v) signifies the greatest common divisor of u and v. Now, using the basis sequence relations described in Part 1 we can express $b_0(2N)$ and $b_1(2N)$ in terms of $b_0(N)$ and $b_1(N)$ as follows

$$b_0(2N)=b_0(N)^2-c_2\,b_1(N)^2$$
$$b_1(2N)=2b_0(N)b_1(N)-c_1\,b_1(N)^2$$

Substituting these expressions into gives

$$2b_0^2 - 2c_1b_0b_1 + \left(c_1^2 - 2c_2\right)b_1^2 \equiv \left(c_1^2 - 2c_2\right)$$

where the argument of each basis sequence symbol, when not explicitly specified, is understood to be N for the remainder of this section. Multiplying through by 2 and substituting the expression for $2b_0$ from equation

$$c_1^2\left(b_1 - 1\right)^2 - 2c_1^2b_1\left(b_1 - 1\right) + D\left(b_1^2 - 1\right) \equiv 0$$

where $D = c_1^2 - 4c_2$, the discriminant of f. Clearly, the congruence is satisfied if $b_1 = 1$ (mod N), meaning that N divides $b_1 - 1$. On the other hand, if N and $b_1 - 1$ are relatively prime, then we can divide the preceding congruence by $b_1 - 1$ to give

$$Db_1 \equiv -D$$

In this case, assuming D is coprime to N, we have $b_1 = -1$ (mod N). Now, since (1) gives $b_0 = (c_1/2)(b_1 - 1)$, it follows that for any prime or symmetric pseudoprime N co-prime to $2c_1D$, and assuming that N either divides or is co-prime to $b_0(N)$, we have

$$\left.\begin{array}{l} b_0(N) \equiv 0 \\ b_1(N) \equiv 1 \end{array}\right\} (\text{mod } N)$$

or

$$\left.\begin{array}{l} b_0(N) \equiv -c_1 \\ b_1(N) \equiv -1 \end{array}\right\} (\text{mod } N)$$

Of course, the divisibility assumptions on N used in this derivation are not implied by the definition of a symmetric pseudoprime, so the congruences are a stronger primality criterion.

Definition: A symmetric pseudoprime N relative to a second degree polynomial f is called *determinate* if $\gcd(2c_1D, N) = 1$ and $\gcd(b_0(N), N) = 1$ or N.

Clearly a determinate symmetric pseudoprime satisfies the equation. To further strenghten the criterion, we can specify which of these two conditions applies to any given N. If N is actually a prime, it follows from the discussion that equation is satisfied if f splits into linear factors in the field Z_N (which implies that D is a square modulo N), and that equation is satisfied if f is irreducible in Z_N (which implies that D is not a square modulo N).

To illustrate the preceeding topics, consider the familiar "Fibonacci polynomial" $f(x) = x^2 - x - 1$. Since this is reversible, we know that a composite integer N is a symmetric pseudoprime relative to f if and only if $s(N) = s(1)$ (mod N). All 86 such composites N less than 10^6 are listed in Table 4, along with the values of $b_0(N)$ and $b_1(N)$ (mod N), and $\gcd(N, b_0(N))$. Nineteen of these symmetric pseudoprimes can immediately be distinguished from primes because they are not determinate (eleven being divisible by the discriminant D = 5, and eighteen having 1 < gcd(N, b_0(N)) < N). Each of the remaining 67 symmetric pseudoprimes is determinate, but eleven of them (listed in parentheses) can be excluded because they exhibit the wrong congruences for their residue classes (mod 5). This leaves 56 composites which cannot be distinguished from primes based on any of the criteria discussed above.

There are second degree polynomials that exhibit a much lower density of symmetric pseudoprimes than the Fibonacci polynomial. For example, the polynomial

$$f(x) = x^2 - 4x - 9$$

has only 4 symmetric pseudoprimes less than 10^6, none of which is determinate. Notice that since this polynomial is not reversible, the single congruence s(N) = s(1) is not sufficient to ensure that N is a symmetric pseudoprime; we must also have s(2N) = s(2).

The first three symmetric pseudoprimes relative to polynomial (5) are listed below.

(mod N)

N	$b_0(N)$	$b_1(N)$	$\gcd(b_0(N), N)$
1105	340	936	85
2465	1479	494	493
15457	10701	2379	1189

The next symmetric pseudoprime relative to polynomial is 38503, making it the smallest determinate symmetric pseudoprime. (In comparison, the Fibonacci polynomial has 18 determinate symmetric pseudoprimes less than 10000.) There are only 69 determinate symmetric pseudoprimes relative to polynomial less than $8(10)^7$. Each of these satisfies the "splitting congruences". It follows that congruence (which is satisfied by half of all primes) is a sufficient primality criteria for integers less than $8(10)^7$.

The seven pseudoprimes appearing in parentheses can be distinguished from primes based on the values of the basis sequence elements b_0((N-1)/2) and b_1((N-1)/2). In a binary algorithm these two elements are always computed in the second-to-the-last step, so it's convenient to use these elements to strengthen the test. Letting m = (N-1)/2 we have

$$b_0(N) = -2c_2b_0(m)b_1(m) + c_1c_2b_1^2(m)$$

and

$$b_1(N) = b_0^2(m) - 2c_1b_0(m)b_1(m) + \left(c_1^2 - c_2\right)b_1^2(m)$$

If N is a prime, it must satisfy the equation. If it satisfies then (using the fact that gcd(N,c_2) = 1 for primes N > c_2) we have

$$b_1(m)\left[c_1b_1(m) - 2b_0(m)\right] \equiv 0$$

Since N is assumed prime, it follows that either N divides $b_1(m)$ or else $\gcd(N, b_1(m)) = 1$. If N divides $b_1(m)$, then clearly

$$\left.\begin{aligned} b_0\left(\frac{N-1}{2}\right) &\equiv 1 \\ b_1\left(\frac{N-1}{2}\right) &\equiv 0 \end{aligned}\right\} \pmod{N}$$

On the other hand, if $\gcd(N, b_1(m)) = 1$, it follows that

$$\left.\begin{aligned} Db_0^2\left(\frac{N-1}{2}\right) &\equiv c_1^2 \\ Db_1^2\left(\frac{N-1}{2}\right) &\equiv 4 \end{aligned}\right\} \pmod{N}$$

The above congruences are sufficient to exclude all 7 of the determinate pseudoprimes listed in parentheses.

Although we have not found any composites relative, they presumably exist, so we will derive the congruence conditions on $b_0((N-1)/2)$ and $b_1((N-1)/2)$ for this case. Again letting $m = (N-1)/2$, we can rearrange the terms and then square to give

$$4c_2^2\, b_0^2(m)\, b_1^2(m) \equiv \left(c_1 + c_1c_2b_1^2(m)\right)^2$$

Also, if we multiply by c_1 and c_2, we can eliminate the term $-2c_1c_2b_0(m)b_1(m)$ from these two equations to give

$$c_2\, b_0^2(m) \equiv c_2^2\, b_1(m) + \left(c_1^2 - c_2\right)$$

Multiplying through by $4c_2b_1(m)^2$, we can then substitute from equation to yield the following quadratic in $b_1(m)^2$:

$$c_2^2\left(c_1^2 - 4c_2\right)b_1^4(m) + c_2\left(4c_2 - 2c_1^2\right)b_1^2(m) + c_1^2 \equiv 0$$

Solving this for $b_1(m)^2$ gives the two possible values

$$c_2Db_1\left(\frac{N-1}{2}\right)^2 \equiv \begin{cases} c_1^2 \\ D \end{cases}$$

which correspond to

$$c_2Db_0\left(\frac{N-1}{2}\right)^2 \equiv \begin{cases} \left(c_1^2 - 2c_2\right)^2 \\ D\, c_1^2 \end{cases}$$

The Legendre symbols $[c_2/N]$ and $[D/N]$ determine which of these two possible solutions applies for any given prime N.

POLYNOMIALS OF DEGREE 3

By Proposition, a composite integer N co-prime to 6 is a symmetric pseudoprime relative to the third degree polynomial

$$f(x) = x^3 + c_1 x^2 + c_2 x + c_3 \qquad c_i \in Z$$

iff all three of the congruences

$$\left.\begin{array}{l} s(N) \equiv s(1) \\ s(2N) \equiv s(2) \\ s(3N) \equiv s(3) \end{array}\right\} \pmod N$$

are satisfied. If the polynomial is reversible (meaning that $c_3 = \pm 1$), then by Proposition an odd composite integer N is a symmetric pseudoprime relative to f iff the two congruences

$$\left.\begin{array}{l} s(N) \equiv s(1) \\ s(2N) \equiv s(2) \end{array}\right\} \pmod N$$

are satisfied. Moreover, if f is reversible, Proposition shows that a composite integer N is a symmetric pseudoprime relative to f iff the two congruences

$$\left.\begin{array}{l} s(N) \equiv s(1) \\ s(-N) \equiv s(-1) \end{array}\right\} \pmod N$$

are satisfied.

The third degree polynomial that has recieved the most attention with respect to pseudoprime tests is

$$f(x) = x^3 - x - 1$$

Following Shanks and Adams, we will refer to this as Perrin's polynomial. This particular polynomial seems to have been selected for study by Lucas and Perrin mainly because $c_1 = 0$, so s(p) is divisible by p for every prime p.

Perrin also remarked that the sequence grows more slowly than the Fibonacci sequence, making some manual computations easier.

The author's attention was drawn because its characteristic recurrence corresponds to the spiral tiling of equilateral triangles, just as the Fibonacci recurrence corresponds to the spiral tiling of perfect squares. These are the only two ways of partitioning the plane into a geometric progression of regular polygons.

Shanks and Adams found that, relative to Perin's polynomial, the smallest composite N such that s(N) (0 (mod N) is 271441.

They also noted that Perrin's sequence is reversible, and that the primality test is strengthened by requiring the reverse congruence s(-N) = s(-1) (mod N). Composite integers N that satisfy s(±N) = s(±1) and four other congruences involving the terms s(±(N+1)) and s(±(N-1)) (mod N) were called "acceptable composites". (It was reported in [4] that, at least up to $50(10)^9$, the congruences involving s(±(N+1)) and s(±(N-1)) did not strengthen of the test.) It is clear that the composite integers N such that s(±N) = s(±1) (mod N) for a reversible cubic are precisely the symmetric pseudoprimes relative to that cubic.

Shanks and Adams found that the smallest symmetric pseudoprime relative to Perin's polynomial is 27664033. Subsequently Shanks, et al, listed all 55 symmetric pseudoprimes less than $50(10)^9$ relative to Perin's polynomial. The algorithm used to compute these results can be deduced. Since Perrin's polynomial has d = 3 and $c_d = -1$, the relations between the coefficient sequences and the symmetric functions with k = 3 give

$$c_0(n)s(2n)+c_1(n)s(n)+c_2(n)s(0)+c_3(n)s(-n)=0$$

Noting that s(0) = 3, that Warring's formula gives $c_0(n) = 1$ and $c_1(n) = -s(n)$, and that

$$c_j(n)=-c_{3-j}(-n)$$

we have

$$s(2n)-s^2(n)+3s(-n)-s(-n)=0$$

which is equivalent to the "Doubling Rule" used by Shanks and Adams:

$$s(2n)=s^2(n)-2s(-n)$$

Furthermore, using the basis sequence relations with r = ±1, we have

$$s(2n\pm1)-s(n)s(n\pm1)+s(-n)s(\pm1)-s(-n\pm1)=0$$

These formulas are valid for all integers n, positive or negative, so they can operate on the "signature"

$$s(n-1) \quad s(n) \quad s(n+1) \quad s(-(n-1)) \quad s(-n) \quad s(-(n+1))$$

to compute s(N) and s(-N) (mod N) in $\log_2(N)$ steps, where each step requires 6 full multiplications.

The basis sequence algorithm requires d(d+1)/2 = 6 full multiplications per step, so in a sense the methods are equally efficient.

However, for this particular case (reversible, third order) the "Doubling Rule" has the advantage of producing both s(N) and s(-N), whereas the basis sequence algorithm must include one additional step to compute s(2N) so as to complete the pseudoprime test. Also, the test based on s(N) and s(2N) is strictly applicable to odd integers only, whereas the test based on s(N) and s(-N) is applicable to all integers.

On the other hand, the "Doubling Rule" has some drawbacks:

1. It requires manipulation and storage of 6 dynamic variables (the "signature"), versus only 3 for the basis sequence algorithm,
2. It applies only to reversible sequences, limiting it to recurrences with only d-1 congruence conditions and only one test per discriminant,
3. It becomes progressively more complicated when generalized to recurrences of higher order. In contrast, the basis sequence algorithm applies to non-reversible as well as reversible sequences, and to recurrences of any order.

For Perrin's sequence, the forward congruence s(N) = s(1) alone is sufficient to exclude most composites other than the symmetric pseudoprimes. The only ordinary pseudoprimes less than the first symmetric pseudoprime are

$$271441=(521)(521) \qquad \text{(Shanks and Adams)}$$
$$904631=(7)(13)(9941)$$
$$16532714=(2)(11)(11)(53)(1289)$$
$$24658561=(19)(271)(4789)$$

These numbers (which have a coincidental affinity for Mersenne prime factors) can be used to determine when certain summations take on integer values. This is based on the fact that the values of s(n) for Perrin's polynomial can be expressed in closed form as

$$s(6N+r)=(6N+r)\sum_{k=0}^{N}\frac{(2N+k+\alpha+\beta-1)!}{(2N-2k+\alpha)!(3k+\beta)!}$$

where r = 0,1,.., or 5, and a,b are the least non-negative integers such that a = r (mod 2) and b = -r (mod 3). Clearly, the summation is an integer iff s(6N+r) = 0 (mod 6N+r). For example, the integer n = 16532714 = 6(2755452) + 2 satisfies the congruence s(n) = 0 (mod n), so 2755452 is the least positive integer N such that the summation

$$\sum_{k=0}^{N}\frac{(2N+k)!}{(2N-2k)!(1+3k)!}$$

is an integer.

Perrin's polynomial is the fundamental reversible cubic with discriminant -23 (which is the smallest negative discriminant for a cubic). As explained by Shanks, et al, there are 12 *reversible* cubics with discriminant -23, but the corresponding recurrences are all of the form s(k) =

$s_{Perrin's}(mk)$ with $m = \pm1, \pm2, \pm3, \pm4, \pm5$, or ±9. Therefore, every symmetric pseudoprime relative to Perrin's polynomial is also a symmetric pseudoprime relative to each of the other reversible polynomials of discriminant -23. However, since we are not restricted to reversible sequences, we can consider other polynomials with discriminant -23, such as

$$f(x) = x^3 + x^2 - 4x - 5$$

Even though the prime types relative to this polynomial are identical to the prime types relative to Perrin's polynomial, the only symmetric pseudoprimes less than $50(10)^9$ relative to both Perrin's polynomial are the four Carmichael numbers. This is because the periods of the two sequences modulo a given prime are not necessarily equal. The smallest symmetric pseudoprime relative to hte equation of the form pq where p and q are both splitting primes is 3042763501 = (27581)(110321).

To give an example with a different discriminant, consider the non-reversible cubic

$$f(x) = x^3 - 9x^2 + 24x - 15$$

which has discriminant -135. Relative to this polynomial the smallest symmetric pseudoprime equal to the product of two splitting primes is 196049701 = (9901)(19801).

An example of a cubic polynomial relative to which the integer 703 satisfies the first and second congruences but not the third. That polynomial was reducible, having the integer roots -17, 23, and 31. The same example can be given for an irreducible cubic by simply taking the coefficients modulo 703. Thus, relative to the irreducible polynomial

$$f(x) = x^3 - 37x^2 - 205x + 170$$

the integer 703 satisfies the first and second congruences but not the third.

POLYNOMIALS OF DEGREE 4

Consider the quartic polynomial

$$f(x) = x^4 - 5x^3 - 2x^2 - 3x - 1$$

By Proposition, a composite integer N co-prime to 6 is a symmetric pseudoprime relative to this polynomial if and only if

$$\left.\begin{aligned} s(N) &\equiv s(1) \\ s(2N) &\equiv s(2) \\ s(3N) &\equiv s(3) \end{aligned}\right\} \pmod N$$

(We could also use the congruences on s(-N), s(N), and s(2N) to test any odd integer N.) The initial values of the coefficient sequences are as follows:

n	$c_0(n)$	$c_1(n)$	$c_2(n)$	$c_3(n)$	$c_4(n)$
0	1	−4	6	−4	1
1	1	−5	−2	−3	−1
2	1	−29	−28	−5	1
3	1	−164	28	−24	−1
4	1	−897	496	−81	1
5	1	−4905	−922	−223	1
.	.	.	.	.	
.	.	.	.	.	

The Galois group is the symmetric group S_4, so all 24 of the permutations of the roots are allowable, and the asymptotic proportion of prime types is as shown below.

Prime Type	***Number of Permulations***	***Asymptotic Proportions***	$k_p \tau_p$
$\{4_1\}$	6	1/4	$2(p^4-1)(p-1)$
$\{1_1 3_1\}$	8	1/3	p^3-1
$\{2_2\}$	3	1/8	p^2-1
$\{1_2 2_1\}$	6	1/4	p^2-1
$\{1_4\}$	1	1/24	$p-1$

Using the basis sequence algorithm, we searched the splitting primes (i.e., primes of type $\{1_4\}$) relative, and found that the smallest symmetric pseudoprime equal to the product of two splitting primes is

$$12282695011 = (78367)(156733)$$

It's easy to show that if p and q are splitting primes then pq is a symmetric pseudoprime if and only if $b_k(p) = b_k(1) \pmod{q}$ and $b_k(q) = b_k(1) \pmod{p}$.

An example of a non-reversible quartic is

$$f(x) = x^4 - 2x^3 + 4x + 2$$

Although this quartic is irreducible (according to Eisenstein's criterion), its Galois group is not the fully symmetric group.

As a result, some root permutations are not allowable. Specifically, there are no primes of type $\{1_1 3_1\}$.

The restricted nature of the roots of this polynomial also shows up in the fact that the general period of the prime type $\{4_1\}$ always divides

$$4\frac{\left(p^4-1\right)}{\left(p+1\right)}=4\left(p^3-p^2+p-1\right)$$

This is because the four roots of (12) satisfy the relation

$$\frac{\theta_1\,\theta_3}{\theta_2\,\theta_4}=\pm i$$

and so, noting that (12) is irreducible, we have

$$\theta^{4\left(p^1-p^2+p-1\right)}\left(\frac{\theta p^1\,\theta p^4}{\theta p^0\,\theta p^2}\right)\equiv\left(\pm i\right)^4=1$$

(The resolvant cubic of this polynomial has one rational root, from which the permutation of the roots in the preceding expressions is determined.)

A complete search of odd integers shows that the smallest symmetric pseudoprime

$$5351537=(1889)(2833)$$

The next symmetric pseudoprime relative is a product of two splitting primes is

$$746331041=(15773)(47317)$$

POLYNOMIALS OF DEGREE 5

A composite integer N co-prime to 120 is a symmetric pseudoprime relative to the fifth degree polynomial

$$f\left(x\right)=x^5+c_1\,x^4+c_2\,x^3+c_3\,x^2+c_4\,x+c_5$$

if and only if all five of the congruences

$$\left.\begin{array}{l} s(N)\equiv s(1)\\ s(2N)\equiv s(2)\\ s(3N)\equiv s(3)\\ s(4N)\equiv s(4)\\ s(5N)\equiv s(5)\end{array}\right\}\pmod N$$

are satisfied. In this section we will focus on the quintic polynomial

$$f\left(x\right)=x^5-x^3-2x^2+1$$

The discriminant of this polynomial is -4511 = -(13)(347), which is the smallest magnitude for any fully symmetric quintic. The initial values of the coefficient sequences for this polynomials are listed below.

n	$c_0(n)$	$c_1(n)$	$c_2(n)$	$c_3(n)$	$c_4(n)$	$c_5(n)$
–3	1	–3	–5	11	–6	1
–2	1	–4	4	–1	2	–1
–1	1	0	–2	–1	0	1
0	1	–5	10	–10	5	–1
1	1	0	–1	–2	0	1
2	1	–2	1	–4	4	–1
3	1	–6	11	–5	–3	1
4	1	–2	–7	–12	8	–1
5	1	–5	–1	–32	–5	1
6	1	–14	55	–79	19	–1
7	1	–7	–57	–135	–14	1
8	1	–18	17	–260	40	–1
9	1	–33	209	–500	–39	1
10	1	–27	–329	–1004	89	–1

Since the Galois group is the fully symmetric S_5, all 120 of the root permutations are allowable, and all 7 of the prime types occur. Their asymptotic proportions are listed in the table below.

Prime Type	*Number of Permulations*	*Asymptotic Proportions*
$\{5_1\}$	24	1/5
$\{1_1 4_1\}$	30	1/4
$\{2_1 3_1\}$	20	1/6
$\{1_1 2_2\}$	15	1/8
$\{1_3 2_1\}$	10	1/12
$\{1_5\}$	1	1/120

The smallest symmetric pseudoprime equal to the product of two splitting primes is

$$2258745004684033 = (27439297)(82317889)$$

The search for this number was carried out using the basis sequence algorithm. With this algorithm the computation of s(N) (mod N) requires $15(\log_2 N)$ full multiplications (mod N).

We could also have used a "Doubling Rule", analagous to that used by Shanks and Adams. For 5th order recurrences the basic doubling rule consists of the two formulas

$$s_1(2n) = s_1^2(n) - 2s_2(n)$$

$$s_2(2n) = s_2^2(n) - 2(j)^n s_1(n) s_2(-n) + 2(j)^n s_1(-n)$$

where $j = -c_5(1) = \pm 1$. The part first comes directly from Warring's formula for $c_2(n)$, which also gives $s_1(4n) = s_1(2n)^2 - 2s_2(2n)$. To deduce the second part, we can set k = 3 in the relevant equation from Section (with appropriate substitutions) to give an expression for $s_1(3n)$ in terms of $s_1(\pm 2n)$ and $s_1(\pm n)$ and $s_2(n)$. Likewise we can set k = 4 to give an expression for $s_1(4n)$ in terms of $s_1(3n)$, $s_1(2n)$, $s_1(\pm n)$, and $s_2(n)$. Then we eliminate $s_1(3n)$ from thes two equations and substitute $s_1(2n)^2 - 2s_2(2n)$ for $s_1(4n)$, and $s_1(n)^2 - 2s_2(n)$ for $s_1(2n)$, to give the second part. The other equations needed to compute terms of the form $s_1(2n+1)$ and $s_2(2n+1)$ can be developed in the same way, although they involve a large number of elements from the $s_1(\pm k)$ and $s_2(\pm k)$ sequences (corresponding to the the "signature").

The type of any prime p with respect to a given polynomial f(x) can easily be determined by checking the number of linear and quadratic factors of f(x) modulo p. To find the number of (distinct) linear factors, we simply count the number of residues u in the range from 1 to p-1 such that $f(u) = 0 \pmod p$. To find the number of quadratic factors, we count the number of residue pairs u,v such that $(x^2 + ux + v)(x^3 + Ax^2 + Bx + C) = f(x)$ modulo p for some integers A,B,C. This implies the conditions

$$\begin{aligned} u + A &= c_1 \\ u + uA + B &= c_2 \\ vA + uB + C &= c_3 \\ vB + uC &= c_3 \\ vB + uC &= c_4 \\ vC &= c_5 \end{aligned}$$

Solving the first three for A, B, and C, and substituting into the last two, we have two necessary and sufficient conditions for $x^2 + ux + v$ to divide f(x) modulo p. For example, with respect to the polynomial (13) we get the two conditions

$$u^4 - u^2 + 2u - 3u^2v + v^2 + v = 0 \pmod p$$

$$2uv^2 - u^3v + uv - 2v - 1 = 0 \qquad \pmod p$$

Of course, if there are n_1 linear factors, we automatically get $n_1(n_1\text{-}1)/2$ non-primitive quadratic factors, so to give the number n_2 of primitive quadratic factors we must subtract $n\text{-}_1(n_1\text{-}1)/2$ from the number of distinct pairs u,v that satisfy the quadratic conditions. Once the linear and quadratic factors have been divided out, what remains is a polynomial of degree m = $5 - n_1 - 2n_2$. Clearly this must be an irreducible factor of degree 3, 4, or 5, so the numbers n_3, n_4, and n_5 of primitive cubic, biquadratic, and quintic factors can immediately be inferred.

With respect to the quintic f(x) given by equation the prime p = 11 is of type $2_1 3_1$, which means that f(x) factors into a quadratic and a cubic in the field $Z_{11}[x]_f$. Specifically we have f(x) = g(x)h(x) where

$$g(x) = (x^2 - 2x - 5) \qquad h(x) = (x^3 - 9x^2 - 3x - 9)$$

The sequence x, x^{11}, x^{121}... has period 6 in $Z_{11}[x]_f$, but the period is 3 in $Z_{11}[x]_h$, and the period is 2 in $Z_{11}[x]_g$. The actual terms of this sequence are listed below:

	In $Z_{11}(x)_f$	*In* $Z_{11}(x)_h$	*In* $Z_{11}(x)_g$
x^{11^0}	$0+1x+0x^2+0x^3\,0x^4$	$0+1x+0x^2$	$0+1x$
x^{11^1}	$6+8x+9x^2+9x^3+4x^4$	$4x+3x+8x^2$	$2+10x$
x^{11^2}	$2+3x+10x^2 1x^3+4x^4$	$5+7x+3x^2$	$0+1x$
x^{11^3}	$10+5x+9x^2+2x^3+4x^4$	$0+1x+0x^2$	$2+10x$
x^{11^4}	$7+4x+0x^2+7x^3+0x^4$	$4+3x+8x^2$	$0+1x$
x^{11^5}	$1+7x+8x^2+3x^3+8x^4$	$5+7x+3x^2$	$2+10x$
x^{11^6}	$0+1x+0x^2+0x^3+0x^4$	$0+1x+0x^2$	$0+1x$

The types of all the primes less than 2000 relative to polynomial are listed in Table. Continuing on to determine the types for all 669 primes less than 5000, we find the following distribution:

Type	*Number of Primes*	*Predicted by*
5_1	127	133.8
$1_1 4_1$	173	167.3
$2_1 3_1$	117	111.5
$1_2 3_1$	106	111.5
$1_1 2_2$	89	83.6
$1_3 2_1$	55	55.8
1_5	2	5.6
Total	669	

The relativity scarcity of splitting primes (i.e., primes of the type 1_5) requires us to examine a larger number of primes in order to get a meaningful estimate of the asymptotic density. We find that, of the 9592 primes less than 100000, there are 75 splitting primes relative, compared with a predicted 79.9. This confirms that the asymptotic density of splitting primes relative to this irreducible quintic is 1/120.

CONGRUENCE CONDITIONS ON THE TERMS OF LINEAR RECURRING SEQUENCES

SECOND ORDER RECURRENCES

To determine congruence conditions on the terms of a linear recurring sequence, we must first determine a closed-form expression for the nth term using only rational functions and root extractions. Consider the second order recurrence

$$x_n \quad c_1 x_{n-1} - c_2 x_{n-2} = 0$$

This recurrence corresponds to a first order recurrence of the form

$$Z_n - \left(\alpha + \beta\sqrt{D}\right) Z_{n-1} = 0$$

where α, β, and D are constants. Let r_n denote the sum of the roots of the characteristic polynomial. The initial value is $r_0 = 1$, and all subsequent values can be generated. Now we define α_n and β_n by the formal relation

$$\frac{\alpha_n + \beta_n\sqrt{D}}{2} = \left(\alpha + \beta\sqrt{D}\right)^n$$

from which we get

$$\alpha_n = \alpha\alpha_{n-1} + D\beta\beta_{n-1}$$
$$\beta_n = \beta\alpha_{n-1} + \alpha\beta_{n-1}$$

The sequences α_n and β_n each individually satisfy the second order recurrence, provided we take the coefficients

$$\alpha = \frac{1}{2}c_1 \qquad \beta = \frac{1}{2}\sqrt{\frac{c_1^2 + 4c_2}{D}}$$

Setting D equal to the determinant, this implies $\alpha = c_1/2$ and $\beta = 1/2$, so we have

$$\frac{\alpha_n + \beta_n\sqrt{D}}{2} = \left(\frac{c_1 + \sqrt{D}}{2}\right)^n$$

By construction the α_n and β_n sequences are integers (assuming c_1 and c_2 are integers) with the initial values

n	α_n	β_n
0	2	0
1	c_1	1

For any prime p we have

$$\alpha_p + \beta_p\sqrt{D} \equiv 2\left(\frac{c_1 + \sqrt{D}}{2}\right)^p \equiv c_1 + D^{(p-1)/2}\sqrt{D} \pmod{p}$$

Therefore

$$\alpha_p \equiv c_1 \qquad \beta_p \equiv D^{(p-1)/2} \equiv [D/p]$$

modulo p, where [D/p] denotes the Legendre symbol. Since the sequences A_n and B_n are linearly independent, we can now express congruence conditions on the pth element of any integer sequence that satisfies the general homogeneous recurrence relation. For example, the basis sequence elements satisfy the congruences

$$b_0(p) \equiv c_1 \frac{1-[D \backslash p]}{2} \qquad b_1(p) \equiv [D \backslash p] \pmod{p}$$

We can also determine quadratic congruence conditions on the elements $b_0(m)$ and $b_1(m)$ where m = (p-1)/2.

THIRD ORDER RECURRENCES

In the quadratic case we found that the roots of the general quadratic polynomial were given by the closed form of the linear recurrence term. In this section we show that the solution of the general cubic is given by the closed form of the third order linear recurrence.

A set of three linearly independent solutions of the recurrence

$$x_n - c_1 X_{n-1} - c_2 X_{n-2} - c_3 X_{n-3} = 0$$

can be found by considering a first order recurrence of the form

$$Z_n - \left(\alpha + \beta D^{1/3} + \gamma D^{2/3}\right) Z_{n-1} = 0$$

where α,β,γ,D are real (but $D^{1/3}$ and $D^{2/3}$ are complex). Letting r(n) denote the nth power of the characteristic root of (4), we have the following initial values

n			r(n)		
0	1	+	$0D^{1/3}$	+	$0D2/3$
1	α	+	$\beta D^{1/3}$	+	$\gamma D^{2/3}$
2	$\left(\alpha^2 + 2\beta\gamma D\right)$	+	$\left(\gamma^2 D + 2\alpha\beta\right) D^{1/3}$	+	$\left(\beta^2 + 2\alpha\gamma\right) D^{2/3}$
3	α_3	+	$\beta_3 D^{1/3}$	+	$\gamma_3 D^{2/3}$

where

$$\alpha_3 = \alpha^3 + \beta^3 D + \gamma^3 D^2 + 6\alpha\beta\gamma D$$

$$\beta_3 = 3\left(\alpha^2 \beta + \beta^2 \gamma D + \alpha\gamma^2 D\right)$$

$$\gamma_3 = 3\left(\alpha^2 \gamma + \alpha\beta^2 + \beta\gamma^3 D\right)$$

Since the three components of each value are incomensurate, these components are each solutions of the same recurrence as the characteristic recurrence of r(n). Identifying this with the general third order recurrence, we have three equations in the three unknown coefficients c_1,c_2,c_3. Solving these equations gives

$$c_1 = 3\alpha$$

$$c_2 = 3\left(\beta\gamma D - \alpha^2\right)$$

$$c_3 = \alpha^3 + \beta^3 D + \gamma^3 D^2 - 3\alpha\beta\gamma D$$

Therefore, we immediately have

$$\alpha = \frac{c_1}{3}$$

Substituting this into the equations for β and γ leads us to define quantities u and v as follows

$$\beta\gamma D = \left(\frac{c_2}{3}\right) = u$$

and

$$\beta^3 D + \gamma^3 D^2 = \left(c_3 + \frac{2c_1^{\ 3}}{27} + \frac{c_1 c_2}{3}\right) = 2v$$

Eliminating either b or g from these equations leads to a quadratic equation with the roots

$$\beta^3 D, \gamma^3 D^2 = v \pm \sqrt{v^2 - u^3}$$

Therefore, we have

$$r(1) = \alpha + \beta D^{1/3} + \gamma D^{2/3}$$
$$= \frac{c_1}{3} + \left(v + \sqrt{v^2 - u^3}\right)^{1/3} + \left(v - \sqrt{v^2 - u^3}\right)^{1/3}$$

By its construction, this is the general form of the roots of the characteristic polynomial, i.e., the general cubic

$$f(x) = x^3 - c_1 x^2 - c_2 x - c_3$$

FOURTH ORDER RECURRENCES

Given any integer k and a sequence of integers s_0, s_1, s_2,... that satisfies a fourth order linear homogeneous recurrence relation for which the resolvant cubic of the characteristic quartic has a rational root, this section describes a method for determining the primes p such that s_p'1 k (mod p).

Consider the general homogeneous fourth order recurrence

$$X_n + k_1 X_{n-1} + k_2 X_{n-2} + k_3 X_{n-3} + k_4 X_{n-4} = 0$$

To analyze this recurrence we first consider the following second order recurrence with complex coefficients

$$Z_n + (a + ib) Z_{n-1} + (c + id) Z_{n-2} = 0$$

where a, b, c, and d are real. Let r(n) denote the sum of the nth powers of the roots of the characteristic polynomial of (6). The initial values are

$$r(0)=2+0i \qquad r(1)=-a-bi$$

and all subsequent r(n) can be generated using (6). If we let A_n and B_n denote, respectively, the real and imaginary coefficients of r(n), then we have

$$A_n = -aA_{n-1} + bB_{n-1} - cA_{n-2} + dB_{n-2}$$
$$B_n = -aB_{n-1} - bA_{n-1} - cB_{n-2} + dA_{n-2}$$

Seperating the real and imaginary components, we find that the sequences A_n and B_n each individually satisfy the fourth order recurrence (5) with the coefficients

$$k_1 = 2a$$
$$k_2 = a^2 + 2c + b^2$$
$$k_3 = 2(ac + bd)$$
$$k_4 = c^2 + d^2$$

These equations imply that $a = k_1/2$ and $c = y/2$ where y is a root of the cubic

$$y_3 + \alpha y^2 + \beta y + \delta = 0$$

with

$$\alpha = -k_2$$
$$\beta = k_1 k_3 - 4k_4$$
$$\delta = -\left(k_3{}^2 + k_1{}^2 k_4 - 4k_2 k_4\right)$$

Equation is called the resolvant cubic of the quartic polynomial corresponding. If x_1, x_2, x_3, x_4 are the roots of the characteristic equation, and y_1, y_2, y_3 are the roots of then

$$y_1 = x_1x_2 + x_3x_4$$
$$y_2 = x_1x_3 + x_2x_4$$
$$y_3 = x_2x_3 + x_1x_4$$

Given the value of c based on any root y, the corresponding values of b and d are

$$b = \pm\sqrt{k_2 - a_2 - 2c}$$
$$b = \pm\sqrt{k_4 - c_2}$$

with the stipulation that

$$bd = \frac{k_3}{2} - ac$$

This last condition relates the signs of b and d, and is a useful identity for ring multiplication to be discussed later.

Example:

To illustrate, consider the fourth order recurrence

$$X_n - 2X_{n-1} + 4X_{n-3} + 2X_{n-4} = 0$$

whose characteristic polynomial

$$f(x) = x^4 - 2x^3 + 4x + 2$$

is irreducible (according to Eisenstein's criterion). For this case we have

$$k_1 = 2 \qquad k_2 = 0 \qquad k_3 = 4 \qquad k_4 = 2$$

which gives $a = -1$ and

$$c^3 - 4c - 3 = 0$$

Taking the root $c = -1$, compute $b = \pm 1$ and $d = \pm 1$. Now, since gives $bd = 1$, it follows that b and d have the same sign. Choosing $b = d = 1$, equation gives the second order complex-valued recurrence

$$Z_n - (1+i)Z_{n-1} - (1+i)Z_{n-2} = 0$$

with the characteristic equation

$$h(z) = z^2 - (1+i)z - (1+i)$$

If we let $r(n) = A_n + iB_n$ denote the sum of the nth powers of the complex roots of h(z), then we have the initial values

n	A_n	B_n
0	2	0
1	1	1
2	2	4
3	–2	8
4	–12	12
5	–34	6
6	–64	–28
.	.	.
.	.	.

By their construction, both of these sequences individually satisfy the original fourth order real-valued recurrence for $n > 4$. If s(n) denotes the sum of the nth powers of the roots, then $s(n) = 2A_n$.

Now, if p is an odd prime, we have the following congruences

$$\begin{aligned} r(p) = \left(A_p + iB_p\right) &\equiv (1+i)^p \\ &\equiv (1)^p + (i)^p \\ &\equiv 1 + i(i)^{p-1} \qquad \pmod{p} \\ &\equiv 1 + i(-1)^{(p-1)/2} \end{aligned}$$

from which it follows that

$$\left.\begin{aligned} A_p &\equiv 1 \\ B_p &\equiv (-1)^{(p-1)/2} \end{aligned}\right\} \pmod{p}$$

To complete the analysis of the general fourth order recurrence we need two more solutions C_n and D_n which, together with A_n and B_n, will form a basis for all possible solutions. That is, any solution X_n will be expressible as

$$x_n = g_1 A_n + g_2 B_n + g_3 C_n + g_4 D_n$$

where the g_m are constants determined by the initial values of X_n. Any four solution sequences constitute a basis provided that they are linearly independent, for which a necessary and sufficient condition is that the determinant

$$\begin{vmatrix} A_0 & B_0 & C_0 & D_0 \\ A_1 & B_1 & C_1 & D_1 \\ A_2 & B_2 & C_2 & D_2 \\ A_3 & B_3 & C_3 & D_3 \end{vmatrix}$$

is non-zero. However, our purpose is to define primality criteria for every possible X_p, which requires that we be able to define congruence conditions on C_n and D_n when n is a prime, similar to the conditions on A_n and B_n given by equations. Our approach will be to derive another second order complex-valued recurring sequence, different from, but with real and imaginary parts that also satisfy the basic fourth-order recurrence.

One possibility would be to use $a = c = -1$ as before, but choose the signs $b = d = +1$. However, this does not lead to independent sequences. Instead, we keep $a = 1$ and return to the cubic for a different value of c. Factoring out the root $c = 1$ leaves

$$c^2 - c - 3 = 0$$

which has the roots $\left(1 \pm \sqrt{13}/2\right)$. To make b and d real valued we choose the root c =, which results in the following set of coefficients

$$a = -1$$

$$b = +\sqrt{-2+\sqrt{13}}$$

$$c = \frac{1-\sqrt{13}}{2}$$

$$d = +\sqrt{\frac{-3+\sqrt{13}}{2}}$$

It is useful to observe that, in view of (10), we have

$$bd = \frac{5-\sqrt{13}}{2}$$

Inserting these values of a, b, c, and d into equation gives a second order complex-valued recurrence with the characteristic polynomial

$$h(z) = z^2 + (a+bi)z + (c+di)$$

Again letting r(n) denote the sum of the nth powers of the roots of h(z), we have the initial values

n	*r(n)*
0	2+0i
1	–1–bi
2	2–2(b+d)i
3	$-2-\left(\frac{7+\sqrt{13}}{2}b+3d\right)i$
.	.
.	.

By their construction, the real and the imaginary parts of r(n), taken seperately, both satisfy the basic fourth order recurrence. Clearly the real part of this r(n) is identical to the A_n sequence already discussed. Our objective is to derive two new and independent integer sequences from the imaginary part of this r(n). Letting I(n) denote the coefficient of the imaginary part of r(n), we have the initial values

n	*I(n)*
0	0
1	–b
2	–2b–2d
3	$-\left(\frac{7+\sqrt{13}}{2}\right)b+3d$
.	.
.	.

As a consequence of equation, the irrational quantities b and d are related by

$$d = b\left(\frac{-1+\sqrt{13}}{6}\right)$$

so every element of the I(n) sequence can be expressed in the form

$$I(n) = b\left(\alpha_n + \beta_n\sqrt{13}\right)$$

where α_n and β_n are rational coefficients. The values of I(0) and I(1) are already of this form. The remaining initial values are

$$I(2) = -b\left(\frac{5+\sqrt{13}}{3}\right) \qquad I(3) = -b\left(3+\sqrt{13}\right)$$

In order to deal only with integer sequences, we will work with 3I(n) instead of I(n), and we will define the two new basis sequences C_n and D_n as follows

$$3I(n) = b\left(C_n + D_n\sqrt{13}\right)$$

The initial values of these sequences are

n	C_n	D_n
0	0	0
1	–3	0
2	–5	–1
3	–9	–3
.	.	.
.	.	.

The four sequences A_n, B_n, C_n, and D_n are linearly independent, as shown by the non-zero determinant

$$\begin{vmatrix} 2 & 0 & 0 & 0 \\ 1 & 1 & -3 & 0 \\ 2 & 4 & -5 & -1 \\ -2 & 8 & -9 & -3 \end{vmatrix} = -12$$

It remains to establish congruence requirements on C_p and D_p for primes p. We know that

$$r(p) \equiv (1-ib)^p \equiv 1^p - (ib)^p \qquad (\text{mod } p)$$

For the remainder of this section, all congruences are understood to be modulo p. Concerning ourselves only with the imaginary part, we have

$$I(p) \equiv -(ib)^p$$
$$\equiv -ib(ib)^{p-1}$$
$$\equiv -\left(-\sqrt{2-\sqrt{13}}\right)^{p-1}$$
$$\equiv -ib\left(2-\sqrt{13}\right)^{(p-1)/2}$$

Therefore, by equation we have

$$C_p + D_p\sqrt{13} \equiv -3\left(2-\sqrt{13}\right)^{(p-1)/2}$$

and also

$$C_p - D_p\sqrt{13} \equiv -3\left(2+\sqrt{13}\right)^{(p-1)/2}$$

from which it follows that

$$C_p{}^2 - 13D_p{}^2 \equiv 9(4-13)^{(p-1)/2}$$
$$\equiv 9(-9)^{(p-1)/2}$$
$$\equiv 9(-1)^{(p-1)/2} = 9[-1/p]$$

where [1/p] is the Legendre symbol. Furthermore, squaring congruences, and multiplying by $\left(2-\sqrt{13}\right)$, we have

$$\left(2-\sqrt{13}\right)\left[C_p{}^2 + 13D_p{}^2 + 2C_pD_p\sqrt{13}\right] \equiv 9\left(2-\sqrt{13}\right)^p$$

$$2\left[C_p{}^2 + 13D_p{}^2 - 13C_pD_p\right] - \left[C_p{}^2 + 13D_p{}^2 - 4C_pD_p\right]\sqrt{13} \equiv 9\left(2-\sqrt{13}\right)^p$$

Expanding the right side using the binomial theorem, we have

$$2\left[C_p{}^2 + 13D_p{}^2 - 13C_pD_p\right] \equiv 9(2)^p$$

and

$$-\left[C_p{}^2 + 13D_p{}^2 - 4C_pD_p\right] \equiv -9(13)^{(p-2)/2}$$

Dividing equation by 2, and equation by 1, and noting that 2P 1'11 and 13(p 1)/2 'I [13/p] (Euler's criterion), we have

$$C_p{}^2 + 13D_p{}^2 - 13C_pD_p \equiv 9$$

and

$$C_p^{\;2} + 13D_p^{\;2} - 4C_pD_p \equiv 9[13/p]$$

Subtracting dividing by 9 gives

$$C_pD_p \equiv [13/p] - 1$$

Substituting this into gives

$$C_p^{\;2} + 13D_p^{\;2} \equiv 13[13/p] - 4$$

Recalling equation

$$C_p^{\;2} + 13D_p^{\;2} \equiv 9[-9/p]$$

we can solve simultaneously for $C_p^{\;2}$ and $13D_p^{\;2}$, which gives

$$C_p^{\;2} \equiv \frac{13[13/p] + p[-1/p]}{2} - 2$$

$$13D_p^{\;2} \equiv \frac{13[13/p] - 9[-1/p]}{2} - 2$$

Letting Φ and θ denote, respectively, the right sides of equations, we have the following complete set of basis element congruences for any prime p:

$$A_p \equiv 1 \qquad B_p \equiv [-1/p] \qquad C_p \equiv \sqrt{\phi} \qquad D_p \equiv \sqrt{\frac{\theta}{13}}$$

along with the condition

$$C_pD_p \equiv [13/p] - 1$$

In view of equation, the residue class of X_p for any sequence of integers X that satisfies the basic recurrence (11) is given by

$$X_p = g_1 + g_2[-1/p] + g_3\sqrt{\phi} + g_4\sqrt{\frac{\theta}{13}}$$

where the g_i may be expressed in terms of the initial values of the X sequence by solving the simultaneous equations

$$\begin{bmatrix} A_0 & B_0 & C_0 & D_0 \\ A_1 & B_1 & C_1 & D_1 \\ A_2 & B_2 & C_2 & D_2 \\ A_3 & B_3 & C_3 & D_3 \end{bmatrix} \cdot \begin{bmatrix} g_1 \\ g_2 \\ g_3 \\ g_4 \end{bmatrix} = \begin{bmatrix} X_0 \\ X_1 \\ X_2 \\ X_3 \end{bmatrix}$$

Since the determinant of the basis matrix is -12, and we want to deal only with integers, we multiply by 12 to clear any fractional values of the g_i, and then define $G_i = 12g_i$. Also, to clear the radicals from we can bring the first two terms on the right over to the left, and then square both sides (bearing in mind that the cross-product of radicals is an integer by equation) to give a quadratic congruence condition on X_p. Since the square of the term $\sqrt{(\theta/13)}$ has a 13 in the denominator, we will multiply through by 13 to give the general form

$$(1872)X_p^2-(312)\left(G_1+G_2[-1/p]\right)X_p+(13)\left(G_1+G_2[-1/p]\right)^2$$
$$-(13)G_3^2\phi+(26)G_3G_4\left(1-[13/p]\right)-G_4^2\theta\equiv 0$$

Notice that if the prime p is not equal to 13, we can divide this congruence by 13 whenever q (mod p) is divisible by 13, which it is when $[1\backslash p]=+1$. Also, if p is not equal to 2 or 3, we may be able to divide by these factors.

To illustrate, consider the particular solution sequence X of recurrence (11) with the initial values $X_0 = 37$, $X_1 = 19$, $X_2 = 43$, and $X_3 = 101$. The values of g_i and G_i for this sequence are

$g_1 = 37/2$ $\quad$ $G_1 = 222$

$g_2 = -121/2$ $\quad$ $G_2 = -726$

$g_3 = -61/3$ $\quad$ $G_3 = -244$

$g_4 = -439/3$ $\quad$ $G_4 = -1756$

Therefore, for any prime p other than 2, 3, or 13, we have the following quadratic congruences on X_p:

(−1/p)	(13/p)	(mod p)
+1	+1	$x_p^2 + 84x_p - 1957 \equiv 0$
+1	−1	$x_p^2 + 84x_p - 36733 \equiv 0$
−1	+1	$13x_p^2 + 84x_p - 1115888 \equiv 0$
−1	−1	$13x_p^2 + 84x_p - 391382 \equiv 0$

These congruences enable us to easily determine the primes p such that X_p Ί y (mod p) for any integer y. For example, the only primes p that divide X_p are 109, 547, and 195691. Also, the only primes p such that X_p Ί 5000 (mod p) are 3 and 157560691.

Chapter 9

Discrete Circular Distributions

Consider a circular arrangement of D bins, into which are placed n marbles randomly with uniform distribution. What is the probability that all the marbles will be located within N consecutive bins? If N is less than or equal to D/2, the answer is easy to see, as discussed in another note.

Briefly, the question is equivalent to whether n randomly placed arc spans of length N all have some common overlap, and we can consider how many of the D^n different arrangements of n such arcs intersect in at least one bin. Focusing on just one particular bin, we know there are N^n distinct arrangements for which every arc includes that bin. Multiplying this by D, the number of bins, gives a first estimate of the total number of arrangements with at least one bin in common. Each arrangement with exactly one bin in common has been counted exactly once in this estimate, but each arrangement with two bins in common has been counted twice (once for each of the common bins).

Likewise each arrangement with three common bins has been counted three times, and so on. To eliminate the duplicates, consider the number of arrangements that have (at least) two particular adjacent bins in common. The number of such arrangements for any particular pair of adjacent bins is clearly $(N–1)^n$, and we can again multiply this by D to give an estimate of the total number of such arrangements.

This counts the arrangements with exactly two common bins just once, but it counts those with three common bins twice, and so on. Therefore, subtracting this number from the previous number, and dividing the result by D^n (the total number of all possible arrangements) gives the probability of overlap between all n of the arcs

$$\Pr\{n,N,D\} = \frac{D(N^n - (N-1)^n)}{D^n}$$

But, again, this applies only if N is less than or equal to D/2. The question is how to generalize this for cases when N is greater than D/2. The solution for continuous circular arcs is well known, but the discrete case seems to be less so. At one extreme, with $N = \text{D}-1$, the answer is easy to see. Letting c_j denote the event that none of the marbles is located in the jth cell (bin), the event of interest is the union of these n events, so the probability of all n marbles being located in an arc of length $N = \text{D-1}$ is

$$\text{pr}\{c_1 + c_2 + \ldots + c_D\} = \sum_{j=1}^{D} \Pr\{c_j\} - \sum_{j=1}^{D-1} \Pr\{c_j c_k\} + \sum_{j=1}^{D-2} \sum_{k=j+1}^{D-1} \sum_{j=1}^{D} \Pr\{c_j c_k c_r\} - \ldots$$

Hence the probability is

$$\Pr\{n, \text{D}-1, \text{D}\} = \sum_{j=1}^{D} (-1)^{j-1} \binom{D}{j} \left(\frac{D-j}{D} \right)^n$$

In the case $N = \text{D} - 2$ we need to assess the probability of the event

$$c_1 c_2 + c_2 c_3 + c_3 c_4 + \ldots + c_{D-1} c_D + c_D c_1$$

The first-order terms are all products of two distinct elements, and there are D such terms, so they contribute $\text{D}[(\text{D}-2)/\text{D}]^n$ to the probability, but the second-order terms are more complicated, because the first-order terms are not all mutually independent.

There are a total of D(D–1)/2 second-order terms, and of these, D consist of products of three distinct elements, and D(D-3)/2 consist of products of four distinct elements.

Proceeding to the third order terms, there are a total of D(D-1)(D-2)/6, consisting of D products of four, D(D–4) products of five, and D(D-4)(D-5)/6 products of six distinct elements. We can continue to find higher order terms, until all combinations have been exhausted, and then consolidate terms, noting that the kth order terms consisting of j products contribute

$$(-1)^{k-1} \left(\frac{D-j}{D} \right)^n$$

to the total probability. The same approach can be taken for any N, by setting the D basic elements to the D products of D–N consecutive elements c_i. As an example, suppose we wish to determine the formula for the probability of all n marbles lying within a span of N=12 cells in a circle of D=16 cells. We seek the probability of the event

$$c_1 c_2 c_3 c_4 + c_2 c_3 c_4 c_5 + c_3 c_4 c_5 c_6 + \ldots$$
$$\ldots + c_{13} c_{14} c_{15} c_{16} + c_{14} c_{15} c_{16} c_1 + c_{15} c_{16} c_1 c_2 + c_{16} c_1 c_2 c_3$$

Applying inclusion-exclusion, we get the following numbers of kth order products of j basic elements:

k / j	1	2	3	4	5	6	7	8	9	10	11	12	13	14	15	16
1	0	0	0	16	0	0	0	0	0	0	0	0	0	0	0	0
2	0	0	0	0	16	16	16	72	0	0	0	0	0	0	0	0
3	0	0	0	0	0	16	32	48	160	128	96	80	0	0	0	0
4	0	0	0	0	0	0	16	48	96	280	384	408	368	168	48	4
5	0	0	0	0	0	0	0	16	64	160	448	784	1024	1024	640	208
6	0	0	0	0	0	0	0	0	16	80	240	680	1360	2016	2160	1456
7	0	0	0	0	0	0	0	0	0	16	96	336	992	2160	3456	4384
8	0	0	0	0	0	0	0	0	0	0	16	112	144	1400	3248	7128
9	0	0	0	0	0	0	0	0	0	0	0	16	128	576	1920	8800
10	0	0	0	0	0	0	0	0	0	0	0	0	16	144	720	7128
11	0	0	0	0	0	0	0	0	0	0	0	0	0	16	160	4192
12	0	0	0	0	0	0	0	0	0	0	0	0	0	0	16	1804
Sum	0	0	0	16	−16	0	0	−56	112	−56	0	16	−48	48	−16	

Naturally, the sum of the terms on the kth row is the binomial coefficient C(16,k). From this table we can infer that, given a circle of 16 cells, the probability of all n marbles lying within 12 consecutive cells is

$$\Pr\{n,12,16\}=16\left[\left(\frac{12}{16}\right)^n-\left(\frac{12}{16}\right)^n\right]-56\left[\left(\frac{8}{16}\right)^n-2\left(\frac{7}{16}\right)^n+\left(\frac{6}{16}\right)^n\right]$$

$$+16\left[\left(\frac{4}{16}\right)^n-3\left(\frac{3}{16}\right)^n+3\left(\frac{2}{16}\right)^n-\left(\frac{1}{16}\right)^n\right]$$

In the same way, we can determine the probabilities with smaller values of D but the same ratio of N/D. This gives

$$\Pr\{n,3,4\}=4\left[\left(\frac{3}{4}\right)^n-\left(\frac{2}{4}\right)^n\right]-2\left[\left(\frac{2}{4}\right)^n-2\left(\frac{1}{4}\right)^n+\left(\frac{0}{4}\right)^n\right]$$

$$\Pr\{n,6,8\}=8\left[\left(\frac{6}{8}\right)^n-\left(\frac{5}{8}\right)^n\right]-12\left[\left(\frac{4}{8}\right)^n-2\left(\frac{3}{8}\right)^n+\left(\frac{2}{8}\right)^n\right]$$

$$\Pr\{n,9,12\}=12\left[\left(\frac{9}{12}\right)^n-\left(\frac{8}{12}\right)^n\right]-30\left[\left(\frac{6}{12}\right)^n-2\left(\frac{5}{12}\right)^n+\left(\frac{4}{12}\right)^n\right]$$

To see the correspondence with the continuous solution, consider the first terms of these expressions, which are multiples of quantities of the form

$$(2q-1)^n-2(2q-1-\alpha)^n+(2q-1-2\alpha)^n=\frac{2\alpha^2}{2q-1}\binom{n}{2}(2q-1)^{n-1}+\ldots$$

where we have put q = N/D and a = 1/D, and we have expanded the expression in powers of a. The coefficient of each of these quantities is simply D = $1/\alpha$, so in the limit as D goes to infinity (and a goes to zero) these terms represent simply the quantity nq^{n-1}.

The second terms are multiples of quantities of the form

$$(2q-1)^n - 2(2q-1-\alpha)^n + (2q-1-2\alpha)^n = \frac{2\alpha^2}{2q-1}\binom{n}{2}(2q-1)^{n-1} + \ldots$$

At this state it would be tempting to think that the coefficients of these terms are

$$? \ \frac{2q-1}{2\alpha^2} = \frac{D}{2}(2N-D) \ ?$$

but in fact to give agreement with the preceding results we must subtract D/2, and we can clearly add or subtract any constant multiple of D/2 without affecting the limiting value, since the ratio of (2N-D–K)/(2N–D) for any constant K still goes to 1 as N and D increase to infinity. Thus the actual coefficients of the second terms are given by

$$\frac{D}{2}(2n-d-1) = \frac{D}{2N-D}\binom{2N-D}{2}$$

The third terms (if present) are multiples of quantities of the form

$$(3q-2)^n - 3(3q-2-\alpha)^n + 3(3q-2-2\alpha)^n - (3q-2-3\alpha)^n$$
$$= \frac{6\alpha^3}{(3q-2)^2}\binom{n}{3}(3q-2)^{n-1} + \ldots$$

Again the temptation is to assume the coefficients of these terms are the reciprocal of the leading factor in this expression, i.e.,

$$? \ \frac{(3q-2)^2}{6\alpha^3} = \frac{D}{6}(3N-2D)^2 \ ?$$

But again, in order to agree with the coefficients derived by inclusion-exclusion, we must deduct a constant from each of the factors in the square. In this case we subtract 1 from the first and 2 from the second, giving the coefficients of the third terms as

$$\frac{D}{6}(3N-2D-1)\,(3N-2D-2) = \frac{D}{3N-2D}\binom{3N-2D}{3}$$

The pattern for higher order coefficients is clear, and of course for any given N and D we will have only k non-zero terms, where kN-(k–1)D must be greater than or equal to k. Hence the number of terms is the greatest integer less than or equal to D/(D–N+1). This explains why, if N is less than or equal to D/2, there is only one term (as we discussed originally). With these coefficients, the continuous case for n points on a circle all lying within an arc of q times the whole perimeter is given by the well-known formula

$$\Pr\{n,q\} = \sum_{j=1}^{j_{\max}} (-1)^{j-1}\binom{n}{j}[1-j(1-q)]^{n-1}$$

where j_{max} is the greatest integer less than but *not* equal to $1/(1-q)$. In our previous example we had N/D = q = 3/4, so j_{max} = 3, and this formula gives

$$\Pr\left\{n, \frac{3}{4}\right\} = \sum_{j=1}^{3} (-1)^{j-1} \binom{n}{j} \left[1 - \frac{j}{4}\right]^{n-1} = \binom{n}{1}\left(\frac{3}{4}\right)^{n-1}$$

$$-\binom{n}{2}\left(\frac{2}{4}\right)^{n-1} + \binom{n}{2}\left(\frac{1}{4}\right)^{n-1}$$

which agrees exactly with the result of taking the limit of the discrete cases with N/D = 3/4 as D goes to infinity.

$$\Pr\{n, N, D\} = \sum_{j=1}^{j_{\max}} \left[\frac{(-1)}{jN-(j-1)D} \binom{jN-(j-1)D}{j} \sum_{k=0}^{j} (-1)^k \binom{j}{k} \left(\frac{jN-(j-1)D-k}{D} \right)^n \right]$$

where j_{max} is the greatest integer less than or equal to D/(D-N+1). Letting M denote D-N, this can be written more succinctly as

$$\Pr\{n, N, D\} = \sum_{j=1}^{j_{\max}} \left[\frac{(-1)^{j-1} D}{D-Mj} \binom{D-Mj}{j} \sum_{k=0}^{j} (-1)^k \binom{j}{k} \left(\frac{D-Mj-k}{D} \right)^n \right]$$

To illustrate the application of this formula, suppose there is a circular arrangement of D = 26 cells, and n marbles are distributed randomly (with uniform density) into these cells. To determine the probability that all n marbles will lie within a span of N = 19 cells, we have j_{max} = 3, because this is the greatest integer less than or equal to 26/(26-19+1). We also have M = 26-19 = 7, so the probability can be expressed as

$$\Pr\{n, 19, 26\} = \sum_{j=1}^{3} \left[(-1)^{j-1} \frac{26}{26-7j} \binom{26-7j}{j} \sum_{k=0}^{j} (-1)^k \binom{j}{k} \left(\frac{26-7j-k}{26} \right)^n \right]$$

$$\Pr\{n, 19, 26\} = \sum_{j=1}^{3} \left[(-1)^{j-1} \frac{26}{26-7j} \binom{26-7j}{j} \sum_{k=0}^{j} (-1)^k \binom{j}{k} \left(\frac{26-7j-k}{26} \right)^n \right]$$

$$= 26\left[\left(\frac{19}{26}\right)^n - \left(\frac{18}{26}\right)^n\right]$$

$$-143\left[\left(\frac{12}{26}\right)^n - 2\left(\frac{11}{26}\right)^n + \left(\frac{10}{26}\right)^n\right]$$

$$+52\left[\left(\frac{5}{26}\right)^n - 3\left(\frac{4}{26}\right)^n + 3\left(\frac{3}{26}\right)^n - \left(\frac{2}{26}\right)^n\right]$$

A plot of this function is shown below.

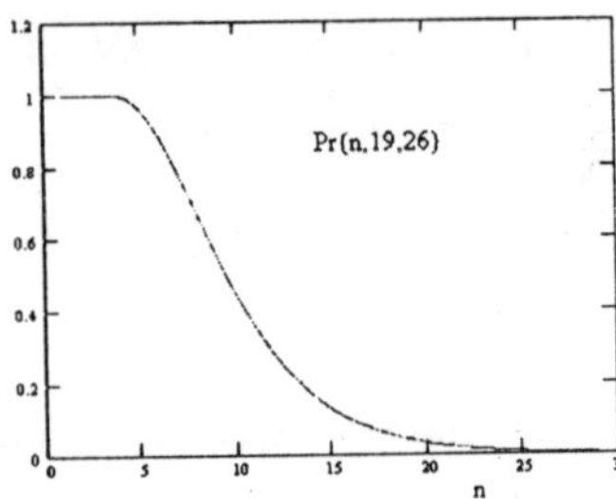

As we would expect, the probability is 1 for $n = 1, 2$, or 3, because any three points on a circular set of 26 cells lie within a span of 19.

A PROCRUSTEAN PROTOCOL

It is sometimes necessary to communicate information by indirect means, and a surprising amount of information can often be conveyed using a pre-established protocol specifically designed for the given circumstances. As an abstract example, consider the natural numbers from 1 to 28, one of which is to be arbitrarily selected as the "winning" number and made known to us. Then four OTHER numbers are arbitrarily drawn from the list (without replacement) and given to us, and our task is to communicate knowledge of the winning number to someone else simply by sending them our four numbers arranged in some particular order. It's easy to establish a protocol for communicating this information on this basis, because the recipient knows the winning number is not one of the four he receives from us, so there are really only 24 possibilities, which can be arranged in ascending order and placed in one-to-one correspondence with the 24 distinct permutations of the four numbers designated a,b,c,d in ascending order.

However, suppose the original list of number is increased from 28 to, say, 42. Can we still establish an effective protocol based on sending four numbers in some order? Clearly not, because there are only 24 distinct "messages" we can send for four given message-numbers, whereas there are 38 possible "winning numbers". Hence we can't possibly distinguish each possible result.

But what if we relax our protocol condition slightly, and permit ourselves to send either all four of the message-numbers in some order OR to send just three of the numbers in some order? In this case we can't rule out the possibility of an effective protocol, because the number of distinct answers we can give with a fixed set of four message-numbers is 24 permutations of four PLUS 6 permutations of three, and there are four distinct ways of choosing three of the four message numbers. Hence we have $24 + 4(6) = 48$ distinct messages that we can send based on any four message-numbers. Of course, the distinctness of some of these may not be evident to the recipient, since they may see only three of our four message numbers, but at least we can't automatically rule out a successful protocol.

One protocol that accomplishes our objective is based on a predefined set of 6-element subsets of the original 42 numbers, defined so as to minimize the overlaps between subsets. For example, we can take the 42 columns from the arrays shown below:

42	41	40	39	38	37	36
1	2	3	4	5	6	7
8	9	10	11	12	13	14
15	16	17	18	19	20	21
22	23	24	25	26	27	28
29	30	31	32	33	34	35
36	37	38	39	40	41	42

35	34	33	32	31	30	29
1	2	3	4	5	6	7
9	10	11	12	13	14	8
17	18	19	20	21	15	16
25	26	27	28	22	23	24
33	34	35	29	30	31	32
41	42	36	37	38	39	40

28	27	26	25	24	23	22
1	2	3	4	5	6	7
10	11	12	13	14	8	9
19	20	21	15	16	17	18
28	22	23	24	25	26	27
30	31	32	33	34	35	29
39	40	41	42	36	37	38

21	20	19	18	17	16	15
1	2	3	4	5	6	7
11	12	13	14	8	9	10
21	15	16	17	18	19	20
24	25	26	27	28	22	23
34	35	29	30	31	32	33
37	38	39	40	41	42	36

14	13	12	11	10	9	8
1	2	3	4	5	6	7
12	13	14	8	9	10	11
16	17	18	19	20	21	15
27	28	22	23	24	25	26
31	32	33	34	35	29	30
42	36	37	38	39	40	41

7	6	5	4	3	2	1
1	2	3	4	5	6	7
13	14	8	9	10	11	12
18	19	20	21	15	16	17
23	24	25	26	27	28	22
35	29	30	31	32	33	34
40	41	42	36	37	38	39

The first array simply lists the numbers 1 through 42 consecutively in horizontal rows. The second array shifts the nth row $1(n-1)$ places to the right (with wrap-around). The third array shifts the nth row 2(n-1) places to the right, and so on. Each column is associated with a unique number from 1 to 42, noted above it.

Our transmission protocol is based on four given "message-numbers" which we will call a,b,c,d in ascending order. There is also a single winning number w. We check each of the four possible sets of three message numbers, (a,b,c), (a,b,d), (a,c,d), and (b,c,d) and compute the sum of the three numbers, and subtract 42 if necessary to place the reduced sum in the range from 1 to 42. (We could more easily express this if we made the range 0 to 41 and worked modulo 42). This will necessarily give us four distinct indices, because no two of these triples differ by more than one number, and the basic elements are distinct.

We then examine the columns in the above arrays corresponding to those four indices. If the winning number w appears in one of those columns, we then transmit the corresponding triple, with its elements arranged in one of the six permutations to pick out the appropriate number from the indexed column.

If, on the other hand, the winning number is NOT in any of the four indexed columns, we will transmit all four message numbers. Of course, these can only be arranged in 24 distinct ways, but these suffice, because we can rule out all the numbers in the four indexed columns, as well as the four message-numbers themselves. If there were no overlap, this would exclude 28 numbers, leaving only 14 to be selected by the 24 permutations. However, there is typically overlap between the numbers in the four indexed columns, and between the column numbers and the message numbers. We can assign the column indices so that no index appears in its own column (to do this in the above arrays we would make the column transpositions (12,13), (20,21), (27,28), (34,35), and (38,39)), but there can still be overlap from one column to another index.

Nevertheless, we are assured that the fours columns (even without the four given incides) will always cover AT LEAST 18 distinct numbers, because every pair of columns shares no more than one number in common, so the most overlap would occur if each of the six pairs of four columns shared a number, thereby reducing the total from 24 down to 24-6 = 18. Hence, from the overall 42 numbers, we are left with no more than 42-18 = 24 possible values of w after excluding those that appear in the four columns tat can be indexed by three of the four given message-numbers. These can therefore be arranged in order and placed in unique correspondence with the permutations of the four message numbers.

At the receiving end, if three numbers are received, we compute the index given by the sum of the three numbers, reduced to the range from 1 to 42, to give us the appropriate column of the table, and then use the ordering to select the winning number w from that column. If FOUR numbers are received, we first compute the four indices from the sets of three, and eliminate all the numbers from the respective columns, as well as eliminating the four message- numbers themselves. We then arrange the remaining numbers into a list in ascending order, and use the arrangement of the four message numbers to select the winning number w.

REGULAR POLYGONAL ARRANGEMENTS

We can distribute six consecutive integers, such as 0 to 5, on the vertices and edges of a triangle in such a way that the sum along each edge is the same. In fact, there are four essentially distinct such arrangements, shown below.

```
     2              4              5              5

    3 4            1 3            0 2            0 1

   1 5 0          2 5 0          3 4 1          4 2 3
```

The common sums of these four triangular arrangements are 6, 7, 8, and 9 respectively. Of course, each of these arrangements has six versions if we count rotations and reflections.

Similarly, arrangements of any 8 consecutive numbers can be constructed on the vertices and edges of a square. In this case there are six essentially distinct arrangements, illustrated below.

```
5 3 1        4 1 5        3 6 2        7 0 5
  4 6          6 2          1 5          3 1
0 7 2        0 7 3        7 0 4        2 4 6
             7 2 1        7 1 3
               3 5          4 2
             0 6 4        0 5 6
```

The common sums of the squares in each of these four columns are 9, 10, 11, and 12 respectively. It's easy to show algebraically that, for each square, the sums of opposite pairs of edges must be equal. If we let V and E denote the sums of the values on the vertices and edges respectively, then the common sum is S = (E+2V)/n for an *n*-sided polygon. We also know that there are a total of $2n$ vertices and edges, so we are distributing the integers 0 to 2n-1, which implies that the sum of all the numbers is $E+V=n(2n-1)$. Solving these equations for E and V, we have

$$E = n[2(2n-1)-S] \quad V = n[S-(2n-1)]$$

This shows that both E and V must be divisible by n, and also that the common sum S must be greater than $(2n-1)$ and less than 2(2n-1). In fact, since there are n edges, each containing n distinct non- negative integers, the least possible value of E is $n(n-1)/2$, so the quantity $2(2n-1)-S$ must not only be positive, it must be greater than or equal to (n–1)/2, and thus S must be less than or equal to $2(2n-1)-(n-1)/2$ = (7n-3)/2. Likewise V must be at least $n(n-1)/2$, so S cannot be less than $(2n-1)+(n-1)/2$ = (5n–3)/2. For $n = 4$ this means S cannot be less than 8.5 and cannot exceed 12.5, consistent with the fact that the smallest common sum is 9 and the largest is 12.

We can place 10 consecutive numbers on the vertices and edges of a pentagon such that the three numbers along any edge have the same sum. This can be done in six essentially distinct ways. Rather than drawing the figures, we can just list the strings of numbers around the perimeter of the figure, beginning with a vertex. The six arrangements are listed below, along with their common sum.

Common sum		
11	1738092546	
13	0765283194	6523814907
14	3476185092	9234716805
16	6283547091	

As in the case of the square arrangements, the alternative patterns for a single common sum differ by just two transpositions of consecutive numbers, and these two transpositions are separated by a single position. As always, the common sum must be no less than (5*n*–3)/2 and no greater than (7*n*–3)/2, which gives the limits of 11 and 16 for *n*=5.

For hexagonal arrangements we have *n*=6, so the only possible common sums are in the range from 13.5 to 19.5, i.e., the set 14, 15, 16, 17, 18, and 19. The essentially distinct solutions for each of these common sums are listed below, along with the corresponding values of V and E, and (in the right-most columns) the sums of the squares of the vertex and edge numbers.

S=14, V=18, E=48

6 3 5 7 2 11 1 9 4 10 0 8 82 424

5 3 6 7 1 11 2 8 4 10 0 9 82 424

8 5 1 11 2 9 3 7 4 10 0 6 94 412

S=15, V=24, E=42

0 6 9 2 4 10 1 11 3 5 7 8 156 350

S=16, V=30, E=36

6 0 10 1 5 7 4 9 3 11 2 8 190 316

7 3 6 8 2 4 10 1 5 11 0 9 214 292

8 5 3 11 2 4 10 0 6 9 1 7 214 292

2 10 4 1 11 5 0 9 7 3 6 8 226 280

0 9 7 5 4 1 11 3 2 8 6 10 226 280

7 9 0 6 10 4 2 11 3 5 8 1 226 280

S=17, V=36, E=30

9 3 5 1 11 2 4 6 7 10 0 8 292 214

11 2 4 8 5 3 9 1 7 10 0 6 292 214

11 2 4 10 3 6 8 0 9 7 1 5 292 214

9 0 8 6 3 4 10 2 5 11 1 7 280 226

9 3 5 8 4 2 11 0 6 10 1 7 280 226

6 4 7 2 8 0 9 3 5 11 1 10 256 250

S=18, V=42, E=24

2 9 7 1 10 0 8 6 4 3 11 5 354 152

S=19, V=48, E=18

11 1 7 2 10 0 9 4 6 8 5 3 412 94

10 0 9 3 7 1 11 2 6 8 5 4 412 94

3 5 11 1 7 4 8 2 9 0 10 6 424 82

These results show some interesting dualities. For example, the first solution with S=14 is the dual of the last solution with S=19, because they have the same split of the numbers into vertex and edge numbers, except that the types are reversed. In other words, the numbers that are assigned to vertices in the first solution are assigned to edges in the last solution, and vice versa. These two solutions are shown pictorially below.

```
      6 3 5          3 5 11
       8 7             6 1
       0 2             10 7
     10 11             0 4
      4 9 1            9 2 8
```

We can generalize the problem by allowing k numbers to be assigned to each edge, where k is greater than 1. For example, we can construct triangles such as the two shown below

```
        1                  1
       6 2                6 10
       8 9                 7 3
    4 5 3 7               8 12
                      9 2 4 11 5
```

In general, to construct a triangle with k numbers on each edge (not counting the vertices), let $N = $ k+2 denote how many numbers are on each edge counting the vertices, and then assign the numbers 1, N, and 2N-1 to the three vertices. Then, to fill in the edge numbers, arrange the remaining numbers into three columns, ascending on each row alternately from the left and the right. For example, to construct a triangle with N=6 numbers on each side, we first assign the numbers 1, 6, and 11 to the vertices, and then we arrange the remaining numbers in three columns as shown below

```
  2  3  4
  8  7  5
  9 10 12
 15 14 13
```

Each column sums to 34, but we actually need columns that sum to 34-5, 34, and 34+5 in order to place these numbers on the edges [6,11], [1,11], and [1,6] of the triangle, respectively. We can do this by leaving the middle column untouched, and making the transpositions (8,5) and (15,13) in the outer columns. This decreases the left-hand column by 5, and increases the right hand column by 5. As an alternate solution, we could make the transpositions (2,4) and (9,12), which would increase the left column by 5 and decrease the right column by 5. Notice that this must always lead to a solution, because the two outer columns will always differ by 2 or 3, so we can form any required adjustment (greater than 1) by transposing pairs.

GOLF TOURNAMENTS, POLYHEDRA, AND LATIN SQUARES

Suppose we need to schedule 32 golfers into foursomes for five rounds of a tournament, in such a way that no one is in the same foursome with anyone else more than once. This is not too hard, because we have enough players to easily avoid repeated pairings. For example, we can use the simple pattern shown below, where the first round has the 32 players listed horizontally, the second round has them listed vertically, and the subsequent rounds skew the successive columns of the vertical listing by 1, 2, and 3 rows per column, respectively.

Four- Round some	1	2	3	4	5
1	1 2 3 4	1 9 17 25	1 16 23 30	1 15 21 27	1 14 19 32
2	5 6 7 8	2 10 18 26	2 9 24 31	2 16 22 28	2 15 20 25
3	9 10 11 12	3 11 19 27	3 10 17 32	3 9 23 29	3 16 21 26
4	13 14 15 16	4 12 20 28	4 11 18 25	4 10 24 30	4 9 22 27
5	17 18 19 20	5 13 21 29	5 12 19 26	5 11 17 31	5 10 23 28
6	21 22 23 24	6 14 22 30	6 13 20 27	6 12 18 32	6 11 24 29
7	25 26 27 28	7 15 23 31	7 14 21 28	7 13 19 25	7 12 17 30
8	29 30 31 32	8 16 24 32	8 15 22 29	8 14 20 26	8 13 18 31

However, suppose we have only 24 players, instead of 32. This is a bit more challenging, since we have to work harder to avoid repeated pairings. This is a standard problem in combinatorial design. One simple approach would be to start with four pair-wise orthogonal latin squares of size 5x5. (For any prime power *N* there exist N-1 pairwise orthogonal Latin squares of size NxN.)

0 1 2 3 4	0 1 2 3 4	0 1 2 3 4	0 1 2 3 4
1 2 3 4 0	2 3 4 0 1	3 4 0 1 2	4 0 1 2 3
2 3 4 0 1	4 0 1 2 3	1 2 3 4 0	3 4 0 1 2
3 4 0 1 2	1 2 3 4 0	4 0 1 2 3	2 3 4 0 1
4 0 1 2 3	3 4 0 1 2	2 3 4 0 1	1 2 3 4 0

Here we've used the integers (mod 5), and the values of the kth array for k=1,2,3,4 are given by

$$A_k(i, j) = k\,(i-1) + (j-1) \pmod 5$$

for i,j = 1,2,3,4,5. These arrays are called "latin squares" because each element appears exacly once in each row and column. They are said to be "pairwise orthogonal" because if you super-impose one on top of the other you get all 25 distinct combinations (i, j) with i, j = 0,1,2,3,4. In other words, each element of one array matches up with each element of the other array exactly once.

So, if we let the elements of the first array represent the first five players, and the elements of the second array represent the next five players, and so on, we can create 25 foursomes from the 20 players by superimposing the four arrays. Since the arrays are pairwise orthogonal, we are assured that no one plays in a foursome with the same person twice.

But we actually need 30 foursomes involving 24 players. This can be done by mixing in four more players, taking advantage of the fact that none of the five players in a given array have yet played with each other. Thus, numbering the players 1 through 24, we have the following five days worth of foursomes. Each of the 24 players plays once per day, and no one plays with the same person twice.

Four-Some	***Day 1***	***Day 2***	***Day 3***	***Day 4***	***Day 5***
1	1 6 11 16	22 8 14 20	3 10 12 19	4 7 15 18	5 9 13 17
2	2 7 12 17	23 9 15 16	4 24 13 20	5 8 24 19	1 10 14 23
3	3 8 13 18	24 10 11 17	5 21 14 16	1 9 21 20	2 6 15 24
4	4 9 14 19	5 6 12 18	1 22 15 17	2 10 22 16	3 7 11 22
5	5 10 15 20	21 7 13 19	2 23 11 18	3 6 23 17	4 8 12 21
6	21 22 23 24	1 2 3 4	6 7 8 9	11 12 13 14	16 18 19 20

This is good, but we can do even better. It's possible to schedule 24 players into foursomes for SEVEN days in such a way that no two players are in the same foursome twice. Obviously this means each player is matched with 21 other players, so he is never matched with 3 of the others. This partitions the players into eight groups of 3, consisting of those who never play together. For convenience, let's say these exclusion groups are i,i+8,i+16 modulo 24 for i=1 to 8. Our arrangement will be symmetrical in these sets. Now, the players 1, 9, 17 never play together, so we can put them in foursomes #1, #2, and #3 respectively on each of the seven days. Then we can fill in the other 21 players in Foursome #1 in such a way we can create foursomes #2 and #3 simply by adding 8 (modulo 24) to each player of the preceding foursome.

Then we can create Foursome #4 using just the 21 players, excluding players 1, 9, and 17, in such a way that (again) Foursomes #5 and #6 are simply the preceding foursome incrementd by 8 (mod 24). The result is summarized below:

Four-some / *Day*	*1*	*2*	*3*	*4*	*5*	*6*
1 1	2 11 21	9 10 19 5	17 18 3 13	4 7 6 24	8 12 14 15	16 20 22 23
2 1	3 12 22	9 11 20 6	17 19 4 14	5 8 7 18	2 13 15 16	10 21 23 24
3 1	4 13 23	9 12 21 7	17 20 5 15	6 2 8 19	3 10 14 16	11 18 22 24
4 1	5 14 24	9 13 22 8	17 21 6 16	7 3 2 20	4 10 11 15	12 18 19 23
5 1	6 15 18	9 14 23 2	17 22 7 10	8 4 3 21	5 11 12 16	13 19 20 24
6 1	7 16 19	9 15 24 3	17 23 8 11	2 5 4 22	6 10 12 13	14 18 20 21
7 1	8 10 20	9 16 18 4	17 24 2 12	3 6 5 23	7 11 13 14	15 19 21 22

For another interesting problem, suppose you're planning a golf tournament in which 12 players will each play one round per day for 4 days. You need to decide how to split up the 12 players into 3 foursomes each day, and you want to choose the foursomes to minimize the number of repeated pairings.

Ideally you'd like to have each pair of players placed together in the same foursome only once, but that's obviously not possible with the conditions of this problem, so the objective is to minimize the repeated pairings AND to make the schedule as symmetrical as possible for the 12 players. What's the "best" (most equally distributed) solution?

In this case it isn't obvious, a priori, how to even precisely characterize the "best" solution, because minimizing the number of repeated pairs doesn't necessarily give a symmetrical solution. One approach is to think of a fairly symmetrical polyhedron with suitable properties. The rhombic dodecahedron comes to mind, because it has 12 faces, each of which has 4 neighbors, and there are 3 dual pairs of mutually exclusive partitions of the 12 faces into three sets of four. Here's the graph of this solid

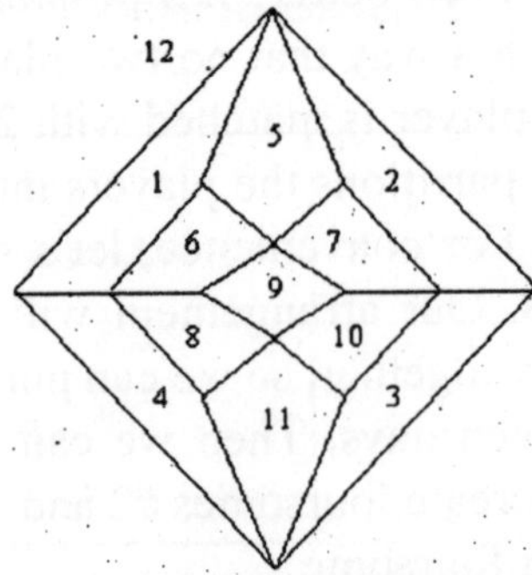

The only pairs of faces whose sets of neighbors are mutually exclusive are [(1,10),(2,8)], [(3,6),(4,7)], [(5,11),(9,12)]. The largest number of faces that are all mutually non-neighbors is four, and each face is a member of exactly three such sets. This gives the partitions

1 2 8 10	1 2 9 11	1 3 7 8
3 4 5 9	3 4 6 7	2 4 6 10
6 7 11 12	5 8 10 12	5 9 11 12

To give a 4th partition, notice that "1" and "10" are each already paired twice with "2" and twice with "8", and these are the dual sets of opposite faces [(1,10),(2,8)]. Also, the opposite faces "1" and "10" are already paired once with each other. Similar pairings exist for the other two dual sets. This suggests that we create three new foursomes by combining non-dual pairs. There are eight distinct ways of doing this. For example, if we combine (1,10) with (4,7), and combine (2,8) with (9,12), and so on, we have

1 4	7 10
2 8	9 12
3 5	6 11

Using this partition for the 4th round, each player's schedule includes four "doubles" and four "singles". No two people are on the same foursome more than twice, and each person plays (at least once) with exactly 8 (of the 11) other players. Is this the best that can be done?

THE COIN IN THREE FOUNTAINS

Draw three intersecting circles and place a coin, heads up, in each of the seven enclosed regions. Two operations are allowed: a FLIP turns over all four of the coins within a single circle, whereas a CLEAR turns all four coins in a single circle heads up. Is it possible to reach a condition in which the coin in the central region (contained within all three of the circles) is tails while all the rest are heads?

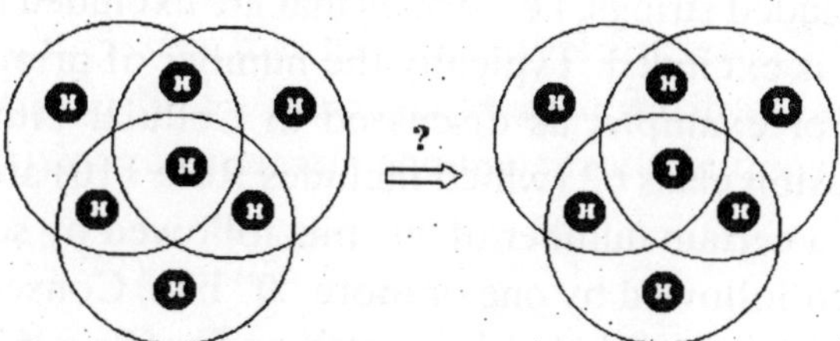

The answer is no. In fact, if, at any stage, all the coins in at least one circle are heads, it's impossible to proceed to the desired final configuration by means of FLIPS and CLEARS. This not only covers a wide range of initial conditions, it also implies that for arbitrary initial configurations no sequence of operations containing a CLEAR can yield the desired final configuration.

PROOF: Assume that at some stage (possibly the initial configuration) all the coins in at least one circle are heads. Therefore, any sequence of FLIPS and CLEARS can be assumed to contain at least one CLEAR. Let A denote the circle that is the subject of the last CLEAR in the sequence, and let B and C denote the other two circles. Every operation following the last CLEAR is a FLIP of either A, B, or C. Let nA, nB, and nC denote the number of these FLIPS respectively. Since the coin in the triple-region was turned heads up by the last CLEAR, the sum (nA+nB+nC) must be odd for the triple-region to end up tails. The double-regions of AB and AC were also turned heads up by the last CLEAR, so the sums (nA+nB) and (nA+nC) must both be even to leave these regions heads up. It follows that nA, nB, and nC must each be odd. However, the oddness of nA implies that the single-region of A, which was turned heads up by the last CLEAR, ends up tails. Thus, the desired final configuration cannot be reached.

IMPOSSIBLE STRINGS FOR CELLULAR AUTOMATA

In the note on Cellular Automata we discussed a class of simple linear two-state automata in which each field value $f_{t+1}(j)$ is either 0 or 1, and is a fixed function of the three prior values $f_t(j-1)$, $f_t(j)$, and $f_t(j+1)$. We saw how the 256 distinct functions ("rules") of this type can be placed into 88 equivalence classes up to direction reversal and field value complementation. Furthermore, we noted that for any given rule there may be impossible (excluded) strings, i.e., strings that are not the successor of any string. For any given rule, we can list the total number of k-bit excluded strings for k = 1, 2,... etc. This gives sequences of integers, and we found that there are only 31 distinct sequences. The members of each of the 88 equivalence classes of rules fall into one of these 31 "exclusion classes". The largest single exclusion class consists of the rules that do not exclude any strings. At the other extreme, Rule 0 excludes all but one string. In between these two extremes are the other 29 exclusion classes, with varying degrees of exclusivity.

Naturally any string that contains an excluded string is also excluded, so we are led to focus on the *primitive* excluded strings, i.e., strings that are excluded but that do not contain any proper substring that is excluded. Typically the number of primitive excluded strings is a partition function. For example, as discussed in Cellular Automata, the primitive excluded strings for exclusion class 62 (which includes Rule 110) are of the form "01*10" where the "*" consists of a certain number of "0" bits followed by some number (possibly zero) of "111" strings each followed by one or more "0" bits. Consequently, letting $P_{62}(k)$ denote the number of primitive excluded k-bit strings for Class 62, we find that this is equal to the number of ordered partitions of k-4-3n into exactly n+1 positive integers, for all n such that k-4-3n is greater than or equal to n+1. Thus the values of n range from 0 to the greatest integer less than or equal to (k-5)/4. The values of $P_{62}(k)$ for k = 1 to 17 are listed below.

0, 0, 0, 0, 1, 1, 1, 1, 2, 3, 4, 5, 7, 10, 14, 19, 26,...

These values satisfy the 4th-order linear recurrence relation

$$P_{62}(k) = P(k-1) + P(k-4)$$

where we have omitted the subscript "62" on the right-hand P terms for convenience. The characteristic polynomial x^4-x^3-1 is irreducible, and has the approximate roots

$$1.380277, -0.819172, 0.219447 \pm 0.914473i$$

Therefore, the sequence asymptotically increases by a factor of about 1.38 as k is incremented. Of course, since any string that contains one of these primitive excluded strings is excluded, the total number of excluded k-bit strings for Class 62 increases more rapidly. The values of this number, which we will denote as $E_{62}(k)$, are listed below for k = 1 to 17.

0, 0, 0, 0, 1, 5, 16, 43, 107, 255, 589, 1328, 2940, 6419, 13862, 29668, 63024,...

These values, too, satisfy a linear recurrence, namely the 6th-order recurrence

$$E_{62}(k) = 5E(k-1) - 9E(k-2) + 8E(k-3) - 4E(k-4)$$
$$-E(k-5) + 2E(k-6)$$

(Again we have omitted the subscript "62" on the right-hand side.) The characteristic polynomial of this recurrence factors

$$x^6 - 5x^5 + 9x^4 - 8x^3 + 4x^2 + x - 2 = (x-1)(x-2)$$
$$(x^4 - 2x^3 + x^2 - x - 1)$$

The total number of k-bit strings is obviously 2^k, so the fraction of k–bit strings that are excluded is $E(k)/2^k$, which approaches 1 as k increases. This isn't surprising, because all it takes is one small primitive excluded string to exclude "almost all" sufficiently long strings. In fact, several of the exclusion classes (0, 3, 4, 7, 14, and 29) have only a single primitive excluded string.

In general, for rule classes that have easily-characterized primitive excluded strings, it seems to be possible to find a linear recurrence for the P and E sequences. Here is a short list of characteristic polynomials and their factorizations for several E-classes:

$$E_0 : x^2 - 3x + 2 = (x-2)(x-1)$$
$$E_2 : x^4 - 3x^3 + 2x^2 - x + 2 = (x-2)(x^3 - x^2 - 1)$$

$$E_3 : x^4 - 4x^3 + 5x^2 - 3x + 2 = (x-2)(x^3 - 2x^2 + x - 1)$$
$$E_4 : x^3 - x^2 + x + 2 = (x-2)(x^2 - x - 1)$$

$$E_7 : x^5 - 4x^4 + 4x^3 + x^2 - 3x + 2 = (x-2)(x^4 - 2x^3 + x - 1)$$
$$E_{14} : x^4 - 3x^3 + x^2 + x + 2 = (x-2)(x^3 - x^2 - x - 1)$$

$E_{18} : x^6 - 5x^5 + 8x^4 - 4x^3 + x - 2 = (x-2)(x^2 - x - 1)$
$(x^3 - 2x^2 + x - 1)$

$E_{25} : x^6 - 4x^5 + 3x^4 + 4x^3 - 4x^2 + x - 2 = (x-2)$
$(x^5 - 2x^4 - x^3 + 2x^2 - 1)$

$E_{29} : x^5 - 4x^4 + 4x^3 + x - 2 = (x-2)(x-1)(x^3 - x^2 - x - 1)$

$E_{46} : x^4 - 4x^3 + 4x^2 + x - 2 = (x-2)(x-1)(x^2 - x - 1)$

$E_{62} : x^6 - 5x^5 + 9x^4 - 8x^3 + 4x^2 + x - 2 = (x-2)(x-1)$
$(x^4 - x^3 + x^2 - x - 1)$

Not surprisingly, each of the characteristic polynomials contains a factor of $(x-2)$, and therefore 2 is the dominant root, causing each sequence (asymptotically) to roughly double on each step. In addition, we see that several other factors are shared by two or more of these polynomials. The factor $(x-1)$ signifies a constant offset in the sequences for classes 0, 29, 46, and 62. We also see the "Fibonacci factor" (x^2-x-1) in the polynomials for classes 4, 18, and 46, the 3rd-order Fibonacci factor (x^3-x^2-x-1) for classes 14 and 29, and the factor (x^3-2x^2+x-1) for classes 3 and 18.

We can also find recurrences for the numbers of primitive excluded strings for some of the rule classes. Here is a short list of the characteristic polynomials and their factorizations for a few of the rule classes with non-trivial P(k) sequences.

$$P_{18} : x^4 - 2x^3 + x^2 - 1 = (x^2 - x - 1)(x^2 - x + 1)$$

$$P_{25} : x^4 - x^2 - x - 1 = (x+1)(x^3 - x^2 - 1)$$

$$P_{62} : x^4 - x^3 - 1 = \text{irreducible}$$

As in the case of Class 62, the primitive excluded strings for Classes 18 and 25 can be expressed in terms of partitions. For Class 18 the primitive excluded strings are "111" plus all the strings of the form 11*11 where "*" signifies an odd number of individual "1" bits separated by at least one "0" from every other "1". Thus the value of $P_{18}(k)$ is the number of ordered partitions of $k-4-n$ into exactly $n+1$ positive parts for odd integers n such that $k-4-n$ is greater than or equal to $n+1$.

Hence n ranges over the odd integers from 1 to the greatest odd integer less than or equal to (k-5)/2. To illustrate, with k=16 the permissible values of n are 1, 3, and 5, so we need to count the numbers of ordered partitions of 11 into two parts, of 9 into four parts, and of 7 into six parts. This gives 10, 56, and 6 respectively, so the total is $P_{18}(16) = 72$. Similarly the primitive excluded strings for Class 25 can be expressed in terms of restricted partitions, although the restriction is slightly more complicated.

The remaining six non-trivial P(k) sequences (i.e., those for the classes 37, 22, 94, 73, 41, and 26) do not appear to satisfy a linear recurrence of order 8 or less. It is not known whether they satisfy any higher-order linear recurrence, or if they have simple generating functions.

HAMILTON CYCLES ON MCCAULEY GRAPHS

Suppose we wish to arrange a string of 8 bits in a circle such that every possible 3-bit string occurs exactly once. One possible solution is 00011101, which (with wrap-around) gives the eight sequences of 3 bits

000	001	011	111	111	110	101	010	100
0	1	3	3	7	6	5	2	4

Each of the eight numbers in this sequence must be given by shifting the preceding number one bit to the left (i.e., multiplying by 2), adding either 0 or 1, and truncating the most significant bit (i.e., taking the residue modulo 8). Thus, beginning with $x = 0$, each succeeding number in the sequence is given by either $2x$ (mod 8) or $2x + 1$ (mod 8). We can express the above solution just in terms of the sequence of eight residues 0, 1, 3, 7, 6, 5, 2, 4. There is only one other solution (up to rotations), namely, 0, 1, 2, 5, 3, 7, 6, 4, corresponding to the string 00010111. These two solutions are just the reflections of each other (up to rotation). Obviously the sequence of bits represents the summand b in the operation $2x + b$ (mod 8) at each step, and this is also the sequence of odd/even numbers in the result.

As noted by Joe McCauley, we can find analogous sequences modulo any even number n (not necessarily a power of 2), such that every residue modulo n appears exactly once, and each number is either 2x (mod n) or 2x + 1 (mod n) relative to the preceding number x. Of course, if n is not a power of 2, the evaluations modulo n will not be pure truncations, so these solutions don't correspond exactly to strings of binary bits taken k bits at a time as they do when n equals 2^k. Thus a more general representation is in terms of a directed graph consisting of n vertices numbered 0 to n-1, with a directed edge from each vertex j to the two vertices $2j$ and $2j + 1$ (mod n). We will call this a McCauley graph, and consider the Hamilton cycles of these graphs. There are no Hamilton cycles for McCauley graphs of odd degree, so we need consider only the graphs for even values of n. The graphs for n = 2, 4, 6, 8, and 10 are shown below.

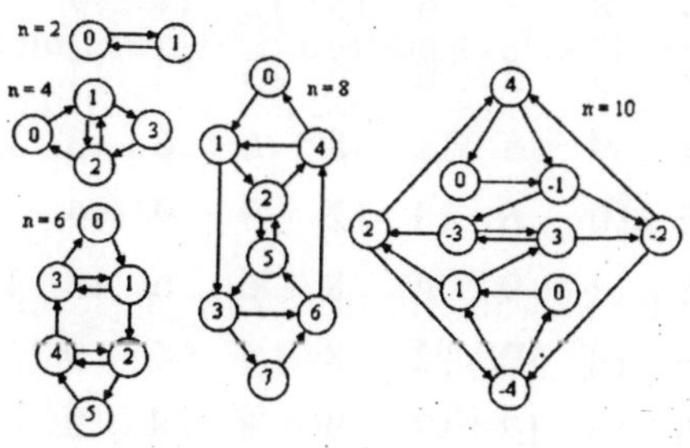

Each of these is a planar graph, meaning that it can be drawn on a two-dimensional surface with no crossing lines. (For clarity we haven't shown the self-edges for the "0" and "n/2" vertices.) It's easy to see that there is only one Hamilton cycle for each of the cases $n = 2$, 4, and 6. For the McCauley graph of degree $n = 8$ there are two distinct Hamilton cycles, as shown in the figures below.

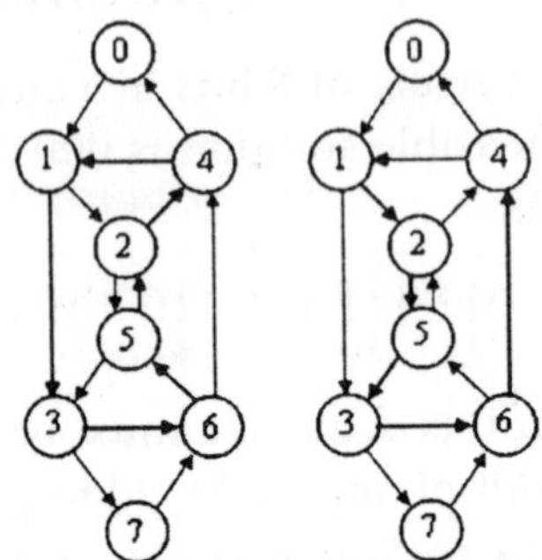

For the McCauley graph of degree $n = 10$ there are three distinct Hamilton cycles, as shown in the figures below.

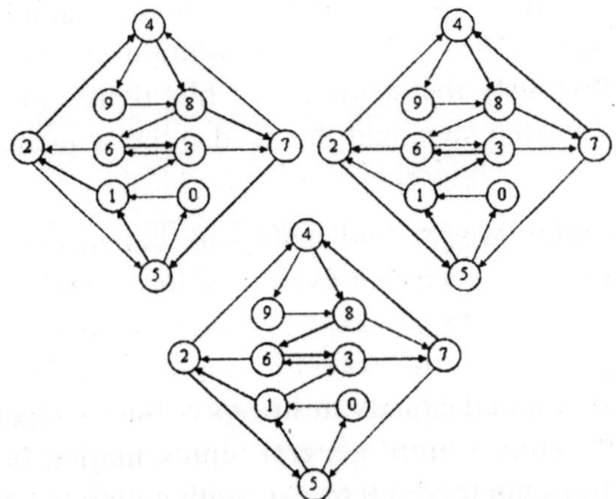

We also note that the graphs are symmetrical if we reduce the residues to the smallest signed residues modulo n–1. For the case $n = 14$ there are seven distinct Hamilton cycles, represented by the sequences of vertices listed below.

0	1	2	4	8	3	6	13	12	11	9	5	10	7
0	1	2	4	9	5	10	6	13	12	11	8	3	7
0	1	2	4	9	5	11	8	6	6	13	12	10	7
0	1	2	5	10	6	13	12	9	9	4	8	3	7
0	1	2	5	11	9	4	8	6	6	13	12	10	7
0	1	3	6	13	12	12	8	4	4	9	5	10	7
0	1	3	6	13	12	12	9	8	8	2	5	10	7

Reducing these to the smallest signed residues modulo 13 shows the symmetries more clearly.

0	1	2	4	−5	3	6	3	6	0	−1	5	−3	−6
0	1	2	4	−4	5	−3	5	−3	6	0	−5	3	−6
0	1	2	4	−4	5	−2	5	−2	−5	3	−1	−3	−6
0	1	2	5	−3	6	0	6	0	−1	−2	−5	3	−6
0	1	2	5	−2	−4	4	−4	4	−5	3	−1	−3	−6
0	1	3	6	0	−1	−2	−1	−2	−5	2	5	−3	−6
0	1	3	6	0	−1	−2	−1	−2	−4	4	5	−3	−6

The number of distinct Hamilton cycles for even values of *n* from 2 to 64 are listed below:

n	cycles	*n*	cyles
2	1	34	3825
4	1	36	5376
6	1	38	13797
8	2	40	24576
10	3	42	27783
12	4	44	95232
14	7	46	182183
16	16	48	262144
18	21	50	629145
20	48	52	1290240
22	93	54	1835001
24	128	56	3670016
26	315	58	9256395
28	448	60	11059200
30	675	62	28629151
32	2048	64	67108864

McCauley noted that the numbers of Hamilton cycles alternate between odd and even, i.e., the number is odd if and only if n/2 is odd. He asked if there are any other patterns in these numbers, and if there is any other way of producing them. It turns out that there is a nice "closed form" number-theoretic expression for these numbers.

First we make a few more observations on the table. Letting H(*n*) denote the number of cycles of size *n*, we note that $H(2(2n)) = 2^{n-1}H(2n)$ for all *n*. For example, H(36) = (256)H(18). This takes care of H(n) for all *n* divisible by 4, so we need only determine the values of H(n) for $n = 2k$ where k is odd. Now we observe that if $n = 2p$ for a prime p we usually have

$$H(2p) = \frac{2^{p-1}}{p} \quad \text{for } p = 3,5,11,13,19,29$$

Of course, by Fermat's theorem, we're assured this quantity is an integer for any prime p. For example, H(2'"11) = $(2^{10} - 1)/11 = 93$. However, this doesn't cover the primes 7, 17, 23, or 31. These are precisely the primes (in the range of the table) for which H(2p) is divisible by p itself. Moreover, for each of these four primes the value of H(2p) is of the form pm^2 for some integer m. In other words, H(2p)/p is a square. Furthermore, in three of these cases (p = 7, 17, and 23) we find that H(2p) = pm^2 = (2p–2)/(2p)–2m, so we can solve for $m = (2^{(p-1)/2}-1)/p$, which implies

$$H(2p) = \frac{\left(2^{(p-1)/2} - 1\right)^2}{p} \quad \text{for} \quad p = 7,17,23$$

The only remaining prime case (in the table) is $p = 31$, for which we have h(2'"31) = 31^5. It's also clear from the table that the values of $h(2n)$ for odd composite n are related to divisors of 2^q–1 for primes q that divide n.

To summarize these observations, let h(n) denote H(2n). We don't lose anything this way, since there are no cycles with odd lengths. With this convention, the observations mentioned above can be summarized as follows:

$$h(2n) = 2^{n-1} h(n) \quad \text{for all positive integers } n$$

$$h(p) = \frac{2^{\phi(p)} - 1}{p} \quad \text{for primes } p = 3,5,\ 11,\ 13,\ 19,\ 29$$

$$h(p) = \frac{\left(2^{\phi(p)/2} - 1\right)^2}{p} \quad \text{for primes } p = 7,\ 17,\ 23$$

where $f(n)$ is Euler's totient function. The prime 31 didn't fit into these patterns, but notice that we have

$$h(p) = \frac{\left(2^{\phi(p)/6} - 1\right)^6}{p} \quad \text{for primes p} = 31,....$$

so we are led to suspect that $h(p)$ for every prime p is of the form

$$h(p) = \frac{\left(2^{\phi(p)/k} - 1\right)^k}{p}$$

for some divisor k of $f(p)$. Furthermore, for each *power* of a prime, we apparently have

$$h(p^r) = \frac{1}{p^r} \prod_{j=1}^{r} \left(2^{\phi(p^j)/k} - 1\right)^k$$

for a suitable divisor k of $f(p)$. Also, notice that for at least some odd primes p,q we have

$$h(pq) = h(p)h(q)\left(2^{\phi(pq)/k} - 1\right)^k$$

where k again is a suitable divisor of $f(n)$. To this point we have left the value of "k" unspecified, other than to say it is a divisor of f, but notice that for each integer N the suitable value of $\phi(N)/k$ is equal to the *order* of 2 modulo N. Letting ord(m,N) denote the least exponent s such that $m^s - 1$ is divisible by N, we have $k = \phi(N)/\text{ord}(2,N)$. This is consistent with the basic operation, which involves cycles generated by iterating $x \to 2x$ (or $2x+1$) modulo N. Combining all these relations, we conjecture that, for any integer $n = 2^r m$ where m is odd we have

$$h(n) = H(2n) = \frac{2^{n-m}}{n} \prod_{d/m} \left(2^{\phi(2,d)} - 1\right)^{\phi(d)/\text{ord}(2,d)}$$

Thus we arrive at an explicit formula for the number of distinct Hamilton cycles of McCauley graphs of degree n. (Note that 2nH(2n) could be regarded as the actual number of Hamilton cycles of degree $2n$, because H(n) was restricted to those beginning with 0, each of which represents 2n others by rotation.) As an example, suppose we wish to determine the value of H(42), which equals h(21). In this case $n = m = 21$, which has the divisors 3, 7, and 21. Using the values $\phi(3) = 2$, $\phi(7) = 6$, $\phi(21) = 12$, and ord(2,3) = 2, ord(2,7) = 3, ord(2,21) = 6, the preceding formula gives

$$h(21) = \frac{1}{21}\left(2^2 - 1\right)\left(2^3 - 1\right)^2\left(2^6 - 1\right)^2 = 27783$$

in agreement with the table value of H(42).

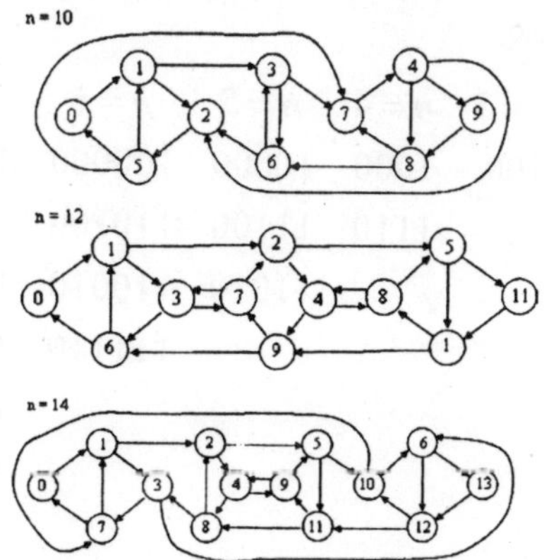

Notice that $h(n)$ is not a simple multiplicative function, because of the cross-product terms involving $2^{\text{ord}(pq)} - 1$. Still, it's an interesting definition based on the divisors. (Can this can be inverted using something like the Mobius function?) This formula is valid for all the values in the table, which is fairly persuasive, but I haven't proven that it gives $h(n)$ for all n. Still, this is an interesting example of how much structure can be inferred (or at least guessed) about a combinatorial function strictly from examining its values, without even considering its definition.

The McCauley graphs are all planar up to $n = 14$, as shown by the figures below. The first non-planar McCauley graph is for $n = 16$, shown in the figure below.

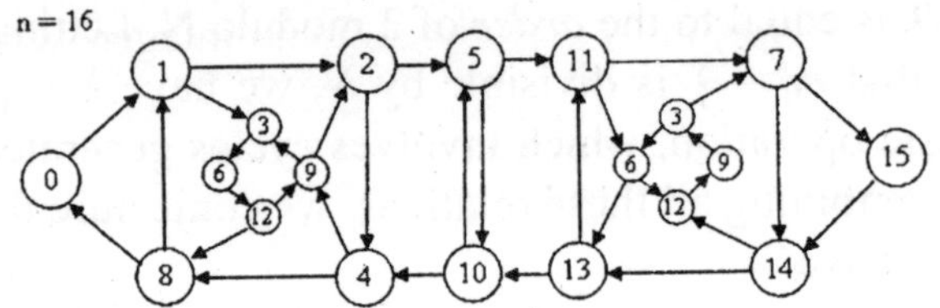

Incidentally, the sequence of integers h(n) is also the number (up to rotation) of $n \times n$ circulant invertible matrices over the field of integers modulo 2. A circulant matrix is one whose rows are all just rotated versions of a single row. For example, the following is a 4 x 4 circulant matrix:

$$\begin{bmatrix} 1 & 2 & 3 & 4 \\ 4 & 1 & 2 & 3 \\ 3 & 4 & 1 & 2 \\ 2 & 3 & 4 & 1 \end{bmatrix}$$

A matrix is invertible if its determinant is non-zero. So, an invertible circulant matrix over the integers modulo 2 is a circulant matrix whose elements are either 0 or 1 and whose determinant is odd (i.e., congruent to 1 modulo 2). Since each row is just a rotated version of one of the rows, we can represent each such matrix (or rather the equivalence class of matrices up to rotation) by just the top row. The matrices for the first several values of n are indicated below.

$n=2$	$n=3$	$n=4$	$n=5$	$n=6$	$n=7$
10	100	1000	10000	100000	1000000
		1110	11100	110100	111000
			11010	110010	1100100
				111110	1010100
					1111100
					1111010
					1110110

We can generalize the above discussion to other moduli. For example, instead of considering all cycles of integers of length n such that each term s_{k+1} is either $2s_k$ or $2s_k$ + 1 modulo n, we could allow each term to be of the form $3s_k$, $3s_k + 1$, or $3s_k + 2$ modulo n. For example, the four solutions (up to rotation) of length 8 are

0 1 4 6 3 2 7 5
0 2 7 6 3 1 4 5
0 1 3 2 7 6 4 5
0 1 4 5 7 6 3 2

Obviously each of these cycles is uniquely characterized by the sequence of "remainders" modulo 3, so they each correspond to an integer with those digits in base 3. The numbers of cycles for n = 2 to 15 are listed below:

n	cycles	n	cycles
2	1	9	14
3	2	10	12
4	2	11	64
5	2	12	64
6	4	13	42
7	4	14	44
8	4	15	512

Recall that with base 2 there were no cycles except when n is divisible by 2. With base 3 there are clearly more cycles when n is divisible by 3 than when it is not, but there do exist cycles in the latter case.

PERMUTATION PUZZLES

Consider five women, Anne, Betsy, Corrine, Donna, and Emily, who live (not necessarily respectively) in the cities Anaheim, Baltimore, Cleveland, Denver, and Enumclaw. Each woman has exactly one son, and their names are Andy, Bob, Chuck, Dave, and Edward (again, not necessarily in that order). The ages of the women are 41, 42, 43, 45, and 46 (again, not necessarily in that order). We're given the following information:

1. Emily lives in Anaheim. She's one year older than Chuck's mother who lives in Cleveland.
2. Edward's mother is Donna. She's one year older than Anne whose son isn't Dave.
3. The woman in Denver (who isn't Betsy) is younger than the woman in Enumclaw.
4. Andy's mother (not Corrine) is 43.
5. The woman living in Baltimore is 45.

How old is each woman, who is her son, and where does she live? Moreover, is there an efficient formal technique for solving problems like this? Perhaps if we think about the thought process we would use to solve it "by hand", and then think about how we might program a computer to do the same thing, the process could be formalized.

The most useful information in the particular puzzle stated above seems to be the ages and their relations to each other. There are two distinct pairs of women whose ages differ by only one, and the list of ages contains only three such possibilities. Furthermore, we know Chuck's mother isn't 45 (from her location), so we must have either [Emily=42,Chuck's=41] or else [Emily=43,Chuck's=42]. Both of these involve the age 42, so we have Donna=46 and Anne=45, hence Anne is in Baltimore. Also, we must put Donna in Enumclaw to make its resident older than the woman in Denver, who must be Corrine. Then we see Andy must be Emily's son, so she is 43, and the rest falls into place. The result is

Emily	***Anne***	***Betsy***	***Corrine***	***Donna***
43 Andy Anaheim	45 Bob Baltimore	42 Chuck Cleveland	41 Dave Denver	46 Edward Enumclaw

There are probably many different and equivalent ways of expressing and formalizing the constraints. One way is in terms of compositions of permutations. We have four groups (names, ages, son's names, and locations) of five elements, and we can arbitrarily assign the numbers 1 to 5 to the elements of each group as shown below

	City	***Woman***	***Son***	***Age***
1	Anaheim	Anne	Dave	41
2	Baltimore	Donna	Chuck	42
3	Cleveland	Emily	Bob	43
4	Denver	Betsy	Andy	45
5	Enumclaw	Corrine	Edward	46

We seek permutations of these four sets that, when aligned, satisfy certain conditions. The solution can be expressed by the following set of four permutations

Cities	12345 I
Women	31452
Sons	43215
Ages	34215

These permutations match each city with the resident woman, her son, and her age. Of course, there are really 120 distinct ways of permuting the five columns, and they all represent the same logical solution. I just arbitrarily chose to arrange the columns so that the cities have the identity permutation 12345. Three other ways of expressing the same solution are

Cities	25134	Cities	43215	Cities	43125
Women	12345 I	Women	54132	Women	54312
Sons	35421	Sons	12345 I	Sons	12435
Ages	45321	Ages	12435	Ages	12345 I

where "I" denotes the category that has the identity permutation. Obviously if the permutation from Cities to Women (for example) is 31452, then the inverse is 25134. Notice that the composition of any permutation and its inverse is the identity permutation. More generally, the composition of any "loop" of permutations must yield the identity, and this fact can be used to infer information about some of the permutations given information about some of the others. To illustrate, consider the three categories Cities, Ages, and Women, and note that we have the permutations

Cities to Women:	31452
Women to Sons:	35421
Sons to Cities:	43215

The composition of these three permutations, in sequence, necessarily gives the identity. To show how this establishes conditions on the solution, recall that we were told explicitly that Emily is in Anaheim, so the permutation from Women to Cities is of the form {**1**}. In addition, as noted above, the numerical clues about the Ages exclude all but two permutations from Cities to Ages, namely, {24135} and {34215}. Taking the second of these, we have

Cities to Ages:	3****
Ages to Women:	*****
Women to Cities:	**1**

The composition of these three permutations must be the identity permutation, which clearly requires that the permutation from Ages to Women must be of the form {**3**}. In other words, we've deduced that the 43 year old woman is Emily, which determines the rest of the solution.

Of course, there's nothing profound going on here. The leading "3" in the permutation from Cities to Ages signifies that the woman in Anaheim is 43, and the middle "1" in the permutation from Women to Cities signifies that Emily is in Anaheim. These two facts in combination obviously imply Emily is 43. This just illustrates how information on one basis is related to information on other bases. It's important to recognize that there are three distinct aspects of puzzles of this kind. The first is just general computation. For example, the particular puzzle above required making use of numerical relations between a given set of numbers.

More generally, conditions could be imposed such as "The age of person A has exactly two prime divisors in common with the age of person B". Before even beginning to solve the puzzle, we must first resolve all the computational "clues" of this kind, and express them as constrants on the allowable permutations. These computational aspects could be arbitrarily difficult–or even practically impossible–to resolve. Fro example, it might be specified that person A lives in city D if and only if the 100 trillionth decimal digit of pi is 7. This is a perfectly deterministic and well-defined "clue", but it is practically useless.

Assuming we can solve the computational aspects of the puzzle, the second task is to determine which permutations between two given sets are allowed based on a set of clues. This has nothing to do with transferring information from one basis to another. It is essentially just the classical "satisfiability problem", i.e., given a set of logical variables feeding into a fixed system of logic gates with a single logical output variable, determine which (if any) input states set the output TRUE. (It is this kind of reasoning that enables us to say the permutation from Cities to Ages must be either {24135} or {34215}, without even taking any other information into account.) This is known to be NP-complete, so we can't expect there to exist any simple recipe that would work in polynomial time in all cases. The third task involved in solving such puzzles is to relate the information that has been provided on different bases. It is only this third task (the simplest of the three) that can be formalized in terms of the compositions of permutations.

CHRISTMAS ORNAMENTS

Given a collection of Christmas ornaments in the shape of arbitrary polyhedra (such as, but not limited to cubes, tetrahedra, icosahedra, etc), suppose each face of each polyhedron is colored depending on the number of edges of the face. For example, every four-sided face (such as a face of a cube) has color number 4, and every triangular face has color number 3, and in general every n-sided face has color number n. Can there be a Christmas Ornament with all sides painted a different color?

If we assume that two faces share no more than one common edge, and that there are a finite number of faces, then the answer is easily seen to be no. Simply list, in order, the numbers of neighbors for the *n* faces, so we have a list of *n* strictly increasing positive integers (because no two faces have the same number of neighbors). Thus at least one of the *n* faces must border on at least *n* other faces, which is impossible because there are only n-1 other faces.

On the other hand, if we remove the restriction to finite-faced polyhedra, the answer is yes, because we can define a fractal structure that has no two surfaces with the same number of "sides". To illustrate this, consider a regular tetrahedron, each of whose four triangular faces has 3 sides. Now draw a triangle on one face, a square on another, a pentagon on another, and a 9-sided polygon on the remaining face. If we exclude these polygonal regions from their respective faces, then the original faces of the tetrahedron have 6, 7, 8, and 12 edges, and the excluded regions have 3, 4, 5, and 9 edges.

Needless to say, these "excluded regions" aren't really distinct faces, because they reside entirely within the planes of the original faces. However, we can fix this by building little pyramids on each of those drawn polygons. This creates a large number of new triangular faces, each of which obviously has three sides. Now we can repeat the trick, drawing a 10-sided polygon on one of those little triangular faces, an 11-sided polygon on another, a 15-sided polygon on another, and so on. Then we build little pyramids on these polygonal bases and repeat the process.

Thus, defining a solid as the limit of this construction process carried on indefinitely, we have a fractal ornament with infinitely many faces, no two of which has the same number of edges. Notice that, even though each face shares only one edge with each of its neighbors, this construction evades the earlier proof simply because it's not a finite construction, i.e., there is no finite integer n equal to the number of faces.

So, in order for the answer to be no, we need to restrict our polyhedra to those with a finite number of faces. The question is whether there exists a finite polyhedron such that no two faces have the same number of edges–where "edge" is defined as a maximal *contiguuous* line segment contained in two specific faces. Thus, two faces can share more than one edge. (Note that a contiguous line segment counts as more than one edge for a given face if it intersects more than one other face.) I'll also assume the surface has genus 2 for the moment. Let's say a face is "normal" if it shares exactly one edge with each of its neighboring faces. We've seen that if all the faces of a solid are normal, it's impossible for all of them to have a different number of edges. Now let's consider a solid with one or more non-normal face(s), (as suggested by David Einstein) such as

Fig. Two Edges Adjoining the Same Face of Neighbour

This implies the existence of two faces X and Y that are adjacent over (at least) two separate segments of their common line, as illustrated below:

Faces X and Y share the common edges AB and CD, but they do not connect to each other on the segment BC. Instead, within the region bounded by the cycle of vertices BcdefCkjihglB we have a set of other surfaces. Let's call a region like this an "anomaly" (borrowed from crystalography).

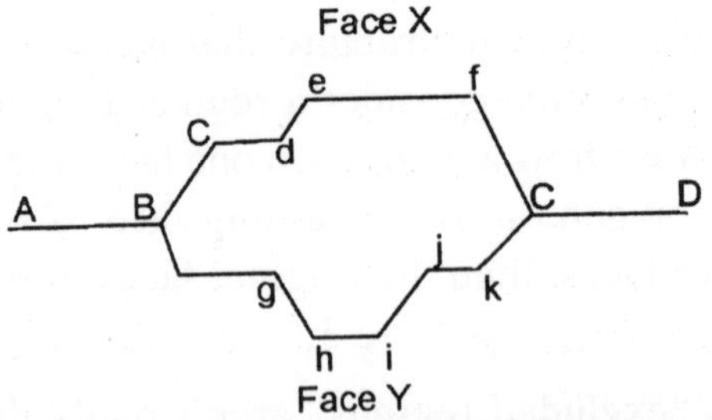

Face X
e f
C d
A B C D
j
g k
h i
Face Y

Suppose all the surfaces enclosed by X and Y in this anomaly are normal. We can then apply our original argument (strengthened by the fact that no face can have fewer than 3 edges) to show that at least one of the n faces in this anomaly must have at least n+2 neighbors, which is impossible because even counting X and Y there are only n+1 available neighbors. Therefore, we conclude that not all the faces in this anomaly can be normal, i.e., there must be two faces that share more than one common edge. In other words, there must be a (strictly smaller) anomaly within the anomaly. We can now repeat the above argument to prove that THIS smaller anomaly must also contain an anomaly, and so on ad infinitum.

This shows again that if we allow the polyhedron to have infinitely many faces the proof doesn't work–fortunately–because the theorem is false in that case. On the other hand, if we consider only polyhedra with a FINITE number of faces, then we must at some stage reach an anomaly that is filled with only normal faces, and so those faces cannot all have a different number of edges. What about the restriction to surfaces of genus 2? I imposed that because of the possibility that for a surface of higher genus an anomaly might "enclose itself", and thereby allow a finite solution, if the anomaly contains a "wormhole". There may be some way to close this loophole (literally!) for surfaces of higher genus, but I don't see it at the moment.

"God hath given you one face,

and you make yourself another."

PYTHAGOREAN GRAPHS

Is it possible to fill an infinite square array with distinct integers such that the sum of the squares of any two adjacent numbers is a square? To illustrate, the following is a 4x4 array with the desired property

1836	105	252	735
1248	100	240	700
936	75	180	525
273	560	1344	3920

Every pair of neighboring numbers (horizontally or vertically) constitutes the legs of a Pythagorean triple. The hypotenuses of these triples are as shown below

	1839		273		777	
2220		145		348		1015
	1252		260		740	
1560		125		300		875
	939		195		555	
975		565		1356		3955
	623		1456		4144	

The key to this problem is to recognize that the row and column conditions are independent. Thus, for any positive integer N we can easily construct a square array of size (N+1)x(N+1) consisting of the values

$$A[m,n] = (12/5)^m\ (24/7)^n\ 35^N$$

$$= 2^{(2m+3n)}\ 3^{(m+n)}\ 5^{(N-m)}\ 7^{(N-n)}$$

for m, n = 0,1,2,..,N. These values are all integers and no two of them have the same number of factors of 5 and 7, so they are all distinct.

Although this gives a recipe for arbitrarily large arrays, it doesn't give an *infinite* array. R. Mentock provided the following nice construction of an infinite array.

For $m = 3M + a$, $n = 3N + b$, define F(m, n) by

$F(m, n) = 6^M * 132^N * f(a, b)$

where

$f(0,0) = 6 * 7$ $\quad f(0,1) = 6 * 24$ $\quad f(0,2) = 6 * 143$

$f(1,0) = 8 * 7$ $\quad f(1,1) = 8 * 24$ $\quad f(1,2) = 8 * 143$

$f(2,0) = 15 * 7$ $\quad f(2,1) = 15 * 24$ $\quad f(2,2) = 15 * 143$

Another, shorter, solution: For m = 2M + a, n = 2N + b, define

F(m, n) by

$F(m, n) = 10^M * 76^N * f(a, b)$

where

$f(0,0) = 7 * 17$ $\quad f(0,1) = 7 * 144$

$f(1,0) = 24 * 17$ $\quad f(1,1) = 24 * 144$

Both of the above "infinite" solutions actually fill only one quadrant of the plane. To completely "tile" the plane with Pythaogrean legs, Mentock notes that

Each of the two solutions are actually combinations of two sequences, each defined along the axes. The other points are just products of these, i.e., G(m, n) = G(m) * G(n). Since we have four of them, we start at zero and assign values to the four semi-axes. G(0, 0) is the LCM of all G(0), and that number is propogated through. Then G(m,n) = G(m) * G(n) like above, *and there we go.*

A slightly different approach was suggested by Dave Radcliffe, who wrote

Given F(m, n) = $10^M * 76^N * f(a,b)$ where

$f(0,0) = 7 * 17$ $\qquad$ $f(0,1) = 7 * 144$

$f(1,0) = 24 * 17$ $\qquad$ $f(1,1) = 24 * 144$

For m, n > = 0 define

G(m, n) = F(m, n) * 11 * 13

G($-m-1$, n) = F(m, n) * 11 * 84

G($-m-1$, $-n-1$) = F(m, n) * 60 * 84

G(m, $-n-1$) = F(m, n) * 60 * 13

Since no F(m,n) is divisible by 11 or 13, G is also one-to-one.

By the way, it's also interesting to see which polyhedral can have distinct numbers at their vertices such that the sum of the squares of any two adjacent numbers is a square. Considering just the Platonic solids, I think there is no solution for the tetrahedron, octahedron, or icosahedrons. Of course there are solutions for the cube, such as

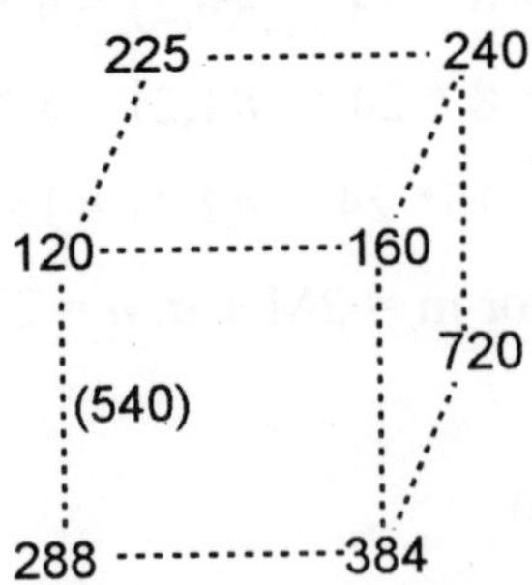

Here the values increase by factors of 4/3, 12/5, and 15/8 in the three principle directions. (Presumably it's possible to fill an infinite 3D orthogonal lattice with distinct numbers in this way.)

This leaves only the dodecahedron. It turns out that it is possible to populate the vertices of a dodecahedron with distinct numbers such that the sum of the squares of each pair of adjacent numbers is a square. Here's an example of a "Pythagorean dodecahedron":

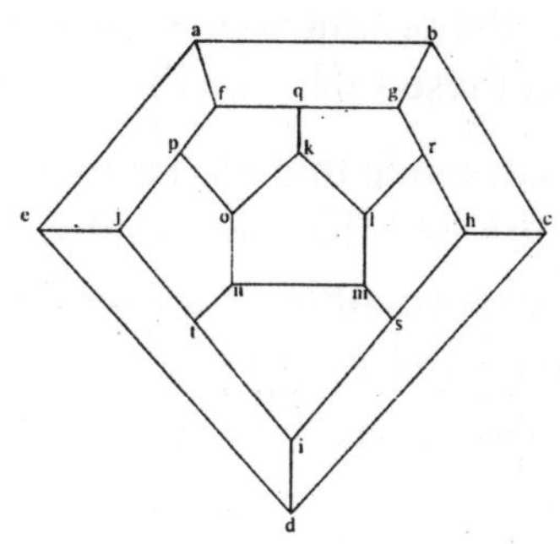

$a = 155040$	$f = 649800$	$k = 802560$	$p = 2006400$
$b = 290700$	$g = 1438965$	$\mathrm{l} = 601920$	$q = 180576$
$c = 897600$	$h = 3762000$	$m = 451440$	$r = 689700$
$d = 1683000$	$i = 2613600$	$n = 240768$	$s = 846450$
$e = 177650$	$j = 1504800$	$\mathrm{o} = 1070080$	$t = 275880$

Another related question, raised by Dave Radcliffe, is whether it possible to connect any two positive integers (greater than 4) by a series of Pythagorean legs. (As Dave put it, "consider a graph G with a vertex for each integer $n > [4]$, and an edge joining m and n if $m^2 + n^2$ is a square. Is this graph connected?") Notice that the numbers 3,4 are a separate graph and don't connect to anything else. But for $n > 4$ it certainly appears to bc truc that the graph is connected.

If you look at this as a special case of the more general "connection criterion" $m^2 + n^2 = k$*square (where k is a square-free integer expressible as a sum of two squares), it seems the case k=1 is unique in giving a connected graph for all the integers (above some number).

In contrast, consider the graph based on $m^2 + n^2 = 2$*square. In this case the naturals split into mutually disjoint graphs, each of which is a multiple of the basic connected graph consisting of the numbers

1, 7, 17, 23, 31, 41, 47, 49, 71, 73, 79, 89, 97, 103, 113, 119,...

These are exactly the numbers whose prime factors are all congruent to +1 or -1 modulo 8. In other words, they are the primes p such that 2 is a square (mod p). The graph based on $m^2 + n^2 = 5$*square splits into mutually disjoint graphs, each of which is a multiple of

1, 2, 4, 8, 11, 16, 19, 22, 29, 31, 32, 38, 41, 44, 58, 59, 61, 62,...

These appear to be exactly the numbers expressible as products of the primes 2, 11, 19, 31, 41, 59, 61, 71, 79, 89, 101,... which I think are the primes p such that 5 is a square (mod p). The graph based on $m^2 + n^2 = 10$*square splits into disjoint graphs that are each multiples of

1, 3, 9, 13, 27, 31, 37, 39, 41, 43, 53, 67, 71, 79, 81, 83,...

Again it appears there are all the products of a particular set of primes, in this case the primes 3, 13, 31, 37, 41, etc., and these look like the primes modulo which 10 is a square.

So, we might be tempted to conclude that the graph of the naturals with the connection criterion $m^2 + n^2 = k$*square splits into disjoint graphs, each of which is a multiple of a graph consisting of the numbers expressible as a product of primes p not dividing k and such that k is a quadratic residue mod p. (This explains why the case k=1 gives just a single totally connected graph.) However, the case k=13 seems to behave differently. It's basic connected set consists of

1, 8, 12, 18, 27, 34, 51, 53, 79, 86, 92, 96, 103, 116, 122,...

The set of primes for this case should be 2,3,13,17,23,29,43,etc. but there seems to be something else going on here.

ELECTORAL DRAWS

The principle of hereditary kingship became firmly established in most countries of western Europe during the middle ages, due partly to the historical accident that, for several centuries, each succeeding king by the time of his death had an adult son who seemed at least nominally capable of becoming the new king. Germany, however, was an exception. During these same centuries, a series of kings of Germany (who also held the title of Holy Roman Emperor) failed to leave any clear heir, either because they had no sons, or because their sons were still children when the king died. As a result, the question of succession had to be decided by other means. Gradually a system relying on a group of Electors was established. The Electors originally consisted of seven German dignitaries, who met after the passing of each king to select his successor. The number of Electors was later increased to nine. They were

Archbishop of Cologne

Archbishop of Mainz

Archbishop of Trier

King of Bohemia

Count Palatine of the Rhine

Duke of Saxony

Margrave of Brandenburg

(Duke of Bavaria)

(Ruler of Hanover)

The odd number of Electors was helpful in avoiding draws. It's interesting that the titles of the Electors themselves (with the exception of the three archbishops) were strictly hereditary, so ultimately the system was still based on the principle of heredity, but allowing some degree of choice.

The Electors chose the German kings for over 500 years, from the 13th century until they were finally abolished in 1806 during the Napoleonic era. Ironically, just 18 years before the final abolition of the German Electors, a new system of Electors was established for selecting the leader of the United States of America. When the constitution of the United States was written in 1788, the power to select a president was given to a group of Electors. However, unlike the German Electors, these Electors were not hereditary, but were to be themselves elected by the legislative bodies of each state of the union. Of course, the members of the legislative bodies were themselves elected, by popular vote of all the citizens. The framers might have considered making citizenship an elective office as well, but decided instead to let citizenship be hereditary. (It's not so easy to completely eliminate the principle of heredity.) New citizens could also be created by naturalization.

In practice the states soon began to allow the presidential Electors to be chosen directly by popular vote, bypassing the state legislatures. The original intent of the framers of the constitution was for the presidential Electors to exercise their own judgment in making their selections for President and Vice President, but the rise of the party system led to the practice of having each candidate Elector public ally declare his allegiance to one party, so by voting for a particular Elector the citizen is essentially voting for the presidential and vice presidential candidates of that party. Another innovation came about when the states realized that they could increase their relative influence by adopting a "winner-take-all" policy, whereby *all* the electoral candidates of the party that gets the most popular votes (in that state) are chosen.

By winning the popular vote in a winner-take-all state with nine Electors, the presidential candidate receives all nine electoral votes, whereas if the Electors were chosen in proportion to the popular vote, the presidential candidate would usually get only five to his opponent's four, giving only a one-vote difference. In the belief that candidates will pay more attention to a "winner-take-all" state than to a proportionate state, nearly all the states quickly adopted this policy. (One or two states still allow their Electors to be split between the parties, in proportion to the popular votes, but this is rare.) Of course, once all the states have adopted the winner-take-all policy, the relative advantage of this policy disappears, leaving only the disadvantage that it allows for the possibility of a presidential candidate winning the overall popular votes but losing the electoral vote. This has actually happened at least twice in United States history.

In presidential elections the fifty states–plus the District of Columbia–have a combined total of 538 Electors. The number of Electors from a state equals the number of the states' representatives in the House and the Senate, which depends (for the House) on the population of the state. Based on the population census of 2000, the number of Electors for each of the states are listed below (applicable to the presidential elections of 2004 and 2008):

55 Californmia	10 Arizona	5 Nevada
34 Texas	10 Maryland	5 New Mexico
31 New Yokr	10 Minnesota	5 Utah
27 Florida	10 Wisconsin	5 West Virginia
21 Illinois	9 Alabama	4 Hawaii
21 Pennsylvania	9 Colorado	4 Idaho
20 Ohio	9 Louisiana	4 Maine
17 Michigan	8 Kentucky	4 New Hampshire
15 Georgia	8 South Carolina	4 Rhode Island
15 New Jersey	7 Connecticut	3 Alaska
15 North Carolina	7 Iowa	3 Delaware
13 Virginia	7 Oklahoma	3 D.C.
12 Massachusetts	7 Oregon	3 Montana
11 Indiana	6 Arkansas	3 North Dakota
11 Missouri	6 Kanasas	3 South Dakota
11 Tennessee	6 Mississippi	3 Vermont
11 Washington	5 Nebraska	3 Wyoming

If we assume every state has a winner-take-all policy, a candidate can either win or fail to win each of the fifty-one regions, so the number of distinct possible outcomes (win patterns) is

$$2^{51} = 2251799813685248 \approx (2.25)10^{15}$$

This represents over two-thousand trillion outcomes.

One possible result of a two-candidate election is an electoral draw, with each candidate receiving exactly 269 electoral votes. This raises the question: In how many distinct ways can a candidate win exactly 269 votes? If we don't distinguish between states with the same number of votes, then we are just dealing with the following multi-set of summands:

3 (8)	4 (5)	5 (5)	6 (5)	7 (4)
8 (2)	9 (3)	10 (4)	11(4)	12 (1)
13 (1)	15 (3)	17 (1)	20 (1)	21 (2)
27 (1)	31 (1)	34 (1)	55 (1)	

A candidate can win 0, 1, ..., or 8 of the three-vote states, and he can win 0, 1, ..., or 5 of the four-vote states, and so on. Thus the total number of distinct win patterns for a candidate (not distinguishing between states with the same number of votes) is

$$(9)(6)^2(5)^3(4)^3(3)^2(2)^8 = 597196800$$

Of these, the number of win patterns giving exactly 269 votes is 40972162 (almost 41 million), which is about one out of 145. This is the coefficient of x^{269} in the expansion of the product

$$\left(1+x^{55}\right)\left(1+x^{34}\right)\left(1+x^{31}\right)\left(1+x^{27}\right)\left(1+x^{21}+x^{2.21}\right)\ldots$$
$$\left(1+x^{3}+x^{2.3}+x^{3.3}+x^{4.3}+x^{5.3}+x^{6.3}+x^{7.3}+x^{8.3}\right)$$

Of all the possible vote totals, from 0 to 538, the one with the most ways of occurring is 269.

If we distinguish between the states with the same number of electoral votes, then the number of distinct outcomes is increased, because each of the roughly 41 million win patterns can occur in multiple distinct ways. For example, a given win pattern may contain 5 of the 8 three-vote states, but there are C(8,5) = 56 ways of choosing five of those eight states. Applying the appropriate factors to each win pattern in the set of 269-winning patterns, we find that the number of distinct ways of winning exactly 269 electoral votes, is

1705744125652

which is just over 17 trillion, about 1/132 of all 2^{51} possible outcomes.

Chapter 10

Recurrence Relations

In mathematics, a recurrence relation is an equation that defines a sequence recursively: each term of the sequence is defined as a function of the preceding terms.

A difference equation is a specific type of recurrence relation.

An example of a recurrence relation is the logistic map:

$$x_{n+1} = rx_n (1 - x_n).$$

Some simply defined recurrence relations can have very complex (chaotic) behaviours and are sometimes studied by physicists and mathematicians in a field of mathematics known as nonlinear analysis.

Solving a recurrence relation means obtaining a closed-form solution: a non-recursive function of *n*.

LINEAR HOMOGENEOUS RECURRENCE RELATIONS

The term *linear* means that each term of the sequence is defined as a linear function of the preceding terms. The *order* of a linear recurrence relation is the number of preceding terms required by the definition—so the relation $a_n = a_{n “ 2}$ is of order two, because there must be at least two preceding terms (whether they are both used or not).

The general form of a linear recurrence relation of order *d* is as follows:

$$a_n = c_1 a_{n-1} + c_2 a_{n-2} + \ldots + c_d a_{n-d} + c$$

where c and c_i (for all i) are allowed to depend on n, but a_i (for all) is not. If is independent of (for all), then the recurrence relation is said to have constant coefficients. Additionally, if then the recurrence relation is homogeneous; such a sequence is also called a linear recursive sequence or LRS.

The linear recurrence, together with *seed values* (initial conditions) for, determines the sequence uniquely.

Fibonacci Numbers

The Fibonacci numbers are defined using the linear recurrence relation

$$F_n = F_{n-1} + F_{n-2}$$

with seed values:

$$F_1 = 1$$

$$F_2 = 1.$$

The sequence of Fibonacci numbers begins:

1, 1, 2, 3, 5, 8, 13, 21, 34, 55, 89...

It can be solved via the matrix solution described below, yielding the closed form expression.

Solving via Linear Algebra

Given an LRS, one can write down its companion matrix, then put it in Jordan normal form (which is diagonal if the eigenvalues are distinct). Expressing the seed in terms of the eigenbasis, say

$$\begin{bmatrix} a_0 \\ \vdots \\ a_{d-1} \end{bmatrix} = b_1 v_1 + \ldots + b_d v_d$$

yields

$$\begin{bmatrix} a_n \\ \vdots \\ a_{n+(d-1)} \end{bmatrix} = C_n \begin{bmatrix} a_0 \\ \vdots \\ a_{d-1} \end{bmatrix} = C^n (b_1 v_1 + \ldots + b_d v_d) = \lambda_1^n b_1 v_1 + \ldots + \lambda_d^n b_d v_d$$

which is a closed form expression (expand on the first coordinate to obtain a closed form expression for a_n).

If the companion matrix is not diagonalizable, then the resulting expression is more complicated, but conceptually the same.

Solving Generally

Solutions to recurrence relations are found by systematic means, often by using generating functions (formal power series) or by noticing the fact that r^n is a solution for particular values of r.

Consider, for example, a recurrence relation of the form

$$a_n = Aa_{n-1} + Ba_{n-2}.$$

Suppose that it has a solution of the form $a_n = r^n$. Substituting this guess in the recurrence relation, we find:

$$r^n = Ar^{n-1} + Br^{n-2}.$$

Dividing through by r^{n-2}, we get

$$r^2 = Ar + B,$$

$$r^2 = Ar - B = 0$$

This is known as the *characteristic equation* of the recurrence relation. Solve for r to obtain the two roots λ_1, λ_2, and if these roots are distinct, we have the solution

$$a_n = C\lambda_1^n + D\lambda_2^n$$

while if they are identical (when $A^2 + 4B = 0$), we have

$$a_n = C\lambda^n + D\lambda^n$$

where constants C and D can be found from the "side conditions" that are often given as $a_0 = a$, $a_1 = b$.

Different solutions are obtained depending on the nature of the roots of the characteristic equation.

GENERAL LINEAR HOMOGENEOUS RECURRENCE RELATIONS

Many linear homogeneous recurrence relations may be solved by means of the hypergeometric series. Special cases of these lead to recurrence relations for the orthogonal polynomials, and many special functions. For example, the solution to

$$J_{n+1} = \frac{2n}{z} J_n - J_{n-1}$$

is given by the $J_n = J_n(z)$, the Bessel function, while

$$(b - n)M_{n-1} + (2n - b - z)M_n - nM_{n+1} = 0$$

is solved by $M_n = M(n,b;z)$ the confluent hypergeometric series.

SOLVING NON-HOMOGENEOUS RECURRENCE RELATIONS

If the recurrence is inhomogeneous, a particular solution can be found by the method of undetermined coefficients and the solution is the sum of the solution of the homogeneous and the particular solutions. Another method to solve an inhomogeneous recurrence is the method of *symbolic differentiation*. For example, consider the following recurrence:

$$a_{n+1} = a_n + 1$$

This is an inhomogeneous recurrence. If we substitute $n \mapsto n + 1$, we obtain the recurrence

$$a_{n+2} = a_{n+1} + 1$$

Subtracting the original recurrence from this equation yields

$$a_{n+1} - a_{n+1} = a_{n+1} - a_n$$

or equivalently

$$a_{n+2} = 2a_{n+1} - a_n$$

This is a homogeneous recurrence which can be solved by the methods explained above. In general, if a linear recurrence has the form

$$a_{n+k} = \lambda_{k-1}a_{n+k-1} + \lambda_{k-2}a_{n+k-2} + \dots + \lambda_1 a_{n+1} + \lambda_0 a_n + p(n)$$

where $\lambda_0, \lambda_1, \dots, \lambda_{k-1}$ are constant coefficients and $p(n)$ is the inhomogeneity, then if $p(n)$ is a polynomial with degree r, then this inhomogeneous recurrence can be reduced to a homogeneous recurrence by applying the method of symbolic differentiation r times.

Example: Let's solve the following inhomogeneous recurrence relation using the method of undetermined coefficients:

$$a_{n+1} = 2a_n + 3^n + 5_n$$

The general solution for the analogous homogeneous relation

$$b_{n+1} = 2b_n$$

is

$$b_n = c_1 2^n$$

The inhomogeneous part $(3^n + 5n)$ leads us to guess the particular solution

$$a_n = c_2 a + c_3 n + c_4$$

(the guess for a^x is a^x, and the guess for a degree-n polynomial is a degree-n polynomial)

Substituting back into the recurrence relation, we get:

$$c_2 3a^{n+1} + c_3(n+1) + c_4 = 2(c_2 3^n + c_3 n + c_4) + 3^n + 5n$$

Looking only at the coefficients of 3^n:

$$3c_2 = 2c_2 + 1$$

$$c_2 = 1$$

Looking only at the coefficients of n:

$$c_3 = 2c_3 + 5$$
$$c_3 = -5$$

Looking only at the coefficients of 1 (constants):

$$c_3 + c_4 = 2c_4$$
$$c_4 = c_3 = -5$$

So the general solution is

$$a_n = c_1 2^n + 3^n - 5n - 5$$

A technique is described for using Newton's identities to compute the coefficients of a multi-step recurrence formulae for simulating the solutions of ordinary linear differential equations of high order. This approach gives the exact root-matched recurrence formula without having to determine the complex roots of the characteristic polynomials.

Let x(t) and y(t) denote continuous functions of time related according to the Nth order differential equation

$$\sum_{k-0}^{N} a_k \frac{d^k y}{dt} = \sum_{k=0}^{N} b_k \frac{d^k x}{dt^k} \qquad a_N, a_0, b_N {}'0$$

where the coefficients a_k and b_k are constants. Simple linear relationships of this form are widely used in numerical simulations of heat transfer, chemical reactions, fluid flows, mechanical dynamics, Markov processes, etc. (Relationships of this form are also used in signal filtering and control-system compensation, but in those applications the frequency response is usually emphasized over the exact time-domain response.) A variety of methods have been used to numerically generate the solutions y(t) of such equations for arbitrary forcing functions x(t). In the digital environment equation is often simulated by means of a multi-step linear recurrence relation of the form

$$h\, Y_n = g\, X_n$$

where Xn and Yn are the column vectors

$$Y_n = \begin{bmatrix} y((n-N)\Delta t) \\ \vdots \\ y((n-N)\Delta t) \\ y(n\Delta t) \end{bmatrix} \quad X_n = \begin{bmatrix} y((n-N)\Delta t) \\ \vdots \\ y((n-N)\Delta t) \\ x(n\Delta t) \end{bmatrix}$$

for some constant Δt (called the *step-size* of the simulation), and g and h are constant row vectors

$$h = [h_0\ h_1 \ldots h_N]\ g = [g_0\ g_1 \ldots g_N]$$

Given the coefficients ak and bk, our objective is to compute, as quickly and efficiently as possible, the components of h and g.

It is well known that if the recurrence is required to match the exact homogeneous response for y(t) when x(t) is constantly zero, then the components of h must be proportional to the elementary symmetric functions (with alternating signs) of the quantities $e^{\alpha_i \Delta t}$, i = 1,2,..,N, where αi are the roots of the characteristic polynomial.

The usual procedure for determining this "root-matched" recurrence is to solve the characteristic polynomial for the roots ai, and then compute the elementary symmetric functions of these roots. For high order relationships the most troublesome aspect of this approach is the determination of all the complex roots of the characteristic polynomials, especially when there are multiple roots.

The most common way of simplifying the calculations is to use a substitution method, e.g., bilinear substitution, but such methods do not reproduce the exact homogeneous response. The purpose of this note is to describe how, using Newton's Identities, we can precede directly from the coefficients to the exact root-matched recurrence coefficients without solving the characteristic polynomials.

DETERMINING THE ROOT MATCHED RECURRENCE

For any integer k, let w(k) denote the sum of the kth powers of the roots $a_1,a_2,..,a_N$. Clearly, w(0) = N. Using Newton's Identities [8,9] the values of w(k) for k = 1,2,... are given by

$a_N\, w(1) + 1a_{N-1} = 0$

$a_N\, w(2) + a_{N-1}\, w(1) + 2a_{N-2} = 0$

$a_N\, w(3) + a_{N-1}\, w(2) + a_{N-2}\, w(1) + 3a_{N-3} = 0$

for k < N, and the recurrence

$a_N\, w(k) + a_{N-1}\, w(k\text{-}1) + ... + a_0\, w(k\text{-}N) = 0$

for all k ≥ N. Now let E(k) denote the sum of the kth powers of the values of $e^{\alpha_i \Delta t}$, i=1,2,..,N. Expanding each of the exponential functions as a Taylor's series and combining terms by powers of Δt gives the following expression for E(k):

$$E(k) = \sum_{j-0}^{\infty} \frac{(k\Delta t)^j}{j!} w(j)$$

Letting σ_k denote (-1)k times the kth elementary symmetric function of the quantities $e^{\alpha_i \Delta t}$, we can now compute σ_0 through σ_N using the identities

$E(1) + 1\sigma_1 = 0$

$E(2) + \sigma_1 E(1) + 2\sigma_2 = 0$

$E(3) + \sigma_1 E(2) + \sigma_2 E(1) + 3\sigma_3 = 0$

The precision of these computations, including the convergence, can be checked by verifying the equality

$$\sigma_N = (-1)^N e^{-a_{N-1}\Delta t / a_N}$$

The general recurrence for the homogeneous equation in y(t) can now be expressed as:

$$S_A Y_n = 0$$

where SA is the row vector

$$S_A = [\sigma_N \; \sigma_{N-1} \ldots \; \sigma_0]$$

Figure presents an implementation of the preceding algorithm in BASIC. The inputs to this program are the order N, the step-size Dt, and the coefficients $a_0, a_1 .., a_N$. The program computes the values of $\sigma_0, \sigma_1, .., \sigma_N$. The parameter "sumto" determines the number of terms to be evaluated in the summation.

By entering the coefficients bk in place of ak into the preceding algorithm we can determine the recurrence for the homogeneous equation in x(t)

$$S_B X_n = 0$$

We can then combine the two sides of the recurrence by applying the appropriate scale factors, which are easily deduced from steady-state conditions. The result is

$$\frac{a_0}{\sum s_A} S_A Y_n = \frac{b_0}{\sum s_B} S_B X_n$$

where ΣS_A and ΣS_B denote the sums of the components of SA and SB respectively.

EQUILATERAL RECURRENCES

If $b_N = 0$ then the characteristic polynomial of the right side has fewer than N roots. In this case the recurrence would involve fewer "past values" of x(t) than of y(t). Note that the coefficients of x_n were uniquely determined by the requirement that x(t) conform to the true homogeneous behavior when y(t) is constantly zero.

However, in practice x(t) is an arbitrary forcing function so we are not really concerned with reproducing the homogeneous "response" of x(t). Instead, we are trying to reconstruct the function x(t) from the discrete samples. Thus it is often considered more appropriate to make use of as many past values of x(t) as is computationally convenient. This typically leads to recurrences that use the same number of past values of x(t) as of y(t), i.e., *equilateral* recurrences. The usual approach in signal processing is to augment the roots of the right-hand characteristic polynomial with fictitious roots of the form ±p i, and then proceed as before. The effect is to add roots equalling -1 to the characteristic polynomial of the recurrence, which essentially mimics bi-linear substitution.

The rationale for the "method of fictitious roots" just described is based on the frequency response characteristics. In the time domain it is often more appropriate to convert into an equilateral equation by a simple change of variables. If we define the variable z(t) such that $y=\lambda(x+z)$ for some constant λ, then (1) can be written

$$\sum_{k-0}^{N} a_k \frac{d^k z}{dt^k} = \sum_{k-0}^{N} \left(\frac{b_k}{\lambda} - a_k \right) \frac{d^k x}{dt^k}$$

For any choice of λ not equal to 0, (b_N/a_N), or (b_0/a_0), this equation is equilateral, so we can apply the method to give the recurrence

$$h\, Z_n = g\lambda\, X_n$$

where $g\lambda$ is based on the coefficients of the right hand side of (5). Reverting back to the y(t) variable gives

$$h\, Y_n = \lambda\, (g\lambda + h)\, X_n$$

For large values of λ the vector $\lambda\,(g\lambda + h)$ is quite insensitive to changes in λ, and it converges in the limit as $|\lambda| \to \infty$. In practice, any λ greater than about 10 times the largest b_i will give very nearly the limiting vector. This "change-of-variables method" enables us to determine equilateral recurrences for non-equilateral differential equations without resorting to fictitious roots.

THE EXPECTED VALUE METHOD

For comparison we briefly describe a method that explicitly treats x(t) as an *independent* variable and y(t) as a *dependent* variable. From this point of view it can be argued that the *expected value* of x(t) based on the N+1 most recent samples is given by the Nth degree polynomial that passes through those samples. Although the method is a straight-forward application of polynomial curve fitting, no explicit formulation of the general Nth-order recurrence formula appears in the literature, so a brief description is given below.

As in our previous recurrence formulas, we require that the homogeneous response of y(t) be matched exactly, so we use the algorithm shown in Figure to determine SA. Combining this with the particular solution based on the Nth order polynomial curve fit of x(t) leads directly to the total recurrence

$$S_A Y_n - S_A T A^{-1} B T^{-1} X_n$$

where A, B, and T are square matrices such that the elements in the jth columns of the kth rows (where j,k range from 0 to N) are given by

$$A_{kj} = \begin{cases} \frac{j!}{k!} a_{j-k} & \text{if } j \geq k \\ 0 & \text{if } j \geq k \end{cases} \quad B_{kj} = \begin{cases} \frac{j!}{k!} b_{j-k} & \text{if } j \geq k \\ 0 & \text{if } j < k \end{cases} \quad T_{kj} = \begin{cases} (k\Delta t)^j & \text{if } k+j>0 \\ 0 & \text{if } k+j=0 \end{cases}$$

This method gives, arguably, the best possible reconstruction of arbitrary forcing functions x(t), although the matrix inversions make it somewhat laborious when applied to high order equations.

Examples: To illustrate the basic algorithm for determining the root-matched recurrence relation for a homogeneous equation, consider the differential equation

$$\frac{d^5y}{dt^5} + 7\frac{d^4y}{dt^4} + 21\frac{d^3y}{dt^3} + 33\frac{d^3y}{dt^2} + 28\frac{dy}{dt} + 10y = 0$$

To simplify the recurrence expressions we will denote $y(n\Delta t)$ by y_n. Taking a time interval of

$\Delta t = 0.1$, the algorithm described gives the following recurrence formula for y_n

$$y_n = -h_4\, y_{n-1} - h_3\, y_{n-2} - h_2\, y_{n-3} - h_1\, y_{n-4} - h_0\, y_{n-5}$$

where

h_0 = -0.4965853038

h_1 = 2.8479524294

h_2 = -6.5428490609

h_3 = 7.5263050469

h_4 = -4.3347524368

(Notice that we've scaled the coefficients to make h_5 equal to 1.) The accuracy of this recurrence can be illustrated by noting that the characteristic roots -1, -2+i, -2-i, -1+i, and -1-i, so one particular analytical solution has the simple form

$$y_n = -8e^{-n\Delta t} + (-6e^{-2n\Delta t} + 14e^{-n\Delta t}) \cos(n\Delta t)$$

Taking y_0 through y_4 from as initial values, it is easily verified that the recurrence numerically reproduces the exact analytical values of y_5, y_6, y_7... etc., very precisely. After 35 iterations of the recurrence formula (carried out to 10 significant decimal digits) the value of y_{40} differs from the analytical value by only 0.0000013, this discrepancy being due entirely to accumulated round-off error. In contrast, using the recurrence relation given by bilinear substitution, the error at y_{40} is 0.0013094. For larger step-sizes the error using bilinear substitution becomes progressively worse, whereas the root-matched recurrence method is equally applicable to any step-size.

We chose a homogeneous equation with simple characteristic roots for the preceding example so that the recurrence values could be easily compared to the analytical solution.

Also, the characteristic roots in the preceding example all have negative real parts, so the solutions are "stable". However, our interest is not limited to "stable" systems. For example, problems involving combustion and thermal "run-away" often lead to "unstable" systems. To illustrate the various methods discussed in this note as applied to one of these "unstable" systems, and to include "numerator dynamics", consider the differential equation

$$\frac{d^5y}{dt^5}-\frac{d^3y}{dt^3}-2\frac{d^2y}{dt^2}+y=\frac{d^3x}{dt^3}-\frac{dx}{dt}-x$$

Taking a time interval of $\Delta t = 1$, the relation between x and y can be simulated using a linear recurrence of the form (2), where the components of h and g are as listed in the table below (normalized for $h_5 = 1$).

i	*All except bilinear substitut*	*Bilinear substitut*	*Expected value method*	*Change of variables method*	*Fictitious roots method*	*Bilinear substit*
0	−1.000000	−1.823529	−0.283164	−0.083311	−0.250012	−0.411765
1	5.060084	8.764706	−0.530963	−0.537182	0.386879	0.529412
2	−11.732310	−18.235294	2.456330	2.489417	0.365333	0.588235
3	14.904837	21.294118	−3.428658	−3.485152	−1.780006	−1.764706
4	−7.146380	−9.117647	0.407632	0.446686	−0.658437	−1.117647
5	1.000000	1.000000	0.092613	0.083311	0.250012	0.294118

All the methods except Bilinear Substitution made use of the BASIC program in Figure to determine the components of h and to assist in determining the components of g. Interestingly, the Change-of-Variables Method gives components of g that closely match those given by the Expected Value Method, even though the former makes no use of a 5th order curve fit of x(t). In contrast, the methods of fictitious roots and bilinear substitution give very different g vectors, and therefore will clearly yield significantly different values of y(t) in response to certain forcing functions x(t).

RECURRENCES FOR HARMONIC SUMS

Let h(n) denote the nth partial sum of the harmonic series, and define $H[k] = h(2^k)$. It's not hard to show that H[n] approaches a linear recurring sequence satisfying

$$H[k] - 2\,H[k-1] + H[k-2] = 0$$

as k increases. The error of this recurrence decreases by a factor of two at each step. Correcting for this error, we arrive at the third- order recurrence

$$2\,H[k] - 5\,H[k-1] + 4\,H[k-2] - H[k-3] = 0$$

Eliminating the main error term from this relation leads to a fourth- order recurrence, and so on. The coefficients of these recurrences up through order six are tabulated below (without signs):

(2)	1	2	1			
(4)	2	5	4	1		
(16)	8	22	21	8	1	
(64)	128	360	358	149	24	1
(256)	8192	23168	23272	9894	1685	88 1

The numbers in parentheses on the left indicate the factor by which the error of the respective recurrence decreases at each step. If c[i, j] denotes the coefficient in the ith row and the jth column, and r_i denotes the "error factor" of the ith row, then the rule of formation for these coefficients can be expressed as

$$c[i, j] = (r_{(i-1)})\, c[i-1, j] + c[i-1, j-1]$$

Note that the sum of the elements of each row, with alternating signs, is zero. Also, the "error factor" increases by a multiple of 4 at each level except the first,

so we have

$$r_1 = 2 \text{ and } r_i = 4^{(i-1)} \text{ for } i>1.$$

This same approach works for other geometric arguments, e.g., if we define G[k] = $h(3^k)$, we can derive a sequence of recurrences with the coefficients listed below (without signs):

(3)	1	2	1		
(9)	3	7	5	1	
(81)	27	66	52	14	1

The rule of formation is the same, except that in this case we have $r_1 = 3$ and $r_i = 9^{(i-1)}$ for i>1.

To generalize this a bit, for any fixed integer m>1 define

$$H[k] = h(m^k)$$

Since h(n) increases logrithmically, it's natural to expect that the above recurrence is a good starting point for any m, because

$$\ln(m^k) + A \ln(m^{(k-1)}) + B \ln(m^{(k-2)}) = k \ln(m) + A (k-1) \ln(m) + B (k-2) \ln(m)$$

Setting this equal to zero and dividing by ln(m) gives

$$k + A (k-1) + B (k-2) = k(1+A+B) - (A+2B) = 0$$

which implies the two conditions

$$1 + A + B = 0 \quad A + 2B = 0$$

so we must have A=-2 and B=1. However, the third order recurrence for $h(m^k)$ gives only two constraints on three coefficients, i.e.,

$$1 + A + B + C = 0 \qquad A + 2B + 3C = 0$$

and similarly for higher order recurrences. Thus, all recurrences for $\ln(m^k)$ of order greater than 2 are underspecified. Anyway, taking the order 2 recurrence we see that the characteristic polynomial splits into linear factors

$$x^2 - 2x + 1 = (x-1)^2.$$

It turns out that the characteristic polynomials of all the asymptotic recurrences for H[k] described previously also split into linear factors. Specifically, for m=2 we have

$$2x^3 - 5x^2 + 4x - 1 = (2x-1)(x-1)^2$$

$$8x^4 - 22x^3 + 21x^2 - 8x + 1 = (4x-1)(2x-1)(x-1)^2$$

$$128x^5 - 360x^4 + 358x^3 - 149x^2 + 24x - 1 = (16x-1)(4x-1)(2x-1)(x-1)^2$$

and so on. Similarly for m=3 we have

$$3x^3 - 7x^2 + 5x - 1 = (3x-1)(x-1)^2$$

$$27x^4 - 66x^3 + 52x^2 - 14x + 1 = (9x-1)(3x-1)(x-1)^2$$

and so on. In general, the characteristic polynomial of the asymptotic Jth-order recurrence (J>3) for H[k] with any fixed m is

$$P(x) = (mx-1)\,(x-1)^2 \prod_{t=1}^{J-3} (m^{(2t)}\,x - 1)$$

so the characteristic roots (eigen values) are

$$1,\ 1,\ 1/m,\ 1/m^2,\ 1/m^4,\ 1/m^6, \ldots,\ 1/m^{(2(J-3))}.$$

This allows us to give an explicit formula for the terms of the Jth-order linear recurring sequence of values that approximate H[k].

Let Q be a JxJ matrix with the components

$Q[i,1] = 1$ for i = 1 to J

$Q[i,2] = i$ for i = 1 to J

$Q[i,3] = 1/m^i$ for i = 1 to J

$Q[i,j] = 1/m^{2i}(j-3)$ for i = 1 to J and j = 4 to J

and let R be a column vector with the components

$R[i] = H[i-1]$ for i=1 to J

Then the asymptotic recurrence for H[k] has the following closed-form solution using the exact values H[0] through H[J-1] as initial values:

$$H[k-1] = [1\ k\ 1/m^{k}\ 1/m^{2k}\ 1/m^{4k} \ldots\ 1/m^{(2k(J-3))}]\ Q^{-1}\ R$$

Clearly, as k increases, all but the first two terms of this expression go to zero, so for large values of k we have approximately

$$H[k-1] = C[1] + C[2]*k$$

where C[1] and C[2] are the first two components of the column vector $C = Q^{-1} R$. The difference between H[k-1] and $h(2^{(k-1)})$ is therefore approximated by

$$C[1] + C[2]\ k - (k-1) \ln(2) = C[1] + \ln(2) + k\ (C[2] - \ln(2))$$

For high order recurrence this value becomes virtually independent of k, which suggests that C[2] approaches ln(2). Making this substitution,

we would expect to find that

$$\text{gamma} \sim = C[1] + C[2]$$

To illustrate, with m = 2 the closed–form solution for the 6th–order recurrence gives

$$C[1] + C[2] = 0.5772175....$$

which is approximately equal to gamma = 0.5772156...

INTERLEAVING FIBONACCI NUMBERS

The sequence of Fibonacci numbers is defined by the initial values $f_0 = 0$, $f_1 = 1$, and the recurrence $f_n = f_{n-1} + f_{n-2}$. The first several values are

n:	0	1	2	3	4	5	6	7	8	9	10	11	...
f_n:	0	1	1	2	3	5	8	12	21	34	55	89	...

Notice that the even-ordered terms satisfy the relations

$$f_2 = 1$$
$$f_4 = 1f_2 + 2$$
$$f_6 = 1f_4 + 2f_2 + 3$$
$$f_8 = 1f_8 + 2f_4 + 3f_2 + 4$$
$$f_{10} = 1f_8 + 2f_6 + 3f_4 + 4f_2 + 5$$

and so on. Similarly the odd-ordered terms satisfy the relations

$$f_1 = 1$$
$$f_3 = 1f_1 + 1$$

$$f_5 = 1f_3 + 2f_1 + 1$$
$$f_7 = 1f_5 + 2f_3 + 3f_1 + 1$$
$$f_9 = 1f_7 + 2f_5 + 3f_3 + 4f_1 + 1$$

and so on. One way of proving that this pattern continues is by induction. We know by inspection up to a certain n that consecutive Fibonacci numbers are given by

$$f_{2n} = 1f_{2(n-1)} + 2f_{2(n-2)} + \ldots + (n-1)f_2 + n$$
$$f_{2n+1} = 1f_{2(n-1)} + 1 + 2f_{2(n-2)+1} + \ldots + (n-1)f_3 + (n)f_1 + 1$$

(Note that this requires $f_1 = f_2 = 1$.) Adding these together, term by term, it follows from the basic Fibonacci recurrence that the next even-ordered value in the sequence is

$$f_{2(n+1)} = 1f_{2n} + 2f_{2(n-1)} + \ldots + (n-1)f_4 + (n)f_2 + (n+1)$$

which is of the same form as the previous even-ordered term, but with n+1 in place of n. Likewise the next odd-ordered value in the sequence can be found by adding this expression for $f_{2(n+1)}$ to the preceding expressing for f_{2n+1}, which gives

$$f_{2(n+1)+1} = 1f_{2n} + 2f_{2(n-1)} + \ldots + (n-1)f_4 + (n)f_2 + (n+1)$$

where we have used the fact that $f_1 = 1$. This is of the same form as the previous odd-ordered term, but with n+1 in place of n, so the induction is complete. There, the interleaving odd and even parts of the Fibonacci numbers satisfy the relations

$$f_{2n} = n + \sum_{k=1}^{n-1} kf_{2(n-k)} \quad f_{2n+1} = 1 + \sum_{k=1}^{n} kf_{2(n-k)+1}$$

Naturally these identities are closely related to the generating functions for the recurrence relations. We know the ratios of successive values of fn asymptotically approach successive powers of the dominant root ϕ of the characteristic equation $x^2 - x - 1 = 0$. Therefore, with n = 6, the even-ordered relation approaches

$$\phi^{12} = 1\phi^{10} + 2\phi^8 + 3\phi^6 + 4\phi^4 + 5\phi^2 + 6$$

and if we divide this by ϕ^{10} we get

$$\phi^2 = 1 + 2\phi^{-2} + 3\phi^{-4} + 3\phi^{-6} + 5\phi^{-8} + 6\phi^{-10}$$

Taking this to the limit as n goes to infinity, and substituting for the derivative of the geometric series, we get

$$\phi^2 = \frac{1}{\left(1-\phi^{-2}\right)^2}$$

which is equivalent to the characteristic equation $\phi^2 - \phi - 1 = 0$. However, this isn't the only possible expansion of the terms of the even (or odd) ordered Fibonacci sequences. For example, given the even-ordered sequence 0, 1, 3, 8, 21, 55, 144, ... we could simply apply the "greedy algorithm" by expressing each term as the sum of the maximum number of the immediately preceding terms. Thus we have

$$2(1)+1=3$$
$$2(3)+1(1)+1=8$$
$$2(8)+1(3)+1(1)+1=21$$
$$2(21)+1(8)+1(3)+1(1)+1=5$$
$$1(55)+1(21)+1(8)+1(3)+1(1)+1=144$$

From this it's easy to see that $f_{2n} = f_{2(n-1)} - f_{2(n-2)} + 2f_{2(n-1)}$, so the even-ordered Fibonacci numbers satisfy the recurrence

$$f_{2n} = 3f_{2(n-1)} - f_{2(n-2)}$$

which has the characteristic equation

$$x^4 - 3x^2 + 1 = \left(x^2 - x - 1\right)\left(x^2 + x - 1\right)$$

This shows that another way of arriving at the recurrence for the kth Fibonacci numbers is to solve for the (k-2)th degree polynomial that gives, when multiplied by the characteristic polynomial of the Fibonacci sequence, a polynomial whose only non-zero coefficients are for powers that are multiples of k. For example, to find the recurrence for every third Fibonacci number, we evaluate the product

$$\left(x^2 - x - 1\right)\left(x^4 + Ax^3 + Bx^2 + Cx + D\right)$$
$$= x^6 + (A-1)x^5 + (B-A-1)x^4 + (C-A-B)x^3 + (D-C-B)x^2 + (-C-D)x - D$$

We need the coefficients of x^5, x^4, x^2, and x to vanish, so we require A = 1, B = 2, C = –1, and D = 1. Hence the characteristic polynomial for every third Fibonacci number is

$$x^4 - 4x^3 - 1 = \left(x^2 - x - 1\right)\left(x^4 + x^3 + 2x^2 - x + 1\right)$$

with the corresponding recurrence relation

$$f_{3n} = 4f_{3(n-1)} + f_{3(n-2)}$$

Naturally the same type of relations can be derived for any linear recurring sequence. For example, consider the "tribonacci" sequence, defined by the initial values $t_0 = t_1 = t_2 = 1$, and the recurrence $t_n = t_{n-1} + t_{n-2} + t_{n-3}$. The first several values are shown below.

n:	0	1	2	3	4	5	6	7	8	9	10	11	12	13	14...
t_n:	1	1	1	3	5	9	17	31	57	105	193	355	653	1201	2209...

To find the recurrence formula for every third term of this sequence, we can apply the greedy algorithm to express the terms 1, 9, 57, 355, 2209, 13745, 85525, ... as follows

$$6(57)+1(9)+4(1)=355$$

$$6(355)+1(57)+2(9)+4(1)=2209$$

$$6(2209)+1(355)+2(57)+2(9)+4(1)=13745$$

$$6(12745)+1(2209)+2(355)+2(57)+2(9)+4(1)=85525$$

From this it's easy to see that $t_{3n} = t_{3(n-1)} - 5t_{3(n-2)} + t_{3(n-3)} + 6t_{3(n-1)}$, so we have the recurrence

$$t_{3n} = 7\, t_{3(n-1)} - 5t_{3(n-2)} + t_{3(n-3)}$$

corresponding to the characteristic polynomial

$$x^9 - 7x^6 + 5x^3 - 1 = (x^3 - x^2 - x - 1)(x^4 + x^5 + 2x^4 - 3x^3 - x + 1)$$

We can relate this back to the greedy recurrence by examining the asymptotic expression in terms of the dominant root f, which can be written in the form

$$\phi^3 = 6 + \phi^{-3} + 2\phi^{-6} + 2\phi^{-9} + \ldots$$

$$= 4 - \phi^{-3} + 2\left(\frac{1}{1-\phi^{-3}}\right)$$

This implies that f is a root of the characteristic polynomial, as expected. Incidentally, the "poles" of this recurrence are 0 and the cube roots of 1.

ROOT-MATCHED RECURRENCES FOR DIFF EQS

Here is a method for determining the "root-matched" recurrence corresponding to a given ordinary differential equation without having to explicitly solve for the roots of the characteristic polynomial. More generally, this method enables us to find the exact root-matched recurrence corresponding to any continuous linear system without having to solve for the eigen values of the system. The following description covers the just simple homogeneous case. Suppose you have an ordinary linear differential equation with constant coefficients:

a5 y''''' + a4 y'''' + a3 y''' + a2 y'' + a1 y' + a0 y = 0

where the (') denotes derivatives with respect to t. The response of y(t) can be reproduced discretely by a recurrence relation of the form

c5 y(5dt) + c4 y(4dt) + c3 y(3dt) + c2 y(2dt) + c1 y(dt) + c0 y(0) = 0

where "dt" is the time increment of the simulation. The question is, given the coefficients ai of the differential equation, how can you compute the coefficients ci of the corresponding recurrence relation.

The coefficients ci must be proportional to the elementary symmetric functions (with alternating signs) of the quantities $e^{(ri\ dt)}$, where ri (i=1,2,..,5) are the roots of the characteristic polynomial. Thus, a straightforward method of computing the ci is to solve the characteristic polynomial for the roots ri, and then compute the elementary symmetric functions of these roots. However, determining all the roots of a high order polynomial can be computationally tricky, particularly for embedded code, where the use of iterative processes can cause problems.

Fortunately, there is a way to compute the ci without having to soive for the characteristic roots of the equation. The technique makes use of "Newton's Identities" and the Taylor's series expansion of the exponential function.

Assume our basic ODE is of order N. For any integer k, let w(k) denote the sum of the kth powers of the roots r1,r2,...,rN. Clearly, w(0)=N. Using Newton's Identities, the values of w(k) for k=1,2,... are given by

$$w(1)a(N) + 1a(N-1) = 0$$

$$w(2)a(N) + w(1)a(N-1) + 2a(N-2) = 0$$

$$w(3)a(N) + w(2)a(N-1) + w(1)a(N-2) + 3a(N-3) = 0 \text{ etc}$$

for k=N. Now let E(k) denote the sum of the kth powers of the values $e^{(ri\ dt)}$, i=1,2,..,N. Expanding each exponential function as a Taylor series and combining terms by powers of dt gives the following expression for E(k):

$$E(k) = \text{SUM from } j=0 \text{ to inf of } \{((k\ dt)^j)\ w(j)/\ j!\}$$

for k=1,2,..,N. The coefficients of the recurrence relation can then be computed using the identities

$$c(N) = 1$$

$$E(1) + 1\ c(N-1) = 0$$

$$E(2) + E(1)\ c(N-1) + 2\ c(N-2) = 0$$

$$E(3) + E(2)\ c(N-1) + E(1)\ c(N-2) + 3\ c(N-3) = 0$$

This gives the exact root-matched recurrence coefficients without ever having to determine the roots. It also has the advantage that no special provisions are needed to cover the case of repeated roots.

THE GOLDEN PENTAGON

The plane can be tiled by a geometric progression of equilateral triangles with the characteristic equation $x^3 - x - 1$ as shown below. In view of the popularity of the "golden rectangle" based on the geometric spiral of squares, it's surprising the analagous construction for equilateral triangles is rarely mentioned. Another picture, this one with the triangles numbered insequence, is shown below.

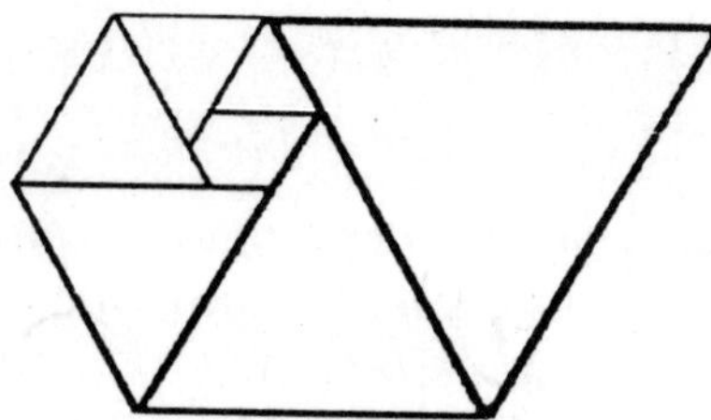

Notice that if h(k) denotes the height of the kth trianglein the spiral, then each of the following recurrences issatisfied

$h(k) = h(k-1) + h(k-5)$

$h(k) = h(k-2) + h(k-3)$

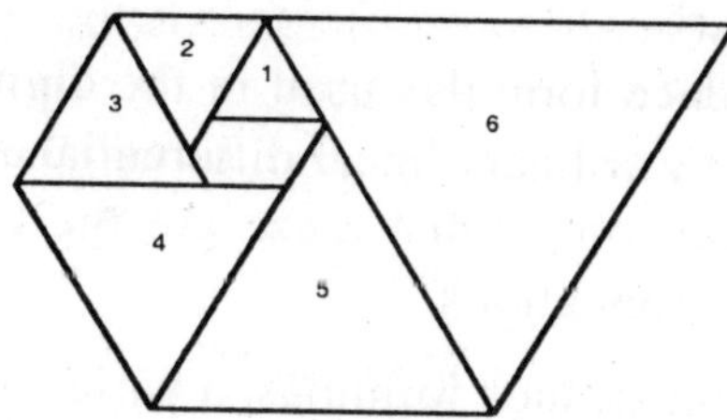

This corresponds to the fact that the sequence has two possiblecharacteristic equations

$x^5 - x^4 - 1 = 0$

$x^3 - x - 1 = 0$

This isn't surprising, because the second polynomial is adivisor of the first, i.e.,

$(x^5 - x^4 - 1) = (x^3 - x - 1)(x^2 - x - 1)$

Each triangle in the sequence is larger than the preceedingtriangle by a factor w, defined as the real root of thecubic factor $x^3 - x - 1$. This value can be expressed as

$$w = \left\langle \frac{1}{2} + \left\langle \frac{1}{2^2} - \frac{1}{3^3} \right\rangle^{1/2} \right\rangle^{1/2} + \left\langle \frac{1}{2} - \left\langle \frac{1}{2^2} - \frac{1}{3^3} \right\rangle^{1.2} \right\rangle^{1/3}$$

The numerical value of this "other golden proportion" is

$$w = 1.324717957244746...$$

Chapter 11

Lead-Lag Algorithms

The derivations of recurrence formulas used in the digital simulation of continuous dynamic response, as modeled by ordinary linear differential equations. Emphasis is placed on first and second order equations, since these are the two most common "transfer functions" used in real-world applications.

Along with the optimum recurrence formulas, a variety of specialized and simplified methods are discussed and evaluated to clarify their underlying assumptions and ranges of applicability. Throughout this document we assume that the time required to perform a recurrence computation is negligible relative to the characteristic times of the dynamic functions being simulated. Also, the "round off" error of the computations is assumed to be negligible. In view of the speed and word size of modern digital processors, both of these assumptions are justified when modeling most large-scale physical phenomena.

DISCRETE-TIME SIMULATION OF FIRST-ORDER RESPONSE

The first order lead/lag transfer function is a commonly used mathematical model for representing simple, continuous, dynamic response. A first order lead/lag coupling of two variables implies the relationship

$$\tau_D \frac{dy}{dt} + y = \tau_N \frac{dx}{dt} + x$$

where x is the independent variable, y is the dependent variable, t is time, and the coefficients t_D and t_N are coupling parameters (not necessarily constant). In the continuous domain this relationship uniquely determines the output response y(t) for any given initial conditions and input function x(t). However, there exists a variety of algorithms for simulating in a digital computer, and these algorithms differ both in their computational complexity and in their computed response characteristics.

There are two reasons for the diversity of digital simulations. First, according to equation the change in the output during any interval of time depends on the value of the input continuously throughout the interval, but in digital systems the input is specified only by a sequence of discrete values. Hence the relationship is underspecified in the discrete-time domain, and can only be solved by making an assumption as to the behavior of the input variable between the discrete values. Second, practical constraints on execution time, execution frequency, and memory storage often lead to the use of modified solutions based on various simplifying assumptions.

OPTIMUM FIRST ORDER SOLUTION

BASIC ASSUMPTIONS

As noted earlier, equation is indeterminate when the input variable is specified only by a sequence of discrete values, unless some assumption is made as to the behavior of the input variable between the discrete values. When only two discrete values (the current and one past value) are available at any one time, as is the case in virtually all first-order algorithms, the best possible assumption is that "x" varies linearly with time between the two discrete values. If x were significantly non-linear over the interval, then clearly two values would be insufficient to specify the function. So, we can represent the input function over an interval Δt by

$$x = x_p + \left(\frac{x_c - x_p}{\Delta t}\right)$$

where the subscripts p and c denote "past" and "current" discrete values, corresponding to the initial and final values for that interval. The derivative of this x(t) during the interval is then just

$$\frac{dx}{dt} = \left(\frac{x_c - x_p}{\Delta t}\right)$$

so equation can be rewritten

$$\tau_D \frac{dy}{dt} + y = \tau_D \left(\frac{x_c - x_p}{\Delta t}\right) + x_p + \left(\frac{x_c - x_p}{\Delta t}\right) t$$

Now, if t_D and t_N are known constants, and we are given the discrete values of x_p, y_p, and x_c for any given interval Dt, then equation can be solved exactly for the current output value y_c. If the coefficients t_D and t_N are variable, the situation is somewhat more complicated. We will consider only the case where these coefficients may be treated as functions of time, so that remains linear. In this case the coefficients are specified for a given interval by their initial and final values (i.e., past and current). As with the input function x(t), this in itself is insufficient specification unless the coefficients can be assumed to vary linearly with time between the current instants. Therefore, we make the assumption that

$$\tau_N = \tau_{NP} + \left(\frac{\tau_{NC} - \tau_{NP}}{\Delta t}\right)t$$

$$\tau_D = \tau_{DP} + \left(\frac{\tau_{DC} - \tau_{DP}}{\Delta t}\right)t$$

On this basis, equation can be written

$$(\dot{\tau}_D t + \tau_{DP})\frac{dy}{dt} + y = (\dot{\tau}_N t + \tau_{NP})\dot{x} + (\dot{x}t + x_p)$$

where

$$\dot{\tau}_D = \left(\frac{\tau_{DC} - \tau_{DP}}{\Delta t}\right) \quad \dot{\tau}_N = \left(\frac{\tau_{NC} - \tau_{NP}}{\Delta t}\right) \quad \dot{x} = \left(\frac{x_c - x_p}{\Delta t}\right)$$

Knowing the initial and final values of x, t_N, and t_D, and the initial value of y for a given Dt, equation can be solved exactly for the final (current) value of y.

GENERAL FIRST ORDER RECURRENCE FORMULAS

Before deriving the actual solutions of equations we will consider the form these solutions must take. We require a formula that can be applied recursively to compute the current value of y based on the current value of x and the past values of x and y. We will find that, in general, the solution can be expressed as a linear combination of the three given values, i.e.,

$$y_c = (f_1)\, y_p + (f_2)\, x_p + (f_3)\, x_c$$

where f_1, f_2, and f_3 are all functions of t_D, t_N, and Dt. However, these three functions are obviously not independent, because if $y_p = x_p = x_c = m$ then clearly $y_c = m$, so we have

$\mu = \mu f_1 + \mu f_2 + \mu f_3$ which implies $1 = f_1 + f_2 + f_3$. Taking advantage of this fact, we can eliminate f_1 from equation to give

$$y_c = (1 - f_2 - f_3)\, y_p + (f_2)\, x_p + (f_3)\, x_c$$

If we then make the substitution $x_c = x_p + (x_c - x_p)$ we have

$$y_c = (1 - f_2 - f_3)\, y_p + (f_2)\, x_p + (f_3)\, x_p + (f_3)(x_c - x_p)$$

which can be written

$$y_c = y_p + A\,(x_p - y_p) + B\,(x_c - x_p)$$

where the coefficients A and B are defined as

$$A = f_2 + f_3;\ B = f_3$$

Any valid lead/lag algorithm that computes y_c as a linear function of y_p, x_p, and x_c must be expressible in the form of equation, regardless of the assumptions made concerning the interpolation of the independent variable. Therefore, we define equation as the standard recurrence formula for digital lead/lag simulations. In subsequent sections we will frequently identify algorithms simply by specifying their standard recurrence coefficients A and B.

OPTIMUM RECURRENCE COEFFICIENTS FOR CONSTANT τ

When t_D and t_N are constant, the optimum recurrence coefficients can be determined by solving equation, which can be rewritten as follows

$$\frac{dy}{dt}+\left(\frac{1}{\tau_D}\right)y=\left(\frac{\dot{x}}{\tau_D}\right)t+\left(\frac{\tau_N\dot{x}+x_p}{\tau_D}\right)$$

Recall that the solution of any equation of the form

$$\frac{dy}{dt}+F(t)y=G(t)$$

where F(t) and G(t) are functions of t is given by

$$y=\frac{1}{e^r}\left(\int G(t)e^r\,dt+K\right)\text{ where } r=\int F(t)dt$$

where K is a constant of integration. Making the substitutions

$$F(t)=\frac{1}{\tau_D}$$

$$F(t)=\left(\frac{\dot{x}}{\tau_D}\right)t+\left(\frac{\tau_D\dot{x}+x_p}{\tau_N}\right)$$

$$r=\int\frac{1}{\tau_D}dt=\frac{t}{\tau_D}$$

we have

$$y=\frac{1}{e^{t/t_D}}\left[\int\left(\frac{\dot{x}}{\tau_D}t+\frac{\tau_D\dot{x}+x_p}{\tau_N}\right)e^{t/t_D}+K\right]$$

Performing the integration and dividing through by $e^{t/tD}$ gives

$$y=\dot{x}t+\dot{x}(\tau_N-\tau_D)+x_p+Ke^{-t/t_D}$$

To determine K we apply the initial condition $y = y_p$ at $t = 0$; which gives

$$K=y_p-x_p-\dot{x}(\tau_N-\tau_D)$$

Inserting this back into the preceding equation, recalling the definition of $\dot{x}$, and noting that $y = y_c$ at $t = \Delta t$, we arrive (after some algebraic manipulation) at the result

$$y_c = (f_1)y_p + (f_2)x_p + (f_3)x_c$$

where

$$f_1 = e^{-\Delta t/\tau_D}$$

$$f_2 = \frac{\tau_N - \tau_D}{\Delta t}\left(e^{-\Delta t/\tau_D} - 1\right) - e^{-\Delta t/\tau_D}$$

$$f_3 = 1 + \frac{\tau_N - \tau_D}{\Delta t}\left(e^{-\Delta t/\tau_D} - 1\right)$$

Note that as required the sum of f_1, f_2, and f_3 is identically 1. Substituting the expressions for f_2 and f_3 into equations gives the optimum recurrence coefficients for constant t:

$$A = 1 - e^{-\Delta t/\tau_D}$$

$$B = 1 + A\frac{\tau_N - \tau_D}{\Delta t}$$

OPTIMUM RECURRENCE COEFFICIENTS FOR VARIABLE τ

When t_N and t_D are variable, we base the optimum recurrence coefficients on equation. For convenience we define the following parameters

$$a = \dot{\tau}_D$$

$$b = \tau_{DP}$$

$$c = \dot{x}\left(1 + \dot{\tau}_N\right)$$

$$d = \dot{x}\tau_{DP} + x_p$$

In these terms equation can be rewritten

$$\frac{dy}{dt} + \frac{1}{at + b}y = \frac{ct + d}{at + b}$$

This is in the form of equation so the general solution is given by equation where

$$F(t) = \frac{1}{at + b} \quad G(t) = \frac{ct + d}{at + b}$$

$$r = \int \frac{1}{at + b}dt = \mathrm{In}\left(\left(at + b\right)^{1/a}\right)$$

Therefore we have

$$y = \frac{1}{\left(at + b\right)^{1/a}}\left[\int\left(\frac{ct + d}{at + b}\right)\left(at + b\right)^{1/a}dt + K\right]$$

Performing the integration and dividing through by $(at+b)^{1/a}$ gives

$$y = \left(\frac{c}{1+a}\right)t + d - \frac{bc}{1+a} + K(at+b)^{-1/a}$$

The constant of integration K is determined by the initial condition $y = y_p$ at t = 0, which gives

$$K = b^{1/a}\left(y_p - d + \frac{bc}{1+a}\right)$$

We can now compute y_c by evaluating equation at t = Dt. If we then replace a,b,c, and d with their respective definitions, we arrive at the result

$$y_c = (f_1)y_p + (f_2)x_p + (f_3)x_c$$

where

$$f_1 = \left(\frac{\tau_{DP}}{\tau_{DC}}\right)^{1/\dot{\tau}_D}$$

$$f_2 = \left[1 - f_1 - \frac{\tau_{NP}}{\Delta t}(1-f_1) + \frac{\tau_{DP}}{\Delta t}\left(\frac{1+\dot{\tau}_N}{1+\dot{\tau}_D}\right)(1-f_1) - \left(\frac{1+\dot{\tau}_N}{1+\dot{\tau}_D}\right)\right]$$

$$f_3 = \left[\frac{\tau_{NP}}{\Delta t}(1-f_1) + \frac{\tau_{NP}}{\Delta t}\left(\frac{1+\dot{\tau}_N}{1+\dot{\tau}_D}\right)(1-f_1) - \left(\frac{1+\dot{\tau}_N}{1+\dot{\tau}_D}\right)\right]$$

Note that, as required, the sum of f_1, f_2, and f_3 is identically 1. Substituting the expressions for f_2 and f_3 into equations gives the optiumum recurrence coefficients for variable τ:

$$A = 1 - \left(\frac{\tau_{DP}}{\tau_{DC}}\right)^{1/\dot{\tau}_D}$$

$$B = \left(\frac{1+\dot{\tau}_N}{1+\dot{\tau}_D}\right) + \frac{A}{\Delta t}\left(\tau_{NP} - \left(\frac{1+\dot{\tau}_N}{1+\dot{\tau}_D}\right)\tau_{DP}\right)$$

When referring to the recurrence formulas the term "optimum" is justified in the following sense: The total solution y(t) equals the solution to the homogeneous equation plus a particular solution for the given forcing function. The only ambiguity is in the particular solution, which depends on how we choose to interpolate the independent function x(t). Note that the forcing function has no effect on the homogeneous solution, and that the particular solution is independent of the actual y(t) values.

It follows that the coefficients of the "y terms" in the recurrence relation must be exactly, *regardless of how the x(t) function is interpolated.* The only ambiguityt in the recurrence formula is in the "x-term" coefficients (f_2 and f_3), and even those have a fully determined sum. In view of this, the term "optimum" is used throughout this document to denote a recurrence formula that has the exact "y-term" coefficient(s), and for which the "x-term" coefficients sum to the correct value.

With regard to the variable-t solution, we would expect to find that it has as a special case the constant-t solution, and in fact if and are zero then clearly equation is equivalent to the constant-t expression for B given by equation. However, it may not be self-evident that equation reduces to the A of equation for constant t. To show that it does, we can rewrite equation as follows $A = 1 - e^{(1/\dot{\tau}_D)\ln(\tau_{DP}/\tau_{DC})}$

If we now recall the series expansion of the natural log

$$\ln(z) = 2\left[\left(\frac{z-1}{z+1}\right) + \frac{1}{3}\left(\frac{z-1}{z+1}\right)^3 + \ldots\right]$$

we see that if the ratio t_{DP}/t_{DC} is close to 1 we can approximate the natural log in equation very accurately by just the first term of the expansion, i.e.,

$$\ln\left(\frac{\tau_{DP}}{\tau_{DC}}\right) \approx 2\left[\frac{\left(\frac{\tau_{DP}}{\tau_{DC}}\right) - 1}{\left(\frac{\tau_{DP}}{\tau_{DC}}\right) + 1}\right] = 2\left(\frac{\tau_{DC} - \tau_{DP}}{\tau_{DC} + \tau_{DP}}\right)$$

in which case equation becomes

$$A \approx 1 - e^{-\frac{\Delta t}{(\tau_{DP} + \tau_{DC})^2}}$$

This makes it clear that as $\dot{\tau}_D$ goes to zero, and t_{DC} approaches t_{DP}, the vairable-t expression for A does in fact reduce to the constant-t case given by equation.

We now demonstrate that the constant-t response approaches the variable-t response if a sufficiently small time interval Δt is used. First, notice that as Δt goes to zero the ratio of t_{DP} to t_{DC} can be made arbitrarily close to 1 for any finite value of $\dot{\tau}_D$. Also, as we have already seen, equation is equivalent to A of equation in the limit as τ_{DP}/τ_{DC} goes to 1. Therefore, as Δt goes to zero the A coefficient for both constant t and variable t is given by equation. Furthermore, if we recall the power series expansion

$$e^z = 1 + \frac{z}{1!} + \frac{z^2}{2!} + \frac{z^3}{3!} + \ldots$$

we see that as Δt goes to zero, A of equation becomes simply $A = \Delta\tau/\tau_D$. Substituting this into the expressions for B in equations, along with the stipulation that $\tau_{NC} \approx \tau_{NP}$ for a sufficiently small Dt, we see that both solutions give $B = t_N/t_D$. Thus, for sufficiently small Δt, the constant-τ and variable-τ solutions both reduce to the form

$$y_c = y_p + \frac{\Delta t}{\tau_D}\left(x_p - y_p\right) + \frac{\tau_N}{\tau_D}\left(x_c - x_p\right)$$

Notice that if Δt is actually zero, but $x_c - x_p$ does not vanish, then this can be written as

$$y_c = y_p + \frac{\tau_N}{\tau_D}\left(x_c - x_p\right)$$

which is the expected response to a step input, viz, an instantaneous change in x yields an instantaneous change in y with a magnitude amplified by the factor τ_N/τ_D.

Examination of equation also shows that the variable-t solution requires t_{DP} and t_{DC} have the same sign, since if they didn't, the ratio τ_{DP}/τ_{DC} would be negative and the result of the exponentiation would, in general, be complex. Another way of stating this restriction is that τ_D can never pass through zero, which effectively prohibits a change of sign for a continuous function. In one sense, the "reason" for this restriction is that we divided by t_D when we wrote equation. More fundamentally, equation shows that when τ_D is zero the differential term in y vanishes and the equation is singular.

It may appear that equation also exhibits a singularity, specifically when $\dot{\tau}_D$ Δt equals -1. However, as long as the absolute value of $\Delta t\left(\dot{\tau}_D / \tau_{DP}\right)$ is less than 1 (which corresponds to the requirement that τ_D never pass through zero during the interval) it can be shown that B remains analytic at $\dot{\tau}_D = -1$, and is given by

$$B = \frac{\tau_{NP}}{\tau_{DP}} + \left(1 + \dot{\tau}_N\right)\left[1 - \left(1 - \frac{\tau_{DP}}{\Delta t}\right)\ln\left(1 - \frac{\Delta t}{\tau_{DP}}\right)\right]$$

SIMPLIFIED SOLUTIONS BASED ON APPROXIMATIONS OF e^z

If τ_N and τ_D are constant (and assuming a constant Δt), then the recurrence coefficients A and B are also constant and can be precomputed off-line using the optimum constant-τ formulas given by equations. However, if τ_N and τ_D are variable, the optimum coefficients given are rather cumbersome, and require storage of past values of τ_D and τ_N. We have shown that the variable-t response can be approximated by the constant-τ solution if Δt is small enough, but in this case the A and B coefficients must be recomputed on-line for each interval, and it may be desirable to simplify the computation. Since the most time-consuming aspect of the computation is the exponential in equation, the most common approach is to substitute some approximation for this exponential.

If $|\Delta t/\tau_D|$ is much less than 1, we can substitute the first terms of the Taylor series expansion for e^z into equation, which leads to the family of approximate recurrence coefficients.

Number of Taylor Series Terms for e^x	*A*	*B*
1	$\frac{\Delta t}{\tau_D}$	$\frac{\tau_N}{\tau_D}$
2	$\frac{\Delta t}{\tau_D}\left(1-\frac{\Delta t}{2\tau_D}\right)$	$\frac{\tau_N}{\tau_D}+\frac{\Delta t}{2\tau_D}\left(1-\frac{\tau_N}{\tau_D}\right)$
3	$\frac{\Delta t}{\tau_D}\left(1-\frac{\Delta t}{2\tau_D}\left(1-\frac{\Delta t}{3\tau_D}\right)\right)$	$\frac{\tau_N}{\tau_d}+\frac{\Delta t}{2\tau_D}\left(1-\frac{\tau_N}{\tau_D}\right)\left(1-\frac{\Delta t}{3\tau_D}\right)$

Many lead/lag algorithms correspond to one of these approximate solutions, or some variation. For example, one simulation program uses an algorithm equivalent to

$$A=\frac{\Delta t}{\tau_D}$$

$$B=\frac{\tau_N+\Delta t}{\tau_D}$$

This corresponds to the lowest order, except that B includes a Δt in the numerator. The effect of this Δt is to introduce a bias in the response to non-constant input functions (because B is the coefficient of Δx in the recurrence formula). This bias was not evident during testing because the only evaluation of the algorithm was based on its response to step inputs, during which Δx was always zero except at a step.

Another example is a commercial embedded software package that uses a lead/lag algorithm with the coefficients

$$A=\frac{\Delta t}{\tau_D}\left(1-\frac{\Delta t}{2\tau_D}\right)\quad B=\frac{\Delta t}{\tau_D}+\frac{\Delta t}{2\tau_D}\left(1-\frac{\tau_N}{\tau_D}-\frac{\Delta t}{2\tau_D}\right)$$

Here, the solution corresponds to the second order approximation, except for the $-\Delta t/2\tau_D$ component in the right hand term of B.

It's interesting to compare all five of the modified B coefficients discussed so far for a particular case. Consider a lead/lag with $\tau_N/\tau_D = 10$ and $\Delta t/\tau_D = r < 1$. The three levels of approximation are

$$B_1=10 \quad B_2=10-\frac{9}{2}r \quad B_3=10-\frac{9}{2}r+\frac{3}{2}r^2$$

whereas the dynamic simulation algorithm of the first example gives B = 10 + r, and the embedded commercial algorithm of the second example gives $B = 10 - (9/2)r - (1/4)r^2$. According to the Taylor series approximation method, the former approximation is wrong in the first order correction for non-vanishing r, while the latter approximation is wrong in the second order. The reason that the designers of these algorithms chose to deviate from the Taylor series approximation as shown here is not known.

Instead of the Taylor series approximation, we could make use of the series expansion of the natural log. If we take just the first term of this series and make the substitution

$$\omega = 2\frac{z-1}{z+1}$$

we have

$$\ln\left(\frac{2+\omega}{2-\omega}\right) \approx w$$

Taking the exponential of both sides gives

$$e^{\omega} \approx \frac{2+\omega}{2-\omega}$$

In the range of interest, this is a much more efficient approximation of the exponential than the truncated Taylor series. Substituting this for the exponential in equations gives the recurrence coefficients

$$A = \frac{\Delta t}{\tau_D + \Delta t/2}$$

$$B = \frac{\tau_N + \Delta t/2}{\tau_D + \Delta t/2}$$

Incidentally, some commercial applications have implemented simple first-order lags which compute y_c based only on x_c and y_p. Since they do not store the past value of x, they effectively assume $x_p = x_c$. With this assumption the value of B is unimportant, because it is the coefficient of $x_c - x_p$ which vanishes. Hence we're only interested in the A coefficient, which these applications take to be $A = \Delta t/(\tau_D + \Delta t)$. Clearly the optimum representation of a first-order lag to this level of approximation would be with $A = \Delta t/(\tau_D + \Delta t/2)$, as can be seen from the preceding formula. It is not known why the application designers chose to neglect the factor of 1/2 on Δt in the denominator. This makes the effective time constant of the algorithm equal to $\tau_D + \Delta t/2$. Typically Δt is about 1/4 of τ_D, so this represents up to 25% error in the time constant.

FINITE DIFFERENCES AND BILINEAR SUBSTITUTION

The most direct approach to determining an approximate recurrence formula for a differential equation of the form $\tau_D \frac{dy}{dt} + y = \tau_N \frac{dx}{dt} + x$

is to substitute simple finite differences for the differentials about the nearest convenient point. The derivative of the function y(t) at the point half-way between the past and current instants can be approximated by

$$\frac{dy}{dt} \approx \frac{y_c - y_p}{\Delta t}$$

The value of the function y(t) at this point is approximately

$$y \approx \frac{y_c + y_p}{2}$$

Similarly for the function x(t) at the same instant we have the approximations

$$\frac{dx}{dt} \approx \frac{x_c - x_p}{\Delta t} \qquad x \approx \frac{x_c + x_p}{2}$$

Making these substutitions into equation and rearranging gives the recurrence formula

$$y_c \approx \left(\frac{2\tau_D - \Delta t}{2\tau_D + \Delta t}\right) y_p + \left(\frac{-2\tau_N + \Delta t}{2\tau_D + \Delta t}\right) x_p + \left(\frac{2\tau_N + \Delta t}{2\tau_D + \Delta t}\right) x_c$$

By equations, the standard form recurrence coefficients are

$$A = \frac{\Delta t}{\tau_D + \Delta t / 2} \qquad B = \frac{\tau_N + \Delta t / 2}{\tau_D + \Delta t / 2}$$

These are identical to equations, which shows that the direct application of finite differences gives the same first-order recurrence formula as results from substituting the exponential approximation of equation into equations.

There is another approach that, called bilinear substitution (or Tustin's algorithm), that leads to the same result as finite differences (and the optimum exponential substitution). To apply this method we first re-write equation in the form

$$(\tau_D D + 1) y_c = (\tau_N D + 1) y_c$$

where D is the differential operator. We then make the substitution

$$D = \frac{2}{\Delta t}\left(\frac{1 - z^{-1}}{1 + z^{-1}}\right)$$

where the symbol z^{-1} is the "past value" operator. For example, $z^{-1} y_c = y_p$. Substituting equation into equation, we have

$$\left[\frac{2\tau_D}{\Delta t}\left(\frac{1-z^{-1}}{1+z^{-1}}\right)+1\right]y_c=\left[\frac{2\tau_N}{\Delta t}\left(\frac{1-z^{-1}}{1+z^{-1}}\right)+1\right]x_c$$

Multiplying through by $\Delta t(1+z^{-1})$ gives

$$\left[2\tau_D\left(1-z^{-1}\right)+\Delta t\left(1+z^{-1}\right)\right]y_c=\left[2\tau_N\left(1-z^{-1}\right)+\Delta t\left(1+z^{-1}\right)\right]x_c$$

Rearranging terms, we have

$$\left(2\tau_D+\Delta t\right)y_c+\left(2\tau_D+\Delta t\right)z^{-1}y_c=\left(2\tau_N+\Delta t\right)x_c+\left(2\tau_N+\Delta t\right)z^{-1}x_c$$

Carrying out the indicated past value operations gives

$$\left(2\tau_D+\Delta t\right)y_c+\left(2\tau_D+\Delta t\right)y_p=\left(2\tau_N+\Delta t\right)x_c+\left(2\tau_N+\Delta t\right)x_p$$

Solving for y_c, we arrive at equation. Thus, when applied to a first-order transfer function, the bilinear substitution method gives the same recurrence formula as the method of finite differences.

PRE-WARPING

A technique called "pre-warping" is sometimes applied to lead/lag algorithms in an attempt to correct the amplitude of the output responding to an oscillating input signal. The standard recurrence coefficients for this method are

$$A=\frac{2}{1+\dfrac{1}{\tan\left(\Delta t/2\tau_D\right)}}\qquad B=\frac{1+\dfrac{1}{\tan\left(\Delta t/2\tau_N\right)}}{1+\dfrac{1}{\tan\left(\Delta t/2\tau_D\right)}}$$

Recalling that as z goes to zero, tan(z) approaches z, we see that the pre-warped coefficients reduct to equations if Δt is small relative to τ_N and τ_D.

The rationale for pre-warping is clarified by the following example. Given a lead/lag simulation with time-step size $\Delta t = 0.25$ seconds and a sinusoidal input function with a period of T = 1 second, each cycle of the input will be described by four discrete values.

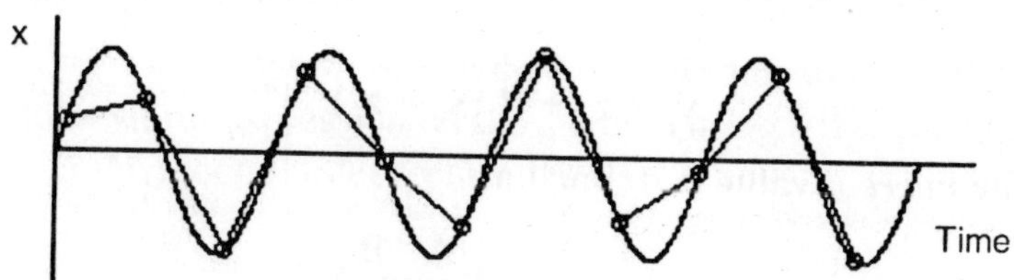

Clearly a sample will rarely be taken at an absolute peak of the wave, even if T is not an exact multiple of Δt and the samples precess through the cycle. Therefore, assuming linear interpolation between the discrete values, we will truncate most of the peaks and under-represent the true amplitude of the input function, which results in a corresponding underestimation of the output amplitude. In an attempt to correct for this, pre-warping biases the recurrence coefficients to reduct the effective lag of the filter, thereby increasing its output amplitude for a given oscillating input. However, the bias necessary to cancel out the under-representation of the input varies with the ration of T to Dt, so the pre-warped coefficients are really only correct for one particular frequency - by convention, the "half power frequency". At lower frequencies, and for most non-oscillatory inputs, the pre-warping coefficients yield less accurate results than the approximate coefficients given by equations. Hence, pre-warping is a highly specialized technique, intended only for a limited range of frequency-response applications.

COMPOSITE PAST VALUE RECURRENCE

All the solution methods discussed so far have assumed the availability of xp and yp, that is, past values of the input and out functions. This requires the dedication of two RAM memory locations for each lead/lag in the system. In contrast, the composite past value (CPV) recurrence scheme required only one stored value. In the constant-t case this recurrence scheme is functionally equivalent to the standard recurrence formula for any given solution coefficients A and B. However, when CPV recurrence is applied in the variable-t case, it does not give exactly the same results as the standard formula.

To derive CPV recurrence formula, we begin by rewriting the standard formula as follows

$$y_c = (1 - A)\, y_p + (A - B)\, x_p + (B)\, x_c$$

Dividing and multiplying the two past-value terms by 1– B gives

$$y_c = (1-B)\left[\frac{1-A}{1-B} y_P + \frac{A-B}{1-B} x_P\right] + (B)x_c$$

If we define the quantity in the brackets as the "composite past value", denoted by z_p, then we can rewrite equation as

$$y_c = (1 - B)\, z_p + (B)\, x_c$$

or

$$y_c = z_p + B\,(x_c - z_p)$$

The composite *current* value is defined as

$$z_c = \frac{1-A}{1-B} y_c + \frac{A-B}{1-B} x_c$$

To simplify this, we substitute the expression for the y_c from equation, which gives

$$z_c = (1-A)z_p + \frac{B(1-B)}{1-B}x_c + \frac{A-B}{1-B}x_c$$

Combining terms in x_c and simplifying, we have

$$z_c = z_p + A(x_c - z_p)$$

Therefore, the CPV recurrence formulas are

$$y_c = z_p + B\ (x_c - z_p)$$
$$z_c = z_p + A\ (x_c - z_p)$$

The derivation of these formulas does not distinguish between the past and current values of A and B, thereby tacitly assuming that these coefficients are constant. However, when τ_N and τ_D are variable, the standard coefficients A and B are also variable, in which case the CPV recurrence formulas are not exactly equivalent to the standard formula on which they are based. To demonstrate this, we can solve equation for z_p and substitute this into equation to give

$$z_c = \frac{1-A_c}{1-B_c}y_c + \frac{A_c - B_c}{1-B_c}x_c$$

A and B are now subscripted with a "c" to denote the current values, since only the current values are available each time the recurrence formulas are executed. Equation is also valid at the past instant, which is to say

$$z_p = \frac{1-A_p}{1-B_p}y_p + \frac{A_p - B_p}{1-B_p}x_p$$

If we substitute this expression for z_p into equation, the resulting equation for y_c can be reduced to the form

$$y_c = y_p + A_{eff}\ (x_p - y_p) + B_{eff}\ (x_c - x_p)$$

where A_{eff} and B_{eff} are the effective standard recurrence coefficients, related to the actual coefficients used in the CPV recurrence formula by

$$A_{eff} = \left(\frac{1-B_c}{1-B_P}\right)A_p + \left(\frac{B_c - B_P}{1-B_P}\right) \qquad B_{eff} = B_c$$

Because of this effect, there are circumstances in which CPV recurrence yields unrealistic output response, even if the optimum formulas for A and B are used. The simplest illustration of this is the case where τ_N and τ_D initially have different values and then become equal. The true first order response is for the output to asymptotically converge on the input function once $\tau_N = \tau_D$, and this can be accuratly simulated using the standard recurrence formula. However, this response cannot be reproduced using CPV recurrence. To understand why, notice that whenever $\tau_N = \tau_D$ the current value of B (denoted by B_c) must equal 1, as is clearly shown by equation. Substituting this into equations gives the effective coefficients $A_{eff} = 1$ and $B_{eff} = 1$. Inserting these into the standard recurrence formula gives the result

$$y_c = y_p + (1)(x_p - y_p) + (1)(x_c - x_p) = x_c$$

Thus, using CPV recurrence, the output undergoes a "step" change and "instantaneously" becomes equal to the current input, regardless of the output's past value.

A possible objection to equation -1) is that it appears to predict divergent behavior when B_p equals 1, whereas in practive CPV recurrence formulas never yield divergent response. The explanation is that B_p can only equal 1 if B_c of the previous iteration was 1. But we have already seen that if B_c was 1 for the previous iteration, then y_c was unconditionally set equal to x_c during that iteration. Therefore, for the present iteration, y_p must equal x_p. Recalling that A is the coefficient $(x_p - y_p)$ in the standard recurrence formula, we see that, in effect, the divergence of A_{eff} at $B_p = 1$ is always "cancelled" by the vanishing of $(x_p - y_p)$.

Digital simulations of continuous dynamic functions have a constraint on their execution frequency imposed by the intrinsic information rate of the input and output signals. The execution frequency must be high enough to resolve and convery the significant characteristics of these signals. However, the simplified methods of simulation described in Sections have another limitation: The approximations on which they are based are valid only if the execution period Δt is small relative to the characteristic times of the function (i.e., t_N and t_D). To illustrate that these two limitations are logically independent, consider the simple lag response

$$\tau \frac{dy}{dt} + y = x$$

where t = 0.05 seconds. Equations give the recurrence formula

$$y_c = y_p + \left(\frac{2\Delta t}{0.1 + \Delta t}\right)(x_p - y_p) + \left(\frac{\Delta t}{0.1 + \Delta t}\right)(x_c - x_p)$$

Now suppose that, based on their maximum information rates, we have determined that x and y can be adequately represented with a sampling frequency of 4 Hz, corresponding to a Δt of 0.25 seconds. Substituting this into equation gives

$$y_c = y_p + (1.4286)(x_p - y_p) + (0.7143)(x_c - x_p)$$

A sample of the response given by this recurrence formula for a single time-step is illustrated.

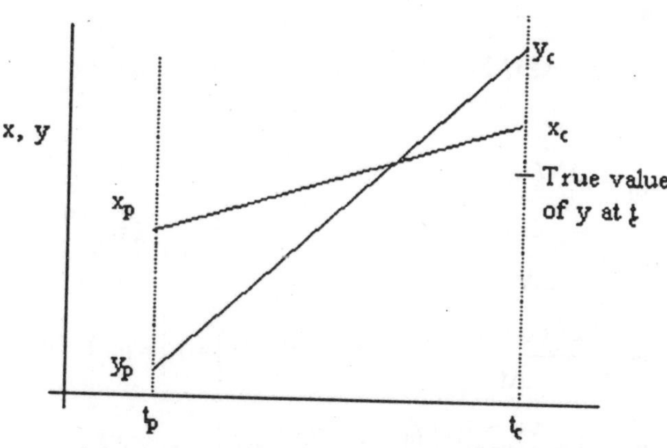

The output computed by equation crosses over the input, which is obviously not representative of a pure lag response. In general, recurrence formulas based on equations will not give an accurate representation of a lead/lag function unless Δt is less than 2t. This means that the execution frequency of the present example would have to be raised from 4 Hz to at least 10 Hz. In contrast, if the optimum coefficients given by equations are used, the recurrence formula is

$$y_c = y_p + (0.9933)(x_p - y_p) + (0.8013)(x_c - x_p)$$

for which an execution frequency of 4 Hz is adequate. Thus the use of non-optimal coefficients imposes unnecessarily high execution frequency on the simulation.

Another way of looking at this is to regard the error in the computed value of y_c for any given interval Δt as consisting of two parts: one part due to the uncertainty in the input function, and one part resulting from the approximations made in determining the coefficients of the recurrence formula. The first part is unavoidable (for a given Δt), but he second part can be eliminated entirely by using the optimum recurrence coefficients.

With the technique of pre-warping, an attempt is made to use these two errors to compensate for each other. These two errors offset each other only under a limited range of conditions. Essentially, pre-warping attempts to account for non-inearity in the input signal over the interval Dt, but there is really no general way of determining the non-linearity of a signal based on just two samples (x_p and x_c). In applying its "correction", pre-warping makes use of information about the input signal beyond what can be inferred from the two given points.

Specifically it depends on the knowledge that the input is periodic with a given frequency. Without such special information, a general method of accounting for non-linearity of the input signal must make use of at least three values of x. With three values we can determine the optimum recurrence formula based on second order interpolation of the input function. Denoting the "past past value" of x by x_{pp}, the expanded recurrence formula can be written

$$y_c = (g_1)\, y_p + (g_2)\, x_{pp} + (g_3)\, x_p + (g_4)\, x_c$$

where

$$g_1 = e^{1\Delta t/\tau_D}$$

$$g_2 = \frac{\tau_N - \tau_D}{2\Delta t}\left[\left(1 - \frac{2\tau_D}{\Delta t}\right) + e^{-\Delta t/\tau_D}\left(1 + \frac{2\tau_D}{\Delta t}\right)\right]$$

$$g_3 = \frac{2(\tau_N - \tau_D)}{\Delta t}\left[\frac{\tau_D}{\Delta t}\left(1 - e^{-\Delta t/\tau_D}\right) - 1\right] - e^{-\Delta t/\tau_D}$$

$$g_4 = 1 + \frac{\tau_N - \tau_D}{2\Delta t}\left[\left(1 - e^{1\Delta/\tau_D}\right) + 2\right]$$

By the same reasoning that led to the relation $f_1 + f_2 + f_3 = 1$, we have the relation

$$g_1 + g_2 + g_3 + g_4 = 1$$

Using this, we can re-write equation in a generalized standard form analagous to equation as follows

$$y_c = y_p + A\,(x_p - y_p) + B\,(x_c - x_p) + C\,(x_c - 2x_p + x_{pp})$$

where

$$A = 1 - e^{-\Delta t/\tau_D}$$

$$B = 1 + A\left(\frac{\tau_N - \tau_D}{\Delta t}\right)$$

$$C = \frac{(\tau_N - \tau_D)}{\Delta t}\left[A\left(\frac{1}{2} + \frac{\tau_D}{\Delta t}\right) - 1\right]$$

If x_{pp}, x_p, and x_c are co-linear vs time, then the quantity multiplied by C vanishes, and equation reduces to equation.

Beginning with equation), we could have defined the composite past value directly as just the sum of the past value terms

$$z_p = (1 - A)\, y_p + (A - B)\, x_p$$

Then the CPV recurrence formulas would have been

$$y_c = z_p + B\, x_c$$

$$z_c = (1 - A)\, y_c + (A - B)\, x_c$$

The advantages of equations is that they can be executed with just two multiplications, whereas equations and involve three multiplications. This is significant because multiplication is a time-consuming operation in most digital devices. However, equations will give better accuracy in the variable-t case because they are equivalent to using the effective standard coefficients

$$A_{eff} = A_p + (B_c - B_p)$$

$$B_{eff} = B_c$$

These coefficients never yield singular values of the coefficients, unlike equations when B_c=1, so no special provisions are needed to account for this computational singularity (recognizing that it cancels when multiplied by $(x_c - x_p)$).

114-B

Encyclopaedia
of
DISCRETE MATHEMATICS

Encyclopaedia
of
DISCRETE MATHEMATICS

Volume 2

Dr. Rakesh Kumar Pandey

ANMOL PUBLICATIONS PVT. LTD.
NEW DELHI - 110 002 (INDIA)

ANMOL PUBLICATIONS PVT. LTD.
Regd. Office: 4360/4, Ansari Road, Daryaganj,
New Delhi-110 002 (India)
Ph.: 23278000, 23261597

Branch Office: No. 1015, Ist Main Road, BSK IIIrd Stage
IIIrd Phase, IIIrd Block,
Bangalore-560 085 (India)
Tel.: 080-41723429
Visit us at: www.anmolpublications.com

Encyclopaedia of Discrete Mathematics

First Edition, 2009

ISBN 978-81-261-4108-1 (Set)

PRINTED IN INDIA

Printed at Mehra Offset Press, Delhi.

Contents

Preface

Discrete Mathematics stands among the *Leading Disciplines of Mathematics and Theoretical Computer Science*. It handles the objects of discrete values. This Encyclopaedia has more than one purpose. Students should learn a particular set of Mathematical Facts and know how to apply them more importantly. It also teaches students how to think Logically and Mathematically. To achieve these goals, the chapters are stressed on *Logical Reasoning, Graph Theory, Set Theory, Binary Relations, Recurrence Relations, Combinatorics, Functions, Matrices*, etc.

Encyclopaedia of Discrete Mathematics has been written for the students who are preparing for an *Advance Degree Programme*. With the background necessary for further study in Theoretical Computer Science. Sufficient topics have been included to accommodate students with varied background. The subject matter has been discussed in such a way that students will find no difficulty to understand it.

Author

List of Symbols

$\pm$	$\not\subset$	$\subset$	$\supset$	$\subseteq$	$\in$	$\notin$	∂
$\neq$	$\equiv$	$\approx$	$\angle$	∇	$\sum$	Π	Ω
ϕ	α	β	χ	ε	γ	η	λ
μ	ν	$\perp$	π	θ	ρ	σ	τ
$\Re$	ω	ξ	ψ	ζ	Υ	$\exists$	$\int$
$\iint$	$\iiint$	$\oint$	$\times$	$\sqrt{\ }$	$\therefore$	$\because$	Δ
$\geq$	$\leq$	$\Rightarrow$	$\Downarrow$	$\Leftrightarrow$	∞	$\propto$	$\mathbb{R}$
$\mathbb{Z}$	$\mathbb{N}$	$\mathbb{C}$	$\mathbb{Q}$				

Chapter 1

Binary Relation

The relation we are going to study here is an abstraction of relations we see in our everyday life such as those between parent and child, between car and owner, among name, social security number, address and telephone number etc. We are going to focus our attention on one key property which all the everyday relations have in common, define everything that has that property as a relation, and study properties of those relations.

One of the places where relation in that sense is used is data base management systems. Along with hierarchical and network models of data, the *relational model* is widely used to represent data in a database. In this model the data in a database are represented as a collection of relations. Informally, each relation is like a table or a simple file. For example, consider the following table.

Employee		
Name	*Address*	*Home Phone*
Amy Angels	35 Mediterranean Av.	224–1357
Barbara Braves	221 Atlantic Av.	301–1734
Charles Cubs	312 Baltic Av.	223–9876

Each row of this table represents a collection of data values such as name, address, and telephone number of a person. Each row is considered an instance of a relation and the table as the collection of the rows is considered a relation, which is the relation we are going to be studying in this chapter.

Operations such as inserting or deleting entries to or from a table, merging two table, finding the intersection of two tables, and searching for certain entries can be described simply and precisely as operations on relations and known mathematical results on relations can be utilized without reinventing them.

The relational model is flexible (easy to expand, easy to modify) and interface to query languages is simple. It is thus widely used today.

DEFINITION OF BINARY RELATION

Here we are going to define relation formally, first binary relation, then general n–ary relation. A relation in everyday life shows an association of objects of a set with objects of other sets (or the same set) such as John owns a red Mustang, Jim has a green Miata etc. The essence of relation is these associations. A collection of these individual associations is a relation, such as the ownership relation between peoples and automobiles. To represent these individual associations, a set of "related" objects, such as John and a red Mustang, can be used. However, simple sets such as {John, a red Mustang} are not sufficient here.

The order of the objects must also be taken into account, because John owns a red Mustang but the red Mustang does not own John, and simple sets do not deal with orders. Thus sets with an order on its members are needed to describe a relation. Here the concept of ordered pair and, more generally, that of ordered n–tuple are going to be defined first. A relation is then defined as a set of ordered pairs or ordered n–tuples.

Ordered pair: An ordered pair is a set of a pair of objects with an order associated with them.

If objects are represented by *x* and *y*, then we write an ordered pair as <*x, y*> or <*y, x*>. In general <*x, y*> is different from <*y, x*>.

Equality of ordered pairs: Two ordered pairs <*a, b*> and <*c, d*> are equal if and only if $a = c$ and $b = d$.

For example, if the ordered pair <*a, b*> is equal to <*1, 2*>, then $a = 1$, and $b = 2$. <*1, 2*> is not equal to the ordered pair <*2, 1*>.

Binary relation: A binary relation from a set *A* to a set *B* is a set of ordered pairs <*a, b*> where *a* is an element of *A* and *b* is an element of *B*.

When an ordered pair <*a, b*> is in a relation *R*, we write *a R b*, or <*a, b*> ∈ *R*. It means that element *a* is related to element *b* in relation *R*.

When $A = B$, we call a relation from *A* to *B* a (binary) relation on *A* .

Cartesian product: The set of all ordered pairs <*a, b*>, where *a* is an element of *A* and *b* is an element of *B*, is called the Cartesian product of *A* and *B* and is denoted by $A \times B$.

Thus a binary relation from *A* to *B* is a subset of Cartesian product $A \times B$.

Examples: If $A = \{1, 2, 3\}$ and $B = \{4, 5\}$, then {<*1, 4*>, <*2, 5*>, <*3, 5*>}, for example, is a binary relation from *A* to *B*.

However, {<*1, 1*>, <*1, 4*>, <*3, 5*>} is not a binary relation from *A* to *B* because *1* is not in *B*.

The parent–child relation is a binary relation on the set of people. <*John, John Jr.*>, for example, is an element of the parent–child relation if John is the father of John Jr.

DEFINITION OF N–ARY RELATION

Here we are going to formally define general n-ary relation using the concept of ordered n-tuple.

Ordered n-tuple: An ordered n-tuple is a set of n objects with an order associated with them. If n objects are represented by $x_1, x_2, ..., x_n$, then we write the ordered n-tuple as $<x_1, x_2, ..., x_n>$.

Cartesian product: Let $A_1, ..., A_n$ be n sets. Then the set of all ordered n-tuples $<x_1, ..., x_n>$, where $x_i \in A_i$ for all i, $1 \leq i \leq n$, is called the Cartesian product of $A_1, ..., A_n$, and is denoted by $A_1 \times ... \times A_n$.

Equality of n-tuples: Two ordered n-tuples $<x_1, ..., x_n>$ and $<y_1, ..., y_n>$ are equal if and only if $x_i = y_i$ for all i, $1 \leq i \leq n$.

For example the ordered 3–tuple $<1, 2, 3>$ can be equal to only $<1, 2, 3>$ and nothing else. It is not equal to the ordered n–tuple $<2, 3, 1>$ for example.

N-ary relation: An n–ary relation on sets $A_1, ..., A_n$ is a set of ordered n–tuples $<a_1, ..., a_n>$ where a_i is an element of A_i for all i, $1 \leq i \leq n$. Thus an n-ary relation on sets $A_1, ..., A_n$ is a subset of Cartesian product $A_1 \times ... \times A_n$.

Example: Let A_1 be a set of names, A_2 a set of addresses, and A_3 a set of telephone numbers. Then a set of 3–tuples < name, address, telephone number > such as {< Amy Angels, 35 Mediterranean Ave, 224–1357 >, < Barbara Braves, 221 Atlantic Ave, 301–1734 >, < Charles Cubs, 312 Baltic Ave, 223–9876 >}, is a 3–ary (ternary) relation over A_1, A_2 and A_3.

Example: Let A_1 be a set of names. Then a set of 1–tuples such as {< Amy >, < Barbara >, < Charles >}, is a 1–ary (unary) relation over A_1.

A unary relation represents a property/characteristic, such as tall, rich etc., shared by the members of A_1 listed in the relation.

Special Relations

The empty set is certainly a set of ordered n–tuples. Therefore it is a relation. It is called the empty relation.

The Cartesian product $A_1 \times ... \times A_n$ of sets $A_1 , ... , A_n$, is also a relation, and it is called the universal relation.

EQUALITY OF RELATIONS

Definition (equality of binary relation): Two binary relations $R_1 \subseteq A_1 \times A_2$ and $R_2 \subseteq B_1 \times B_2$ are equal if and only if $A_1 = B_1$, $A_2 = B_2$, and $R_1 = R_2$ as a set.

For example, let $R_1 = \{<1, 2> , <2, 2>\} \subseteq \{1, 2\} \times \{1, 2\}$, and $R_2 = \{<a, b> , <b, b>\} \subseteq \{a, b\} \times \{a, b\}$. Then $R_1 = R_2$ if and only if $a = 1$ and $b = 2$.

Definition (equality of n–ary relation): An n-ary relation $R_1 \subseteq A_1 \times ... \times A_n$ and an m-ary relation $R_2 \subseteq B_1 \times ... \times B_m$ are equal if and only if $m = n$, $A_i = B_i$ for each i, $1 \leq i \leq n$, and $R_1 = R_2$ as a set of ordered n-tuples.

RECURSIVE DEFINITION OF RELATION

Certain relations can be defined recursively. Note that a relation is a set. Therefore a recursive definition of a relation follows the same format as that of sets. Here only examples are given.

Example: Let us define recursively the relation "less than" denoted by $R_<$ on the set of natural numbers N.

Note that $R_< = \{<a, b> \mid a \in N \wedge b \in N \wedge a < b\} = \{<0, 1>, <0, 2>,..., <1, 2>, ...\}$.

Basis Clause: $<0, 1> \in R_<$ (or $0 \; R_< \; 1$), meaning 0 is less than 1.

Inductive Clause: For all x and y in N, if $<x, y> \in R_<$, then $<x, y + 1> \in R_<$, and $<x + 1, y + 1> \in R_<$.

Extremal Clause: Nothing is in $R_<$, unless it is obtained from the Basis and Inductive Clauses.

Informally one can see that this definition is correct as follows:

First, $<0, 1>$ is the "simplest" element in $R_<$.

Next by arranging the ordered pairs in $R_<$ as follows, one can see that the two operations in the Inductive Clause generate them all:

<0, 1>, <0, 2>, <0, 3>, ..., <0, n>, ...

<1, 2>, <1, 3>, <1, 4>, ..., <1, n>, ...

<2, 3>, <2, 4>, <2, 5>, ..., <2, n>, ...

....

....

In each row, the first (leftmost) ordered pair gives the "smallest" ordered pair with the first number. For example $<2, 3>$ is the smallest with 2 as the first component. It is obtained from the first ordered pair of the row immediately above it by using $<x + 1, y + 1>$ of the Inductive Clause, apply that to $<1, 2>$ to get $<2, 3>$, for example.

Within each row, the ordered pairs are obtained from the first one by using $<x, y + 1>$ of the Inductive Clause successively.

Thus all the ordered pairs of $R_<$ are generated from $<0, 1>$ by following the Inductive Clause.

Example: Let $R_{a+b=c}$ be the set of triples of natural numbers <*a, b, c*> which satisfy $a + b = c$. Then $R_{a+b=c}$ on the set of natural numbers N can be defined recursively as follows.

Basis Clause: $<0, 0, 0> \in R_{a+b=c}$.

Inductive Clause: For all *x, y* and *z* in *N*, if $<x, y, z> \in R_{a+b=c}$, then $<x+1, y, z+1>$ and $<x, y+1, z+1> \in R_{a+b=c}$.

Extremal Clause: Nothing is in $R_{a+b=c}$ unless it is obtained from the Basis and Inductive Clauses.

PROPERTIES OF BINARY RELATION AND OPERATIONS

DIGRAPH

A digraph is short for directed graph, and it is a diagram composed of points called vertices (nodes) and arrows called arcs going from a vertex to a vertex.

For example the figure below is a digraph with 3 vertices and 4 arcs.

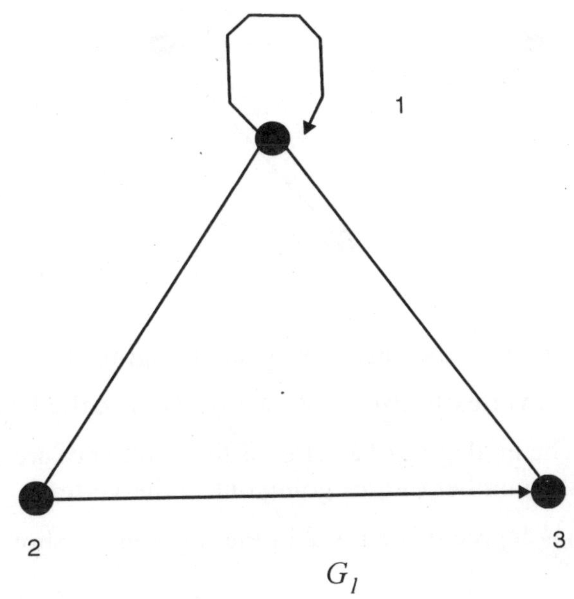

G_1

In this figure the vertices are labeled with numbers *1, 2,* and *3*.

Mathematically a digraph is defined as follows.

Digraph: A digraph is an ordered pair of sets $G = (V, A)$, where V is a set of vertices and A is a set of ordered pairs (called arcs) of vertices of V.

In the example, G_1, given above, $V - \{1, 2, 3\}$, and $A = \{<1, 1>, <1, 2>, <1, 3>, <2, 3>\}$.

Digraph representation of binary relations: A binary relation on a set can be represented by a digraph.

Let R be a binary relation on a set A, that is R is a subset of $A \times A$.

Then the digraph, call it G, representing R can be constructed as follows:

1. The vertices of the digraph G are the elements of A, and
2. $<x, y>$ is an arc of G from vertex x to vertex y if and only if $<x, y>$ is in R.

Example: The less than relation R on the set of integers $A = \{1, 2, 3, 4\}$ is the set $\{<1, 2>, <1, 3>, <1, 4>, <2, 3>, <2, 4>, <3, 4>\}$ and it can be represented by the following digraph.

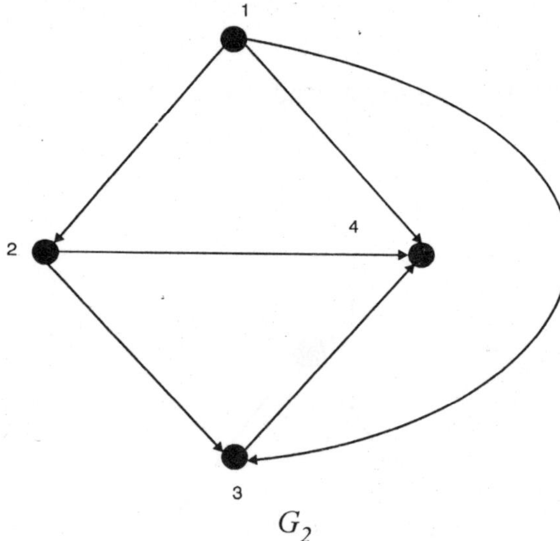

G_2

Let us now define some of the basic concepts on digraphs.

Loop: An arc from a vertex to itself such as $<1, 1>$, is called a loop (or self–loop).

Degree of vertex: The in–degree of a vertex is the number of arcs coming to the vertex, and the out–degree is the number of arcs going out of the vertex.

For example, the in–degree of vertex 2 in the digraph G_2 shown above is 1, and the out–degree is 2.

Path: A path from a vertex x_0 to a vertex x_n in a digraph $G = (V, A)$ is a sequence of vertices $x_0, x_1, \ldots, x_n$ that satisfies the following:

for each i, $0 \le i \le n - 1$, $<x_i, x_{i+1}> \in A$, or $<x_{i+1}, x_i> \in A$, that is, between any pair of vertices there is an arc connecting them.

x_0 is the initial vertex and x_n is the terminal vertex of the path.

A path is called a directed path if $<x_i, x_{i+1}> A \in$, for every i, $0 \le i \le n - 1$.

If the initial and the terminal vertices of a path are the same, that is, $x_0 = x_n$, then the path is called a cycle .

If no arcs appear more than once in a path, the path is called a simple path. A path is called elementary if no vertices appear more than once in it except for the initial and terminal vertices of a cycle. In a simple cycle one vertex appears twice in the sequence: once as the initial vertex and once as the terminal vertex.

Note: There are two different definitions for "simple path". Here we follow the definition of Berge[1], Liu[2], Rosen[3] and others. A "simple path" according to another group (Cormen et al[4], Stanat and McAllister[5] and others) is a path in which no vertices appear more than once.

Connected graph: A digraph is said to be connected if there is a path between every pair of its vertices.

Example: In the digraph G_3 given below,

1, 2, 5 is a simple and elementary path but not directed,

1, 2, 2, 5 is a simple path but neither directed nor elementary.

1, 2, 4, 5 is a simple elementary directed path,

1, 2, 4, 5, 2, 4, 5 is a directed path but not simple (hence not elementary),

1, 3, 5, 2, 1 is a simple elementary cycle but not directed, and

2, 4, 5, 2 is a simple elementary directed cycle.

This digraph is connected.

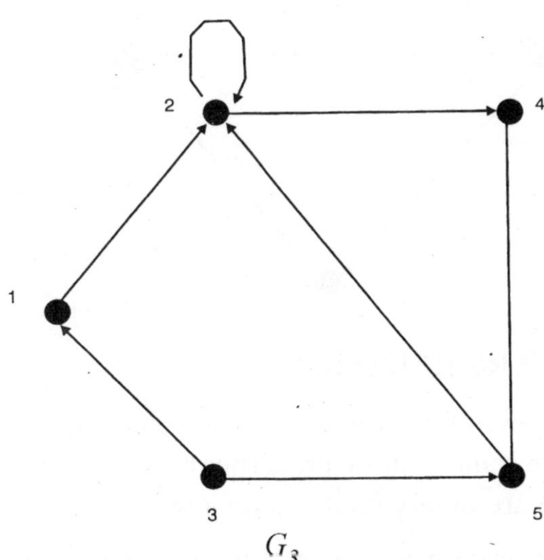

G_3

Sometimes we need to refer to part of a given digraph. A partial digraph of a digraph is a digraph consisting of arbitrary numbers of vertices and arcs of the given digraph, while a subdigraph is a digraph consisting of an arbitrary number of vertices and all the arcs between them of the given digraph. Formally they are defined as follows:

Subdigraph, partial digraph: Let $G = (V, A)$ be a digraph. Then a digraph (V', A') is a partial digraph of G, if $V' \subseteq V$, and $A' \subseteq A \cap (V' \times V')$. It is a subdigraph of G, if $V' \subseteq V$, and $A' = A\ (V' \times V')$.

DIGRAPH REPRESENTATION OF BINARY RELATION

A binary relation on a set can be represented by a digraph. Let R be a binary relation on a set A, that is R is a subset of $A \times A$. Then a digraph representing R can be constructed as follows:

Let the elements of A be the vertices of the digraph G, and let <*x, y*> be an arc of G from vertex x to vertex y if and only if <*x, y*> is in R.

Example: The less than relation R on th e set of integers $A = \{1, 2, 3, 4\}$ is the set {<*1*, *2*> , <*1*, *3*>, <*1*, *4*>, <*2*, *3*> , <*2*, *4*> , <*3*, *4*>} and it can be represented by the following digraph.

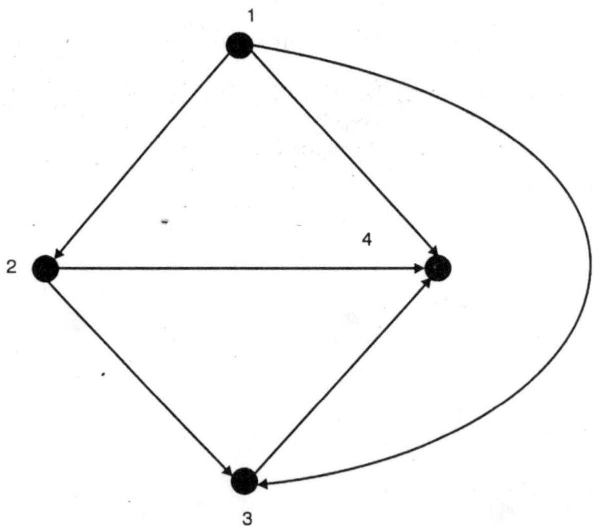

PROPERTIES OF BINARY RELATION

Certain important types of binary relation can be characterized by properties they have. Here we are going to learn some of those properties binary relations may have. The relations we are interested in here are binary relations on a set.

reflexive relation: A relation R on a set A is called reflexive if and only if < a, a > $\in R$ for every element a of A.

Example: The relation $\leq$ on the set of integers $\{1, 2, 3\}$ is $\{<1, 1>, <1, 2>, <1, 3>, <2, 2>, <2, 3>, <3, 3>\}$ and it is reflexive because $<1, 1>$, $<2, 2>$, $<3, 3>$ are in this relation. As a matter of fact $\leq$ on any set of numbers is also reflexive. Similarly $\geq$ and $=$ on any set of numbers are reflexive. However, $<$ (or $>$) on any set of numbers is not reflexive.

Example: The relation $\subseteq$ on the set of subsets of $\{1, 2\}$ is $\{<\emptyset, \emptyset>, <\emptyset, \{1\}>, <\emptyset, \{2\}>, <\emptyset, \{1, 2\}>, <\{1\}, \{1\}>, <\{1\}, \{1, 2\}>, <\{2\}, \{2\}>, <\{2\}, \{1, 2\}>, <\{1, 2\}, \{1, 2\}>\}$

and it is reflexive. In fact relation on any collection of sets is reflexive.

Irreflexive relation: A relation R on a sct A is called irreflexive if and only if $<a, a> \notin R$ for every element a of A.

Example: The relation $>$ (or $<$) on the set of integers $\{1, 2, 3\}$ is irreflexive. In fact it is irreflexive for any set of numbers.

Example: The relation $\{<1, 1>, <1, 2>, <1, 3>, <2, 3>, <3, 3>\}$ on the set of integers $\{1, 2, 3\}$ is neither reflexive nor irreflexive.

Symmetric relation: A relation R on a set A is called symmetric if and only if for any a, and b in A, whenever $<a, b> \in R$, $<b, a> \in R$.

Example: The relation $=$ on the set of integers $\{1, 2, 3\}$ is $\{<1, 1>, <2, 2> <3, 3>\}$ and it is symmetric. Similarly $=$ on any set of numbers is symmetric. However, $<$ (or $>$), $\leq$ (or $\geq$ on any set of numbers is not symmetric.

Example: The relation "being acquainted with" on a set of people is symmetric.

Antisymmetric relation: A relation R on a set A is called antisymmetric if and only if for any a, and b in A, whenever $<a, b> \in R$, and $<b, a> \in R$, $a = b$ must hold. Equivalently, R is antisymmetric if and only if whenever $<a, b> \in R$, and $a \neq b$, $<b, a> \notin R$. Thus in an antisymmetric relation no pair of elements are related to each other.

Example: The relation $<$ (or $>$) on any set of numbers is antisymmetric. So is the equality relation on any set of numbers.

Transitive relation: A relation R on a set A is called transitive if and only if for any a, b, and c in A, whenever $<a, b> \in R$, and $<b, c> \in R$, $<a, c> \in R$.

Example: The relation $\leq$ on the set of integers $\{1, 2, 3\}$ is transitive, because for $<1, 2>$ and $<2, 3>$ in $\leq$, $<1, 3>$ is also in , for $<1, 1>$ and $<1, 2>$ in $\leq$, $<1, 2>$ is also in , and similarly for the others. As a matter of fact $\leq$ on any set of numbers is also transitive. Similarly and $= \geq$ on any set of numbers are transitive.

The following figures show the digraph of relations with different properties.

1. Is reflexive, antisymmetric, symmetric and transitive, but not irreflexive.
2. Is neither reflexive nor irreflexive, and it is antisymmetric, symmetric and transitive.

3. Is irreflexive but has none of the other four properties.
4. Is irreflexive, and symmetric, but none of the other three.
5. Is irreflexive, antisymmetric and transitive but neither reflexive nor symmetric.

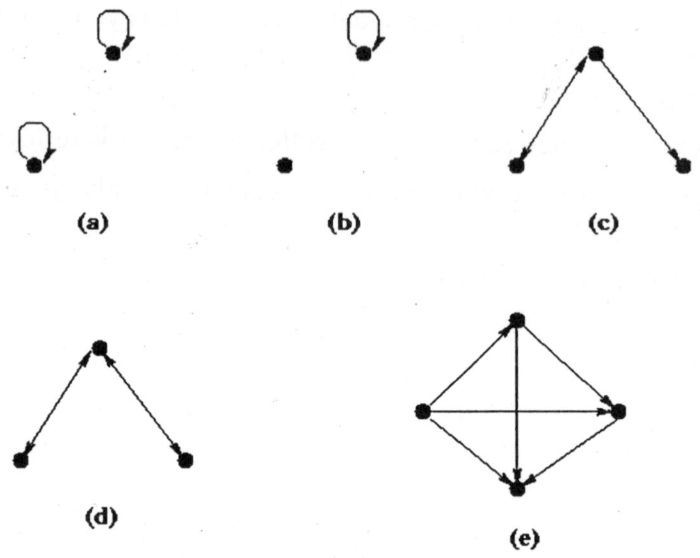

(a) (b) (c)

(d) (e)

OPERATIONS ON RELATIONS

SET OPERATIONS

A relation is a set. It is a set of ordered pairs if it is a binary relation, and it is a set of ordered *n*–tuples if it is an *n*–ary relation. Thus all the set operations apply to relations such as $\cup$, $\cap$, and complementing.

For example, the union of the "less than" and "equality" relations on the set of integers is the "less than or equal to" relation on the set of integers. The intersection of the "less than" and "less than or equal to" relations on the set of integers is the "less than" relation on the same set. The complement of the "less than" relation on the set of integers is the "greater than or equal to" relation on the same set.

COMPOSITE RELATIONS

If the elements of a set A are related to those of a set B, and those of B are in turn related to the elements of a set C, then one can expect a relation between A and C. For example, if Tom is my father(parent–child relation) and Sarah is a sister of Tom (sister relation), then Sarah is my aunt (aunt–nephew/niece relation). Composite relations give that kind of relations.

Composite relation: Let R_1 be a binary relation from a set A to a set B, R_2 a binary relation from B to a set C. Then the composite relation from A to C denoted by R_1R_2(also denoted by R_1 O R_2 is defined as

$$R_1R_2 = \{<a, c> \mid a \in A \wedge c\ C \wedge \exists\ b\ [b \in B < \wedge\ a, b> \in R_1 \wedge <b, c> \in R_2]\}.$$

In English, this means that an element a in A is related to an element c in C if there is an element b in B such that a is related to b by R_1 and b is related to c by R_2. Thus R_1R_2 is a relation from A to C via B in a sense. If R_1 is a parent–child relation and R_2 is a sister relation, then R_1R_2 is an aunt–nephew/niece relation.

Example: Let $A = \{a_1, a_2\}$, $B = \{b_1, b_2, b_3\}$, and $C = \{c_1, c_2\}$. Also let $R_1 = \{<a_1, b_1>, <a_1, b_2>, <a_2, b_3>\}$, and $R_2 = \{<b_1, c_1>, <b_2, c_1>, <b_2, c_2>, <b_3, c_1>\}$. Then $R_1R_2 = \{<a_1, c_1>, <a_1, c_2>, <a_2, c_1>\}$.

This is illustrated in the following figure. The dashed lines in the figure of R_1R_2 indicate the ordered pairs in R_1R_2, and dotted lines show ordered pairs that produce the dashed lines. (The lines in the left figure are all supposed to be solid lines.)

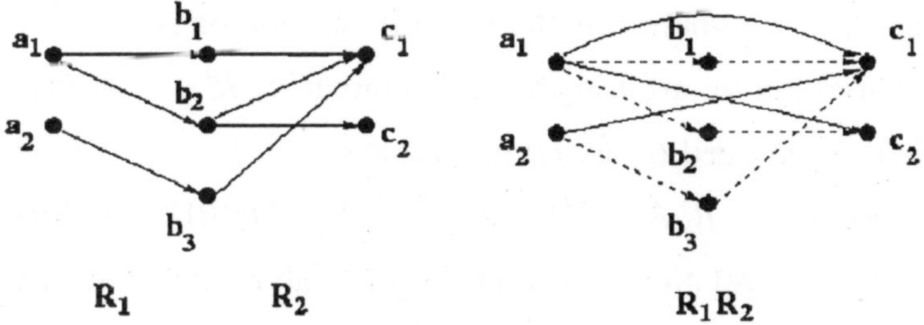

Example: If R is the parent–child relation on a set of people A, then RR, also denoted by R^2, is the grandparent–grandchild relation on A.

More examples: The digraphs of R^2 for several simple relations R are shown below:

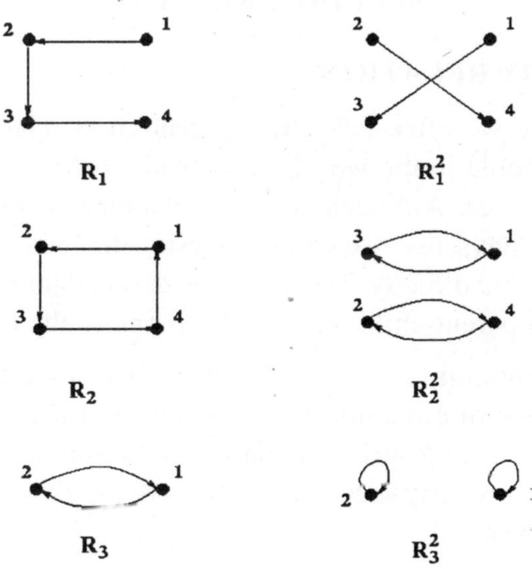

PROPERTIES OF COMPOSITE RELATIONS

Composite relations defined above have the following properties. Let R_1 be a relation from A to B, and R_2 and R_3 be relations from B to C. Then

1. $R_1(R_2R_3) = (R_1R_2)R_3$
2. $R_1(R_2 \cup R_3) = R_1R_2 \cup R_1R_3$
3. $R_1(R_2 \cap R_3) \subseteq R_1R_2 \cap R_1R_3$

Proofs for these properties are omitted.

POWERS OF RELATION

Let R be a binary relation on A. Then R^n for all positive integers n is defined recursively as follows:

Definition(power of relation):

Basis Clause: $R^0 = E$, where E is the equality relation on A.

Inductive Clause: For an arbitrary natural number n , $R^{n+1} = R^nR$.

Note that there is no need for extremal clause here.

Thus for example $R^1 = R$, $R^2 = RR$, and $R^3 = R^2R = (RR)R = R(RR) = RRR$.

The powers of binary relation R on a set A defined above have the following properties.

1. $R^{m+n} = R^mR^n$,
2. $(R^m)^n = R^{mn}$.

SPECIAL RELATIONS

CLOSURE OF BINARY RELATION

In our everyday life we often talk about parent-child relationship. This is a binary relation on the set of people in the world, dead or alive. Also we are often interested in ancestor-descendant relations. Although the latter relation can be obtained from the former, hence it is redundant in that sense, we do use ancestor-descendant relations which give us necessary information more directly. This ancestor-descendant relation relates two people if there is a sequence of parent-child relations from one to the other.

It includes the parent-child relation as a subset. The ancestor-descendant relation is an example of the closure of a relation, in particular the transitive closure of the parent-child relation. In general, the closure of a relation is the smallest extension of the relation that has a certain specific property such as the reflexivity, symmetry or transitivity. Formally they are defined as follows:

Reflexive closure: A relation R' is the reflexive closure of a relation R if and only if

1. R' is reflexive,
2. R R', and
3. For any relation R'', if $R \subseteq R''$ and R'' is reflexive, then $R' \subseteq R''$, that is, R' is the smallest relation that satisfies (1) and (2).

Example: Let R be the less-than relation on the set of integers I, that is $R = \{< a, b> \mid a \in I \wedge b \in I \wedge a < b\}$.

Then the reflexive closure $r(R)$ of R is the union of R and the equality relation on I, that is $r(R) = \{<a, b> \mid a \in I \wedge b \in I\ a \in b\}$

The digraph of the reflexive closure of a relation is obtained from the digraph of the relation by adding a self-loop at each vertex if one is already not there.

Symmetric and transitive closures can be defined similarly.

Symmetric closure: A relation R' is the symmetric closure of a relation R if and only if

1. R' is symmetric,
2. R R', and
3. For any relation R'', if $R \subseteq R''$, and R'' is symmetric, then $R' \subseteq R''$, that is, R' is the smallest relation that satisfies (1) and (2).

Example: Let R be the less–than relation on the set of integers I. Then the symmetric closure of R, denoted by $s(R)$ is $s(R) = \{<a, b> \mid a\ I\ b\ I\ [a < b\ a > b]\}$ that is $\{<a, b> \mid a\ I\ b\ I\ a\ b\}$

The digraph of the symmetric closure of a relation is obtained from the digraph of the relation by adding for each arc the arc in the reverse direction if one is already not there.

Transitive closure: A relation R' is the transitive closure of a relation R if and only if

1. R' is transitive,
2. $R \subseteq R'$, and
3. For any relation R'', if $R \subseteq R''$ and R'' is transitive, then $R' \subseteq R''$, that is, R' is the smallest relation that satisfies (1) and (2).

Examples: The transitive closure of a parent–child relation is the ancestor–descendant relation as mentioned above, and that of the less–than relation on I is the less–than relation itself.

The digraph of the transitive closure of a relation is obtained from the digraph of the relation by adding for each directed path the arc that shunts the path if one is already not there.

Two more examples of closures are given below in terms of digraphs.

The arrows with two heads represent arrows going in opposite directions.

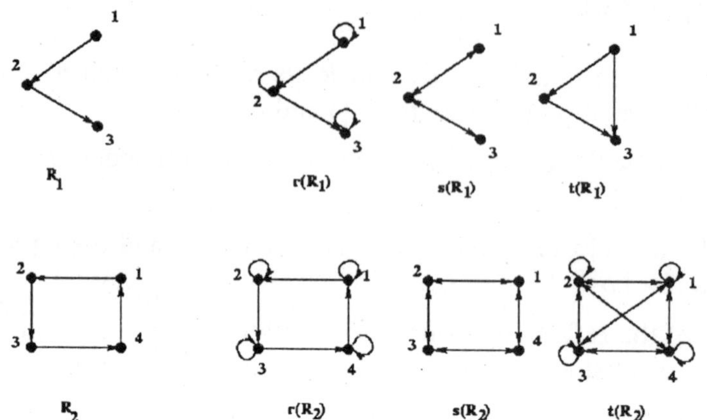

PROPERTIES OF CLOSURE

The closures have the following properties. They are stated here as theorems without proof.

Theorem: Let E denote the equality relation, and R^c the inverse relation of binary relation R, all on a set A, where $R^c = \{< a, b > | < b, a \in R\}$. Then

1. $r(R) = R \cup E$
2. $s(R) = R \cup R^c$
3. $t(R) = \cup_{i=1}^{\infty} R^i = \cup_{i=1}^{n} R^i$, if $|A| = n$.
4. R is reflexive if and only if $r(R) = R$.
5. R is symmetric if and only if $s(R) = R$.
6. R is transitive if and only if $t(R) = R$.

EQUIVALENCE RELATION

On the face of most clocks, hours are represented by integers between 1 and 12. However, since a day has 24 hours after 12 hours, a clock goes back to hour 1, and starts all over again from there. Thus each pair of hours such as 1 and 13, 2 and 14, etc. share one number 1, 2, ...etc., respectively.

The same applies when we are interested in more than 24 hours. 25th hour is 1, so are 37th, 49th etc. What we are doing here essentially is that we consider the numbers in each group such as 1, 13, 25, ..., equivalent in the sense that they all are represented by one number. Being representable by one number such as we see on clocks is a binary relation on the set of natural numbers and it is an example of equivalence relation we are going to study here.

The concept of equivalence relation is characterized by three properties as follows:

Definition(equivalence relation): A binary relation *R* on a set *A* is an equivalence relation if and only if

- *R* is reflexive
- *R* is symmetric, and
- *R* is transitive.

Example: The equality relation (=) on a set of numbers such as {*1, 2, 3*} is an equivalence relation.

Example: The congruent modulo *m* relation on the set of integers i.e. {<*a, b*>| *a b* (mod *m*)}, where *m* is a positive integer greater than *1*, is an equivalence relation.

Note that the equivalence relation on hours on a clock is the congruent mod *12*, and that when *m* = *2*, i.e. the congruent mod *2*, all even numbers are equivalent and all odd numbers are equivalent. Thus the set of integers are divided into two subsets: evens and odds.

Example: Taking this discrete structures course together this semester is another equivalence relation.

Equivalence relations can also be represented by a digraph since they are a binary relation on a set. For example the digraph of the equivalence relation congruent mod *3* on {*0, 1, 2, 3, 4, 5 , 6*} is as shown below. It consists of three connected components.

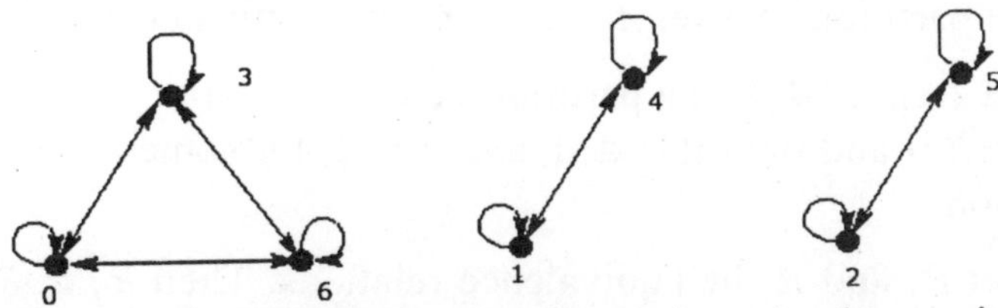

The set of even numbers and that of odd numbers in the equivalence relation of congruent mod *2*, and the set of integers equivalent to a number between 1 and 12 in the equivalence relation on hours in the clock example are called an equivalence class. Formally it is defined as follows:

Equivalence class: For an equivalence relation *R* on a set *A*, the set of the elements of *A* that are related to an element, say *a*, of *A* is called the equivalence class of element *a* and it is denoted by [*a*].

Example: For the equivalence relation of hours on a clock, equivalence classes are

$[1] = \{1, 13, 25, ...\} = \{1+ 12n: n \in N\}$,

$[2] = \{2, 14, 26, ...\} = \{2+ 12n: n \in N\}$,

........,

where N is the set of natural numbers. There are altogether twelve of them.

For an equivalence relation R on a set A, every element of A is in an equivalence class. For if an element, say b , does not belong to the equivalence class of any other element in A, then the set consisting of the element b itself is an equivalence class. Thus the set A is in a sense *covered* by the equivalence classes. Another property of equivalence class is that equivalence classes of two elements of a set A are either disjoint or identical, that is either $[a] = [b]$ or $[a] \cap [b] = \emptyset$ for arbitrary elements a and b of A. Thus the set A is partitioned into equivalence classes by an equivalence relation on A. This is formally stated as a theorem below after the definition of partition.

Partition: Let A be a set and let $A_1, A_2, ..., A_n$ be subsets of A. Then $\{A_1, A_2, ..., A_n\}$ is a partition of A, if and only if

1. $\cup_{i=1}^{n} A_i = A$, and
2. $A_i \cap A_j = \emptyset$, if $A_i \neq A_j$, $1 \leq i, j \leq n$.

Example: Let $A = \{1, 2, 3, 4, 5\}$, $A_1 = \{1, 5\}$, $A_2 = \{3\}$, and $A_3 = \{2, 4\}$. Then $\{A_1, A_2, A_3\}$ is a partition of A. However, $B_1 = \{1, 2, 5\}$, $B_2 = \{2, 3\}$, and $B_3 = \{4\}$ do not form a partition for A because $B_1 \cap B_2 \neq \emptyset$, though $B_1 \neq B_2$.

Theorem: The set of equivalence classes of an equivalence relation on a set A is a partition of A.

Conversely, a partition of a set A determines an equivalence relation on A.

Theorem: Let $\{A_1, ..., A_n\}$ be a partition of a set A. Define a binary relation R on A as follows: $<a, b> \in R$ if and only if $a \in A_i$ and $b \in A_i$ for some i, $1 \leq i \leq n$. Then R is an equivalence relation.

Theorem : Let R_1 and R_2 be equivalence relations. Then $R_1 \cap R_2$ is an equivalence relation, but $R_1 \cup R_2$ is not necessarily an equivalence relation.

ORDER RELATION

Shoppers in a grocery store are served at a cashier on the first–come–first–served basis. When there are many people at cashiers, lines are formed. People in these lines are ordered for service: Those at the head of a line are served sooner than those at the end. Cars waiting for the signal to change at an intersection are also ordered similarly. Natural numbers can also be ordered in the increasing order of their magnitude. Those are just a few examples of order we encounter in our daily lives. The order relations we are going to study here are an abstraction of those relations. The properties common to orders we see in our daily lives have been extracted and are used to characterize the concepts of order. Here we are going to learn three types of order: partial order, total order, and quasi order.

Partial order: A binary relation *R* on a set *A* is a partial order if and only if it is :

1. Reflexive,
2. Antisymmetric, and
3. Transitive.

The ordered pair <*A*, *R*> is called a poset (partially ordered set) when *R* is a partial order.

Example: The less-than-or-equal-to relation on the set of integers *I* is a partial order, and the set *I* with this relation is a poset.

Example: The subset relation on the power set of a set, say {1, 2} , is also a partial order, and the set {1, 2} with the subset relation is a poset.

Total order: A binary relation *R* on a set *A* is a total order if and only if it is

1. A partial order, and
2. For any pair of elements *a* and *b* of *A*, $< a, b > \in R$ or $< b, a > \in R$.

That is, every element is related with every element one way or the other.

A total order is also called a *linear order*.

Example: The less-than-or-equal-to relation on the set of integers *I* is a total order.

The strictly-less-than and proper-subset relations are not partial order because they are not reflexive. They are examples of some relation called quasi order.

Quasi order: A binary relation *R* on a set *A* is a quasi order if and only if it is

1. Irreflexive, and
2. Transitive.

A quasi order is necessarily antisymmetric as one can easily verify.

Caution Like many other definitions there is another fairly widely used definition of quasi order in the literature. According to that definition a quasi order is a relation that is *reflexive* and transitive. That is something quite different from "quasi order" we have defined here.

Example: The less-than relation on the set of integers *I* is a quasi order.

Example: The proper subset relation on the power set of a set, say {1, 2} , is also a quasi order.

The concept of *least/greatest number* in a set of integers can be generalized for a general poset. We start with the concepts of minimal/maximal elements.

minimal/maximal element: Let $< A, \preceq >$ be a poset, where $\preceq$ represents an arbitrary partial order. Then an element $b \in A$ is a minimal element of A if there is no element $a \in A$ that satisfies $a \preceq b$. Similarly an element $b \in A$ is a maximal element of A if there is no element $a \in A$ that satisfies $b \preceq a$.

Example: The set of $\{\{1\}, \{2\}, \{1, 2\}\}$ with $\subseteq$ has two minimal elements $\{1\}$ and $\{2\}$. Note that $\{1\}$, and $\{2\}$ are not related to each other in $\subseteq$. Hence we can not say which is "smaller than" which, that is, they are not comparable.

least/greatest element: Let $< A, \preceq >$ be a poset. Then an element $b \in A$ is the least element of A if for every element $a \in A$, $b \preceq a$.

Note that the least element of a poset is unique if one exists because of the antisymmetry of $\preceq$.

Example: The poset of the set of natural numbers with the less–than–or–equal–to relation has the least element 0.

Example: The poset of the powerset of $\{1, 2\}$ with $\in$ has the least element $\emptyset$.

Well order: A total order R on a set A is a well order if every non–empty subset of A has the least element.

Example: The poset of the set of natural numbers with the less–than–or–equal–to relation is a well order, because every set of natural numbers has the least element.

The poset of the set of positive real numberes with the less–than–or–equal–to relation is not a well order, because the set itself does not have any least element (0 is not in the set).

A digraph of a binary relation on a set can be simplified if the relation is a partial order. Hasse diagrams defined as follows are such graphs.

Hasse diagram: A Hasse diagram is a graph for a poset which does not have loops and arcs implied by the transitivity. Further, it is drawn so that all arcs point upward eliminating arrowheads.

To obtain the Hassse diagram of a poset, first remove the loops, then remove arcs $<$ a, b $>$ if and only if there is an element c that $<$ a, c $>$ and $<$ c, b $>$ exist in the given relation.

Example: For the relation $\{< a, a >, < a, b >, < a, c >, < b, b >, < b, c >, < c, c >\}$ on set $\{a, b, c\}$, the Hasse diagram has the arcs $\{< a, b >, < b, c >\}$ as shown below.

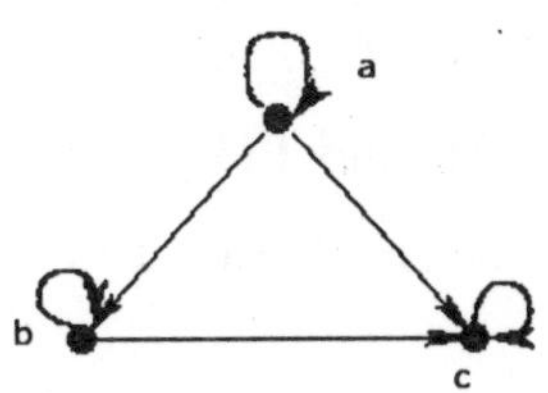

Digraph for Portial Order

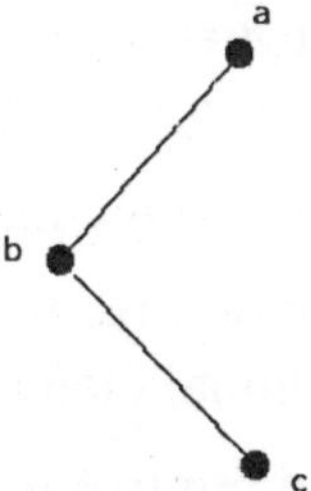

Hasse Digram

TOPOLOGICAL SORTING

The elements in a finite poset can be ordered linearly in a number of ways while preserving the partial order. For example {Ø, {1}, {2}, {1, 2}} wit h the partial order ⊆, can be ordered linearly as Ø, {1}, {2}, {1, 2}, or Ø, {2}, {1}, {1, 2}. In these orders a set appears before (to the left of) another set if it is a subset of the other. In real life, tasks for manufacturing goods in general can be partially ordered based on the prerequisite relation, that is certain tasks must be completed before certain other tasks can be started. For example the arms of a chair must be carved before the chair is assembled. Scheduling those tasks is essentially the same as arranging them with a linear order (ignoring here some possible concurrent processing for simplicity's sake).

The topological sorting is a procedure to find from a partial order on a finite set a linear order that does not violate the partial order. It is based on the fact that a finite poset has at least one minimal element. The basic idea of the topological sorting is to first remove a minimal element from the given poset, and then repeat that for the resulting set until no more elements are left. The order of removal of the minimal elements gives a linear order. The following algorithm formally describes the topological sorting.

Algorithm Topological Sor

Input: A finite poset <*A*, *R*>.

Output: A sequence of the elements of *A* preserving the order *R*.

integer *i*;

i : = 1;

while (*A*) {

pick a minimal element *b* from *A*;

A : = *A* – {*b*};

i : = *i* + 1;

output *b*

}

Example: Let $A = \{\{1\}, \{2\}, \{3\}, \{1, 2\}, \{1, 3\}, \{2, 3\}, \{1, 2, 3\}\}$ with the partial order $\subseteq$.

This given A has three minimal elements $\{1\}$, $\{2\}$, and $\{3\}$.

Select $\{2\}$ and remove it from A. Let A denote the resultant set i.e. $A := A - \{2\}$. The new A has two minimal elements $\{1\}$, and $\{3\}$.

Select $\{1\}$ and remove it from A. Denote by A the resultant set, that is $A = \{\{3\}, \{1, 2\}, \{1, 3\}, \{2, 3\}, \{1, 2, 3\}\}$.

This new A has two minimal elements $\{3\}$ and $\{1, 2\}$.

Select $\{1, 2\}$ and remove it from A.

Proceeding in like manner, we can obtain the following linear order:

$\{\{2\}, \{1\}, \{1, 2\}, \{3\}, \{1, 3\}, \{2, 3\}, \{1, 2, 3\}\}$.

FUNCTION

A function is something that associates each element of a set with an element of another set (which may or may not be the same as the first set). The concept of function appears quite often even in nontechnical contexts. For example, a social security number uniquely identifies the person, the income tax rate varies depending on the income, the final letter grade for a course is often determined by test and exam scores, homeworks and projects, and so on.

In all these cases to each member of a set (social security number, income, tuple of test and exam scores, homeworks and projects) some member of another set (person, tax rate, letter grade, respectively) is assigned.

As you might have noticed, a function is quite like a relation. In fact, formally, we define a function as a special type of binary relation.

Definition (function): A function, denote it by f, from a set A to a set B is a relation from A to B that satisfies:

1. For each element a in A, there is an element b in B such that $<a, b>$ is in the relation, and
2. If $<a, b>$ and $<a, c>$ are in the relation, then $b = c$.

The set A in the above definition is called the domain of the function and B its codomain.

Thus, f is a function if it *covers* the domain (maps every element of the domain) and it is *single valued.*

The relation given by f between a and b represented by the ordered pair $<a, b>$ is denoted as $f(a) = b$, and b is called the image of a under f.

The set of images of the elements of a set *S* under a function *f* is called the image of the set *S* under *f*, and is denoted by $f(S)$, that is,

$f(S) = \{f(a) \mid a\ S\}$, where *S* is a subset of the domain *A* of *f*.

The image of the domain under *f* is called the range of *f*.

Example: Let *f* be the function from the set of natural numbers *N* to *N* that maps each natural number *x* to x^2 . Then the domain and codomain of this *f* are *N*, the image of, say *3*, under this function is *9*, and its range is the set of squares, i.e. $\{0, 1, 4, 9, 16, ...\}$.

Sum and product: Let *f* and g be functions from a set *A* to the set of real numbers *R*.

Then the sum and the product of *f* and g are defined as follows:

For all *x*, $(f + g)(x) = f(x) + g(x)$, and

for all *x*, $(f*g)(x) = f(x)*g(x)$,

where $f(x)*g(x)$ is the product of two real numbers $f(x)$ and $g(x)$.

Example: Let $f(x) = 3x + 1$ and $g(x) = x^2$. Then $(f + g)(x) = x^2 + 3x + 1$, and $(f*g)(x) = 3x^3 + x^2$

One-to-one: A function *f* is said to be one–to–one (injective) , if and only if whenever $f(x) = f(y)$, $x = y$.

Example: The function $f(x) = x^2$ from the set of natural numbers *N* to *N* is a one–to–one function. Note that $f(x) = x^2$ is not one–to–one if it is from the set of integers(negative as well as non–negative) to *N* , because for example $f(1) = f(-1) = 1$.

Onto: A function *f* from a set *A* to a set *B* is said to be onto(surjective) , if and only if for every element *y* of *B* , there is an element *x* in *A* such that $f(x) = y$, that is, *f* is onto if and only if $f(A) = B$.

Example: The function $f(x) = 2x$ from the set of natural numbers *N* to the set of non–negative even numbers *E* is an onto function. However, $f(x) = 2x$ from the set of natural numbers *N* to *N* is not onto, because, for example, nothing in *N* can be mapped to *3* by this function.

Bijection: A function is called a bijection , if it is onto and one–to–one.

Example: The function $f(x) = 2x$ from the set of natural numbers *N* to the set of non–negative even numbers *E* is one–to–one and onto. Thus it is a bijection.

Every bijection has a function called the inverse function.

These concepts are illustrated in the figure below. In each figure below, the points on the left are in the domain and the ones on the right are in the codomain, and arrows show $< x, f(x) >$ relation.

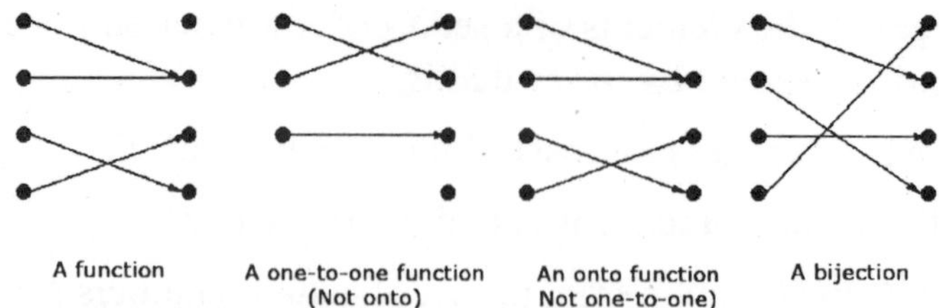

Inverse: Let f be a bijection from a set A to a set B. Then the function g is called the inverse function of f, and it is denoted by f^{-1}, if for every element y of B, $g(y) = x$, where $f(x) = y$.. Note that such an x is unique for each y because f is a bijection.

For example, the rightmost function in the above figure is a bijection and its inverse is obtained by reversing the direction of each arrow.

Example: The inverse function of $f(x) = 2x$ from the set of natural numbers N to the set of non–negative even numbers E is $f^{-1}(x) = 1/2\ x$ from E to N. It is also a bijection.

A function is a relation. Therefore one can also talk about composition of functions.

Composite function: Let g be a function from a set A to a set B, and let f be a function from B to a set C. Then the composition of functions f and g, denoted by fg, is the function from A to C that satisfies

$fg(x) = f(g(x))$ for all x in A.

Example: Let $f(x) = x^2$, and $g(x) = x + 1$. Then $f(g(x)) = (x + 1)^2$.

GROWTH OF FUNCTIONS

One of the important criteria in evaluating algorithms is the time it takes to complete a job. To have a meaningful comparison of algorithms, the estimate of computation time must be independent of the programming language, compiler, and computer used; must reflect on the size of the problem being solved; and must not depend on specific instances of the problem being solved. The quantities often used for the estimate are the worst case execution time, and average execution time of an algorithm, and they are represented by the number of some key operations executed to perform the required computation.

As an example for an estimate of computation time, let us consider the sequential search algorithm.

Example: Algorithm for Sequential Search

Algorithm SeqSearch(L, n, x)

L is an array with n entries indexed $1, \ldots, n$, and x is the key to be searched for in L.

Output: if x is in L, then output its index, else output 0.

Index := *1;*

while (*index* $\leq n$ and $L[index] \neq x$)

Index := *index* + *1* ;

if (*index > n)* , then *index := 0*

return *index* .

The worst case time of this algorithm, for example, can be estimated as follows: First the key operation is comparison of keys comparing $L[index]$ with x . Most search algorithms (if not all) need "comparison of keys". The largest number of execution of this comparison is n , which occurs when x is not in L or when x is at $L[n]$, and the while loop is executed n times. This quantity n thus obtained is used as an estimate of the worst case time of this sequential search algorithm.

Note that in the while loop two comparisons and one addition are performed. Thus one could use $3n$ as an estimate just as well. Also note that the very first line and the last two lines are not counted in. The reasons for those are firstly that differences in implementation details such as languages, commands, compilers and machines make differences in constant factors meaningless, and secondly that for large values of n , the highest degree term in n dominates the estimate. Since we are mostly interested in the behavior of algorithms for large values of n , lower terms can be ignored compared with the highest term. The concept that is used to address these issues is something called big–oh, and that is what we are going to study here.

BIG – OH

The following example gives the idea of one function growing more rapidly than another. We will use this example to introduce the concept the big–Oh.

Example: $f(n) = 100\,n^2$, $g(n) = n^4$, the following table and figure show that $g(n)$ grows faster than $f(n)$ when $n > 10$. We say f is big–Oh of g.

n	$f(n)$	$g(n)$
10	10,000	10,000
50	250,000	6,250,000
100	1,000,000	100,000,000
150	2,250,000	506,250,000

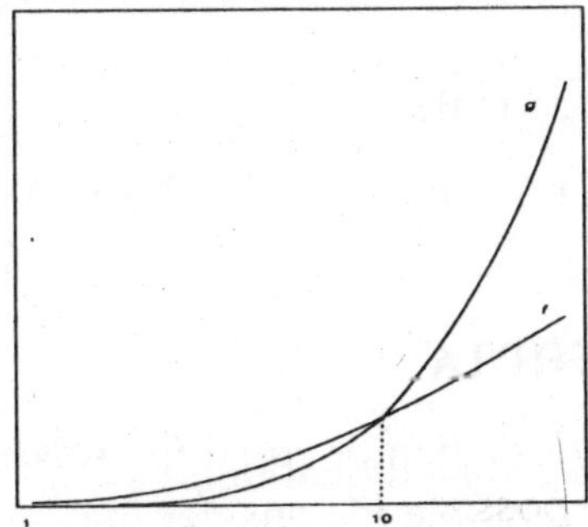

Big-oh: Let f and g be functions from the set of integers (or the set of real numbers) to the set of real numbers. Then $f(x)$ is said to be $O(g(x))$, which is read as $f(x)$ is big-oh of $g(x)$, if and only if there are constants C and n_0 such that

$$| f(x) | C | g(x) |$$

whenever $x > n_0$.

Note that big–oh is a binary relation on a set of functions (What kinds of properties does it have ? reflexive ? symmetric ? transitive ?).

For example, $5x + 10$ is big–oh of x^2, because $5x + 10 < 5x^2 + 10x^2 = 15x^2$ for $x > 1$.

Hence for $C = 15$ and $n_0 = 1$, $| 5x + 10 | \leq C | x^2 |$.

Similarly it can be seen that $3x^2 + 2x + 4 < 9x^2$ for $x > 1$. Hence $3x^2 + 2x + 4$ is $O(x^2)$.

In general, we have the following theorem:

Theorem: $a_n x^n + ... + a_1 x + a_0$ is $O(x^n)$ for any real numbers a_n , ..., a_0 and any nonnegative number n .

Note: Let $f(x) = 3x^2 + 2x + 4$, $g(x) = x^2$, from the above illustration, we have that $f(x)$ is $O(g(x))$. Also, since $x^2 < 3x^2 + 2x + 4$, we can also get $g(x)$ is $O(f(x))$. In this case, we say these two functions are of the *same order*. Please refer to *big–theta* for more information.

GROWTH OF COMBINATIONS OF FUNCTIONS

Big–oh has some useful properties. Some of them are listed as theorems here.

Let use start with the definition of max function.

Definition(max function): Let $f_1(x)$ and $f_2(x)$ be functions from a set A to a set of real numbers B. Then $\max(f_1(x) , f_2(x))$ is the function from A to B that takes as its value at each point x the larger of $f_1(x)$ and $f_2(x)$.

Theorem: If $f_1(x)$ is $O(g_1(x))$, and $f_2(x)$ is $O(g_2(x))$, then $(f_1 + f_2)(x)$ is $O(\max(g_1(x) , g_2(x)))$.

From this theorem it follows that if $f_1(x)$ and $f_2(x)$ are $O(g(x))$, then $(f_1 + f_2)(x)$ is $O(g(x))$, and

$(f_1 + f_2)(x)$ is $O(\max(f_1(x) , f_2(x)))$.

Theorem: If $f_1(x)$ is $O(g_1(x))$, and $f_2(x)$ is $O(g_2(x))$, then $(f_1 * f_2)(x)$ is $O(g_1(x) * g_2(x))$.

BIG – OMEGA AND BIG – THETA

Big–oh concerns with the "less than or equal to" relation between functions for large values of the variable. It is also possible to consider the "greater than or equal to" relation and "equal to" relation in a similar way. Big–Omega is for the former and big–theta is for the latter.

- *Big–omega*: Let *f* and *g* be functions from the set of integers (or the set of real numbers) to the set of real numbers. Then $f(x)$ is said to be $\Omega(g(x))$, which is read as $f(x)$ is big–omega of $g(x)$, if there are constants C and n_0 such that

  $$| f(x) | \geq C \, | g(x) |$$

  whenever $x > n_0$.

Definition (big–theta): Let *f* and *g* be functions from the set of integers (or the set of real numbers) to the set of real numbers. Then $f(x)$ is said to be $(g(x))$, which is read as $f(x)$ is big–theta of $g(x)$, if $f(x)$ is $\Theta(g(x))$, and $(g(x))$. We also say that $f(x)$ is of order $g(x)$. (We have seen this at Big–Oh when introducing the concept of *same order*.)

For example, $3x^2 - 3x - 5$ is $\Omega(x^2)$, because $3x^2 - 3x - 5 \geq x^2$ for integers $x > 2$ $(C = 1, n_0 = 2)$.

Hence by Theorem it is (x^2).

In general, we have the following theorem:

Theorem: $a_n x^n + \ldots + a_1 x + a_0$ is $\Theta(x^n)$ for any real numbers $a_n, \ldots, a_0$ and any nonnegative number n.

LITTLE – OH AND LITTLE – OMEGA

If $f(x)$ is $O(g(x))$, but not $\Theta(g(x))$, then $f(x)$ is said to be $o(g(x))$, and it is read as $f(x)$ is little–oh of $g(x)$. Similarly for little–omega (ω).

For example x is $o(x^2)$, x^2 is $o(2^x)$, 2^x is $o(x!)$, etc.

CALCULATION OF BIG – OH

Basic knowledge of limits and derivatives of functions from calculus is necessary here. Big–oh relationships between functions can be tested using limit of function. as follows:

Let $f(x)$ and $g(x)$ be functions from a set of real numbers to a set of real numbers.

Then

1. If , then $f(x)$ is $o(g(x))$. Note that if $f(x)$ is $o(g(x))$, then $f(x)$ is $O(g(x))$.
2. If , then $g(x)$ is $o(f(x))$.
3. If , then $f(x)$ is $(g(x))$.
4. If , then $f(x)$ is $O(g(x))$.

For example,

$(4/x + 3/x^2 + 5/x^4)/(1 - 3/x - 5/x^3 - 4/x^4) = 0$.

Hence

$(4x^3 + 3x^2 + 5)$ is $o(x^4 - 3x^3 - 5x - 4)$,

or equivalently, $(x^4 - 3x^3 - 5x - 4)$ is $(4x^3 + 3x^2 + 5)$.

Let us see why these rules hold. Here we give a proof for 4. Others can be proven similarly.

Proof: Suppose.

By the definition of limit this means that, n_0 such that whenever $x > n_0$.

Hence

Let =, then whenever $x > n_0$.

Since we are interested in non–negative functions f and g , this means that | $f(x)$ | C | $g(x)$ |

Hence $f(x) = O(g(x))$.

L'Hospital (L'Hτpital)'s Rule

$f(x)/g(x)$ is not always easy to calculate. For example take $x^2/3^x$. Since both x^2 and 3^x go to as x goes to and there is no apparent factor common to both, the calculation of the limit is not immediate. One tool we may be able to use in such cases is L'Hospital's Rule, which is given as a theorem below.

Theorem (L'Hospital):

If $f(x) =$ and $g(x) =$, and $f(x)$ and $g(x)$ have the first derivatives, $f'(x)$ and $g'(x)$, respectively, then $f(x)/g(x) = f'(x)/g'(x)$.

This also holds when $f(x) = 0$ and $g(x) = 0$, instead of $f(x) =$ and $g(x) =$.

For example, $x/e^x = 1/e^x = 0$, because $(e^x)' = e^x$, where e is the base for the natural logarithm.

Similarly $\ln x/x = (1/x)/1 = 1/x = 0$.

Note that this rule can be applied repeatedly as long as the conditions are satisfied.

So, for example, $x^2/e^x = 2x/e^x = 2/e^x = 0$.

SUMMARY OF BIG – OH

Sometimes, it is very necessary to compare the order of some common used functions including the following:

1 log$n$ n $n$log$n$ n^2 2^n $n!$ n^n

Now, we can use what we've learned above about the concept of big–Oh and the calculation methods to calculate the order of these functions. The result shows that each function in the above list is big–oh of the functions following them. The figure below displays the graphs of these funcions, using a scale for the values of the functions that doubles for each successive marking on the graph.

Chapter 2

Composition of Functions

In mathematics, a composite function, formed by the composition of one function on another, represents the application of the former to the result of the application of the latter to the argument of the composite. The functions $f : X \to Y$ and $g : Y \to Z$ can be composed by first applying f to an argument x and then applying g to the result. Thus one obtains a function $(g \circ f) : X \to Z$ defined by $(g \circ f)(x) = g(f(x)$ for all x in X. The notation g o f is read as "g circle f" or "g composed with f".

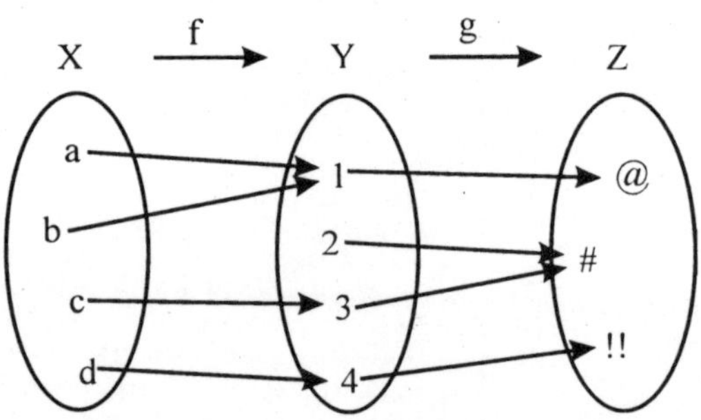

Fig. The Composition of f and g

The composition of functions is always associative. That is, if f, g, and h are three functions with suitably chosen domains and codomains, then $f \circ (g \circ h) = (f \circ g) \circ h$. Since there is no distinction between the choices of placement of parentheses, they may be safely left off.

The functions g and f commute with each other if $g \circ f = f \circ g$. In general, composition of functions will not be commutative. Commutativity is a special property, attained only by particular functions, and often in special circumstances. For example,

$|x| + 3 = |x + 3|$ only when $x \geq 0$. But inverse functions always commute to produce the identity mapping.

Derivatives of compositions involving differentiable functions can be found using the chain rule. "Higher" derivatives of such functions are given by Faa di Bruno's formula. As an example, suppose that an airplane's elevation at time *t* is given by the function *h*(*t*) and that the oxygen concentration at elevation *x* is given by the function *c*(*x*). Then (c o h) (*t*) describes the oxygen concentration around the plane at time *t*.

Functional Powers

If $Y \subseteq X$ then $f: X \rightarrow Y$ may compose with itself; this is sometimes denoted f^2 . Thus:

(f o f)(x) – $f(f(x))$ =f^2 (x)

(f o f o)(x) = $f(f(f(x)))$ = $f^3(x)$

Repeated composition of a function with itself is called function iteration.

The functional powers f of f n = f n o f = f n + 1 for natural n follow immediately.

By convention, *f*(0) = idD(*f*) (the identity map on the domain of *f*).

If $f: X \rightarrow X$ admits an inverse function, negative functional powers f^{-k} ($k > 0$) are defined as the opposite power of the inverse function, $(f^{-1})k$.

Note: If *f* takes its values in a ring (in particular for real or complex-valued *f*), there is a risk of confusion, as *f n* could also stand for the *n*-fold product of *f*, e.g. *f* 2(*x*) = *f* (*x*) . *f* (*x*).

(For trigonometric functions, usually the latter is meant, at least for positive exponents. For example, in trigonometry, this superscript notation represents standard exponentiation when used with trigonometric functions: sin2(*x*) = sin(*x*). sin(*x*)

However, for negative exponents (especially –1), in nevertheless usually refers to the inverse function, e.g., tan–1 = arctan ($\neq$ 1/tan).

In some cases, an expression for *f* in *g*(*x*) = *f r*(*x*) can be derived from the rule for g given non-integer values of *r*. This is called fractional iteration. A simple example would be that where *f* is the successor function, *f r*(*x*) = *x* + *r*.

Iterated functions occur naturally in the study of fractals and dynamical systems.

Composition Mapping

Suppose one has two (or more) functions $f: X \rightarrow Y$, $g: X \rightarrow X$ having the same domain and range. Then one can form long, potentialy complicated chains of these functions composed together, such as f o f o g o f. Such long chains have the algebraic structure of a maping, sometimes called the composition mapping. In general, composition mappings can have remarkably complicated structure. One particular notable example is the de Rham curve. The set of all functions $f: X \rightarrow X$ is called the full transformation semigroup on *X*.

If the functions are bijective, then the set of all possible combinations of these functions form a group; and one says that the group is generated by these functions.

The set of all bijective functions $f: X \rightarrow X$ form a group with respect to the composition operator; this is sometimes called the composition group.

Composition Operator

Given a function g, the composition operator Cg is defined as that operator which maps functions to functions as

$$C_g f = f \circ g$$

Composition operators are studied in the field of operator theory.

Definition: Let $f: S \rightarrow T$ and $g: T \rightarrow U$ be functions. The composition g o f of f and g is the function from S to U defined by the formula (g o f)(x) = $g(f(x)$ for all $x \in S$.

Proposition: Composition of functions is associative.

Definition: Let f: $S \rightarrow T$ be a function then f is said to map S onto T if for each element $y \in T$ there exists an element $x \in S$ with $f(x) = y$.

If $f(x_1) = f(x_2)$ implies $x_1 = x_2$ for all element x_1 $x_2 \in S$, then f is said to be a one-to-one function.

If f is both one-to-one and onto, then it is called a one-to-one correspondence from S to T.

Proposition: Let $f: S \rightarrow T$ be a function. Suppose that S and T are finite sets with the same number of elements. Then f is one-to-one if and only if it is onto.

Proposition: Let $f: S \rightarrow T$ and $g: T \rightarrow U$ be functions.

(*a*) If f and g are one-to-one, the g o f is one-to-one.

(*b*) If f and g are onto, then g o f is onto.

Definition: Let S be a set. The identity function $1_s : S \rightarrow S$ is defined by the formula

$$1_s(x) = x \text{ for all } x \in S.$$

If $f: S \rightarrow T$ is a function, then a function $g: T \rightarrow S$ is called an inverse for f if

$$g \circ f = 1_s \text{ and } f \circ g = 1_T$$

Proposition: Let $f: S \rightarrow T$ be a function. If f has an inverse, then it must be one-to-one and onto. Conversely, if f is one-to-one and onto, then it has a unique inverse.

Examples: If the graph of a function is known, how can you decide whether a function is one-to-one (injective) or onto (surjective)?

Solution:

Injection: If a horizontal line intersects the graph of a function in more than one place, then there are two different points a and b for which $f(a) = f(b)$, but a # b. Then the function is not one-to-one.

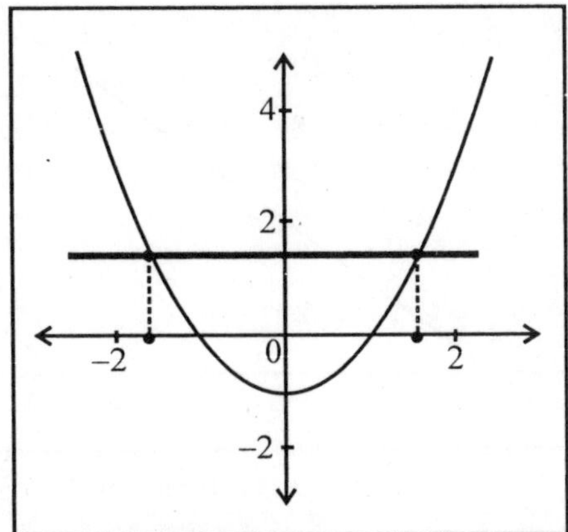

Not a one-to-one Function

Surjection: If you place a light on the left and on the right hand side of the coordinate system, then the shadow of the graph on the y axis is the image of the domain of the function. If that shadow covers the range of the function, then the function is onto.

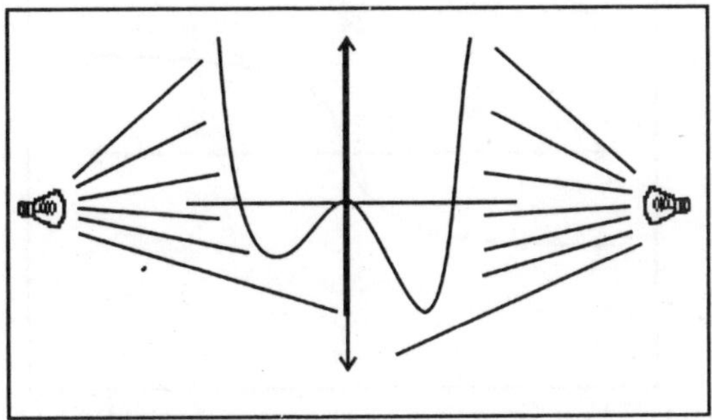
Not an onto Function (if Range is R)

Note that every function can be modified to be onto by setting its range to be the image of its domain.

Example: Which of the following functions are one-one, onto, or bijections? The domain for all functions is R.

1. $f(x) = 2x + 5$
2. $g(x) = \arctan(x)$
3. $g(x) = \sin(x)$
4. $h(x) = 2x^3 + 5x^2 - 7x + 6$

Solution:

1. *Linear Function:* This function is linear. The equation $y = 2x + 5$ has a unique solution for every x, so that the function is one-one and onto, i.e. a bijection. In fact, all linear functions are bijections.

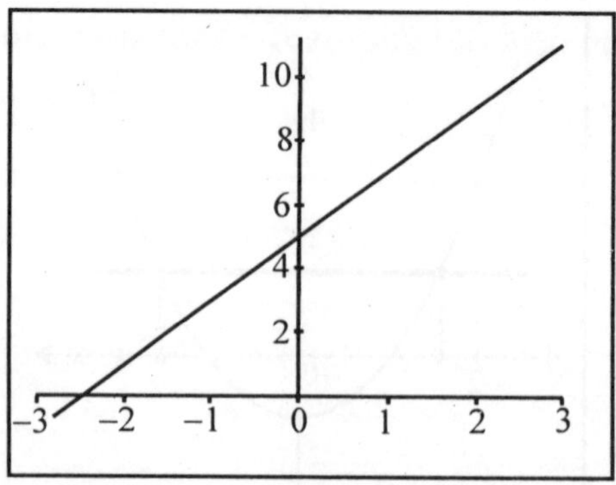

$f(x) = 2x + 5$ is a Bijection.

2. *Inverse Functions:* This function is the inverse function of the tangent function. A such, it must be one-to-one. It is onto if the range is $[-\pi/2, \pi/2]$, but not if the range is *R*. In fact, all inverse functions are one-to-one.

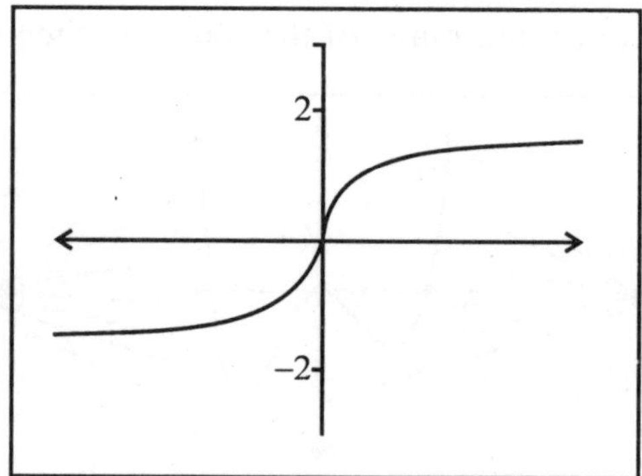

$g(x) = Arctan(x)$ is Injective, not Surjective on *R*.

3. *Periodic Function:* Since this function is periodic, it can not be one-to-one. It is onto, if the range is the interval [–1, 1], but not onto if the range is *R*. In fact, every periodic function is not one-to-one.

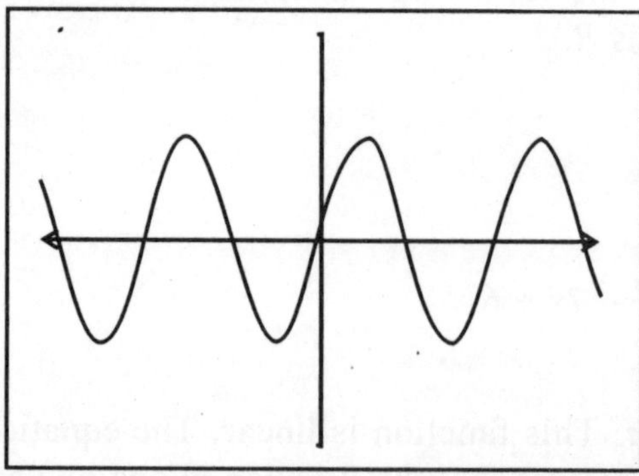

$g(x) = \sin(x)$ is Neither one-to-one nor onto *R*.

4. *Odd Degree Polynomial:* This is an odd-degree polynomial. Hence, the limit as x approaches plus or minus infinity must be plus or minus infinity, respectively. That means that the function is onto. Since most third degree equations have three zero, this function is probably not one-to-one. A look at the graph confirms this. In fact, every odd-degree polynomial is onto while no even degree polynomial is onto.

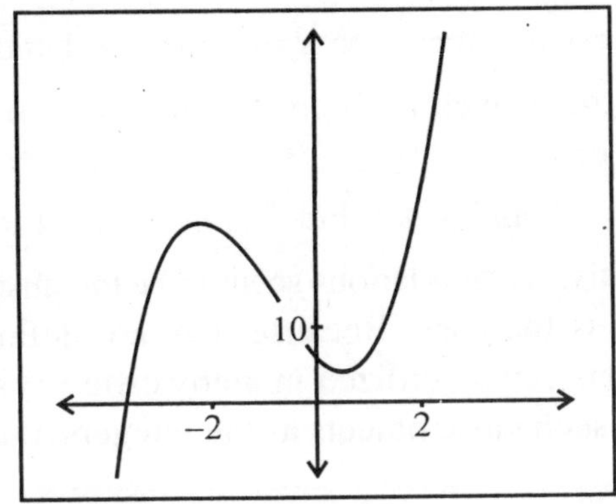

$h(x) = 2x^3 - 7x + 6$ is onto, not one-to-one

Equivalence Relations and Classes

Just as there were different classes of functions (bijections, injections, and surjections), there are also special classes of relations. One of the most useful kind of relation (besides functions, which of course are also relations) are those called equivalence relations.

Definition: Equivalence Relation

1. Let S be a set and r a relation between S and itself. We call r and equivalence relation on S if r has the following three properties:

   (a) Reflexivity: Every element of S is related to itself

   (b) Symmetry: If s is related to t then t is related to s

   (c) Transitivity: If s is related to t and t is related to u, then s is related to u.

Examples: Let $A = \{1, 2, 3, 4\}$ and $B = \{a, b, c)$ and define the following two relations:

1. $r : \{(a, a), (b, b), (a, b), (b, a)\}$

   $s : 1\sim 1, 2 \sim 2, 3 \sim 3, 4 \sim 4, 1 \sim 4, 4 \sim 1, 2 \sim 4, 4 \sim 2$

3. Which one is an equivalence relation, if any?

Solution: Consider the relation r: $\{(a, a), (b, b), (a, b), (b, a)\}$. Then this is, technically speaking, not an equivalence relation, for the simple reason that the element c has no association. However, if we restrict our relation to a new domain that does not include the element c, then it might be an equivalence relation. We need to check:

1. *Reflexive:* every element that has a relation does also have a relation with itself
2. *Symmetric:* clearly true
3. *Transitive:* follows from the first two in this case.
   (*a*) Consider the relation 1~ 1, 2 ~ 2, 3 ~ 3, 4 ~ 4, 1 ~ 4, 4 ~ 1, 2 ~ 4, 4 ~ 2. Then this is not an equivalence relation. We need to check:
   (*b*) *Reflexive:* every element with a relation is related to itself.
   (*c*) *Symmetric:* true as well (3 is only related to 3, so symmetry condition does not apply here)
3. *Not transitive:* 1 ~ 4 and 4 ~ 2, but 1 is not related to 2.

   At first glance equivalence relations seem to be too abstract to be useful. However, just the opposite is the case. Because they are defined in an abstract fashion, equivalent relations can be utilized in many different situations. In fact, they can be used to define such basic objects as the integers, the rational numbers, and the real numbers.

   The main result about an equivalence relation on a set A is that it induces a partition of A into disjoint sets. This property is the one that will allow us to define new mathematical objects based on old ones in the next section.

Theorem: Equivalence Classes

1. Let r be an equivalence relation on a set A. Then A can be written as a union of disjoint sets $A = \bigcup_a A_a$ with the following properties:
2. If a, b are in A then $a \sim b$ if and only if a and b are in the same set A_a
3. The subsets A_a are non-empty and pairwise disjoint.
4. The sets Aa are called equivalence classes.

Proof: We have to decide what the equivalence classes Aa should be: Since by property (1) two elements a and b are supposed to be related if and only if they are in the same class, it seems natural to define:

- $A_a = \{b \in A : b \sim a\}$

To simplify notation a little, we denote this class by $A(a)$ Now we need to check whether this is a good definition, and whether it satisfies the above properties.

Because of reflexivity of the equivalence relation, the class $A(a)$ contains the element a and is therefore not empty.

Next, we will show that two classes that are different can bot have any elements in common, i.e. they are disjoint. Recall that disjoint means that two sets have either an empty intersection or are the same sets. Take two elements a and a' and suppose that $A(a)$ and $A(a')$ have a non-empty intersection; say both classes contain the element c. Then

- $c \sim a$, because c is in $A(a)$
- $a \sim c$, by symmetry and the above line
- $c \sim a$ because c is in $A(a')$
- Therefore, by transitivity, $a \sim a'$

Now take any element b in $A(a')$. Then $b \sim a'$ and $a' \sim a$ so that $b \sim a$ by transitivity, and hence be is contained in $A(a)$. Therefore $A(a')$ is a subset of $A(a)$. The same argument works the other way around as well (by symmetry), showing that $A(a)$ is a subset of $A(a')$. Hence, $A(a) = A(a')$. This shows that two different classes must be disjoint.

Finally, we need to show that all classes $A(a)$ make up the original set A. But that is clear, because every element a in A belongs to the class $A(a)$, hence the union of all classes $A(a)$ must be contained in A, and therefore must be equal to A.

Example: Consider the set Z of all integers. Define a relation r by saying that x and y are related if their difference $y = x$ is divisible by 2. Then

1. Check that this relation is an equivalence relation
2. Find the two equivalence classes, and name them appropriately.
3. How would you add these equivalence classes, if at all?

Solution:

1. *Equivalence Relation:* reflexive: $x - x$ is equal to zero, which divisible by two. Hence, every element is related to itself.

   Symmetry: If $x \sim y$, the $y - x$ is divisible by 2. But then $-(y - x) = x - y$ is divisible by two.

   Hence, $y \sim x$

   Transitivity:

   (*a*) If $x \sim y$ then $y - x = 2n$ for some integer n

   (*b*) If $y \sim z$ then $z - y = 2m$ for some integer m

   But then $z - x = (2m + y) - (y - 2n) = 2\ (m + n)$, so that $z - x$ is divisible by 2. In other words, x and z are related.

2. Equivalence Classes: Two elements are in the same equivalence class if and only if they are related. If x and y are in the same class, then y – x = 2m for some integer n.

   (*a*) If y was even, then y = 2m for some integer m, and x = 2m – 2n must also be even.

(*b*) If y was odd, then y = 2m + 1 for some integer m, and x = 2m + 1 – 2n must also be odd

Therefore, there are two equivalence classes, and they are appropriately labeled:

(*i*) E = even numbers: E = [(2)] contains all even numbers

(*ii*) O = odd numbers: O = [(3)] contains all odd numbers.

3. *Adding Equivalence Classes:* Define $[x] + [y] = [x + y]$. We need to show that this is well- defined, i.e. independent of the particular representative of the equivalence classes of $[x]$ and $[y]$. Take $x \sim x'$ and $y \sim y'$.

(*a*) Then $x' - x = 2m$ and $y' - y = 2n$ for some integers n and m.

(*b*) Then $(x' + y') - (x + y) = x' - x + y' - y = 2m + 2n = 2(m + n)$

That means $(x' + y')$ and $(x + y)$ are related if $x \sim x'$ and $y \sim y'$. But then

(*c*) $[x] + [y] = [x + y] = [x' + y'] = [x'] + [y']$

Hence, addition does not depend on the particular representative from a class, so that addition as defined above is indeed a well-defined operation.

A better would be the following: Defined two classes 0 and 1 by saying:

(*d*) All numbers divisible by 2 with no remainder are in class 0

(*e*) All numbers divisible by 2 with remainder 1 are in class 1

Then add equivalence classes by adding numbers modulo 2.

Example: What kind of equivalence classes do you get when x and y are defined to be related if their difference is divisible by m? How could you add those?

Solution:

1. This relation is an equivalence relation (i.e. the three conditions are satisfied)
2. There are m equivalence classes:

(*a*) All numbers divisible by m with no remainder are in class 0

(*b*) All numbers divisible by m with remainder 1 are in class 1

(*c*) All numbers divisible by m with remainder 2 are in class 2

(*d*) All numbers divisible by m with remainder m-1 are in class $m - 1$

3. Addition can be defined by adding modulo m. That is, if we consider the equivalence classes obtained by dividing the differences by, say, 5, then we have, as an example:

(*a*) [(2)] + [(1)] = [(3)]

(*b*) [(2)] + [(4)] = [(1)]

(*c*) [(3)] + [(4)] = [(3)]

(*d*) etc...

Here is another, more complicated example:

Example: Consider the set $R \times R \setminus \{(0, 0)\}$ of all points in the plane minus the origin. Define a relation between two points (x, y) and (x', y') by saying that they are related if they are lying on the same straight line passing through the origin. Then, check that this relation is an equivalence relation and find a graphical representation of all equivalence classes by picking an appropriate member for each class.

Definition: Let S be a set. A subset R of $S \times S$ is called an equivalence relation on S if

1. For all $a \in S$, $(a, a) \in R$;
2. For all $a, b \in S$, if $(a, b) \in R$ then $(b, a) \in R$;
3. For all $a, b, c \in S$ if (a, b) R and $(b, c) \in R$, then $(a, c) \in R$.

We will write $a \sim b$ to denote the fact that $(a, b) \in R$.

Definition: Let ~be an equivalence relation on the set S. For a given element $a \in S$ we define the equivalence class of a to be the set of all elements of S that are equivalent to a.

We will use the notation $[a]$. In symbols, $[a] = \{x \in S \mid x \sim a\}$.

The notation $S/\sim$ will be used for the collection of all equivalence classes of S under ~.

Example: Let $f : S -> T$ ..... be any function. For $x_1, x_2 \in S$ we define $x_1 \sim x_2$ if $f(x_1) = f(x_2)$. Then for all $x_1, x_2, x_3 \in S$ we have

1. $f(x_1) = f(x_1)$;
2. if $f(x_1) = f(x_2)$, then $f(x_2) = f(x_1)$;
3. if $f(x_1) = f(x_2) = f(x_3)$, then $f(x_1) = (x_3)$.

This shows that we have defined an equivalence relation on the set S. The proof of this is easy because the equivalence relation is defined in terms of equality of the images $f(x)$, and equality is the most elementary equivalence relation.

The collection of all equivalence classes of S under ~will be denoted by S/f.

Proposition: Let S be a set, and let ~ be an equivalence relation on S. Then each element of **S** belongs to exactly one of the equivalence classes of S determined by the relation~

Definition: Let S be any set. A family P of subsets of S is called a partition of S if each element of S belongs to exactly one of the members of P.

Proposition: Any partition P of a set S determines an equivalence relation.

Theorem: If $F: S \rightarrow T$ is any function, and ~ is the equivalence relation defined on S by letting

$x_1 \sim x_2$ if $f(x_1) = f(x_2)$ for all $x_{1,} x_2 S$, then there is a one-to-one correspondence between the elements of the image $f(S)$ of S under f and the equivalence classes S/f of the relation ~.

If $f : S \rightarrow T$ is a function and y belongs to the image $f(S)$, then the inverse image of y is

$$f^{-1}(y) = \{x \in S \mid f(x) = y\}.$$

The inverse images of elements of $f(S)$ are the equivalence classes in S/f. (Note carefully that we are not implying that f has an inverse function.)

Permutations

Definition: Let S be a set. A function $\sigma : S -> S$ is called a permutation of S if σ is one-to-one and onto.

The set of all permutations of S will be denoted by Sym(S).

The set of all permutations of the set $\{1, 2, ..., n\}$ will be denoted by S_n.

Propositio shows that the composition of two permutations in Sym(S) is again a permutation. It is obvious that the identity function on S is one-to-one and onto.

Propositions shows that any permutation in Sym(s) has an inverse function that is also one-to-one and onto. We can summarize these important properties as follows:

(i) If $\sigma, \tau \in$ Sym(S), then $\sigma\ \tau \in$ Sym(S);

(ii) $1_s \in$ Sym(S);

(iii) if, $\sigma \in$ Sym(S) then $\sigma^{-1} \in$ Sym(S).

Definition: Let S be a set, and let $\sigma, \in$ Sym(S). Then is called a cycle of length k if there exist elements $a_1, a_2, ..., a_k \in S$ such that

$\sigma(a_1) = a_2, \sigma(a_2) = a_3 ,....., \sigma(a_{k-1}) = a_k, \sigma(a_k) = a_1$, and

$\sigma(x) = x$ for all other elements $x \in S$ with $x \neq a_1$ for $i = 1, 2, ..., 1, 2, ..., k$.

In this case we write $\sigma = (a_1, a_2, ... a_k)$.

We can also write $\sigma = (a_1, a_2, ... a_k)$ or $= (a_3, a_k, a_1, a_2)$, etc. The notation for a cycle of length k can thus be written in k different ways, depending on the starting point. The notation (1) is used for the identity permutation.

Definition: Let $\sigma = (a_1, a_2, \ldots.. a_k)$ and $\tau = (b_1, a_2, \ldots.. b_m)$ be cycles in Sym(S), for a set *S*. Then σ and τ are said to be **disjoint** if $a_i \neq b_j$ for all *i, j*.

Proposition: Let *S* be any set. If σ and τ are disjoint cycle in Sym (S), then $\sigma\tau = \tau\sigma$.

Theorem: Every permutation in S_n can be written as a product of disjoint cycles. The cycles that appear in the product are unique.

Definition: Lct $\sigma \in S_n$ The least positive integer m such that $\sigma^m = (1)$ is called the order of σ.

Proposition: Let $\sigma \in S_n$ have order *m*. Then for all integers *j, k* we have

$$\sigma^j = \sigma^k \text{ if and only if } j \equiv k \pmod{m}.$$

Proposition: Let $\sigma \in S_n$ be written as a product of disjoint cycles. Then the order of is the least common multiple of the lengths of its cycle.

Definition: A cycle (a_1, a_2) of length two is called a transposition.

Proposition: Any permutation in S_n where $n \geq 2$ can be written as a product of transpositions.

Theorem: If a permutation is written as a product of transpositions in two ways, then the number of transpositions is either even in both cases or odd in both cases.

Definition: A permutation σ is called even if it can be written as a product of an even number of transpositions, and odd if it can be written as a product of an odd number of transpositions.

RELATION

After knowing some of the basic elements of set theory (sets), we will move on to the second most elementary concept, the concept of relations and functions. Let *A* and *B* two sets. A relations between *A* and *B* is a collection of ordered pairs (*a*, *b*) such that $a \in A$ and $b \in B$. Often we use the notation $a \sim b$ to indicated that a and b are related, rather then the order pair notation (*a*, *b*). Note that this does not mean that *each* element from A needs to be associated with one (or more) elements from B. It is sufficient if *some* associations between elements of *A* and *B* are defined. In contrast, there is the definition of a function:

Thus, a set of ordered pairs is called a relation.

For example, (1, 2), (3, 5), (7, 10) ... are relations.

The first components in the ordered pairs (*x*-coordinate) is called the domain, and the second components (*y*-coordinate of *f*(*x*) coordinate) is called the range.

Functions: Domain and Range

Let's return to the subject of domain and range. When functions are first introduced, you will probably have some slightly pathetic "functions" and relations to deal with, being just sets of points. These won't be terribly useful or interesting functions and relations, but your text wants you to get the idea of what the domain and range are.

The first components in the ordered pairs (x-coordinate) is called the domain, and the second components (y-coordinate or $f(x)$ coordinate) is called the range.

Definition: Let A and B be two sets. Since, a function f from A to B is a relation between A and B such that for each $a \in A$ there is one and only one associated $b \in B$. The set A is called the domain of the function, B is called its range. Often a function is denoted as $y = f(x)$ or simply $f(x)$ or simply $f(x)$, indicating the relation $\{(x, f(x)\}$.

Definition: Let S and T be sets. A function from S into T is a subset F of S X T such that each element $x \in S$ there is exactly one element $y \in T$ such the $(x, y) \in F$. The set S is called the domain of the function, and the set T is called the codomain of the function. The subset $\{y \in T \mid (x, y) \in F$ for some $x \in S\}$ of the codomain is called the image of the function.

Example: Let $S = \{1, 2, 3\}$ and $T = \{u, v, w\}$. The subsets

$F_1 = \{(1, u), (2, v), (3, w)\}$ and

$F_2 = \{(1, u), (2, v), (3, w)\}$

Of S X T both define functions since in both cases each element of S occurs exactly once among the ordered pairs. The subset

$$F_3 = \{(1, u), (3, w)\}$$

does not define a function with domain S because the element $2 \in S$ does not appear as the first component of any ordered pair.

Note the F_3 is a function if the domain is changed to the set $\{1, 3\}$.

Unlike the conventions used in calculus, the domain and codomain must be specified as well as the "rule of correspondence" (list of pairs) when you are presenting a function.

The subset

$F_4 = \{(1, u), (2, u), (2, v), (3, w)\}$

does not define a function since 2 appears as the first component of two ordered pairs.

When a candidate such as F_4 fails to be a function in this way, we say that it is not "well-defined."

For example, find the range; given the domain $\{2, 4, 6\}$ of $f(x) = 3x - 1$

$f(2) = 3(2) - 1 = 6 - 1 = 5$

$f(4) = 3(4) - 1 = 12 - 1 = 11$

$f(6) = 3(6) - 1 = 18 - 1 = 17$

the range is {5, 11, 17} and the ordered pairs are (2, 5); (4, 11); (6, 17)

The outcomes of a function (i.e. the elements from the range associated to elements in the domain) do not only depend on the rule of the function (such as x is associated with sin(x) but also on the domain of the function. therefore, we need to specify those outcomes that are possible for a given rule and given domain:

Functions Versus Relations

There are different ways of looking at functions. We will consider a few. But first, we need to discuss some terminology.

A "relations" is just a relationship between sets of information. Think of all the people in one of your classes, and think of their heights. The pairing of names and heights is a relation. In relations and functions, the pairs of names and heights are "ordered", which means one comes first and the other comes second. To put it another way, we could set up this pairing so that either you give me a name, and then I give you that person's height, or else you give me a height, and I give you the names of all the people who are that tall. (By the way, the set of all the starting points is called "the domain" and the set of all the ending points is called "the range." The domain is where you start; the range is where you go. The domain is the x' s; the range is the y' s. I'll explain more on the subject of determining domains and ranges later.)

A function is a well-behaved relation. Just as with members of your own family, some members of the relation family are better behaved than other. (This means that, while all functions are relations, since they pair information, not all relations are functions. Functions are a sub-classification of relations.) When we say that a function is "a well-behaved relation", we mean that, given a starting point, we know exactly where to go. That is, given an x, we get only and exactly one y.

Let's return to our relation of your classmates and their heights, and let's suppose that the domain is the set of everybody's heights. And let's suppose that there's a pizza-delivery guy waiting in the hallway. And all the delivery guy knows is that the pizza is for the student in your classroom who is five-foot-five. Now let the guy in. Who does he go to? What if you are five-five? And what if you're out of cash? You see, the relation "height indicates name" is not well-behaved. It is not a function. Given the relationship (x, y) = (five-foot-five person, name), there might be six different possibilities for y = "name". For a relation to be a function, there must be only and exactly one y that corresponds to a given x. Here are some pictures of this:

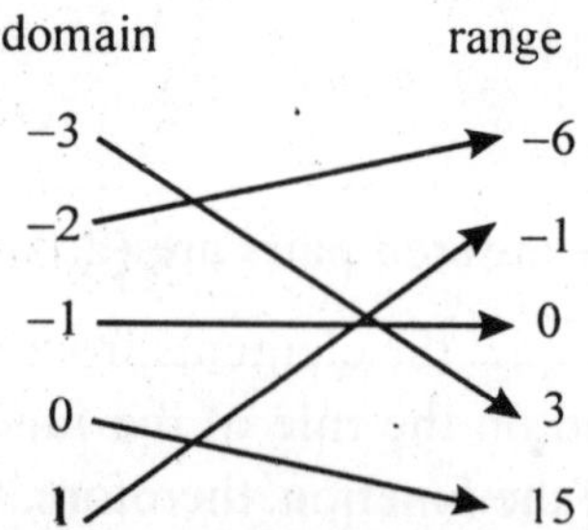

This is a function. You can tell by tracing from each x to each y. There is only one y for each x; there is only one arrow coming from each x.

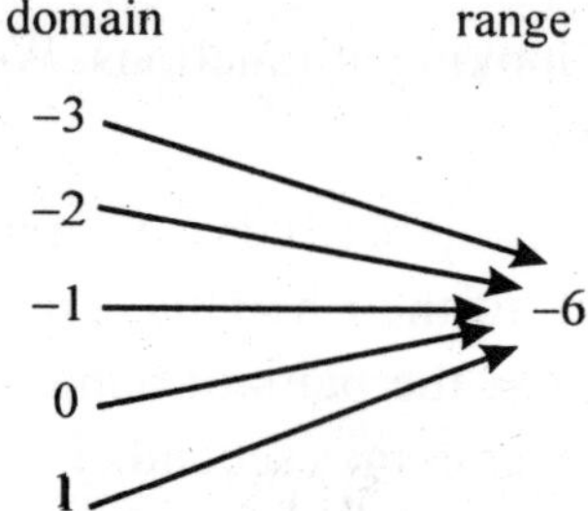

This is a function, There is only one arrow coming from each x; there is only one y for each x. It just so happens that it's always the same y, but it is only one y. So this is a function; it's just an extremely boring function!

domain range

–3 → –6
–2 → –1
–1 → 0
0 → 3
1 → 3
1 → 15

This one is not a function. If you'll notice, there are two arrow coming from the number 1; the number 1 is associated with two different range elements. So this is a relation, but it is not a function.

domain range

–3 → –6
–2 → –1
–1 → 0
0 → 3
1 → 3
1 → 15

Each element of the domain that has a pair in the range is nicely well-behaved. But what about that 16? It is in the domain, but it has no range element that corresponds to it! This won't work. So then this is not a function.

Vertical Line Test

Given a graph, it can be determined if it is a function or a relation.

In order for the graph to be a function, the vertical line must only intersect the graph at one and only one point.

Looking at this function stuff graphically, what if we had the relation {(2, 3), (2, – 2)}?

We already know that this is not a function, since $x = 2$ goes to both $y = 3$ and $y = -2$.

If we graph this relation, it looks like:

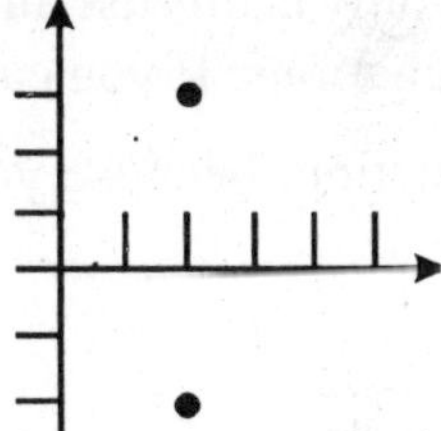

Notice that you can draw a vertical line through the two points, like this:

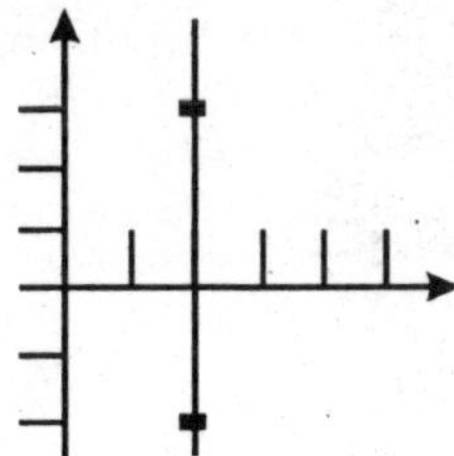

This characteristic of non-functions by I-don't-know-who, and was codified in The Vertical Line Test: Given the graph of a relation, if you can draw a vertical line that crosses the graph in more than one place, then the relation is not a function. Here are a couple examples:

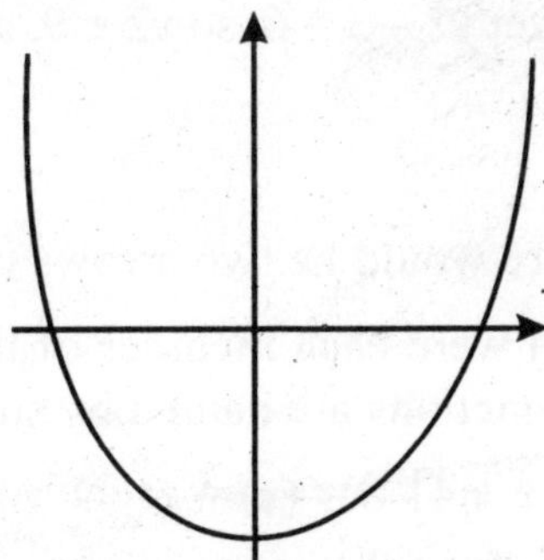

This one is a function. There is no vertical line that will cross this graph twice.

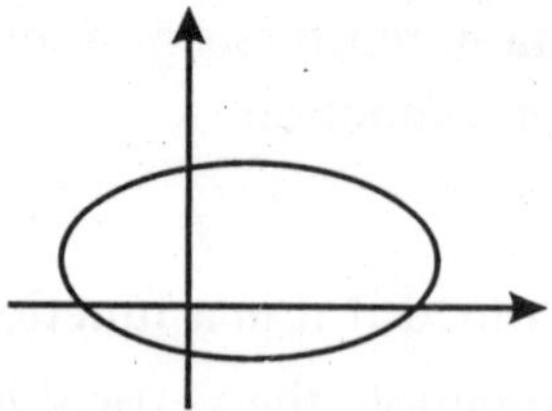

This one is not a function. Any number of vertical lines will intersect this intersect this oval twice. For instance, the y-axis intersects twice.

"Is it a function?" – Quick answer without the graph.

Think of all the graphing that you've done so far. The simplest method is to solve for y. make a T-chart, pick some values for x, solve for the corresponding values of y, plot your points, and connect the dots. Not only is this useful for graphing, but this methodology gives yet another way of identifying functions: If you can solve for "y". then it's a function.

For example, $2y + 3x = 6$ is a function, because you can solve for y:

$$2y + 3x = 6$$

$$2y = -3x + 6$$

$$y = (-3/2)x + 3$$

On the other hand, $y^2 + 3x = 6$ is not a function, because you can not solve for a unique y:

$$y^2 + 3x = 6$$

$$y2 = -3x + 6$$

$$y = \pm\sqrt{-3x+6}$$

This is solved for "y", but it's not unique. Do you take the positive square root,

or the negative? So, in this case, the relation is not a function. (You can also check this by using our first definition from above.

Think of "x = –1". Then we get $y2 - 3 = 6$, so $y2 = 9$, and then y can be either –3 or + 3.

That is,

if we did an arrow chart, there would be two arrows coming from $x = -1$.)

Thus, a function is a relation were each member of the domain is reserved one and only one member of the range. Functions are named by suing single letters.

For example, $f = \{(x, y) | y = x + 4\}$ or $f(x) = x + 4$

The equations above are representations of functions. The second example is the most commonly used representation; it is called the function notation.

The ordered pair for functions is $(x, f(x))$.

Consider the example; $f(x) = x + 4$ at $x = 2$.

$f(2) = 2 + 4 = 6$, thus the ordered pair is (2, 6)

Example: Find the domain and range of: $f(x) = 2/(x-3)$

Solution: Since x cannot be zero; the domain is $D = \{x|\ x^1\ 3\}$ and $f(x)$ can have any value; thus the range is $R = \{f(x)|$ any real number$\}$

Example: State the domain and range of the following relation. Is the relation a function?

$$\{(2, -3), (4, 6), (3, -1), (6, 6), (2, 3)\}$$

This list of points, being a relationship between certain *x*'s and certain *y*'s, is a relation. The domain is all the *x*-values, and the range is all the *y*-values. you list the values without duplication:

Domain: {2, 3, 4, 6}

Range: {–3, –1, 3, 6}

(It is customary to list these values in numerical order, but it is not required. Sets are called "unordered lists", so you can list the numbers in any order you feel like. Just don't duplicate.)

While this is a relation (because *x*'s and *y*'s are being related to each other), you have two points with the same *x*-values: (2, –3) and (2, 3). Since $x = 2$ gives you two possible destinations, then this relation is not a function.

Note that all I had to do to check whether the relation was a function was to look for duplicate *x*- values. If you find a duplicate *x*-value, then the different *y*-values mean that you do not have a function.

Example: State the domain and range of the following relation. Is the relation a function?

$$\{(-3, 5), (-2, 5), (-1, 5), (0, 5), (1, 5), (2, 5)\}$$

List the *x*-values for the domain and the *y*-values for the range:

Domain: {–3, –2, –1, 0, 1, 2}

Range: {5}

This is another example of a "boring" function (like we saw in the first chart higher up in this page), because all the *x*-values go to the exact same *y*-values. But each *x*-value is different, so, while boring, this relation is indeed a function. (In point of fact, these points lie on the horizontal line $y = 5$.)

There is one other case for finding the domain and range of functions. They will give you a function and ask you to find the domain (and maybe the range, too). I have only ever seen (or can even think of) two things at this stage in your mathematical career that you'll have to check in order to determine the domain of the function they'll give you, and those two things are denominators and square roots.

Example: Determine the domain and range of the given function

$$\frac{x^2+x-2}{x^2-x-2}$$

Solution: The domain is all the values that x is allowed to take on. The only problem you have with this function is that you have to be careful not to divide by zero. So the only values that x can not take on are those which would cause division by zero. So set the denominator equal to zero and solve; the domain will be everything else.

$$x^2 - x - 2 = 0$$
$$(x - 2)(x + 1) = 0$$
$$x = 2 \text{ or } x = -1$$

Then the domain is "all x not equal to – 1 or 2".

The range is a bit trickier, which is why they may not ask for it. In general, though, they'll want you to graph the function and find the range from the picture. In this case:

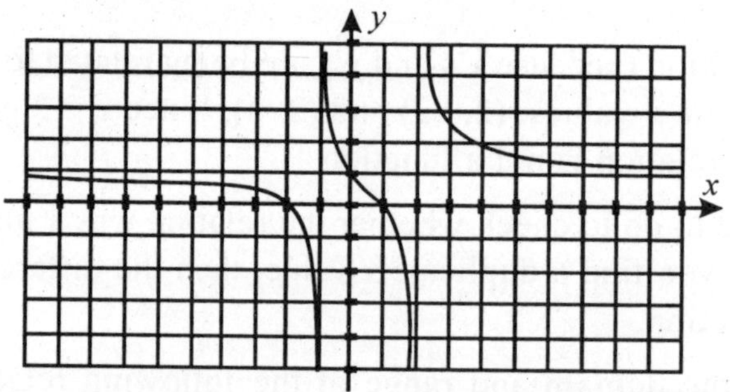

As you can see from the picture, the graph covers all y-values (that is, the graph will go as low as you like, and will also go as high as you like). Since the graph will eventually cover all possible values of y, then the range is "all real numbers".

Example: Determine the domain and range of the given function:

$$-\sqrt{-2x+3}$$

The domain is all values that x can take on. The only problem you have with this function is that you cannot have a negative inside the square root. So set the insides greater-than-or-equal-to zero, and solve. This will be the domain:

$-2x + \geq 30$

$-2x \geq -3$

$2x \leq 3$

$x \leq 3/2 = 1.5$

Then the domain is "all $x \leq 3/2$".

The range requires a graph. Be sure to be careful when graphing radicals:

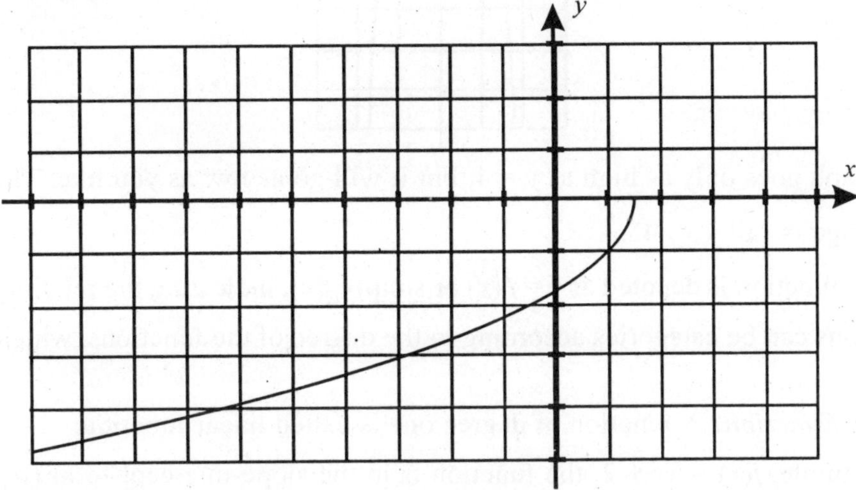

The graph starts at $y = 0$ and goes down from there. While the graph goes down very slowly, you know that, eventually, you can go as low as you like (by picking an x that is sufficiently big). Also, from your experience with graphing, you know that the graph will never start coming back up. Then the range is "$y \leq 0$".

Example: Determine the domain and range of the given function:

$y = -x^4 + 4$

This is just a garden-variety polynomial. There are no denominators (so no division-by-zero problems) and no radicals (so no square-root-of-a-negative problems). There are no problems with a polynomial. There are no values that you can't plug in for x. When you have a polynomial, the answer is always that the domain is "all x".

The range will vary from polynomial to polynomial, and they probably won't even ask, but, if they do, look at the picture:

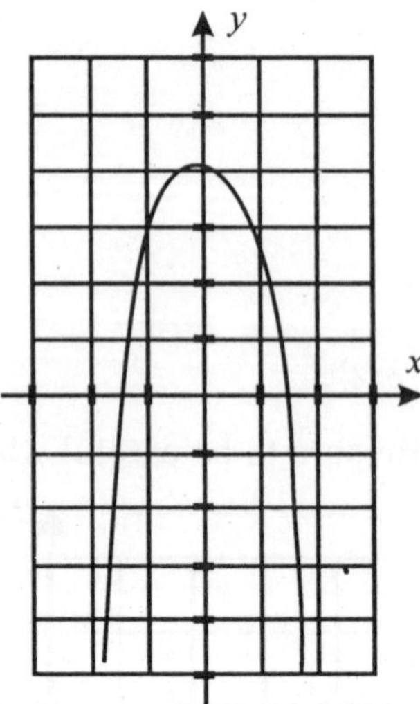

The graph goes only as high as $y = 4$, but it will go as low as you like. Then:

The range is "all $y \leq 4$".

Often a function is denoted as $y = f(x)$ or simply $f(x)$, indicating the relation $\{(x, f(x))\}$.

Functions can be categories according to the degree of the functions, which are listed below:

Linear Function: A function of degree one is called linear function.

For example, $f(x) = 3x + 2$, the function is in the slope-intercept form ($y = mx + b$) m(slope) = 3, y-intercept = 2

Quadratic Function: A function of degree two is called quadratic function. [$f(x) = ax^2 + bx + c$]

The same as graphing quadratic equations. For example, graph: $f(x) = 2x^2 + 8x + 9$

$y = 2x^2 + 8x + 9$ replace $f(x)$ with y

$y - 9 = 2x + 8x$ complete the square

$y - 9 + 4 = 2\,(x^2 + 4x + 4)$

$y - 1 = 2(x + 2)^2$ parabola with vertex at $(-2, 1)$

$y = 2(x + 2)^2 + 1$ solve for y and replace y with $f(x)$

$f(x) = 2(x + 2)^2 + 1$

Cubic Function: A function of degree three is called cubic function. [$f(x) = ax^3 + bx^2 + cx + d$]

Biquadratic Function: A function of degree four is called biquadratic function. [$f(x) = ax^4 + bx^3 + cx^2 + dx + e$]

Graph of Function: To graph functions, just change y into $f(x)$ from graphing equations. Graphing functions is the same as graphing equations with the exception of naming the domain and the range.

Rules for Graphing Other Function:

1. Determine the domain of the graph.
2. Check for symmetry.
   if $f(-x) = f(x)$, then symmetric about the y-axis.
   if $f(-x) = -f(x)$, then symmetric about the origin.
   (functions rule out the possibility that the graph has an x-axis symmetry)
3. Find the x-intercept and y-intercept of the graph
   Evaluate $f(0)$ to find the y-intercept.
   To find the x-intercept, find the value or values that will make $f(x) = 0$
4. Plot some points for the graph and reflect, according to the symmetry test.
   For example, graph: $f(x) = | x |$ the domain is $D = \{x \mid x \text{ is any real number}\}$
   $f(-x) = | -x | = | x | = f(x)$ symmetric about the y-axis.
   $f(0) = | 0 | = 0$, $f(x) =$ at $x = 0$ $x = 0$ the intercepts are tat the origin.

x	$f(x)$
1	1
–2	2
–1	1
1	2

Examples: Let $A = \{1, 2, 3, 4)$, $B = (14, 7, 234)\}$, $C = \{a, b, c\}$, and $R =$ real numbers.

Define the following relations:

1. r is the relation between A and B that associates the pairs $1 \sim 234$, $2 \sim 7$, $3 \sim 14$, $4 \sim 234$, $2 \sim 234$
2. f is the relation between A and C that relates the pairs $\{(1, c), (2, b), (3, a), (r, b)\}$
3. g is the relation between A and C consisting of the associations $\{(1, a), (2, a), (3, a)\}$
4. h is the relation between R and itself consisting of pairs $\{(x, \sin(x))\}$

Solution:

1. The relation r is not a function, because the element 2 from the set A is associated with two elements from B

2. The relation f is a function, because every element from A has exactly one relation from the set C.
3. The relation g is not a function, because the element $\{4\}$ from the domain A has no element associated with it.
4. The relation h is a function with domain R, because every element $\{x\}$ in R has, exactly one element $\{\sin(x)\}$ associated with it.

The outcomes of a function (i.e. the elements from the range associated to elements in the domain) do not only depend on the rule of the function (such as x is associated with $\sin(x)$) but also on the domain of the function. Therefore, we need to specify those outcomes that are possible for a given rule and a given domain:

Definition: Image and Preimage

Let A and B be two sets and f a function from A to B. Then the image of f is defined as

1. Image $(f) = \{b \in B$: there is an $a \in A$ with $f(a) = b\}$.

   Let A and B be two sets and f a function from A To B. If C is a subset of the range B then the preimage, or inverse image, of C under the function f is the set defined as

2. $f^{-1}(C) = \{x \in A : f(x) \in C\}$

   Function can be classified into three groups: those for which every element in the image has one preimage, those for which the range is the same as the image, and those which have both of these properties. Accordingly, we make the following definitions:

   As an example, consider the following functions:

Example: Let $f(x) = 0$ if x is rational and $f(x) = 1$ if x is irrational. This function is called Dirichlet's Function. The range for f is R

1. Find the image of the domain of the Dirichlet Function when:
   (*a*) The domain of f is Q
   (*b*) The domain of f is R
   (*c*) The domain of f is [0, 1] (the closed interval between 0 and 1]
2. What is the preimage of R? What is the preimage of [–1/2, 1/2]?
3. What is the image of [0, 2] and the preimage of [1, 4]?
4. Find the image and the preimage of [–2, 2].

Solution:

1. (*a*) When the domain is Q, we have that $f(x) = 0$ for any x, because x must be a rational number. Hence, the image of the domain Q is the set consisting of the single element $\{0\}$

   (*b*) When the domain is R, we have that $f(x)$ could be 0 or 1, because x could be rational or irrational. $f(x)$ can not be any other number. Hence, the image of the domain R is the set consisting of the two element $\{0, 1\}$.

   (*c*) The interval $[0, 1]$ contains irrationals as well as rational numbers. Therefore, $f(x)$ could be equal to 0 or 1, and the image of $[0, 1]$ under f is the set with the two elements $\{0, 1\}$.

2. The preimage of R is the set of all elements such that $f(x)$ is contained in R. Since R includes the numbers 0 and 1, the preimage of R under f is everything in the domain, i.e. R.

   The preimage of $[-1/2, 1/2]$ consists of all elements such that $f(x)$ is contained in $[-1/2, 1/2]$. This set does not contain 1, so the preimage of that set can not contain any irrational number. hence, the preimage of $[-1/2, 1/2]$ is the set of rational umbers Q.

   Let $f(x) - x^2$, with domain and range being R. Then use the graph of the function to determine:

3. If we draw the graph of the function, and shade the interval $[0, 2]$ on the x axis in red, then its projection onto the y axis will be the image (in green).

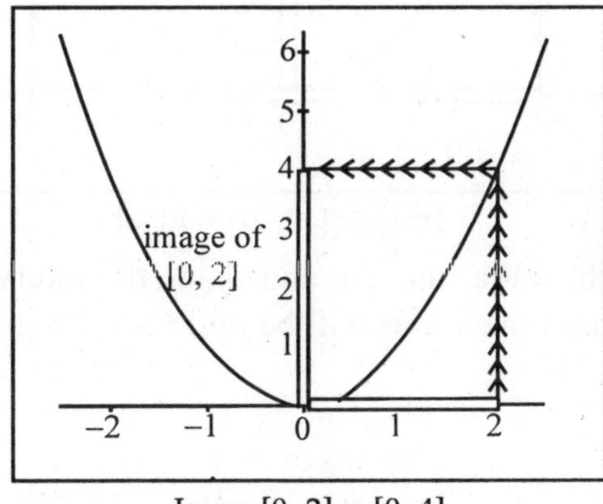

Imag $[0, 2] = [0, 4]$

If we draw the graph of the function, and shade the interval $[1, 4]$ on the y axis in red, then its projection onto the x axis will be the preimage (in green).

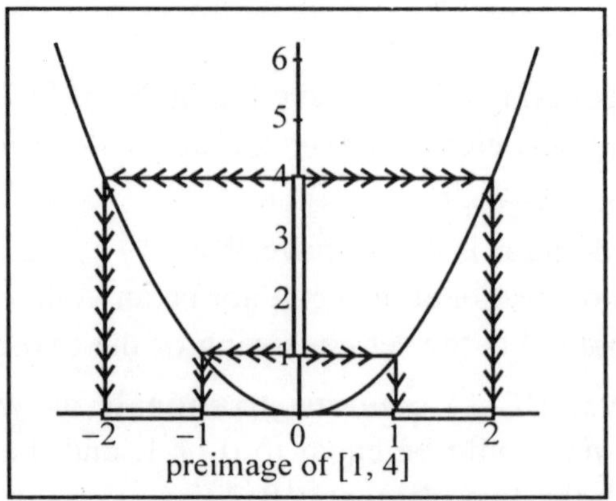

Preimage ([1, 4] = [-2, -1] ∪ [1, 2]

(*iv*) If we draw the graph of the function and shade the interval [–2, 2] on the *x* axis (in red) then its projection onto the *y* axis will be the image (in green).

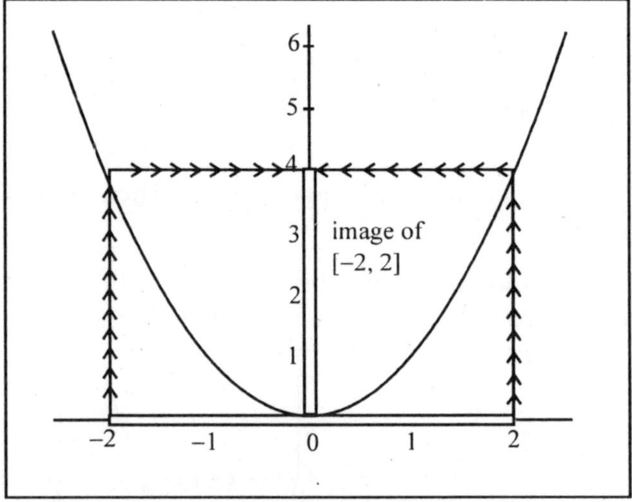

Image ([-2, 2]) = [0, 4]

If we draw the graph of the function and shade the interval [–2, 2] on the *y* axis (in red) then its projection onto the *x* axis will be preimage (in green).

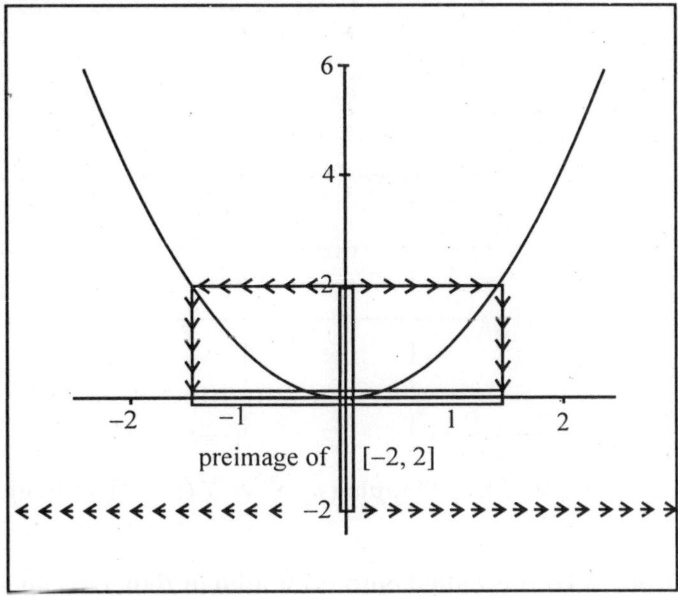

Preimage ([−2, 2]) = [−sqrt (2), sqrt (2)]

Function can be classified into three groups: those for which every element in the image has one preimage, those for which the range is the same as the image, and those which have both of these properties. Accordingly, we make the following definitions:

Types of Mapping

Mappings can be classified into a number of different types.

Injective

A function f from A to B is called one to one (or one-one) in whenever $f(a) = f(b)$ then a = b. Such functions are also called injections.

Thus, the mapping $a : X \rightarrow Y$ is called injective or one-one if for each $x \in X$ there exists at most $y \in Y$ one such that $a(x) = y$. (Note that no such x may exist). Another way of saying that a is injective is $a(x)$ implies y where x. A graphical representation of an injective mapping is show in figure 1.

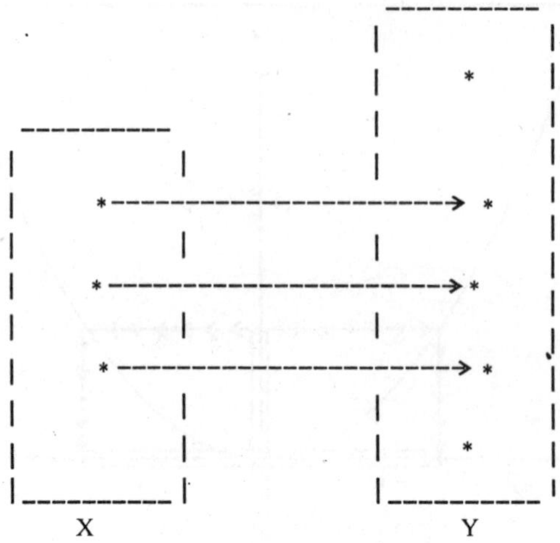

Fig. An Injective Mapping $a : X \rightarrow Y$ (not Surjective)

Surjective

A function *f* from *A* To *B* is called onto if for all *b* in *B* there is an *a* in *A* such that *f*(*a*) = *b*. Such functions are also called surjections.

Thus, the mapping $a : X \rightarrow Y$ is called surjective or onto if for each $x \in X$, there exists at least one $y \in Y$ such that $a(x) = y$. This type of mapping is illustrated in fig.

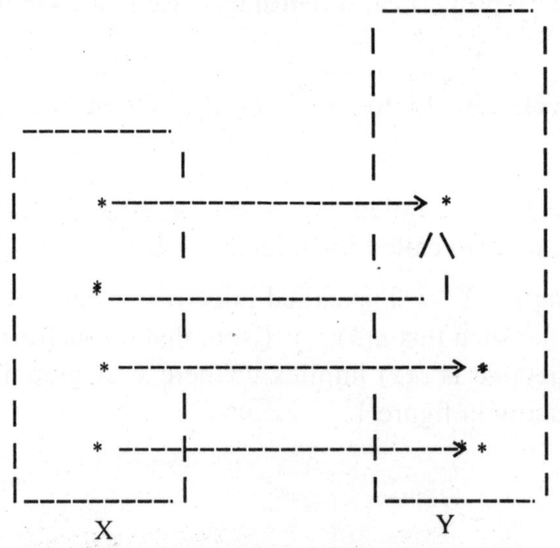

Fig. A Surjective Mapping $a : X \rightarrow Y$ (not Injective)

Bijective

A function *f* from *A* to *B* is called a bijection if it is one to one and onto, i.e. bijections are functions that are injective and surjective.

Thus, the mapping which is both injective and surjective is called bijective or one-to-one and such a mapping is illustrated in Figure. For such mapping, there is a one-to-one correspondence between the members of X and Y.

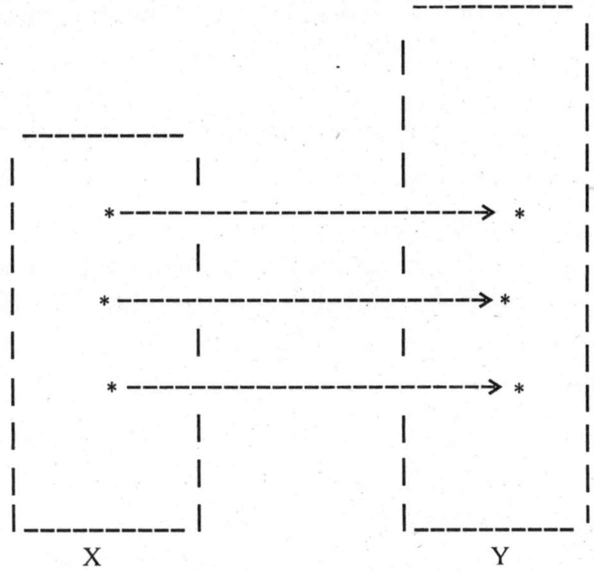

Fig. A bijective mapping $a : X \rightarrow Y$ (both injective and surjective)

Horizontal Line Test:

The horizontal line test implies a one to one function, meaning that there is only one value of x associated with each value of $f(x)$.

Examples: Let $A = \{1, 2, 3, 4\}$, $B = \{14, 7, 234\}$, $C = \{a, b, c\}$, and R = real numbers. Define the following relations:

1. r is the relation between A and B that associates the pairs $1 \sim 234 \sim 7$, $3 \sim 14$, $4 \sim 234$, $2 \sim 234$
2. f is the relation between A and C that relates the pairs $\{(1, c), (2, b), (3, a), (4, b)\}$
3. g is the relation between a and C consisting of the associations $\{(1, a), (2, a), (3, a)\}$
4. h is the relation between R and itself consisting of pairs $\{x, \sin(x)\}$

   Which of those relations are functions?

Solution:

1. The relation r is not a function, because the element 2 from the set A is associated with two elements from B.
2. The relation f is a function, because every element from A has exactly one relation from the set C.

3. The relation g is not a function, because the element $\{4\}$ from the domain A has no element associated with it.
4. The relation h is a function with domain R, because every element $\{4\}$ from the domain A has no element associated with it.
5. The relation h is a function with domain R, because every element $\{x\}$ in R has exactly one element $\{\sin(x)\}$ associated with it.

Abstract Definition of a Function

A fairly precise, but somewhat informal definiton of a function follows. Let A and B be sets, A function from A to B is determined by any association of a unique element of B with each element of A. The set A is called the domain of the function; the set B is called the codomain.

Functions Versus Maps

A map is usually a (total) function as defined here, especially for authors using function as synonym for partial functions with a domain (of definition) not necessarily containing all points of the source or "departure set". (For them, $R \to R; x \to 1/x$ is a function with domain $R\backslash\{0\}$.)

Mathematical Definition of a Function

The informal idea of a function as a rule has been used since ancient times and is still used as the definition of a function in informal contexts such as introductory calculus textbooks. A typical example of this informal definition, is that a function is a rule that assigns to each element of a set D a single element of set C. This informal definition is sufficient for many purposes, but it relies on the undefined concept of a "rule". In the late 1900s, the question of what constitutes a valid rule defining a function came to the forefront. The consensus of modern mathematicians is that the word "rule" should be interpreted in the most general sense possible: as an arbitrary binary relation.

Thus it is common in advanced mathematics for an example to formally define a function f from a set D to a set C to be a set Gf of ordered pairs (x, y) in the Cartesian product D X C. It is required that for each x in D there is at most one pair (x, y) in the set Gf. The "rule" of the function is: given x in D, if there is a pair (x, y) in Gf then $f(x) = y$, and otherwise $f(x)$ is not defined. The set of x for which f is defined is called the domain of f; if the domain of f is all of D then f is called total and the notation $f : D \to C$ is used. The set C is called the codomain of the function; this must be specified because it is not determined by Gf. The set Gf is called the graph of the function.

In most areas of mathematics, the word function is used to mean total function, although non-total functions (called partial functions) are important in functional analysis, mathematical logic, and category theory.

Definition of Functions in Set Theory

Variations of this formal definition are sometimes more convenient for specific disciplines. In some contexts of category theory, even if a function f from a set D is not defined for every element of D, the set D is still called the domain of f. I set theory, it is common to identify the function f with its graph Gf; this identification removes the need to specify either D or C in the formal definition.

Total and Partial Mappings

If a mapping $a : X \rightarrow Y$ is defined for every $x \in X$, then the mapping a is said to be total. The mappings illustrated in Figures are all total (that is every element in X maps to an element in Y). If a mapping a is not defined for all $x \in X$, then the mapping a is said to be partial. Expressed another way, we can say that a mapping a is partial if not all possible inputs have a defined output. An example of a partial mapping is illustrated below in Figure.

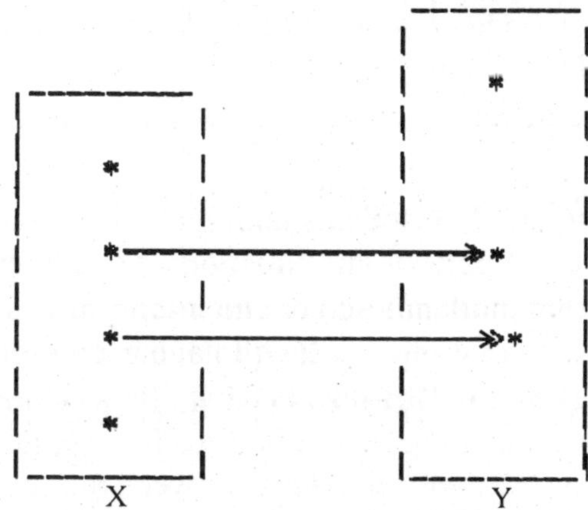

Fig. Example of a Partial Mapping

This mapping is partial since there are elements in X that are not mapped to a corresponding value in Y.

We should emphasise again that we treat mappings as single-valued, that is for any mapping a: $X \rightarrow Y$, we assume that for any $x \in Y$, there is at most one $y \in Y$ such that $y = a(x)$. This is an essential property for computer science applications since, for example, it would be most undesirable for an array element to have to distinct values.

Note also that the terminology is by no means standard in this area. In mathematics, functions (mappings) are by definition, total, that is for each $x \in X$ there is a unique $y \in Y$ such that $y = a(x)$. In computer science applications however, the function (mapping) a from X to Y is usually defined by stating that a maps an element from X onto at most one element of Y. Hence there is no necessity for every member of X to be mapped to a member of Y.

Real Life Examples of Mappings

In order to capture the various concepts associated with mappings and to summaries, it will be helpful to look briefly at two examples of mappings.

Consider firstly a class of 25 students who sit an examination where marks are awarded in the range) to 100. If we denote the set of student names by S and the set of possible marks by M (so that M will be the set $\{m \mid m \in N, 0 \leq m)$, then the "examinations-results-list" constitutes a total mapping since every student has an examination mark and that mark is unique. We can write this mapping as

$$\textit{examination-results-}\ S \rightarrow M$$

If every student whose name appears in S sat the examination, the mapping is total. Note that not every possible examination mark will appear on the results list because there are 101 possible marks but only 25 students. Hence for each mark, it is not the case there is at least one student name, who has attained that mark and so the mapping is not surjective. If it transpired that each of the 25 students achieved a different mark in the examination, then the mapping would be injective (since for each m, there is at most one student with that mark).

Suppose now that there are 200 students taking the examination and that each of the 101 possible marks has been scored. In this situation, for each mark, there is at least one student who has achieved that mark and so now the mapping is sujective. The mapping is not injective however since at least one mark will have been scored by more two or more students. In the (unlikely) event of 101 students taking the examination and each scoring a different mark, the resulting mapping would be bijective (both injective and surjective) and we would have a one-to-one correspondence between the set of student names and the set of examination marks.

As a second example, consider a simple memory manager in an operating system. The heap contains memory available to applications programs and the space in the heap zone is subdivided into a number of contiguous blocks, say. Each block in the zone is either allocated to a user or is free. The storage manager is responsible for recording which blocks are currently owned by which user. If we denote the set of user-names by U and the set of blocks By H (so that H will be the set $\{m \mid m \in N, 0 \leq m \leq h_N\}$, then the storage manager constitutes a mapping from the set of block numbers to the set of users which we can express as

Storage-manager: Note that this constitutes a partial mapping in the situation where one or more blocks is not owned by any user. Also, since, in general, some users may not own any blocks while other users may own several, then the mapping will be neither surjective nor injective.

Chapter 3

Set Theory

The concept of set is fundamental to mathematics and computer science. Everything mathematical starts with sets. For example, relationships between two objects are represented as a set of ordered pairs of objects, the concept of ordered pair is defined using sets, natural numbers, which are the basis of other numbers, are also defined using sets, the concept of function, being a special type of relation, is based on sets, and graphs and digraphs consisting of lines and points are described as an ordered pair of sets.

Though the concept of set is fundamental to mathematics, it is not defined rigorously here. Instead we rely on everyone's notion of "Set" as a collection of objects or a container of objects. In that sense "Set" is an undefined concept here. Similarly we say an object "belongs to "or "is a member of" a set without rigorously defining what it means. "An object(element) x belongs to a set A" is symbolically represented by "x A". It is also assumed that sets have certain (obvious) properties usually asssociated with a collection of objects such as the union of sets exists, for any pair of sets there is a set that contains them etc.

This approach to set theory is called "naive set theory "as opposed to more rigorous "axiomatic set theory". It was first developed by the German mathematician Georg Cantor at the end of the 19th century. For more on naive and axiomatic set theories click here which is not required for this course. Though the naive set theory is not rigorous, it is simpler and practically all the results we need can be derived within the naive set theory. Thus we shall be following this naive set theory in this course.

REPRESENTATION OF SET

A set can be described in a number of different ways. The simplest is to list up all of its members if that is possible. For example $\{1, 2, 3\}$ is the set of three numbers 1, 2, and 3. {indicates the beginning of the set, and} its end. Every object between them separated by commas is a member of the set. Thus $\{\{1, 2\}, \{\{3\}, 2\}, 2\}, \{1\}\}$ is the set of the elements $\{1, 2\}$, $\{\{3\}, 2\}$ and $\{1\}$.

A set can also be described by listing the properties that its members must satisfy. For example, $\{x|\ 1 \leq x \leq 2$ and x is a real number.$\}$ represents the set of real numbers between 1 and 2, and $\{x|\ x$ is the square of an integer and $x \leq 100\}$ represents the set $\{0, 1, 4, 9, 16, 25, 36, 49, 64, 81; 100\}$.

A third way to describe a set is to give a procedure to generate the members of the set. The recursive/inductive definition is an example and it is going to be studied later.

In this representation, first, basic elements of the set are presented. Then a method is given to generate elements of the set from known elements of the set. Thirdly a statement is given that excludes undesirable elements (which may be included in the set otherwise) from the set. for example the set of natural numbers N can be defined recursively as the set that satisfies the following (1), (2), and (3):

1. $0 \in N$
2. For any number x if $x \in N$, then $x + 1 \in N$.
3. Nothing is in N unless it is obtained from (1) and (2).

Following this definition, the set of natural numbers N can be obtained as follows:

First by (1), 0 is put into N.

Then by (2), since 0 is in N, $0 + 1\ (= 1)$ is in N.

Then by (2) again, $1 + 1\ (= 2)$ is in N.

Proceeding in this manner all the natural numbers are put into N.

Note that if we don't have (3), $0.5, 1.5, 2.5, \ldots$ can be included in N, which is not what we want as the set of natural numbers.

BASICS OF SET

Definition (Equality of sets): Two sets are equal if and only if they have the same elements.

More formally, for any sets A and B, $A = B$ if and only if $x\ [x \in A \leftrightarrow x \in B]$.

Thus for example $\{1, 2, 3\} = \{3, 2, 1\}$, that is the order of elements does not matter, and $\{1, 2, 3\} = \{3, 2, 1, 1\}$, that is duplications do not make any difference for sets.

Definition (Subset): A set A is a subset of a set B if and only if everything in A is also in B.

More formally, for any sets A and B, A is a subset of B, and denoted by $A \subseteq B$, if and only if $\forall x\ [x \in A \rightarrow x\ B]$.

If $A \subseteq B$, and $A \neq B$, then A is said to be a proper subset of B and it is denoted by $A \subset B$.

For example $\{1, 2\} \subseteq \{3, 2, 1\}$.

Also $\{1, 2\} \subset \{3, 2, 1\}$.

Cardinality: If a set S has n distinct elements for some natural number n, n is the cardinality (size) of S and S is a finite set. The cardinality of S is denoted by $|S|$.

. For example the cardinality of the set $\{3, 1, 2\}$ is 3.

Empty set: A set which has no elements is called an empty set.

More formally, an empty set, denoted by ØØ, is a set that satisfies the following:

$$\forall x\ x \notin \varnothing,$$

where $\notin$ means "is not in" or "is not a member of".

Note that Ø and {Ø} are different sets. {Ø} has one element namely in it. So {Ø} is not empty. But Ø has nothing in it.

Definition(Universal set): A set which has all the elements in the universe of discourse is called a universal set.

More formally, a universal set, denoted by U, is a set that satisfies the following:

$\forall x\ x \in \text{U}$.

Three subset relationships involving empty set and universal set are listed below as theorems without proof. Their proofs are found elsewhere.

Note that the set A in the next four theorems are arbitrary. So A can be an empty set or universal set.

Theorem: For an arbitrary set A $A \subseteq U$.

Theorem: For an arbitrary set A $\varnothing \subseteq A$.

Theorem: For an arbitrary set A $A \subseteq A$.

Definition(Power set): The set of all subsets of a set A is called the power set of A and denoted by 2^A or $p(A)$.

For example for $A = \{1, 2\}$, $p(A) = \{ \varnothing, \{1\}, \{2\}, \{1, 2\}\}$.

For $B = \{\{1, 2\}, \{\{1\}, 2\}, \varnothing\}$, $p(B) = \{ \varnothing, \{\{1, 2\}\}, \{\{\{1\}, 2\}\}, \{\varnothing\}, \{\{1, 2\}, \{\{1\}, 2\}\varnothing\}, \{\{1, 2\},\}, \{\{\{1\}, 2\},\}, \{\{1, 2\}, \{\{1\}, 2\}, \varnothing\}\}$.

Also $p(\varnothing) = \{\varnothing\}$ and $(\{\varnothing\}) = \{\varnothing, \{\varnothing\}\}$.

Theorem: For an arbitrary set A, the number of subsets of A is $2^{|A|}$.

MATHEMATICAL REASONING

Mathematical theories are constructed starting with some fundamental assumptions, called axioms, such as "sets exist" and "objects belong to a set" in the case of naive set theory, then proceeding to defining concepts(definitions) such as "equality of sets", and "subset", and establishing their properties and relationships between them in the form of theorems such as "Two sets are equal if and only if each is a subset of the other", which in turn causes introduction of new concepts and establishment of their properties and relationships. Proofs are the arguments for establishing those properties and relationships. At the bottom level these arguments follow the inference rules of propositional and predicate logic, that is the conclusion of the theorem being proved must be derived from its hypotheses, axioms, definitions, and proven theorems using inference rules.

However, at the bottom level they become tedious and inefficient as one can easily imagine. Thus in actual proofs short–cuts are taken using already proven theorems, using multiple inference rules in one step without explicitly mentioning them individually, omitting "obvious" proofs, and so on.

Finding a proof is in general an art. There is no single method that works for all cases. However, at this level the most important thing to remember is to know and understand definitions of concepts involved. The next important thing to keep in mind is to look up relevant facts and try to use them. Even if you don't see the entire path to the goal, if you move one step forward from where you are, you get a new perspective and it often gives you some good ideas to pursue. Needless to say that you must not forget the inference rules. It is not a bad idea to review "Problem Solving" we studied earlier here.

There are also some well used and often very useful proof techniques such as trivial proof, vacuous proof, direct proof, proof by contradiction, proving the contrapositive, and proof by induction. These are explained below with proofs of the theorems on subset relation as examples

Theorem: $A \subseteq U$.

Proof: By the definition of $\subseteq$, we need to show that $\forall x[x \in A \rightarrow x \in U]$.

For that, what we need is to show that for an arbitrary x, $x \in A \rightarrow x \in U$ holds according to the inference rule "universal generalization".

Since x is an object of the universe of discourse, $x \in U$ is true for any arbitrary object by the Universal Instantiation. Hence $x \in A \rightarrow x \in U$ is true for any arbitrary object x ($p \rightarrow q$ is always true if q is true regardless of what p is). Thus by the Universal Generalization $\forall x[x \in A \rightarrow x \in U]$, that is $A \subseteq U$, by the definition of subset.

We say $p \rightarrow q$ is *trivially true* if q is true, and this kind of proof (i.e. showing q is true for $p \rightarrow q$ without referring to p) is called a trivial proof.

Theorem: $\emptyset \subseteq A$.

Proof: By the definition of $\subseteq$, we need to show that $\forall x[x \in \emptyset \rightarrow x \in A]$. For that, what we need is to show that for an arbitrary x, $x \in \emptyset \rightarrow x \in A$ holds. Then apply the Universal Generalization.

Since $\forall x \notin \emptyset$, for any arbitrary x, $x \notin \emptyset$ is false by the Universal Instantiation. Hence

$x\in\emptyset \to x\in A$ is true for any arbitrary x ($p \to q$ is always true if p is false regardless of what q is). Hence by the Universal Generalization $\forall x[x\in\emptyset \to x\in A]$. Thus $\emptyset \subseteq A$ by the definition of subset.

We say $p \to q$ is *vacuously true* if p is false, and this kind of proof (i.e. showing p is false for $p \to q$) is called a vacuous proof.

Proof: Variations for Theorem

Theorem: Like most others, can be proven in a number of other ways. Here we try to prove it in two other ways.

Proof by Contrapositive: In this method, to prove $p \to q$ we prove its contrapositive $\neg p \to \neg q$, , instead.

So to prove $x\in\emptyset \to x\in A$ for an arbitrary x, try to prove $x\notin A \to x\notin \emptyset$.

In this case since $x\notin \emptyset$ is true for any arbitrary x, $x\notin A \to x\notin \emptyset$ is trivially true. Hence its contraposiive is also true.

Proof by Contradiction: In this method, to prove p we assume $\neg p$ and derive a contradiction from that. Then since $\neg p$ implies a contradiction, it can not hold true. Hence p must be true.

So to prove $\forall x[x\in\emptyset \to x\in A]$ we assume that $\neg\forall x[x\in\emptyset \to x\in A]$.

$\neg\forall x[x\in\emptyset \to x\in A]$ is equivalent to $\exists x\neg[x\in\emptyset \to x\in A]$.

This in turn is equivalent to $\exists x[x\in\emptyset \wedge x\notin A]$.

However, $\exists x[x\in\emptyset \wedge x\notin A]$ implies $\exists x[x\in\emptyset]$ by formula 4 of the relationships between the connectives and quantifiers from predicate logic and "simplification" from the implications of propositional logic. But $x\notin\emptyset$ for any x, which contradicts $\exists x[x\in\emptyset]$.

Hence $\neg\forall x[x\in\emptyset \to x\in A]$, can not be true.

Hence, $\emptyset \subseteq A$.

More Proofs

Theorem: $A = B$ iff $A \subseteq B$ and $B \subseteq A$.

Proof: By the definition of $A = B$,

$$\begin{aligned} A = B &\Leftrightarrow \forall x[x\in A \leftrightarrow x\in B] \\ &\Leftrightarrow \forall x[x\in A \to x\in B] \wedge (x\in B \to x\in A] \\ &\Leftrightarrow \forall x[x\in A \to x\in B] \wedge \forall x(x\in B \to x\in A] \end{aligned}$$

Since $A \subseteq B \Leftrightarrow \forall x[x \in A \rightarrow x \in B]$, this means that $A \subseteq B$ and $B \subseteq A$.

Hence, $A = B$ iff $A \subseteq B$ and $B \subseteq A$.

SET OPERATIONS

Sets can be combined in a number of different ways to produce another set. Here four basic operations are introduced and their properties are discussed.

Union: The union of sets A and B, denoted by $A \cup B$, is the set defined as

$A \cup B = \{x \mid x \in A\ x \in B\}$

Example: If $A = \{1, 2, 3\}$ and $B = \{4, 5\}$, then $A \cup B = \{1, 2, 3, 4, 5\}$.

Example: If $A = \{1, 2, 3\}$ and $B = \{1, 2, 4, 5\}$, then $A \cup B = \{1, 2, 3, 4, 5\}$.

Note that elements are not repeated in a set.

Intersection: The intersection of sets A and B, denoted by $A \cap B$, is the set defined as

$A\ B = \{x \mid x\ A\ x\ B\}$

Example: If $A = \{1, 2, 3\}$ and $B = \{1, 2, 4, 5\}$, then $A\ B = \{1, 2\}$.

Example: If $A = \{1, 2, 3\}$ and $B = \{4, 5\}$, then $A \cap B = \emptyset$.

Difference: The difference of sets A from B, denoted by $A - B$, is the set defined as

$$A - B = \{x \mid x \in A\ x \notin B\}$$

Example: If $A = \{1, 2, 3\}$ and $B = \{1, 2, 4, 5\}$, then $A - B = \{3\}$.

Example: If $A = \{1, 2, 3\}$ and $B = \{4, 5\}$, then $A - B = \{1, 2, 3\}$.

Note that in general $A - B \neq B - A$

Complement: For a set A, the difference $U - A$, where U is the universe, is called the complement of A and it is denoted by $\overline{A}$.

Thus $\overline{A}$ is the set of everything that is not in A.

The fourth set operation is the Cartesian product We first define an ordered pair and Cartesian product of two sets using it. Then the Cartesian product of multiple sets is defined using the concept of *n*–tuple.

Orderedpair: An ordered pair is a pair of objects with an order associated with them.

If objects are represented by x and y, then we write the ordered pair as $<x, y>$.

Two ordered pairs $<a, b>$ and $<c, d>$ are equal if and only if $a = c$ and $b = d$. For example the ordered pair $<1, 2>$ is not equal to the ordered pair $<2, 1>$.

Cartesian product: The set of all ordered pairs $<a, b>$, where a is an element of A and b is an element of B, is called the Cartesian product of A and B and is denoted by $A \times B$.

Example: Let $A = \{1, 2, 3\}$ and $B = \{a, b\}$. Then

$A \times B = \{<1, a>, <1, b>, <2, a>, <2, b>, <3, a>, <3, b>\}$.

Ordered n–tuple: An ordered *n*–tuple is a set of *n* objects with an order associated with them (rigorous definition to be filled in). If *n* objects are represented by $x_1, x_2, ..., x_n$, then we write the ordered *n*–tuple as $<x_1, x_2, ..., x_n>$.

Cartesian product: Let $A_1, ..., A_n$ be *n* sets. Then the set of all ordered *n*–tuples $<x_1, ..., x_n>$, where $x_i \in A_i$ for all *i*, $1 \le i \le n$, is called the Cartesian product of $A_1, ..., A_n$, and is denoted by $A_1 \times ... \times A_n$.

Example: Let $A - \{1, 2\}$, $B = \{a, b\}$ and $C = \{5, 6\}$. Then

$A \times B \times C = \{<1, a, 5>, <1, a, 6>, <1, b, 5>, <1, b, 6>, <2, a, 5>, <2, a, 6>, <2, b, 5>, <2, b, 6>\}$.

Definition (equality of n–tuples): Two ordered *n*–tuples $<x_1, ..., x_n>$ and $<y_1, ..., y_n>$ are equal if and only if $x_i = y_i$ for all *i*, $1 \le i \le n$.

For example the ordered *3*–tuple $<1, 2, 3>$ is not equal to the ordered *n*–tuple $<2, 3, 1>$.

PROPERTIES OF SET OPERATION

Basic properties of set operations are discussed here. 1 – 6 directly correspond to identities and implications of propositional logic, and 7 – 11 also follow immediately from them as illustrated below.

1. $A \cup \emptyset = A$

   $A \cap U = A$

   .... Identity Laws

2. $A \cup U = U$

   $A \cap \emptyset = \emptyset$

   .... Domination Laws

3. $A \cup A = A$

   $A \cap A = A$

   .... Idempotent Laws

4. $A \cup B = B \cup A$

   $A \cap B = B \cap A$

   .... Commutative Laws

5. $(A \cup B) \cup C = A \cup (B \cup C)$

   $(A \cap B) \cap C = A \cap (B \cap C)$

   .... Associative Laws

6. $A \cup (B \cap C) = (A \cup B) \cap (A \cup C)$

   $A \cap (B \cup C) = (A \cap B) \cup (A \cap C)$

   .... Distributive Laws

7. If $A \subseteq B$ and $C \subseteq D$, then $A \cup C \subseteq B \cup D$, and $A \cap C \subseteq B \cap D$.

8. If $A \subseteq B$, then $A \cup B = B$ and $A \cap B = A$.

9. $A \cup (B - A) = A \cup B$

10. $A \cap (B - A) = \emptyset$

11. $A - (B \cup C) = (A - B) \cap (A - C)$ (cf. $\overline{A \cup B} = \overline{A} \cap \overline{B}$)

    $A - (B \cap C) = (A - B) \cup (A - C)$ (cf. $\overline{A \cap B} = \overline{A} \cup \overline{B}$)

    .... De Morgan's Laws

12. $B = \overline{A}$ if and $A \cup B = U$ only if and $A \cap B = \emptyset$

13. $\overline{\overline{A}} = A$.

    Additional properties:

14. $A \subseteq A \cup B$

15. $A \cap B \subseteq A$

The properties $1 \sim 6$, and 11 can be proven using equivalences of propositional logic. The others can also be proven similarly by going to logic, though they can be proven also using some of these properties (after those properties are proven, needless to say). Let us prove some of these properties.

Proof for: We are going to prove this by showing that every element that is in $A \cup B$ is also in $B \cup A$ and vice versa.

Consider an arbitrary element x. Then by the definition of set union

$x \in A \cup B \Leftrightarrow x \in A \vee x \in B$

$\Leftrightarrow x \in A \vee \in B$ by the commutativity of

$\Leftrightarrow x \in B \cup A$ by the definition of set union.

Hence by Universal Generalization, every element is in $A \cup B$ is also in $B \cup A$.

Hence $A \cup B = B \cup A$.

Note here the correspondence of the commutativity of $\cup$ and that of $\vee$ This correspondence holds not just for the commutativity but also for others.

Furthermore a similar correspondence exists between $\cap$ and $\wedge$, and between $\subseteq$ and $\rightarrow$.

Proof for: By the definition of the equality of sets, we need to prove that $\forall x[x \in A \cup (B \cap C)$ if and only if $x \in (A \cup B) \cap (A \cup C)]$

For that, considering the Univesal Generalization rule, we need to show that for an arbitrary element in the universe x, $x \in A \cup (B \cap C)$ if and only if $x \in (A \cup B) \cap (A \cup C)$

Here the *only if* part is going to be proven. The *if* part can be proven similarly. $x \in A \cup (B \cap C) \Leftrightarrow x \in (A \cup B) \cap (A \cup C)$ by the definition of $\cup$.

$\Leftrightarrow x \in A \vee (x \in B \wedge x \in C)$ by the definition of $\cap$.

$\Leftrightarrow (x \in A \vee x \in B) \wedge (x \in A \vee x \in C)$ by the distribution from the equivalences of propositional logic.

$\Leftrightarrow (x \in A \cup B) \wedge (x \in A \cup C)$ by the definition of $\cup$.

$\Leftrightarrow x \in (A \cup B) \cap (A \cup C)$ by the definition of $\cap$.

Proof for:

1. If $A \subseteq B$ then $A \cup B = B$.

   Let x be an arbitrary element in the universe.

   Then $x \in A \cup B \Leftrightarrow x \in A \vee x \in B$.

   Since $A \subseteq B$, $x \in A \rightarrow x \in B$.

   Also $x \in B \rightarrow x \in B$.

   Hence $x \in A \cup B \rightarrow x \in B$.

   Hence $A \cup B \subseteq B$.

   Since $B \subseteq A \cup B$ (use "addition" rule), $A \cup B = B$ follows.

2. Similarly for $A \cap B = A$.

   Alternative proof:

   For example, (b) can be proven as follows:

   First $A \cap B \subseteq A$.

   Then since $A \subseteq A$, and $A \subseteq B$, by 7 $A \cap A \subseteq A \cap B$.

   Since $A \cap A = A$ by 3, $A \subseteq A \cap B$.

Proof for: Let x be an arbitrary element in the universe.

Then $[x \in A \cup (B - A)] \Leftrightarrow [x \in A \vee (x \in \wedge x \notin A)]$

$\Leftrightarrow [(x \in A \vee x \in B) \wedge (x \in A \vee x \notin A)]$

$\Leftrightarrow [(x \in A \vee x \in B) \wedge \text{True}]$

$\Leftrightarrow [x \in A \vee x \in B]$

Hence $A \cup (B - A) = A \cup B$.

Alternative proof

This can also proven using set properties as follows.

$A\ (B - A) = A \cup (\mathrm{B} \cup \bar{A})$ by the definition of $(B - A)$.

$= (\mathrm{A} \cup \mathrm{B}) \cap (\mathrm{A} \cup \bar{A})$ by the distribution.

$= (\mathrm{A} \cup \mathrm{B}) \cap \mathrm{U}$

$= (\mathrm{A} \cup \mathrm{B})$ by 1.

Proof for: Suppose $A \cap (B - A) \Leftrightarrow x \in A \wedge B - A$.

Then there is an element x that is in $A \cap (B - A)$, i.e.

$x \in A \cap (B - A) \Leftrightarrow x \in A \wedge x \in B - A$

$\Leftrightarrow x \in A \wedge (x \in B \wedge x \notin A)$

$\Leftrightarrow (x \in A \wedge \notin A) \wedge x \in B)$

$\Leftrightarrow F$

Hence $A \cap (B - A) \neq \emptyset$ does not hold.

Hence $A \cap (B - A) = \emptyset$.

This can also be proven in the similar manner to 9 above.

Proof for: Let x be an arbitrary element in the universe.

Then $x \in A - (B \cup C) \Leftrightarrow x \in A \wedge x \notin B \cup C \Leftrightarrow x \in A \wedge \neg (x \in B \vee x \in C)$

$\Leftrightarrow x \in A \wedge (x \notin B \wedge x \notin C)$

$\Leftrightarrow (x \in A \wedge x \notin B) \wedge (x \in A \wedge x \notin C)$

$\Leftrightarrow x \in A - B \wedge x \in A - C$

$\Leftrightarrow x \in (A - B) \cap (A - C)$

Hence $A - (B \cup B) = (A - B) \cap (A - C)$

Proof for: (a) $A \cup B = U \wedge A \cap B = \emptyset \Rightarrow B = A$?

Try to prove $B \subseteq \overline{A}$ and $\overline{A} \subseteq B$.

Let x be an arbitrary element in the universe.

Then if $x \in B$, then $x \notin A$ since $A \cap B = \emptyset$. Hence $x \in \overline{A}$.

Hence $B \subseteq \overline{A}$.

If $x \in \overline{A}$, then $x \notin A$. Since $x \in A \vee x \in B$ (from $A \cup B = U$), $x \in B$ must hold. Hence $\overline{A} \subseteq B$.

Hence $B = A$.

(b) $B = \overline{A} = \Rightarrow (A \cup B = U \wedge A \cap B = \emptyset)$

Since $B = \overline{A}$, $A \cup B = A \cup (U - A) = A \cup U$ since $A \subseteq U$.

Also $A \cap B = A \cap (U - A) = \emptyset$ by 10 above.

Proof for: Since $A \cup A = A \cup A$, $A \cup A = U$.

Also since $A \cap \overline{A} = \overline{A} \cap A$, $\overline{A} \cap A = \emptyset$.

Hence A satisfies the conditions for the complement of A.

Hence $A = \overline{\overline{A}}$.

RECURSIVE DEFINITION

Sets which have too many elements to list them up, and for which there are no convenient or obvious predicates to specify their elements can often be defined using a recursive definition (also called inductive definition). It essentially gives a procedure to generate the members of the set one by one starting with some subset of its elements.

In this type of definition, first a collection of elements to be included initially in the set is specified. These elements can be viewed as the seeds of the set being defined. Next, the rules to be used to generate elements of the set from elements already known to be in the set (initially the seeds) are given. These rules provide a method to construct the set element by element starting with the seeds. These rules can also be used to test elements for the membership in the set.

A recursive definition of a set always consists of three distinct clauses:

1. The basis clause (or simply basis) of the definition establishes that certain objects are in the set. This part of the definition specifies the "seeds" of the set from which the elements of the set are generated using the methods given in the

inductive clause. The set of elements specified here is called basis of the set being defined.

2. The inductive clause (or simply induction) of the definition establishes the ways in which elements of the set can be combined to produce new elements of the set. The inductive clause always asserts that if objects are elements of the set, then they can be combined in certain specified ways to create other objects. Let us call the objects used to create a new object the parents of the new object, and the new object is their child .
3. The extremal clause asserts that unless an object can be shown to be a member of the set by applying the basis and inductive clauses a finite number of times, the object is not a member of the set.

The set you are trying to define recursively is the set that satisfies those three clauses.

There are a number of other ways of expressing the extremal clause that are equivalent to the extremal clause given above.

EXAMPLES OF RECURSIVE DEFINITION OF SET

Example: Definition of the Set of Natural Numbers N

The set N is the set that satisfies the following three clauses:

Basis Clause: $0 \in N$

Inductive Clause: For any element x in N, $x + 1$ is in N.

Extremal Clause: Nothing is in unless it is obtained from the Basis and Inductive Clauses.

The basis for this set N is $\{0\}$. The $x + 1$ in the Inductive Clause is the parent of x, and x is the child of $x + 1$.

Following this definition, the set of natural numbers N can be obtained as follows:

First by the Basis Clause, 0 is put into N.

Then by the Inductive Clause, since 0 is in N, $0 + 1$ $(= 1)$ is in N. 0 is the parent of 1, and 1 is the child of 0.

Then by the Inductive Clause again, $1 + 1$ $(= 2)$ is in N. 1 is the parent of 2, and 2 is the child of 1.

Proceeding in this manner all the "natural numbers" are put into N.

Note that if we don't have the Extremal Clause, 0.5, 1.5, 2.5, ... can be included in N, which is not what we want as the set of natural numbers.

Example: Definition of the Set of Nonnegative Even Numbers *NE*

The set *NE* is the set that satisfies the following three clauses:

Basis Clause: $0 \in NE$

Inductive Clause: For any element *x* in NE, $x + 2$ is in *NE*.

Extremal Clause: Nothing is in unless it is obtained from the Basis and Inductive Clauses.

Example: Definition of the Set of Even Integers *EI*

The set *EI* is the set that satisfies the following three clauses:

Basis Clause: $0 \in EI$

Inductive Clause: For any element *x* in *EI*, $x + 2$, and $x - 2$ are in *EI*.

Extremal Clause: Nothing is in *EI* unless it is obtained from the Basis and Inductive Clauses.

Example: Definition of the Set of Strings *S* over the alphabet $\{a, b\}$ excepting empty string.

This is the set of strings consisting of *a*'s and *b*'s such as *abbab*, *bbabaa*, etc.

The set *S* is the set that satisfies the following three clauses:

Basis Clause: $a \in S$, and $b \in S$.

Inductive Clause: For any element *x* in *S*, $ax \in S$, and $bx \in S$.

Here *ax* means the concatenation of *a* with *x*.

Extremal Clause: Nothing is in *S* unless it is obtained from the Basis and Inductive Clauses.

Tips for recursively defining a set:

For the "Basis Clause", try simplest elements in the set such as smallest numbers (0, or 1), simplest expressions, or shortest strings. Then see how other elements can be obtained from them, and generalize that generation process for the "Inductive Clause".

The set of propositions (propositional forms) can also be defined recursively.

GENERALIZED SET OPERATIONS

As we saw earlier, union, intersection and Cartesian product of sets are associative. For example

$(A \cup B) \cup C = A \cup (B \cup C)$

To denote either of these we often use $A \cup B \cup C$.

This can be generalized for the union of any finite number of sets as $A_1 \cup A_2 .... \cup CA_n$, which we write as

$$\cup_{i=1}^{n} A_i$$

This generalized union of sets can be rigorously defined as follows:

Definition ($\cup_{i=1}^{n} A_i$):

Basis Clause: For $n = 1$, $\cup_{i=1}^{n} A_i = A_1$.

Inductive Clause: $\cup_{i=1}^{n+1} A_i = (\cup_{i=1}^{n} A_i) \cup A_{n+1}$

Similarly the generalized intersection $\cap_{i=1}^{n} A_i$ and generalized Cartesian product $x_{i=1}^{n} A_i$ can be defined.

Based on these definitions, De Morgan's law on set union and intersection can also be generalized as follows:

Theorem: *(Generalized De Morgan)*

$\overline{\cup_{i=1}^{n} A_i} = \cap_{i=1}^{n} \overline{A_i}$, and

$\overline{\cap_{i=1}^{n} A_i} = \cup_{i=1}^{n} \overline{A_i}$

Proof: These can be proven by induction on n and are left as an exercise.

MORE EXAMPLES ON RECURSIVE DEFINITION OF SET

Example: Recursively define the set $S = \{n^2 \mid n$ is a natural number greater than $0\}$, that is $S = \{1, 4, 9, 16, ...\}$.

S is the set that is defined by the following clauses:

Basis Clause: 1 is in S.

Inductive Clause: If x in S, then $x + 2x^{1/2} + 1$ is in S.

Extremal Clause: Nothing is in S unless it is obtained from the Basis and Inductive Clauses.

How do we get $x + 2x^{1/2} + 1$ in the Inductive Clause ?

Since $S = \{1, 4, 9, 16, ...\}$and its elements can be ordered 1, 4, 9, ..., n^2 , for the n–th element, which is n^2, the next element is $(n+1)^2$. Hence if we let x be the n–th element, then $x = n^2$. Hence $n = x^{1/2}$. But the element next to x is $(n+1)^2$ which is equal to $n^2 + 2n + 1$. By substituting $n = x^{1/2}$ into that we obtain $x + 2x^{1/2} + 1$.

Example: Recursively define the set $T = \{2^n - 2 \mid n$ is a natural number greater than 0$\}$, that is $T = \{0, 2, 6, 14, 30, ...\}$.

T is the set that is defined by the following clauses:

Basis Clause: 0 is in T.

Inductive Clause: If x is in T, then $2x + 2$ is in T.

Extremal Clause: Nothing is in T unless it is obtained from the Basis and Inductive Clauses.

How do we get $2x + 2$ in the Inductive Clause ?

Since $T = \{0, 2, 6, 14, ...\}$ and its elements can be ordered 0, 2, 6, ..., $2^n - 2$, for the n–th element, which is $2^n - 2$, the next element is $2^{n+1} - 2$. Hence if we let x be the n–th element, then $x = 2^n - 2$. Now $2^{n+1} - 2 = 2*2^n - 2 = 2*(2^n - 2) + 2 = 2x + 2$.

Example: Recursively define the set of strings $U = \{a^n bc^n \mid$ n is a natural number$\}$, that is U is the set $\{b, abc, aabcc, ...\}$.

Since the simplest element in U is b, let us select b as the basis element. Then from b we can obtain *abc* by prepending a and appending c to it. Similarly *aabcc* can be obtained from *abc* by prepending a and appending c to it. Similarly for the rest of the elements of the set U. Thus U is the set that satisfy the following clauses:

Basis Clause: b is in U.

Inductive Clause: If x is in U, then *axc* is in U.

Extremal Clause: Nothing is in U unless it is obtained from the Basis and Inductive Clauses.

Example: Recursively define the set $V = \{x \mid x$ is a natural number and $\lfloor x/2 \rfloor$ is even$\}$, that is $x = \{0, 4, 8, 12, 16, ...\}$. Note that $\lfloor x \rfloor$ denotes the largest integer that is not larger than x. If x is nonnegative as in this example, it is the integer part of x.

Since 0 is the smallest element of x, let us select 0 as the basis element of V. For the inductive clause note that for $\lfloor x \rfloor$ to increase by 2 to the next even number, x needs to grow by 4. Hence we have the following definition for V:

V is the set that is defined by the following clauses:

Basis Clause: 0 is in V.

Inductive Clause: If x is in V, then $x + 1$ is in V.

Extremal Clause: Nothing is in V unless it is obtained from the Basis and Inductive Clauses.

We have defined four sets recursively. However, we are at this point not exactly sure whether or not those definitions correctly define the sets. To verify those definitions, that is, to check whether or not the elements of each set have the specified properties, we can use proof by induction. Let us try to verify the definition of Example.

Verification of Definition of Set S:

To verify the definition of Example we are going to prove by (general) induction that the elements of the set S is in the form n^2, where n is a natural number greater than 0.

In the Basis Step we prove that the element in the basis, that is 1 in this case, has the property. Then in the Inductive Step we prove that if an arbitrary element,say x, has the property, then the element generated from x has the property.

Basis Step: Since $1 = 1^2$, the basis element 1 of the set S is of the form n^2, where n is a natural number greater than 0.

Inductive Step: Let x be an arbitrary element of S and $x = n^2$ for some natural number n. Then the element generated from x by the Inductive Clause of the definition of S is $x + 2x^{1/2} + 1$.

We are trying to show that $x + 2x^{1/2} + 1$ is of the form n^2.

Since $x = n^2$, $x^{1/2} = n$. Hence $x + 2x^{1/2} + 1 = n^2 + 2n + 1 = (n + 1)^2$, which is the square of a natural number $n + 1$.

End of the proof.

Next let us try to verify the definition of Example .

Verification of Definition of set U:

To verify the definition of Example we are going to prove by (general) induction that the elements of the set U is in the form $a^n bc^n$, where n is a natural number.

In the Basis Step we prove that the element in the basis, that is b in this case, has the property. Then in the Inductive Step we prove that if an arbitrary element, say x, has the property, then the element generated from x has the property.

Basis Step: b is in the form $a^n bc^n$ with $n = 0$.

Inductive Step: Let x be an arbitrary element of U and $x = a^n bc^n$ for some natural number n. Then the element generated from x by the Inductive Clause of the definition of U is axc which is equal to $aa^n bc^n c$, which in turn is equal to $a^{(n+1)} bc^{(n+1)}$. Hence if an arbitrary element x has the property, then its children also have the property.

RECURSIVE DEFINITION OF FUNCTION

Condition: The domain of the function you wish to define recursively must be a set defined recursively.

How to define function recursively: First the values of the function for the basis elements of the domain are specified. Then the value of the function at an element, say *x*, of the domain is defined using its value at the parent(s) of the element *x*.

A few examples are given below.

They are all on functions from integer to integer except the last one.

Example: The function $f(n) = n!$ for natural numbers *n* can be defined recursively as follows:

The function *f* is the function that satisfies the following two clauses:

Basis Clause: $f(0) = 0! = 1$

Inductive Clause: For all natural number *n*, $f(n+1) = (n+1) f(n)$.

Note that here Extremal Clause is not necessary, because the set of natural numbers can be defined recursively and that has the extremal clause in it. So there is no chance of other elements to come into the function being defined.

Using this definition, *3!* can be found as follows:

Since $0! = 1$, $1! = 1 * 0! = 1 * 1 = 1$,

Hence $2! = 2 * 1! = 2 * 1 = 2$.

Hence $3! = 3 * 2! = 3 * 2 * 1 = 6$.

Example: The function $f(n) = 2n + 1$ for natural numbers *n* can be defined recursively as follows:

The function *f* is the function that satisfies the following two clauses:

Basis Clause: $f(0) = 1$

Inductive Clause: For all natural number *n*, $f(n+1) = f(n) + 2$.

See above for the extremal clause.

Example: The function $f(n) = 2^n$ for natural numbers *n* can be defined recursively as follows:

The function *f* is the function that satisfies the following two clauses:

Basis Clause: $f(0) = 1$

Inductive Clause: For all natural number *n*, $f(n+1) = 2 f(n)$.

Example: The function *L* from the set *S* of strings over $\{a, b\}$ to the set of natural numbers that gives the length of a string can be defined recursively as follows:

The function *L* is the function that satisfies the following two clauses:

Basis Clause: For symbols *a* and *b* of the alphabet, $L(a) = 1$ and $L(b) = 1$.

Inductive Clause: For any string x and y of S, $L(xy) = L(x) + L(y)$, where xy is the concatenation of strings x and y.

This function L gives the number of a's and b's

RECURSIVE ALGORITHM

A recursive algorithm is an algorithm which calls itself with "smaller (or simpler)" input values, and which obtains the result for the current input by applying simple operations to the returned value for the smaller (or simpler) input. More generally if a problem can be solved utilizing solutions to smaller versions of the same problem, and the smaller versions reduce to easily solvable cases, then one can use a recursive algorithm to solve that problem. For example, the elements of a recursively defined set, or the value of a recursively defined function can be obtained by a recursive algorithm.

If a set or a function is defined recursively, then a recursive algorithm to compute its members or values mirrors the definition. Initial steps of the recursive algorithm correspond to the basis clause of the recursive definition and they identify the basis elements. They are then followed by steps corresponding to the inductive clause, which reduce the computation for an element of one generation to that of elements of the immediately preceding generation.

In general, recursive computer programs require more memory and computation compared with iterative algorithms, but they are simpler and for many cases a natural way of thinking about the problem.

Example: Algorithm for finding the k–th even natural number

Note here that this can be solved very easily by simply outputting $2*(k - 1)$ for a given k. The purpose here, however, is to illustrate the basic idea of recursion rather than solving the problem.

Algorithm: Even(positive integer k)

Input: k, a positive integer.

Output: k–th even natural number (the first even being 0)

Algorithm:

if $k = 1$, then return 0;

else return Even(k–1) + 2 .

Here the computation of Even(k) is reduced to that of Even for a smaller input value, that is Even(k–1). Even(k) eventually becomes Even(1) which is 0 by the first line. For example, to compute Even(3), Algorithm Even(k) is called with $k = 2$. In the computation of Even(2), Algorithm Even(k) is called with $k = 1$. Since Even(1) = 0, 0 is returned for the computation of Even(*2*), and Even(*2*) = Even(*1*) + *2* = *2* is obtained. This value *2* for Even(2) is now returned to the computation of Even(*3*), and Even(*3*) = Even(2) + *2* = *4* is obtained.

As can be seen by comparing this algorithm with the recursive definition of the set of nonnegative even numbers, the first line of the algorithm corresponds to the basis clause of the definition, and the second line corresponds to the inductive clause.

By way of comparison, let us see how the same problem can be solved by an iterative algorithm.

Algorithm 1–a: Even(positive integer *k*)

Input: *k*, a positive integer

Output: *k*–th even natural number (the first even being *0*)

Algorithm:

```
int i, even;
i := 1;
even := 0;
while(i < k) {
even := even + 2;
i := i + 1;
}
return even .
```

Example: Algorithm for computing the *k*–th power of *2*

Algorithm 2 Power of 2(natural number k)

Input: *k* , a natural number

Output: *k*–th power of *2*

Algorithm: If *k* = *0*, then return *1*;

else return 2*Power of 2(*k* – *1*) .

By way of comparison, let us see how the same problem can be solved by an iterative algorithm.

Algorithm 2–a Power of 2(natural number k)

Input: *k* , a natural number

Output: *k*–th power of *2*

Algorithm:

int *i*, *power*;

i := *0*;

power := *1*;

while(*i* < *k*) {

power := *power* * *2*;

i := *i* + *1*;

}

return *power* .

The next example does not have any corresponding recursive definition. It shows a recursive way of solving a problem.

Example: Recursive Algorithm for Sequential Search

Algorithm 3 SeqSearch(*L*, *i*, *j*, *x*)

Input: *L* is an array, *i* and *j* are positive integers, $i \leq j$, and *x* is the key to be searched for in *L*.

Output: If *x* is in *L* between indexes *i* and *j*, then output its index, else output *0*.

Algorithm:

if $i \leq j$, then

{

if *L*(*i*) = *x*, then return *i* ;

else return SeqSearch(*L*, *i*+*1*, *j*, *x*)

}

else return *0*.

Recursive algorithms can also be used to test objects for membership in a set.

Example: Algorithm for testing whether or not a number *x* is a natural number

Algorithm 4 Natural(a number *x*)

Input: A number *x*

Output: "Yes" if *x* is a natural number, else "No"

Algorithm:

if *x* < *0*, then return "No"

else

if $x = 0$, then return "Yes"

else return Natural($x - 1$)

Example: Algorithm for testing whether or not an expression w is a proposition(propositional form) .

Algorithm Proposition(a string w)

Input: A string w

Output: "Yes" if w is a proposition, else "No"

Algorithm:

if w is 1(true), 0(false), or a propositional variable, then return "Yes"

else if $w = \sim w_1$, then return Proposition(w_1)

else

if ($w = w_1\ w_2$ or $w_1\ w_2$ or $w_1\ w_2$ or $w_1\ w_2$) and

Proposition(w_1) = Yes and Proposition(w_2) = Yes

then return Yes

else return No

end

PROOF BY INDUCTION

MATHEMATICAL INDUCTION FIRST PRINCIPLE

As we have seen in recursion , the set of natural numbers can be defined recursively, and its elements can be generated one by one starting with *0* by adding *1*. Thus the set of natural numbers can be described completely by specifying the basis element (*0*), and the process of generating an element from a known element in the set.

Taking advantage of this, natural numbers can be proven to have certain properties as follows:

First it is proven that the basis element, that is *0*, has the property in question (basis step). You prove that the seeds (the first generation elements) have the property.

Then it is proven that if an arbitrary natural number, denote it by *n*, has the property in question, then the next element, that is $n + 1$, has that property (inductive step). Here you prove that the property is inherited from one generation (n) to the next generation ($n + 1$).

When these two are proven, then it follows that all the natural numbers have that property. For since 0 has the property by the basis step, the element next to it, which is 1 has the same property by the inductive step. Then since 1 has the property, the elemen next to it, which is 2, has the same property again by the inductive step. Proceeding likewise any natural number can be shown to have the property. This process is somewhat analogous to the knocking over a row of dominos with knocking over the first domino corresponding to the basis step.

More generally mathematical statements involving a natural number n such as $1 + 2 + \ldots + n = n(n + 1)/2$ can be proven by mathematical induction by the same token.

To prove that a statement $P(n)$ is true for all natural number $n \leq n_0$, where n_0 is a natural number, we proceed as follows:

Basis Step: Prove that $P(n_0)$ is true.

Induction: Prove that for any integer $k \geq n_0$, if $P(k)$ is true (called induction hypothesis), then $P(k+1)$ is true.

The first principle of mathematical induction states that if the basis step and the inductive step are proven, then $P(n)$ is true for all natural number $k \geq n_0$.

As a first step for proof by induction, it is often a good idea to restate $P(k+1)$ in terms of $P(k)$ so that $P(k)$, which is assumed to be true, can be used.

Example: Prove that for any natural number n, $0 + 1 + \ldots + n = n(n + 1)/2$.

Proof:

Basis Step: If $n = 0$, then $LHS = 0$, and $RHS = 0 * (0 + 1) = 0$.

Hence $LHS = RHS$.

Induction: Assume that for an arbitrary natural number n, $0 + 1 + \ldots + n = n(n + 1)/2$ . ... *Induction Hypothesis*

To prove this for $n+1$, first try to express LHS for $n+1$ in terms of LHS for n, and somehow use the induction hypothesis.

Here let us try

$$LHS \text{ for } n + 1 = 0 + 1 + \ldots + n + (n + 1) = (0 + 1 + \ldots + n) + (n + 1) .$$

Using the induction hypothesis, the last expression can be rewritten as

$$n(n + 1)/2 + (n + 1) .$$

Factoring $(n + 1)$ out, we get

$$(n + 1)(n + 2)/ 2 ,$$

which is equal to the RHS for $n+1$.

Thus $LHS = RHS$ for $n+1$.

PROGRAM CORRECTNESS USING INDUCTION

Loops in an algorithm/program can be proven correct using mathematical induction. In general it involves something called "loop invariant" and it is very difficult to prove the correctness of a loop. Here we are goin to give a few examples to convey the basic idea of correctness proof of loop algorithms.

First consider the following piece of code that computes the square of a natural number:

(We do not compute the square this way but this is just to illustrate the concept of loop invariant and its proof by induction.)

SQUARE FUNCTION: SQ(N)

$S \leftarrow 0$

$i \leftarrow 0$

while $i < n$

$S \leftarrow S + n$

$i \leftarrow i + 1$

return S

Let us first see how this code computes the square of a natural number. For example let us compute 3^2 using it.

First S ← 0 and i ← 0 give S = 0 and i = 0 initially.

Since i < 3, the while loop is entered.

$S \leftarrow 0 + 3$

$i \leftarrow 0 + 1$

producing S = 3 and i = 1.

Since i < 3, the while loop is entered the second time.

$S \leftarrow 3 + 3$

$i \leftarrow 1 + 1$

producing S = 6 and i = 2.

Since i < 3, the while loop is entered the third time.

$S \leftarrow 6 + 3$

$i \leftarrow 2 + 1$

producing S = 9 and $i = 3$.

Since $i = 3$, the while loop is not entered any longer, $S = 9$ is returned and the algorithm is terminted.

In general to compute n^2 by this algorithm, n is added n times.

To prove that the algorithm is correct, let us first note that the algorithm stops after a finite numbe of steps. For i increases one by one from 0 and n is a natural number. Thus i eventually becomes equal to n.

Next, to prove that it computes n^2, we show that after going through the loop k times, S = k*n and i = k hold. This statement is called a loop invariant and mathematical induction can be used to prove it.

Proof by induction.

Basis Step: k = 0. When k = 0, that is when the loop is not entered, S = 0 and i = 0. Hence S = k*n and i = k hold.

Induction Hypothesis: For an arbitrary value m of k, S = m * n and i = m hold after going through the loop m times.

Inductive Step: When the loop is entered (m + 1)–st time, S = m*n and i = m at the beginning of the loop. Inside the loop,

$S \leftarrow m*n + n$

$i \leftarrow i + 1$

producing S = (m + 1)*n and i = m + 1.

Thus $S = k*n$ and $i = k$ hold for any natural number k.

Now, when the algorithm stops, $i = n$. Hence the loop will have been entered n times. Thus $S = n*n = n^2$. Hence the algorithm is correct.

The next example is an algorithm to compute the factorial of a positive integer.

FACTORIAL FUNCTION: FAC(N)

$i \leftarrow 1$

$F \leftarrow 1$

while i $\Leftarrow n$

$F \leftarrow F * i$

$i \leftarrow i + 1$

return F

Let us first see how this code computes the factorial of a positive integer. For example let us compute 3 !.

First $i \leftarrow 1$ and $F \leftarrow 1$ give $i = 1$ and $F = 1$ initially.

Since $i < 3$, the while loop is entered.

$F \leftarrow 1 * 1$

$i \leftarrow 1 + 1$

producing F = 1 and i = 2.

Since i < 3, the while loop is entered the second time.

$F \leftarrow 1 * 2$

$i \leftarrow 2 + 1$

producing F = 2 and i = 3.

Since i = 3, the while loop is entered the third time.

F ← 2 * 3

i ← 3 + 1

producing F = 6 and i = 4.

Since i = 4, the while loop is not entered any longer, F = 6 is returned and the algorithm is terminted.

To prove that the algorithm is correct, let us first note that the algorithm stops after a finite numbe of steps. For i increases one by one from 1 and n is a positive integer. Thus i eventually becomes equal to n.

Next, to prove that it computes n !, we show that after going through the loop k times, F = k ! and i = k + 1 hold. This is a loop invariant and again we are going to use mathematical induction to prove it.

Proof by induction.

Basis Step: k = 1. When k = 1, that is when the loop is entered the first time, F = 1 * 1 = 1 and i = 1 + 1 = 2. Since 1! = 1, F = k! and i = k + 1 hold.

Induction Hypothesis: For an arbitrary value m of k, F = m! and i = m + 1 hold after going through the loop m times.

Inductive Step: When the loop is entered (m + 1)–st time, F = m! and i = (m+1) at the beginning of the loop. Inside the loop,

F ← $m!$* (m + 1)

i ← $(m + 1) + 1$

producing F = (m + 1)! and i = (m + 1) + 1.

Thus F = k! and i = k + 1 hold for any positive integer k.

Now, when the algorithm stops, i = n + 1. Hence the loop will have been entered n times. Thus F = n! is returned. Hence the algorithm is correct.

MATHEMATICAL INDUCTION-SECOND PRINCIPLE

There is another form of induction over the natural numbers based on the second principle of induction to prove assertions of the form $\forall x\, P(x)$. This form of induction does not require the basis step, and in the inductive step $P(n)$ is proved assuming $P(k)$ holds for all $k < n$. Certain problems can be proven more easily by using the second principle than the first principle because $P(k)$ for all $k < n$ can be used rather than just $P(n - 1)$ to prove $P(n)$.

Formally the second principle of induction states that if $\forall n\, [\forall k\, [k < n \rightarrow P(k)] \rightarrow P(n)]$, then $\forall n\, P(n)$ can be concluded.

Here $\forall k\, [k < n \rightarrow P(k)]$ is the induction hypothesis.

The reason that this principle holds is going to be explained later after a few examples of proof.

Example: Let us prove the following equality using the second principle:

For any natural number n, $1 + 3 + \ldots + (2n + 1) = (n + 1)^2$.

Proof: Assume that $1 + 3 + \ldots + (2k + 1) = (k + 1)^2$ holds for all k, $k < n$.

Then $1 + 3 + \ldots + (2n + 1) = (1 + 3 + \ldots + (2n - 1)) + (2n + 1)$

$= n^2 + (2n + 1) = (n + 1)^2$ by the induction hypothesis.

Hence by the second principle of induction $1 + 3 + \ldots + (2n + 1) = (n + 1)^2$ holds for all natural numbers.

Example: Prove that for all positive integer n, $\Sigma_{i=1}^{n} i\,(i!) = (n + 1)! - 1$

Proof: Assume that

$1 * 1! + 2 * 2! + \ldots + k * k! = (k + 1)! - 1$ for all k, $k < n$.

Then $1 * 1! + 2 * 2! + \ldots + (n - 1) * (n - 1)! + n * n!$

$= n! - 1 + n * n!$ by the induction hypothesis.

$= (n + 1)n! - 1$

Hence by the second principle of induction $\Sigma_{i=1}^{n}\ i\,(i!) = (n + 1)! - 1$ holds for all positive integers.

Example: Prove that any positive integer n, $n > 1$, can be written as the product of prime numbers.

Proof: Assume that for all positive integers k, $n > k > 1$, k can be written as the product of prime numbers.

We are going to prove that n can be written as the product of prime numbers.

Since n is an integer, it is either a prime number or not a prime number. If n is a prime number, then it is the product of 1, which is a prime number, and itself. Therefore the statement holds true.

If n is not a prime number, then it is a product of two positive integers, say p and q. Since both p and q are smaller than n, by the induction hypothesis they can be written as the product of prime numbers (Note that this is not possible, or at least very hard, if the First Principle is being used). Hence n can also be written as the product of prime numbers.

Chapter 4

Matrix

In mathematics, a matrix (plural matrices) is a rectangular table of *elements* (or *entries*), which may be numbers or, more generally, any abstract quantities that can be added and multiplied. Matrices are used to describe linear equations, keep track of the coefficients of linear transformations and to record data that depend on multiple parameters. Matrices are described by the field of matrix theory. They can be added, multiplied, and decomposed in various ways, which also makes them a key concept in the field of linear algebra.

In this article, the entries of a matrix are real or complex numbers unless otherwise noted.

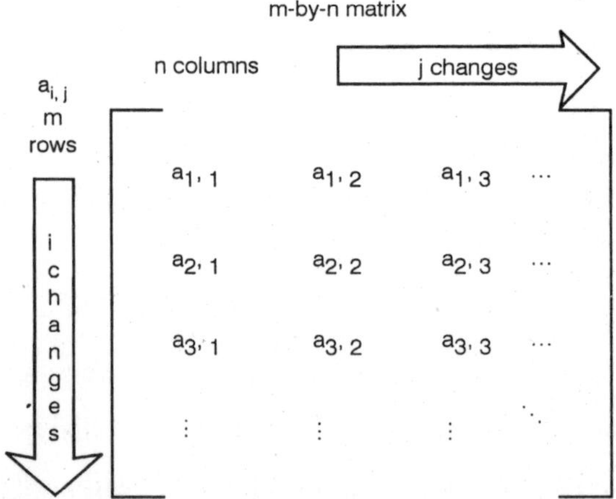

TERMINOLOGY AND NOTATION

The horizontal lines in a matrix are called rows and the vertical lines are called columns. A matrix with *m* rows and *n* columns is called an *m*-by-*n* matrix (written *m* X *n*) and *m* and *n* are called its dimensions. The dimensions of a matrix are always given with the number of rows first, then the number of columns. It is commonly said that an *m*-by-*n* matrix has an order of *m* X *n* ("order" meaning size). Two matrices of the same order whose corresponding entries are equivalent are considered equal.

The entry that lies in the *i*-th row and the *j*-th column of a matrix is typically referred to as the *i,j*, or (*i,j*), or *(i,j)*-th, or $(i,j)^{th}$ entry of the matrix. Again, the row is always noted first, then the column.

Almost always upper-case letters denote matrices, while the corresponding lower-case letters, with two subscript indices, represent the entries. For example, the $(i,j)^{th}$ entry of a matrix A is most commonly written as $a_{i,j}$. Alternative notations for that entry are A[*i,j*] or $A_{i,j}$. In addition to using upper-case letters to symbolize matrices, many authors use a special typographical style, commonly boldface upright (non-italic), to further distinguish matrices from other variables. Following this convention, A is a matrix, distinguished from *A*, a scalar. An alternate convention is to annotate matrices with their dimensions in small type underneath the symbol, for example, $A_{m\times n}$ for an *m*-by-*n* matrix.

We often write $A := (a_{i,j})_{i=1,m;\ j=1,...,n}$ or $A := (a_{i,j})_{m\times n}$ to define an *m* X *n* matrix A. In this case, the entries $a_{i,j}$ are defined separately for all integers $1 \le i \le m$ and $1 \le j \le n$. In some programming languages, the numbering of rows and columns starts at zero. Texts which use any such language extensively frequently follow that convention, so we have $0 \le i \le m\text{-}1$ and $0 \le j \le n-1$.

A matrix where one of the dimensions equals one is often called a *vector*, and interpreted as an element of real coordinate space. An *m* X 1 matrix (one column and *m* rows) is called a column vector and a 1 X *n* matrix (one row and *n* columns) is called a row vector.

MATHEMATICAL DEFINITION

An *m* X *n* ($m, n \in \mathbb{N}$) matrix *A* is a function $A : \{1, 2, ..., m\} \text{ X } \{1, 2, ..., n\} \to S$, where *S* is any non-empty set. ($\{1, 2, ..., m\}$ X $\{1, 2, ..., n\}$ is the Cartesian product of sets $\{1, 2, ..., m\}$ and ($\{1, 2, ..., n\}$.)

We say that matrix *A* is a matrix over the set . Important thing to note is that, if we want to have matrix algebra, the set must be a ring and matrix must be a square matrix (see Square matrices and related definitions below for further explanation). Since the set of all square matrices over a ring is also a ring, matrix algebra is usually called matrix ring.

Since this article mainly considers matrices over real numbers, matrices shown here are actually functions $A : (\{1, 2, ..., m\} \text{ X } \{1, 2, ..., n\} \to \mathbb{R}$.

Example: The matrix

$$A = \begin{bmatrix} 9 & 8 & 6 \\ 1 & 2 & 7 \\ 4 & 9 & 2 \\ 6 & 0 & 5 \end{bmatrix} \text{ or } A = \begin{pmatrix} 9 & 8 & 6 \\ 1 & 2 & 7 \\ 4 & 9 & 2 \\ 6 & 0 & 5 \end{pmatrix}$$

is a 4 X 3 matrix. The element $a_{2,3}$ or A [2, 3] is 7. In terms of the mathematical definition given above, this matrix is a function A: {1, 2, 3, 4} X {1, 2, 3}→ ℝ and, for example, $A((2, 3)) = 7$ and $A((3, 1)) = 4$.

The matrix

$$R= \begin{bmatrix}1 & 2 & 3 & 4 & 5 & 6 & 7 & 8 & 9\end{bmatrix}$$

is a 1 X 9 matrix, or 9-element row vector.

BASIC OPERATIONS

SUM

Two or more matrices of identical dimensions m and n can be added. Given m-by-n matrices A and B, their sum A+B is the m-by-n matrix computed by adding corresponding elements:

$$A + B= (a_{i,j})_{1\leq i\leq m;\, 1\leq j\leq n} + (b_{i,j})_{1\leq i\leq m;\, 1\leq j\leq n}$$

$$= (a_{i,j} + b_{i,j})_{1\leq i\leq m;\, 1\leq j\leq n}\,.$$

For example:

$$\begin{bmatrix}1 & 3 & 1\\ 1 & 0 & 0\\ 1 & 2 & 2\end{bmatrix} + \begin{bmatrix}0 & 0 & 5\\ 7 & 5 & 0\\ 2 & 1 & 1\end{bmatrix} = \begin{bmatrix}1+0 & 3+0 & 1+5\\ 1+7 & 0+5 & 0+0\\ 1+2 & 2+1 & 2+1\end{bmatrix} = \begin{bmatrix}1 & 3 & 6\\ 8 & 5 & 0\\ 3 & 3 & 3\end{bmatrix}.$$

Another, much less often used notion of matrix addition is the direct sum.

SCALAR MULTIPLICATION

Given a matrix A and a number c, the scalar multiplication cA is computed by multiplying every element of A by the scalar c (i.e.) $(cA)i,j = c \cdot a_{i,j}$). For example:

$$2\cdot\begin{bmatrix}1 & 8 & -3\\ 4 & -2 & 5\end{bmatrix} = \begin{bmatrix}2\cdot 1 & 2\cdot 8 & 2\cdot -3\\ 2\cdot 4 & 2\cdot -2 & 2\cdot 5\end{bmatrix} = \begin{bmatrix}2 & 16 & -6\\ 8 & -4 & 10\end{bmatrix}$$

Matrix addition and scalar multiplication turn the set $M(m, n, \mathbb{R})$ of all m-by-n matrices with real entries into a real vector space of dimension $m \cdot n$.

MATRIX MULTIPLICATION

Multiplication of two matrices is well-defined only if the number of columns of the left matrix is the same as the number of rows of the right matrix. If A is an m-by-n matrix and B is an n-by-p matrix, then their matrix product AB is the m-by-p matrix given by:

$$(AB)_{i,j}= a_{i,1}b_{1,j} + a_{i,2}b_{2,j} + \ldots + a_{i,n}b_{n,j}$$

for each pair (i,j). For example:

$$\begin{bmatrix} 1 & 0 & 2 \\ -1 & 3 & 1 \end{bmatrix} \times \begin{bmatrix} 3 & 1 \\ 2 & 1 \\ 1 & 0 \end{bmatrix} = \begin{bmatrix} (1\times 3+0\times 2+2\times 1) & (1\times 1+0\times 1+2\times 0) \\ (-1\times 3+3\times 2+1\times 1) & (-1\times 1\times 3+1+1\times 0) \end{bmatrix} = \begin{bmatrix} 5 & 1 \\ 4 & 2 \end{bmatrix}.$$

Matrix multiplication has the following properties:

1. (AB)C = A(BC) for all k-by-m matrices A, m-by-n matrices B and n-by-p matrices C ("associativity").
2. (A+B)C = AC+BC for all m-by-n matrices A and B and n-by-k matrices C ("right distributivity").
3. C(A+B) = CA+CB for all m-by-n matrices A and B and k-by-m matrices C ("left distributivity").

Matrix multiplication is not commutative; that is, given matrices A and B and their product defined, then generally AB ≠ BA. It may also happen that AB is defined but BA is not defined.

Besides the ordinary matrix multiplication just described, there exist other operations on matrices that can be considered forms of multiplication, such as the Hadamard product and the Kronecker product.

LINEAR TRANSFORMATIONS

Matrices can conveniently represent linear transformations (also known as "linear maps") between finite-dimensional vector spaces. Let R^n be an n-dimensional vector space, and let the vectors in this space be represented in matrix format as column vectors (n-by-1 matrices). For every linear map $f: R^n \to R^m$ there exists a unique m-by-n matrix A such that

$$f(x) = A_x$$

for each vector x in R^n.

We say that the matrix A "represents" the linear map f, or that A is the "transformation matrix" of f.

Matrix multiplication neatly corresponds to the composition of maps. If the k-by-m matrix B represents another linear map $g : R^m \to R^k$, then the composition $g \circ f$ is represented by BA:

$$(g \circ f)(x) = g(f)(x))\ g(Ax) = B(Ax) = (BA)x.$$

This follows from the above-mentioned associativity of matrix multiplication.

More generally, a linear map from an n-dimensional vector space to an m-dimensional vector space is represented by an m-by-n matrix, provided that bases have been chosen for each. This property makes matrices powerful data structures in high-level programming languages.

RANKS

The rank of a matrix A is the dimension of the image of the linear map represented by A; this is the same as the dimension of the space generated by the rows of A, and also the same as the dimension of the space generated by the columns of A. It can also be defined without reference to linear algebra as follows: the rank of an m-by-n matrix A is the smallest number k such that A can be written as a product BC where B is an m-by-k matrix and C is a k-by-n matrix (although this is not a practical way to compute the rank).

TRANSPOSE

The transpose of an m-by-n matrix A is the n-by-m matrix A^T (also sometimes written as A^{tr}, or ${}^tA$, or A') formed by turning rows into columns and columns into rows, i.e., $A^T[i, j] = A[j, i]$ for all indices i and j. If A describes a linear map with respect to two bases, then the matrix A^T describes the transpose of the linear map with respect to the dual bases, see dual space.

We have $(A + B)^T = A^T + B^T$ and $(AB)^T = B^T A^T$.

SQUARE ROOT

Given two bounded matrices T and B, B is a square root of T if T = B*B.

SQUARE MATRICES AND RELATED DEFINITIONS

A square matrix is a matrix which has the same number of rows and columns. The set of all square n-by-n matrices, together with matrix addition and matrix multiplication is a ring. Unless $n = 1$, this ring is not commutative.

M(n, R), the ring of real square matrices, is a real unitary associative algebra. M(n, C), the ring of complex square matrices, is a complex associative algebra.

The unit matrix or identity matrix I_n of size n is the n-by-n matrix in which all the elements on the main diagonal are equal to 1 and all other elements are equal to 0. The identity matrix is named thus because it satisfies the equations $MI_n = M$ and $I_nN = N$ for any m-by-n matrix M and n-by-k matrix N. For example, if $n = 3$:

$$I_3 = \begin{bmatrix} 1 & 0 & 0 \\ 0 & 1 & 0 \\ 0 & 0 & 1 \end{bmatrix}.$$

The identity matrix is the identity element in the ring of square matrices.

Invertible elements in this ring are called invertible matrices or non-singular matrices. An *n*-by-*n* matrix A is invertible if and only if there exists a matrix B such that

$$AB = I_n\ (= BA).$$

In this case, B is the inverse matrix of A, denoted by A^{-1}. The set of all invertible *n*-by-*n* matrices forms a group (specifically a Lie group) under matrix multiplication, the general linear group.

If λ is a number and v is a non-zero vector such that $Av = \lambda v$, then we call v an eigenvector of A and λ the associated eigenvalue. (*Eigen* means "own" in German and in Dutch.) The number λ is an eigenvalue of A if and only if $A-\lambda I_n$ is not invertible, which happens if and only if $p_A(\lambda) = 0$, where $p_A(x)$ is the characteristic polynomial of A. $p_A(x)$ is a polynomial of degree *n* and therefore has *n* complex roots if multiple roots are counted according to their multiplicity. Every square matrix has at most *n* complex eigenvalues.

The determinant of a square matrix A is the product of its *n* eigenvalues, but it can also be defined by the *Leibniz formula*. Invertible matrices are precisely those matrices with a nonzero determinant.

The Gaussian elimination algorithm is of central importance: it can be used to compute determinants, ranks and inverses of matrices and to solve systems of linear equations.

The trace of a square matrix is the sum of its diagonal entries, which equals the sum of its *n* eigenvalues.

The matrix exponential function is defined for square matrices, using power series.

VECTOR AND MATRIX

EUCLIDEAN VECTOR AND EUCLIDEAN SPACE

The *n*–dimensional (real) euclidean space R^n is the collection of *n*–tuples of numbers, called vectors. Vectors can be either presented horizontally,

$$x = (x_1, x_2, \ldots, x_n)$$

or vertically.

$$x = \begin{bmatrix} x_1 \\ x_2 \\ \vdots \\ x_n \end{bmatrix}$$

The only thing you need to keep in mind is that, when vectors are mixed with matrices, it is often more convenient to use vertical vectors.

The number n is the dimension of the vector x. The numbers $x_1, x_2,..., x_n$ are called the coordinates of x. Strictly speaking, these are the coordinates with respect to the standard basis vectors

$$\begin{aligned} e_1 &= (1, 0,..., 0), \\ e_2 &= (0, 1,..., 0), \\ &\vdots \\ e_n &= (0, 0,..., 1). \end{aligned}$$

We also mention the special zero vector

$$0 = (0, 0,..., 0).$$

Example: The 1–dimensional euclidean space

$R^1 = \{(x): x \text{ is a real number}\}$

can be identified with all real numbers R and be visualized as a straight line.

The 2–dimensional euclidean space

$R^2 = \{(x_1, x_2): x_1, x_2 \text{ are real numbers}\}$

can be visualized as the coordinate plane.

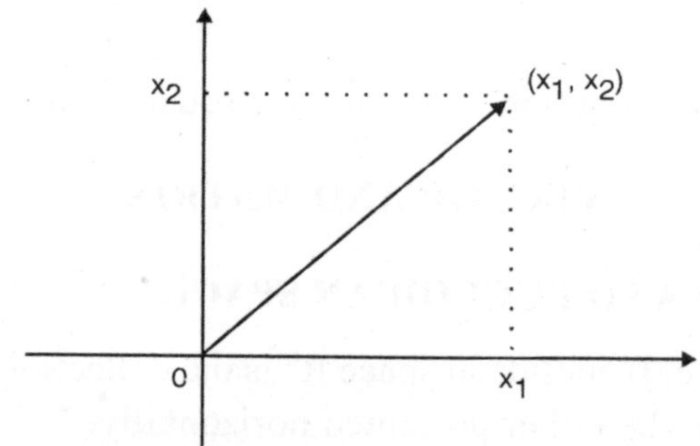

The 3–dimensional euclidean space

$R^3 = \{(x_1, x_2, x_3): x_1, x_2, x_3 \text{ are real numbers}\}$

can be visualized as the world we are living in.

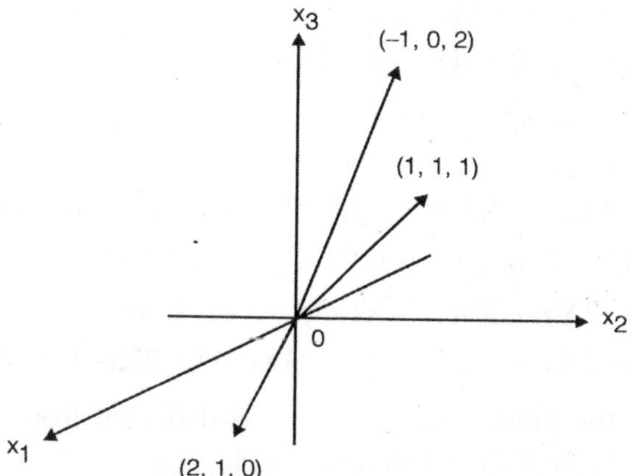

The 0–dimensional euclidean space R^0 consists of a single point – the origin.

The example illustrates that the euclidean vectors can be geometrically presented (in a coordinate system) as an arrow starting from the origin. If an arrow does not start from the origin, we need to parallel shift it so that it begins at the origin.

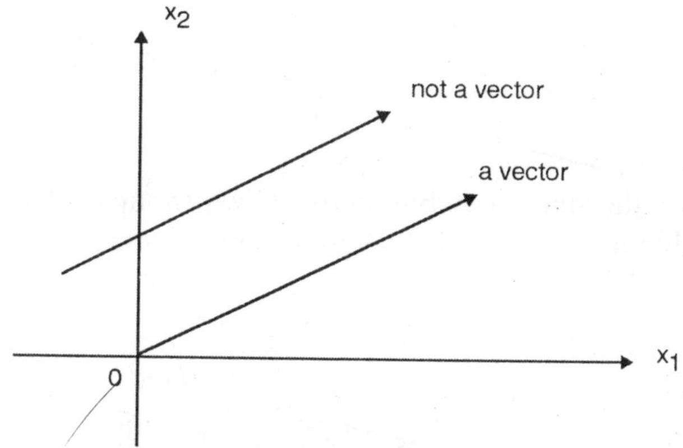

Vectors of the same dimension can be added

$(x_1, x_2,..., x_n) + (y_1, y_2,..., y_n) = (x_1 + y_1, x_2 + y_2,..., x_n + y_n).$

A number (scalar) can also be multiplied to vectors

$c\ (x_1, x_2,..., x_n) = (cx_1, cx_2,..., cx_n).$

The addition and scalar multiplication can also be combined to form linear combinations $c_1x_1 + c_2x_2 +... + c_kx_k$.

Example: Let $u = (1, 2)$, $v = (3, 4)$, $w = (5, 6)$, $x = (1, 0, -3)$, $y = (2, -1, 5)$. Then

$$u + v = (1 + 3, 2 + 4) = (4, 6)$$
$$3u = (3X1, 3X2) = (3, 6)$$
$$u + v + w = (1 + 3 + 5, 2 + 4 + 6) = (9, 12)$$
$$3u - 4v + w = (3X1 - 4X3 + 5, 3X2 - 4X4 + 6) = (-4, -4)$$
$$x + y = (1 + 2, 0 - 1, -3 + 5) = (3, -1, 2)$$
$$-2x = (-2X1, -2X0, -2X(-3)) = (-2, 0, 6)$$
$$-2x + 3y = (-2X1 + 3X2, -2X0 + 3X(-1), -2X(-3) + 3X5) = (4, -3, 21)$$

Moreover, because the dimension of u, v, w is different from the dimension of x, y, expressions such as $u + x$, $2v + 3y$ are meaningless.

Geometrically, the addition is described by parallelogram, and the scalar multiplication is the stretching/shrinking and sometimes reversing (for negative scalars) of vectors.

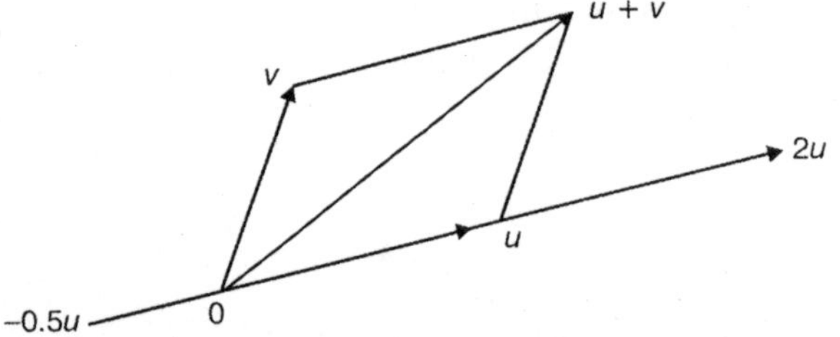

This picture shows the linear combinations of two (nonparallel) vectors. It suggests that all the linear combinations of the two vectors form a plane passing through the origin.

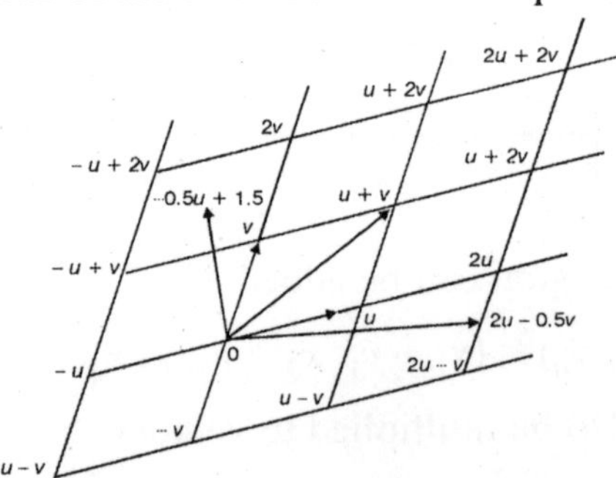

Finally, the operations have the following obvious properties.

commutative

$$u + v = v + u$$

associative

$$(u + v) + w = u + (v + w)$$

zero

$$u + 0 = u = 0 + u$$

distributive

$$r(u + v) = ru + rv$$

distributive

$$(r + s)u = ru + su$$

associative

$$(rs)u = r(su)$$

VECTOR ADDITION IS GIVEN BY PARALLELOGRAM

Let us look at the addition of two vectors (red and blue in the picture) in R^2. The green vector is constructed from the parallelogram.

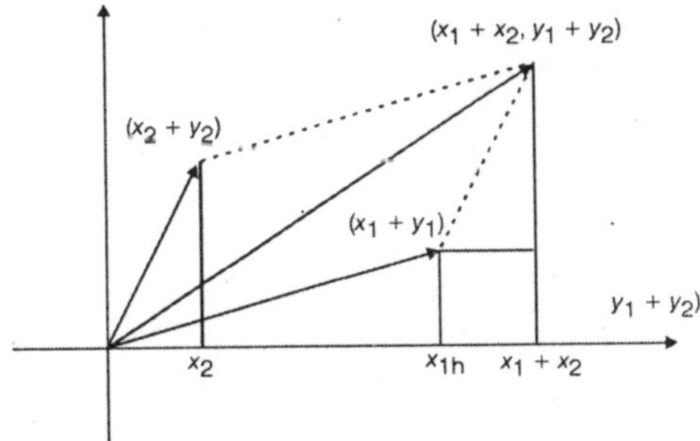

The parallelogram tells us that the two right triangles (in gray shade) are identical. In particular, the two pink segments of the x–axis have the same length. This implies that the x–coordinate of the green vector is $x_1 + x_2$. By similar argument, the y–coordinate of the green vector is $y_1 + y_2$.

MATRIX

Let $M(m, n)$ be the collection of all m by n matrices. The zero matrix is

$$O = \begin{bmatrix} 0 & 0 & \ldots & 0 \\ 0 & 0 & \ldots & 0 \\ \vdots & \vdots & & \vdots \\ 0 & 0 & \ldots & 0 \end{bmatrix}$$

Like vectors, matrices of the same size may be added.

$$\begin{bmatrix} a_{11} & a_{12} & \cdots & a_{1n} \\ a_{21} & a_{22} & \cdots & a_{2n} \\ \vdots & \vdots & & \vdots \\ a_{m1} & a_{m2} & \cdots & a_{mn} \end{bmatrix} + \begin{bmatrix} b_{11} & b_{12} & \cdots & b_{1n} \\ b_{21} & b_{22} & \cdots & b_{2n} \\ \vdots & \vdots & & \vdots \\ b_{m1} & b_{m2} & \cdots & b_{mn} \end{bmatrix}$$

$$= \begin{bmatrix} a_{11}+b_{11} & a_{12}+b_{12} & \cdots & a_{1n}+b_{1n} \\ a_{21}+b_{21} & a_{22}+b_{22} & \cdots & a_{2n}+b_{2n} \\ \vdots & \vdots & & \vdots \\ a_{m1}+b_{m2} & a_{m2}+b_{m2} & \cdots & a_{mn}+b_{mn} \end{bmatrix}$$

A number (scalar) can also be multiplied to matrices

$$c\begin{bmatrix} a_{11} & a_{12} & \cdots & a_{1n} \\ a_{21} & a_{22} & \cdots & a_{2n} \\ \vdots & \vdots & & \vdots \\ a_{m1} & a_{m2} & \cdots & a_{mn} \end{bmatrix} = \begin{bmatrix} ca_{11} & ca_{12} & \cdots & ca_{1n} \\ ca_{21} & ca_{22} & \cdots & ca_{2n} \\ \vdots & \vdots & & \vdots \\ ca_{m1} & ca_{m2} & \cdots & ca_{mn} \end{bmatrix}$$

Moreover, we may combine the two operations and form the linear combinations of matrices (of the same size).

Example: Let

$$A = \begin{bmatrix} 1 & 1 & -2 \\ 3 & 0 & 1 \end{bmatrix}, B = \begin{bmatrix} -1 & 0 & 1 \\ -2 & 1 & 1 \end{bmatrix}, C = \begin{bmatrix} 0 & 1 & 0 \\ 1 & 0 & 0 \end{bmatrix}, D = \begin{bmatrix} 0 & 1 \\ 1 & 2 \end{bmatrix}$$

Then

$$A + B = \begin{bmatrix} 1+(-1) & 1+0 & (-2)+1 \\ 3+(-2) & 0+1 & 1+1 \end{bmatrix} = \begin{bmatrix} 0 & 1 & -1 \\ 1 & 1 & 2 \end{bmatrix}$$

$$A + B + C = \begin{bmatrix} 1+(-1)+0 & 1+0+1 & (-2)+1+0 \\ 3+(-2)+1 & 0+1+0 & 1+1+0 \end{bmatrix} = \begin{bmatrix} 0 & 2 & -1 \\ 2 & 1 & 2 \end{bmatrix}$$

$$2A = \begin{bmatrix} 2\times 1 & 2\times 1 & 2\times(-2) \\ 2\times 3 & 2\times 0 & 2\times 1 \end{bmatrix} = \begin{bmatrix} 2 & 2 & -4 \\ 6 & 0 & 2 \end{bmatrix}$$

$$-3D = \begin{bmatrix} (-3)\times 0 & (-3)\times 1 \\ (-3)\times 1 & (-3)\times 2 \end{bmatrix} = \begin{bmatrix} 0 & -3 \\ -3 & -6 \end{bmatrix}$$

$$3A - 4B + 2C = \begin{bmatrix} 3+4+0 & 3+0+2 & -6-4+0 \\ 9+8+2 & 0-4+0 & 3-4+0 \end{bmatrix} = \begin{bmatrix} 7 & 5 & -10 \\ 19 & -4 & -1 \end{bmatrix}$$

On the other hand, expressions such as $A + D$, $B + D$, $A + D$ are meaningless.

Euclidean vectors can be considered as matrices of special sizes. Specifically, a vertical n–dimensional euclidean vector is an n by 1 matrix, and the euclidean space R^n is identified with $M(n, 1)$. On the other hand, if we present vectors horizontally, then R^n is identified with $M(1, n)$.

Although euclidean vectors may be presented either vertically or horizontally, you are strongly advised to use vertical vectors when they are mixed with matrices. For example, in expressing a system of linear equation as $Ax = b$, because of the presence of the matrix A, we should consider x and b as vertical vectors.

Conversely, matrices can be considered as an (ordered) collection of vectors. Specifically, an m by n matrix is equivalent to n euclidean column vectors of dimension m.

$$A = [a_1\ a_2 \ldots\ a_n] \Leftrightarrow a_1, a_2, \ldots, a_n \in R^n.$$

Similarly, we may also consider row vectors, so that an m by n matrix is equivalent to m euclidean row vectors of dimension n.

Example: The matrix

$$A = \begin{bmatrix} 1 & 2 & 3 \\ 4 & 5 & 6 \end{bmatrix}$$

can be considered as three 2–dimensional column vectors

$$A = [a_1\ a_2\ a_3],\ a_1 = \begin{bmatrix} 1 \\ 4 \end{bmatrix},\ a_2 = \begin{bmatrix} 2 \\ 5 \end{bmatrix},\ a_3 = \begin{bmatrix} 3 \\ 6 \end{bmatrix}$$

It can also be considered as two 3–dimensional row vectors.

$$A = \begin{bmatrix} b_1 \\ b_2 \end{bmatrix},\ b_1 = (1, 2, 3),\ b_2 = (4, 5, 6)$$

The relations between the vectors and the matrices are compatible with addition and scalar multiplication. For example, for three column matrices $A = [a_1\ a_2\ a_3]$, $B = [b_1\ b_2\ b_3]$, we have

$$A + B = [a_1 + b_1\ a_2 + b_2\ a_3 + b_3],\ cA = [ca_1\ ca_2\ ca_3].$$

Because of this compatibility, the properties for the addition and scalar multiplication of vectors are also hold for matrices.

SOME MATRIX NOTATIONS AND TERMINOLOGIES

In theoretical discussions about matrices, we sometimes use the notation $A = (a_{ij})$ to denote a matrix, where a_{ij} is the entry at the i–th row and j–th column (the ij–entry). Sometimes we even indicate the size by writing $A = (a_{ij})_{mXn}$. In terms of this compact notation, we have

$$(a_{ij}) + (b_{ij}) = (a_{ij} + b_{ij}),\ c(a_{ij}) = (ca_{ij}).$$

Example:

$$(i+j)_{3\times 3} = \begin{bmatrix} 1+1 & 1+2 & 1+3 \\ 2+1 & 2+2 & 2+3 \\ 3+1 & 3+2 & 3+3 \end{bmatrix} = \begin{bmatrix} 2 & 3 & 4 \\ 3 & 4 & 5 \\ 4 & 5 & 6 \end{bmatrix}$$

$$(ij)_{4\times 5} = \begin{bmatrix} 1\times 1 & 1\times 2 & 1\times 3 & 1\times 4 & 1\times 5 \\ 2\times 1 & 2\times 2 & 2\times 3 & 2\times 4 & 2\times 5 \\ 3\times 1 & 3\times 2 & 3\times 3 & 3\times 4 & 3\times 5 \\ 4\times 1 & 4\times 2 & 4\times 3 & 4\times 4 & 4\times 5 \end{bmatrix} = \begin{bmatrix} 1 & 2 & 3 & 4 & 5 \\ 2 & 4 & 6 & 8 & 10 \\ 3 & 6 & 9 & 12 & 15 \\ 4 & 8 & 12 & 16 & 20 \end{bmatrix}$$

$$(j^2)_{4\times 3} = \begin{bmatrix} 1^2 & 2^2 & 3^2 \\ 1^2 & 2^2 & 3^2 \\ 1^2 & 2^2 & 3^2 \\ 1^2 & 2^2 & 3^2 \end{bmatrix} = \begin{bmatrix} 1 & 4 & 9 \\ 1 & 4 & 9 \\ 1 & 4 & 9 \\ 1 & 4 & 9 \end{bmatrix}$$

For an n by n square matrix $A = (a_{ij})$, the terms $a_{11}, a_{22}, \ldots, a_{nn}$ are the diagonal entries of the matrix. A is a diagonal matrix if all the off–diagonal entries are zero. In other words, $a_{ij} = 0$ for $i \neq j$. The following are all the 2 by 2 and 3 by 3 diagonal matrices.

$$\begin{bmatrix} a & 0 \\ 0 & b \end{bmatrix}$$

$$\begin{bmatrix} a & 0 & 0 \\ 0 & b & 0 \\ 0 & 0 & c \end{bmatrix}$$

A is an upper triangular matrix if all the entries below the diagonal are zero. In other words, $a_{ij} = 0$ for $i > j$. The following are all the 2 by 2 and 3 by 3 upper triangular matrices.

$$\begin{bmatrix} a & b \\ 0 & c \end{bmatrix}$$

$$\begin{bmatrix} a & b & c \\ 0 & d & e \\ 0 & 0 & f \end{bmatrix}$$

Similarly, A is a lower triangular matrix if all the entries above the diagonal are zero. In other words, $a_{ij} = 0$ for $i < j$. The following are all the 2 by 2 and 3 by 3 upper triangular matrices.

$$\begin{bmatrix} a & 0 \\ b & c \end{bmatrix}$$

$$\begin{bmatrix} a & 0 & 0 \\ b & c & 0 \\ d & e & f \end{bmatrix}$$

Clearly, A is upper and lower triangular at the same time if and only if it is diagonal.

Exercise: Draw the vectors (1, 3), (0, 1), (0, 0), (–1, –3), (2, –2).

Answer:

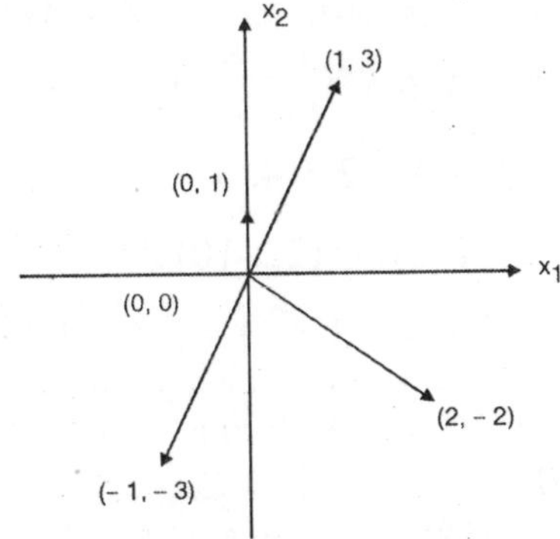

Exercise: Draw the collections of vectors.

$X = \{(0, y): y \text{ is any number}\}$

$Y = \{(x, 0): x \text{ is any number}\}$

$S = \{(x, y): x^2 + y^2 = 1\}$

$D = \{(x, x): x \text{ is any number}\}$

$E = \{(x, y): x + 3y = 0\}$

Answer:

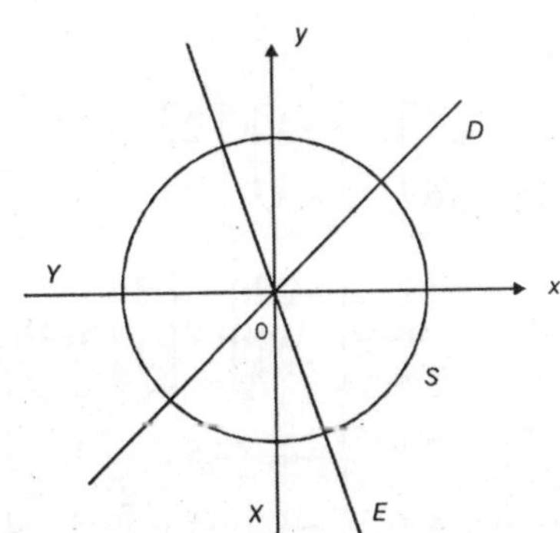

Exercise: Let $u = (1,-5)$, $v = (2, 4,1)$, $w = (3, 0, 1, -1, 4)$, $x = (2, -4, 3)$, $y = (-7, 2, 1)$, $z = (1, 2, 0, -3, 4)$. Compute.

$$u + v,\ v + y,\ y + v,\ w + z,\ z + w,\ x + y + v,$$

$$2u,\ -3v,\ -w,\ 0x,$$

$$2u + 5v,\ -2y + 3v,\ w - 3z,\ 4x + 2y + v.$$

Answer: $u + v$ is meaningless

$$v + y = (2 - 7, 4 + 2, 1 + 1) = (-5, 6, 2)$$

$$y + v = (-7 + 2, 2 + 4, 1 + 1) = (-5, 6, 2)$$

$$w + z = (3 + 1, 0 + 2, 1 + 0, -1 - 3, 4 + 4) = (4, 2, 1, -4, 8)$$

$$z + w = (1 + 3, 2 + 0, 0 + 1, -3 - 1, 4 + 4) = (4, 2, 1, -4, 8)$$

$$x + y + v = (2 - 7 + 2, -4 + 2 + 4, 3 + 1 + 1) = (-3, 2, 5)$$

$$2u = (2 \times 1, 2 \times (-5)) = (2, -10)$$

$$-3v = (-3 \times 2, -3 \times 4, -3 \times 1) = (-6, -12, -3)$$

$$-w = -(3, 0, 1, -1, 4) = (-3, 0, -1, 1, -4)$$

$$0x = (0, 0, 0)$$

$2u + 5v$ is meaningless

$$-2y + 3v = -2(-7, 2, 1) + 3(2, 4,1) = (-2 \times (-7) + 3 \times 2, -2 \times 2 + 3 \times 4, -2 \times 1 + 3X1) = (20, 8, 1)$$

$$w - 3z = (3, 0, 1, -1, 4) - 3(1, 2, 0, -3, 4) = (3 - 3 \times 1, 0 - 3 \times 2, 1 - 3 \times 0, -1 - 3 \times (-3), 4 - 3 \times 4) = (0, -6, 1, 8, -8)$$

$$4x + 2y + v = 4(2, -4, 3) + 2(-7, 2, 1) + (2, 4,1)$$
$$= (4 \times 2 + 2 \times (-7) + 2, 4 \times (-4) + 2 \times 2 + 4, 4 \times 3 + 2 \times 1 + 1)$$
$$= (-4, 8, 15)$$

Exercise: Compute.

Answer:

$$\begin{bmatrix}1\\3\end{bmatrix} - \begin{bmatrix}5\\6\end{bmatrix} = \begin{bmatrix}-4\\-3\end{bmatrix}, \begin{bmatrix}2\\4\end{bmatrix} - \begin{bmatrix}7\\8\end{bmatrix} = \begin{bmatrix}-5\\-4\end{bmatrix}, 3\begin{bmatrix}1\\3\end{bmatrix} - 5\begin{bmatrix}5\\6\end{bmatrix}$$

$$= \begin{bmatrix}-22\\-21\end{bmatrix}, 3\begin{bmatrix}2\\4\end{bmatrix} - 5\begin{bmatrix}7\\8\end{bmatrix} = \begin{bmatrix}-29\\-28\end{bmatrix}$$

$$\begin{bmatrix}1 & 2\\3 & 4\end{bmatrix} - \begin{bmatrix}5 & 7\\6 & 8\end{bmatrix} = \begin{bmatrix}-4 & -5\\-3 & -4\end{bmatrix}, 3\begin{bmatrix}1 & 2\\3 & 4\end{bmatrix} - 5\begin{bmatrix}5 & 7\\6 & 8\end{bmatrix}$$

$$\begin{bmatrix} 0 & -2 & 3 \\ 1 & 0 & -3 \\ 2 & -2 & 2 \end{bmatrix} + \begin{bmatrix} 2 & 0 & -1 \\ 0 & 4 & -3 \\ 3 & -3 & 0 \end{bmatrix} = \begin{bmatrix} 2 & -2 & 2 \\ 1 & 4 & -6 \\ 5 & -5 & 2 \end{bmatrix}, -2\begin{bmatrix} 0 & -2 & 3 \\ 1 & 0 & -3 \\ 2 & -2 & 2 \end{bmatrix}$$

$$+3\begin{bmatrix} 2 & 0 & -1 \\ 0 & 4 & -3 \\ 3 & -3 & 0 \end{bmatrix} = \begin{bmatrix} 4 & 4 & -9 \\ -2 & 12 & -15 \\ 5 & -5 & -4 \end{bmatrix}$$

$$a\begin{bmatrix} 1 & 0 & 0 \\ 0 & 0 & 0 \\ 0 & 0 & 0 \end{bmatrix} + b\begin{bmatrix} 0 & 0 & 0 \\ 0 & 1 & 0 \\ 0 & 0 & 0 \end{bmatrix} + c\begin{bmatrix} 0 & 0 & 0 \\ 0 & 0 & 0 \\ 0 & 0 & 1 \end{bmatrix} = \begin{bmatrix} a & 0 & 0 \\ 0 & b & 0 \\ 0 & 0 & c \end{bmatrix}$$

$$a_{11}\begin{bmatrix} 1 & 0 \\ 0 & 0 \end{bmatrix} + a_{12}\begin{bmatrix} 0 & 1 \\ 0 & 0 \end{bmatrix} + a_{21}\begin{bmatrix} 0 & 0 \\ 1 & 0 \end{bmatrix} + a_{22}\begin{bmatrix} 0 & 0 \\ 0 & 1 \end{bmatrix} = \begin{bmatrix} a_{11} & a_{12} \\ a_{21} & a_{22} \end{bmatrix}$$

$$\begin{bmatrix} 1 & -3 & 5 & 10 \\ 0 & 5 & 1 & -7 \\ 1 & 0 & 9 & -3 \\ 2 & 0 & 1 & 0 \\ -6 & 3 & 0 & 0 \end{bmatrix} + \begin{bmatrix} 4 & -9 & 8 \\ 4 & 0 & 3 \\ 1 & 4 & 0 \\ 0 & 0 & 4 \\ 0 & 5 & 3 \end{bmatrix} = ?, 12\begin{bmatrix} 0 & 0 & 0 & 0 & 0 \\ 0 & 0 & 0 & 0 & 0 \\ 0 & 0 & 0 & 0 & 0 \end{bmatrix} - 13\begin{bmatrix} 0 & 0 & 0 & 0 & 0 \\ 0 & 0 & 0 & 0 & 0 \\ 0 & 0 & 0 & 0 & 0 \\ 0 & 0 & 0 & 0 & 0 \\ 0 & 0 & 0 & 0 & 0 \end{bmatrix}$$

The last two are meaningless because the sizes do not match.

Exercise: Take the standard basis vectors $e_1 = (1, 0, 0)$, $e_2 = (0, 1, 0)$, $e_3 = (0, 0, 1)$ of R^3. Compute the linear combination $x_1e_1 + x_2e_2 + x_3e_3$. Then generalize your result to other dimensions.

Answer: $x_1e_1 + x_2e_2 + x_3e_3 = x_1(1, 0, 0) + x_2(0, 1, 0) + x_3(0, 0, 1) = (x_1, 0, 0) + (0, x_2, 0) + (0, 0, x_3) = (x_1, x_2, x_3)$. In general, for the standard basis $e_1 = (1, 0,..., 0)$, $e_2 = (0, 1,..., 0),..., e_n = (0, 0,..., 1)$ of R^n, we have $(x_1, x_2,..., x_n) = x_1e_1 + x_2e_2 + ... + x_ne_n$.

Exercise: For $u = (1, 2)$, $v = (3, 4)$, $w = (5, 6)$, compute the linear combination $x_1u + x_2v + x_3w$. Do the same for $a_1 = (1, 2, 3)$, $a_2 = (4, 5, 6)$, $a_3 = (7, 8, 9)$. Express your computation in vertical vectors and draw a general observation.

Answer:

$$x_1\begin{bmatrix} 1 \\ 2 \end{bmatrix} + x_2\begin{bmatrix} 3 \\ 4 \end{bmatrix} + x_3\begin{bmatrix} 5 \\ 6 \end{bmatrix} = \begin{bmatrix} x_1 \\ 2x_1 \end{bmatrix} + \begin{bmatrix} 3x_2 \\ 4x_2 \end{bmatrix} + \begin{bmatrix} 5x_3 \\ 6x_3 \end{bmatrix} = \begin{bmatrix} x_1 + 3x_2 + 5x_3 \\ 2x_1 + 4x_2 + 6x_3 \end{bmatrix}$$

$$x_1\begin{bmatrix} 1 \\ 2 \\ 3 \end{bmatrix} + x_2\begin{bmatrix} 4 \\ 5 \\ 6 \end{bmatrix} + x_3\begin{bmatrix} 7 \\ 8 \\ 9 \end{bmatrix} = \begin{bmatrix} x_1 \\ 2x_1 \\ 3x_1 \end{bmatrix} + \begin{bmatrix} 4x_2 \\ 5x_2 \\ 6x_2 \end{bmatrix} + \begin{bmatrix} 7x_3 \\ 8x_3 \\ 9x_3 \end{bmatrix} = \begin{bmatrix} x_1 + 4x_2 + 7x_3 \\ 2x_1 + 5x_2 + 8x_3 \\ 3x_1 + 6x_2 + 9x_3 \end{bmatrix}$$

The results are of the form Ax, with the matrix A being respectively

$$\begin{bmatrix} 1 & 3 & 5 \\ 2 & 4 & 6 \end{bmatrix} = [u\ v\ w]$$

$$\begin{bmatrix} 1 & 4 & 7 \\ 2 & 5 & 8 \\ 3 & 6 & 9 \end{bmatrix} = [a_1\ a_2\ a_3]$$

In general, we have

$$x_1a_1 + x_2a_2 + \ldots + x_na_n = Ax$$

where $A = [a_1\ a_2 \ldots a_n]$ is a matrix with the vectors $a_1, a_2, \ldots, a_n$ as columns.

Exercise: Express the vectors $b = (4, 6)$, $c = (1, 1)$, $e_1 = (1, 0)$, $e_2 = (0, 1)$ as linear combinations of the vectors $u = (1, 2)$, $v = (3, 4)$, $w = (5, 6)$. Can you express as linear combinations of the vectors u, v only?

Answer: We would like to write b, c, e_1, e_2 in the form $x_1(1, 2) + x_2(3, 4) + x_3(5, 6)$. From the previous exercise, we have

$$x_1\begin{bmatrix} 1 \\ 2 \end{bmatrix} + x_2\begin{bmatrix} 3 \\ 4 \end{bmatrix} + x_3\begin{bmatrix} 5 \\ 6 \end{bmatrix} = \begin{bmatrix} x_1 + 3x_2 + 5x_3 \\ 2x_1 + 4x_2 + 6x_3 \end{bmatrix}$$

Therefore the problem is to solve the system of equations

$$x_1 + 3x_2 + 5x_3 = b_1$$
$$2x_1 + 4x_2 + 6x_3 = b_2$$

For $b_1 = 4$, $b_2 = 6$ on the right side, we find $x_1 = 1$, $x_2 = 1$ is one solution. Therefore $b = u + v$. In fact, the solution is not unique. For example, $x_1 = 2$, $x_2 = -1$, $x_3 = 1$ is another solution, from which we get $b = 2u - v + w$.

Incidentally, we have already expressed b as a linear combination of u and v: $b = u + v$. Such expressions are equivalent to solutions of

$$x_1 + 3x_2 = b_1$$
$$2x_1 + 4x_2 = b_2$$

We do the similar computation for the other vectors and find $c = -1/2\ u + 1/2\ v$, $e_1 = -2u + v$, $e_2 = 3/2\ u - 1/2\ v$.

Exercise: Express the vectors $b = (1, 1, 1)$, $c = (-1, 0, 1)$, $e_1 = (1, 0, 0)$ as linear combinations of the vectors $a_1 = (1, 2, 3)$, $a_2 = (4, 5, 6)$, $a_3 = (7, 8, 9)$. Is the expression (i.e., the coefficients) unique?

Answer: We would like to write b, c, e_1 in the form $x_1(1, 2, 3) + x_2(4, 5, 6) + x_3$ $(7, 8, 9)$. The computation is similar to the previous exercise. From this exercise, we have

$$x_1\begin{bmatrix}1\\2\\3\end{bmatrix} + x_2\begin{bmatrix}4\\5\\6\end{bmatrix} + x_3\begin{bmatrix}7\\8\\9\end{bmatrix} = \begin{bmatrix}x_1 + 4x_2 + 7x_3\\2x_1 + 5x_2 + 8x_3\\3x_1 + 6x_2 + 9x_3\end{bmatrix}$$

Therefore the problem is to solve the system.

$$x_1 + 4x_2 + 7x_3 = b_1$$
$$2x_1 + 5x_2 + 8x_3 = b_2$$
$$3x_1 + 6x_2 + 9x_3 = b_3$$

For $b_1 = b_2 = b_3 = 1$ on the right side, we find $x_1 = -1/3$, $x_2 = 1/3$, $x_3 = 0$ is one solution. Therefore $b = -1/3\ a_1 + 1/3\ a_2$. Since the solution is not unique, the expression is not unique. Another possibility is $b = -1/6\ a_1 + 1/6\ a_3$.

For $b_1 = -1$, $b_2 = 0$, $b_3 = 1$ on the right side, we find $x_1 = 4/3$, $x_2 = 0$, $x_3 = -1/3$ is one solution. Therefore $c = 4/3\ a_1 - 1/3\ a_3$. Again the expression is not unique.

For $b_1 = 1$, $b_2 = 0$, $b_3 = 0$ on the right side, there is no solution. Therefore e_1 is not a linear combination of a_1, a_2, a_3.

Exercise: Express the matrix

$$X = \begin{bmatrix}3 & 4\\2 & 3\end{bmatrix}$$

as linear combinations of the matrices.

$$A = \begin{bmatrix}1 & 2\\0 & 1\end{bmatrix},\ B = \begin{bmatrix}1 & 0\\2 & 1\end{bmatrix},\ C = \begin{bmatrix}0 & 1\\-1 & 0\end{bmatrix}$$

Is the expression (i.e., the coefficients) unique? Is it true that any 2 by 2 matrix is a linear combination of A, B and C?

Answer: By comparing the four entries on the both side of the equation $x_1A + x_2B + x_3C = X$,

$$x_1\begin{bmatrix}1 & 2\\0 & 1\end{bmatrix} + x_2\begin{bmatrix}1 & 0\\2 & 1\end{bmatrix} + x_3\begin{bmatrix}0 & 1\\-1 & 0\end{bmatrix} = \begin{bmatrix}3 & 4\\2 & 3\end{bmatrix}$$

we have the following system of equations.

$$x_1 + x_2 = 3$$
$$2x_1 + x_3 = 4$$

$$2x_2 - x_3 = 2$$
$$x_1 + x_2 = 3$$

From the row echelon form of the augmented matrix,

$$\begin{bmatrix} 1 & 1 & 0 & 3 \\ 0 & 2 & -1 & 2 \\ 0 & 0 & 0 & 0 \\ 0 & 0 & 0 & 0 \end{bmatrix}$$

we see that the system has non–unique solutions. One solution is $x_1 = 2$, $x_2 = 1$, $x_3 = 0$. Therefore $X = 2A + 1B + 0C$. Moreover, the expression is not unique.

For a general 2 by 2 matrix $X = (b_{ij})$ (i.e., b_{ij} is the entry of X at the i–th row and j–th column), the corresponding system is

$$x_1 + x_2 = b_{11}$$
$$2x_1 + x_3 = b_{12}$$
$$2x_2 - x_3 = b_{21}$$
$$x_1 + x_2 = b_{22}$$

Then from the row echelon form above (or by the fact that there are more equations than variables), we see that solutions do not always exist. In other words, not any X can be expressed as a linear combination of A, B and C.

Remark: In an earlier exercise, we have seen how the problem of expressing a vector as a linear combination of some other vectors is equivalent to solving a system of linear equations. This exercise shows that we have similar relation for matrices.

LINEAR TRANSFORMATION

TRANSFORMATION

Transformations are rules that convert one data into another data. We may denote a transformation by

$T: X \to Y$,

where T is the transformation (also called map, function), X is the source set (or domain), and Y is the target set (or codomain). Taking any element $x \in X$ as input, the transformation produces an element $y = T(x) \in Y$ as output. We also call y the image (or value) of x under the transformation T. We may also include more specific information about the transformation by writing $T: X \to Y$, $x \to$

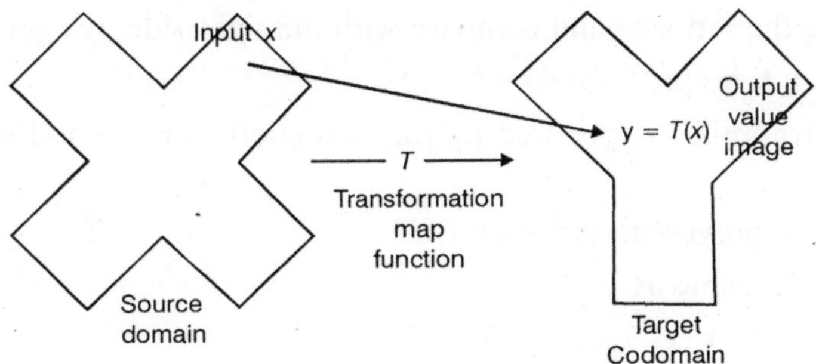

A more rigorous theory of transformations can be found here.

Example: The following are some transformations in everyday life.

Age: people → number

ID_s: student → number

ID_p: professor → number

Instructor: course → professor

Maker: product → manufacturer

Capital City: country → city

Population: city → number

Example In mathematics, transformations are often given by formulae.

Square: $R \to R$, $x \to x^2$

Reflection in x–axis: $R^2 \to R^2$, $(x, y) \to (x, -y)$

Polar–Cartesian: $R^2 \to R^2$, $(r, \theta) \to (r\cos\theta, r\sin\theta)$

Root–Polynomial: $R^2 \to R^2$, $(x_1, x_2) \to (x_1 + x_2, x_1x_2)$

T: $R^3 \to R^2$, $(x_1, x_2, x_3) \to (x_1 + 2x_2 + 3x_3, 4x_1 + 5x_2 + 6x_3)$

Here are two examples of mathematical transformations given by descriptions instead of formulae.

Sign: R → R, positive number → 1; negative number → –1; 0 → 0

Dirichlet function: R → R, rational number → 1; irrational number → 0

ROOT-POLYNOMIAL TRANSFORMATION

If x_1, x_2 are the roots of a quadratic polynomial $x^2 + ax + b$, then we have

$(x - x_1)(x - x_2) = x^2 + ax + b$.

By expanding the left side and compare with the right side, we get

$a = -(x_1 + x_2),\ b = x_1 x_2.$

Since the polynomial is specified by the two coefficients a and b, we see that the transformation

roots $(x_1, x_2) \rightarrow$ polynomial $x^2 + ax + b$

is essentially the same as

$R^2 \rightarrow R^2,\ (x_1, x_2) \rightarrow (x_1 + x_2, x_1 x_2)$

As an exercise for yourself, how would you define the transformation *Root–Polynomial*: $R^3 \rightarrow R^3$ for three roots of polynomials of degree 3?

Finally, you may try the general case *Root–Polynomial*: $R^n \rightarrow R^n$.

MATRIX TRANSFORMATION

We may think of the left side of a system of linear equations as a transformation. For example, the left side of the system

$$x_1 + 2x_2 + 3x_2 = b_1$$
$$4x_1 + 5x_2 + 6x_4 = b_2$$

can be considered as the linear transformation T in the example above. By expressing vectors in vertical form, we also see that

$$T\begin{bmatrix} x_1 \\ x_2 \\ x_3 \end{bmatrix} = \begin{bmatrix} x_1 + 2x_2 + 3x_3 \\ 4x_1 + 5x_2 + 6x_3 \end{bmatrix} = \begin{bmatrix} 1 & 2 & 3 \\ 4 & 5 & 6 \end{bmatrix}\begin{bmatrix} x_1 \\ x_2 \\ x_3 \end{bmatrix}$$

In particular, the formula for the transformation T is given by a matrix A:

$$T(x) = Ax.$$

In general, for any m by n matrix A, we may construct a transformation T: $R^n \rightarrow R^m$ by the above formula. Such transformations are called matrix transformations.

Example: The following matrices

$$\begin{bmatrix} 1 & -1 & 1 \\ 3 & 1 & -1 \\ 2 & 2 & 1 \end{bmatrix}$$

$$\begin{bmatrix} 1 & 3 & 2 & 0 & 1 \\ -1 & -1 & -1 & 1 & 0 \\ 0 & 4 & 2 & 4 & 3 \\ 1 & 3 & 2 & -2 & 0 \end{bmatrix}$$

give us matrix transformations,

$$T_1\begin{bmatrix} x_1 \\ x_2 \\ x_3 \end{bmatrix} = \begin{bmatrix} x_1 - x_2 + x_3 \\ 3x_1 + x_2 - x_3 \\ 2x_1 + x_2 + x_3 \end{bmatrix} : R^3 \to R^3$$

$$T_1\begin{bmatrix} x_1 \\ x_2 \\ x_3 \\ x_4 \\ x_5 \\ x_6 \end{bmatrix} = \begin{bmatrix} x_1 + 3x_2 + 2x_3 + x_5 \\ -x_1 - x_2 - x_3 + x_4 + x_6 \\ 4x_2 + 2x_3 + 4x_4 + 3x_5 + 3x_6 \\ x_1 + 3x_2 + 2x_3 - 2x_4 \end{bmatrix} : R^6 \to R^4$$

or written horizontally,

T_1: $R^3 \to R^3$, $(x_1, x_2, x_3) \to (x_1 - x_2 + x_3, 3x_1 + x_2 - x_3, 2x_1 + 2x_2 + x_3)$

T_2: $R^6 \to R^4$, $(x_1, x_2, x_3, x_4, x_5) \to (x_1 + 3x_2 + 2x_3 + x_5, -x_1 - x_2 - x_3 + x_4 + x_6, 4x_2 + 2x_3 + 4x_4 + 3x_5 + 3x_6, x_1 + 3x_2 + 2x_3 - 2x_4)$

Example The transformations

S: $R^3 \to R^3$, $(x_1, x_2, x_3) \to (x_1 - x_2 + x_3, 3x_1 + x_2 - x_3 - 2, 2x_1 + 2x_2 + x_3)$

T: $R^2 \to R^2$, $(x, y) \to (xy, x^2)$

are not matrix transformations because they cannot be expressed as Ax for a *constant* matrix A.

LINEAR TRANSFORMATION

The following example should convince you that the expression Ax is linear:

$A(x + y) = Ax + Ay$, $A(cx) = cAx$.

A more general proof can be found here.

Example: Let

$$A = \begin{bmatrix} a & b \\ c & d \end{bmatrix}$$

The corresponding linear transformation is

$$T\begin{bmatrix} x_1 \\ x_2 \end{bmatrix} = \begin{bmatrix} ax_1 + bx_2 \\ cx_1 + dx_2 \end{bmatrix} : R^2 \to R^2$$

We have

$$T\begin{bmatrix} x_1 + y_1 \\ x_2 + y_2 \end{bmatrix} = \begin{bmatrix} a(x_1 + y_1) + b(x_2 + y_2) \\ c(x_1 + y_1) + d(x_2 + y_2) \end{bmatrix} = \begin{bmatrix} ax_1 + bx_2 \\ cx_1 + dx_2 \end{bmatrix} + \begin{bmatrix} ay_1 + by_2 \\ cy_1 + dy_2 \end{bmatrix}$$

$$= T\begin{bmatrix} x_1 \\ x_2 \end{bmatrix} + T\begin{bmatrix} y_1 \\ y_2 \end{bmatrix}$$

and (we use k instead of c as the scalar to avoid confusion in notations)

$$T\begin{bmatrix} kx_1 \\ kx_2 \end{bmatrix} = \begin{bmatrix} akx_1 + bkx_2 \\ ckx_2 + dkx_2 \end{bmatrix} = \begin{bmatrix} k(ax_1 + bkx_2) \\ k(cx_1 + dx_2) \end{bmatrix} = k\begin{bmatrix} ay_1 + by_2 \\ cy_1 + dy_2 \end{bmatrix} kT\begin{bmatrix} x_1 \\ x_2 \end{bmatrix}$$

You are strongly encouraged to verify the linearity for matrices of other sizes.

The discussion suggests us to introduce the following concept.

A transformation $T: R^n \to R^m$ is a linear transformation if it satisfies

$$T(u + v) = T(u) + T(v),\ T(cu) = cT(u).$$

By taking $c = 0$ in the definition, we have

$$T(0) = 0.$$

Moreover, by combining the two properties in the definition, we see that linear transformations preserve linear combinations.

$$T(c_1u_1 + c_2u_2 + \dots + c_ku_k) = c_1T(u_1) + c_2T(u_2) + \dots + c_kT(u_k).$$

Example Suppose $T: R^2 \to R^2$ is a linear transformation. Suppose we also know (recall $e_1 = (1, 0)$, $e_2 = (0, 1)$)

$$T(e_1) = (1, 2),\ T(e_2) = (3, 4)$$

Then by

$$(5, -3) = 5\ (1, 0) - 3(0,1) = 5e_1 - 3e_2,$$

$$(-2, 4) = -2e_1 + 4e_2,$$

we have

$$T(5, -3) = 5T(e_1) - 3T(e_2) = 5(1, 2) - 3(3, 4) = (-4, -2),$$

$$T(-2, 4) = -2T(e_1) + 4T(e_2) = -2(1, 2) + 4(3, 4) = (10, 12).$$

In general, by

$$(x_1, x_2) = x_1e_1 + x_2e_2,$$

we have

$$T(x_1, x_2) = x_1T(e_1) + x_2T(e_2) = x_1(1, 2) + x_2(3, 4) = (x_1 + 3x_2, 2x_1 + 4x_2).$$

We remark that T is a matrix transformation given by

$$\begin{bmatrix} 1 & 3 \\ 2 & 4 \end{bmatrix} = [T(e_1)\ T(e_2)]$$

Example: The transformation

$$S(x_1, x_2, x_3) = (2x_1 + 3x_2 - 7x_3 + 1,\ 10x_1 - 4x_2 - 8x_3,\ -4x_1 + x_2 - 5)$$

is not a linear transformation because

$$S(0) = S(0, 0, 0) = (1, 0, -5) \text{“‘‘} 0$$

The transformation

$$T(x, y) = (x + y, xy)$$

does satisfy $T(0) = 0$. However, we have

$T(1, 0) + T(0, 1) = (1 + 0, 1X0) + (0 + 1, 0X1) = (1, 0) + (1, 0) = (2, 0)$.

This is different from

$T((1, 0) + (0, 1)) = T(1, 1) = (1 + 1, 1X1) = (2, 1)$.

Therefore T is not linear.

Since the transformations are not linear, they are not matrix transformations.

The linearity of Ax helps us to construct solutions of systems of linear equations. The key idea is that, if x and x' are respectively solutions of $Ax = b$ and $Ax = b'$, then $cx + c'x'$ is a solution of $Ax = cb + c'b'$.

Example: The vectors $\boldsymbol{x} = (1, 1, 2)$ and $x' = (0, 1, -1)$ are the respective solutions of the following systems.

$$x_1 - x_2 + x_3 = 2$$
$$3x_1 + x_2 - x_3 = 2$$
$$2x_1 + 2x_2 + x_3 = 6$$
$$x_1 - x_2 + x_3 = -2$$
$$3x_1 + x_2 - x_3 = 2$$
$$2x_1 + 2x_2 + x_3 = 1$$

Note that the two systems have the same left side. The following systems

$$x_1 - x_2 + x_3 = 0$$
$$3x_1 + x_2 - x_3 = 4$$
$$2x_1 + 2x_2 + x_3 = 7$$
$$x_1 - x_2 + x_3 = 1$$

$$3x_1 + x_2 - x_3 = 1$$
$$2x_1 + 2x_2 + x_3 = 3$$
$$x_1 - x_2 + x_3 = 8$$
$$3x_1 + x_2 - x_3 = -4$$
$$2x_1 + 2x_2 + x_3 = 3$$

still have the same left side but different right sides. By

$$(0, 4, 7) = (2, 2, 6) + (-2, 2\ 1),$$
$$(1, 1, 3) = 1/2\ (2, 2, 6),$$
$$(8, -4, 3) = (2, 2, 6) - 3(-2, 2\ 1),$$

we see that

$$x + x' = (1, 1, 2) + (0, 1, -1) = (1, 2, 1),$$
$$1/2\ x = 1/2\ (1, 1, 2) = (1/2, 1/2, 1),$$
$$x - 3x' = (1, 1, 2) - 3(0, 1, -1) = (1, -2, 5),$$

are the respective solutions of the three new systems.

MATRIX TRANSFORMATIONS ARE LINEAR

Theorem $A(x + y) = Ax + Ay$, $A(cx) = cAx$.

Proof By an earlier exercise, we have

$x_1a_1 + x_2a_2 + ... + x_na_n = Ax$

where $A = [a_1\ a_2 ...\ a_n]$ is a matrix with the vectors $a_1, a_2, ..., a_n$ as columns, and $x_1, x_2, ..., x_n$ are the coordinates of x. Thus

$$\begin{aligned} Ax + Ay &= x_1a_1 + x_2a_2 + ... + x_na_n + y_1a_1 + y_2a_2 + ... + y_na_n \\ &= (x_1a_1 + y_1a_1) + (x_2a_2 + y_2a_2) + ... + (x_na_n + y_na_n) \\ &= (x_1 + y_1)a_1 + (x_2 + y_2)a_2 + ... + (x_n + y_n)a_n \end{aligned}$$

Since $x_1 + y_1, x_2 + y_2, ..., x_n + y_n$ are the coordinates of $x + y$, we see that the summation is $A(x + y)$.

The proof for $A(cx) = cAx$ is similar.

LINEARITY IN PICTURE

We can often visualize transformations between euclidean spaces. For example, the name of transformation *Reflection in x–axis*: $R^2 \to R^2$, $(x, y) \to (x, -y)$ is inspired by the picture.

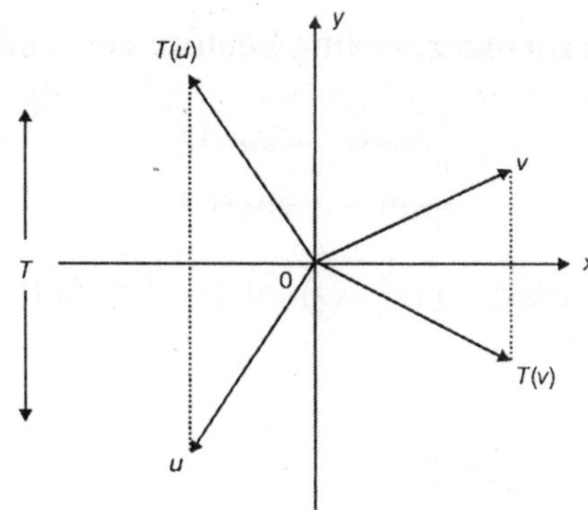

Example: The transformation *Reflection in y–axis*: $R^2 \to R^2$, $(x, y) \to (-x, y)$

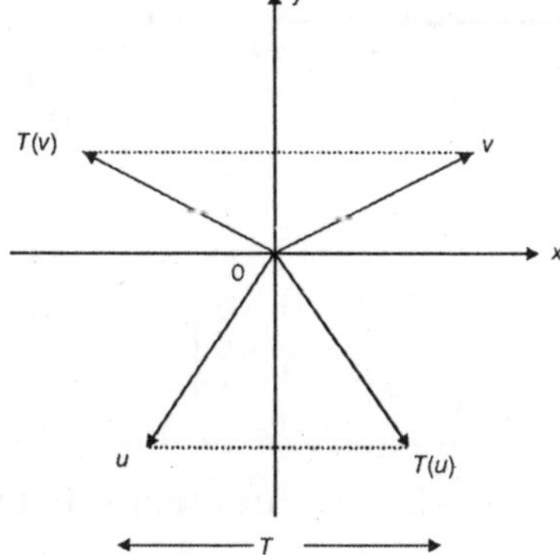

is given by the matrix

$$\begin{bmatrix} -1 & 0 \\ 0 & 1 \end{bmatrix}$$

The transformation *Rotation by &\theta;*: $R^2 \to R^2$, $(x, y) \to (x\cos\theta + y\sin\theta, -x\sin\theta + y\cos\theta)$

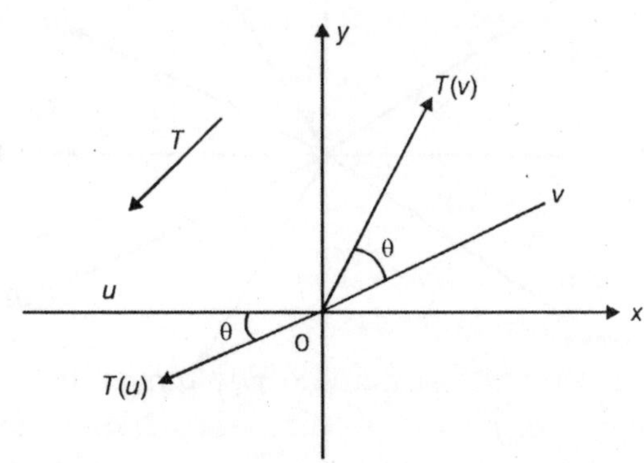

is given by the matrix (remember writing vectors vertically)

$$\begin{bmatrix} \cos\theta & -\sin\theta \\ \sin\theta & \cos\theta \end{bmatrix}$$

The transformation *Projection to* (x_1, x_3)*–plane*: $R^3 \to R^2$, $(x_1, x_2, x_3) \to (x_1, x_3)$

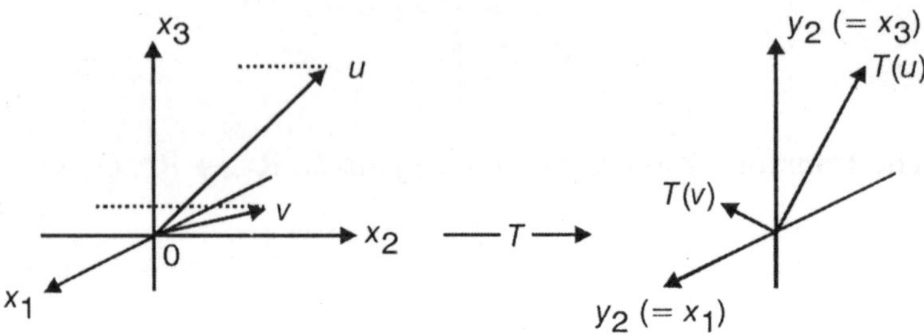

is given by the matrix

$$\begin{bmatrix} 1 & 0 & 0 \\ 0 & 0 & 1 \end{bmatrix}$$

Linear transformations are characterized by the property that addition and scalar multiplication are preserved. Geometrically, this means that parallelograms and stretching/ shrinking are preserved.

Example: The transformation *Reflection in x–axis* clearly preserves parallelograms and stretching/shrinking.

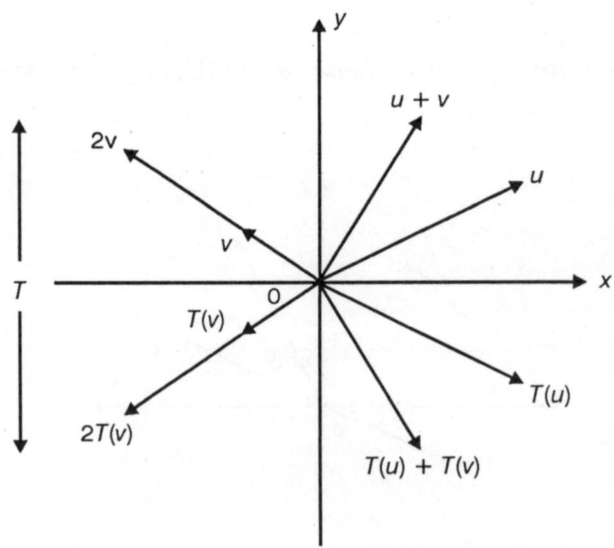

The rotation transformation also preserves the two operations.

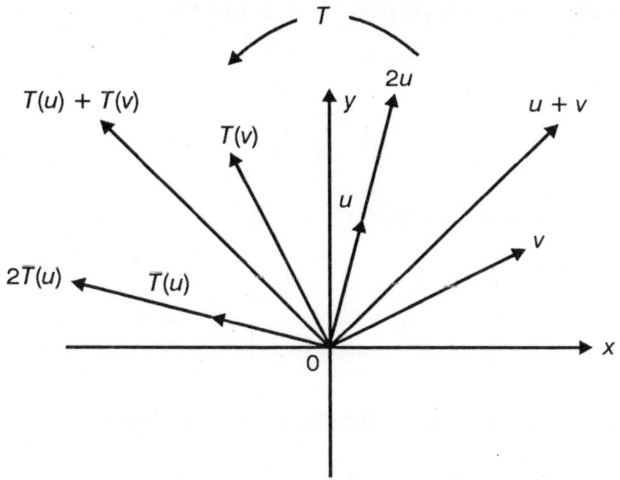

The projection transformation is also a linear transformation.

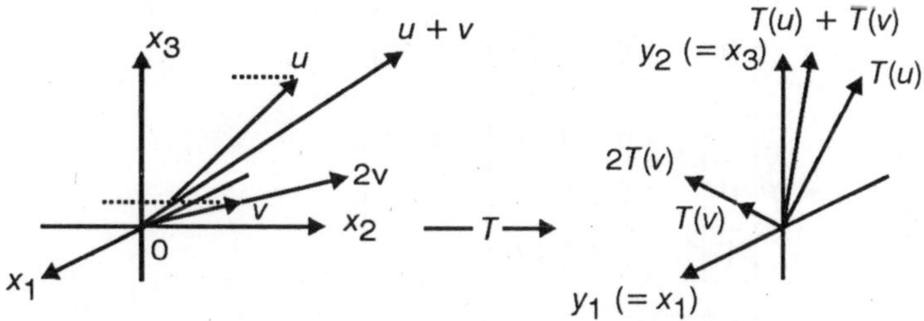

LINEAR TRANSFORMATION = MATRIX TRANSFORMATION

We have seen matrix transformations are linear. Conversely, this example suggests the converse is also true.

Any linear transformation T: $\mathrm{R}^n \to \mathrm{R}^m$ is also a matrix transformation, with the matrix given by

$$A = [T(e_1)\ T(e_2) \ldots\ T(e_n)].$$

In other words, the columns of the matrix A are the values of the linear transformation T at the standard basis. Thus the concepts of matrix transformations and linear transformations (between euclidean spaces) are equivalent.

Example: The identity transformation id: $\mathrm{R}^n \to \mathrm{R}^n$ is clearly linear. The corresponding matrix is

$$[id(e_1)\ id(e_2) \ldots\ id(e_n)] = [e_1\ e_2 \ldots\ e_n] = \begin{bmatrix} 1 & 0 & \ldots & 0 \\ 0 & 1 & \ldots & 0 \\ \vdots & \vdots & & \vdots \\ 0 & 0 & \ldots & 1 \end{bmatrix}$$

called the identity matrix and denoted by I_n.

Example: In this example, we started from a matrix

$$A = \begin{bmatrix} 1 & -1 & 1 \\ 3 & 1 & -1 \\ 2 & 2 & 1 \end{bmatrix}$$

and produced a matrix transformation (which is also a linear transformation).

$$T\begin{bmatrix} x_1 \\ x_2 \\ x_3 \end{bmatrix} = \begin{bmatrix} x_1 - x_2 + x_3 \\ 3x_1 + x_2 - x_3 \\ 2x_1 + 2x_2 + x_3 \end{bmatrix} \quad R^3 \to R^3$$

Conversely, we may recover the columns of A as follows

$$[\text{col } 1] \text{ of } A = T\begin{bmatrix} 1 \\ 0 \\ 0 \end{bmatrix} = \begin{bmatrix} 1-0+0 \\ 3+0-0 \\ 0-1+0 \end{bmatrix} = \begin{bmatrix} 1 \\ 3 \\ 2 \end{bmatrix}$$

$$[\text{col } 2] \text{ of } A = T\begin{bmatrix} 0 \\ 0 \\ 1 \end{bmatrix} = \begin{bmatrix} 0-0-0 \\ 0+1-0 \\ 0-2+0 \end{bmatrix} = \begin{bmatrix} -1 \\ 1 \\ 2 \end{bmatrix}$$

$$[\text{col } 3] \text{ of } A = T\begin{bmatrix} 0 \\ 0 \\ 1 \end{bmatrix} = \begin{bmatrix} 0-0+1 \\ 0+0-1 \\ 0-0+1 \end{bmatrix}$$

Example: The reflections and rotations on R^2 are linear transformations <u>because</u> they preserve parallelogram and stretching/shrinking. To find their matrices, we simply need to apply reflections and rotations to the standard basis.

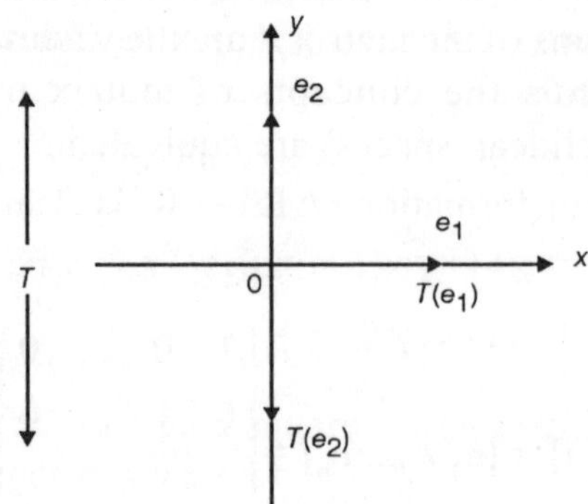

The picture shows that the reflections of e_1, e_2 in x–axis are $e_1 = (1, 0)$, $-e_2 = (0, -1)$. Therefore the matrix for the reflection is

$$\begin{bmatrix} 1 & 0 \\ 0 & -1 \end{bmatrix}$$

Moreover, we also see from picture that for the rotation by angle θ, $T(e_1) = (\cos\theta, \sin\theta)$, $T(e_2) = (-\sin\theta, \cos\theta)$.

Therefore the matrix for the rotation is

$$\begin{bmatrix} \cos\theta & -\sin\theta \\ \sin\theta & \cos\theta \end{bmatrix}$$

Exercise: Find the corresponding transformations, and indicate the source and target euclidean spaces.

1. [1 2]

Ans. $T(x_1, x_2) = x_1 + 2x_2$: $R^2 \to R^1$

2. $\begin{bmatrix} 1 \\ 2 \end{bmatrix}$

Ans. $T(x_1) = (x_1, 2x_1)$: $R^1 \to R^2$

3. $\begin{bmatrix} 2 & 0 & 2 & 1 \\ 1 & 1 & -1 & 0 \\ 0 & 1 & -2 & 1 \end{bmatrix}$

Ans. $T(x_1, x_2, x_3, x_4) = (x_1 + 2x_3 + x_4, x_1 + x_2 - x_3, x2 - 2x_3 + x_4)$: $R^4 \to R^3$

4. $\begin{bmatrix} 1 & 2 & 1 \\ 2 & 1 & 1 \\ 0 & 3 & 1 \\ 3 & 0 & 1 \end{bmatrix}$

Ans. $T(x_1, x_2, x_3) = (x_1 + 2x_2 + x_3, 2x_1 + x_2 + x_3, 3x2 + x_3, 3x_1 + x_3)$: $R^3 \to R^4$

5. $\begin{bmatrix} 1 & -1 & 1 & 0 & 2 \\ 2 & -2 & 0 & 2 & 2 \\ -1 & 1 & 2 & -3 & 1 \\ -2 & 2 & 1 & -3 & -1 \end{bmatrix}$

Ans. $T(x_1, x_2, x_3, x_4, x_5) = (x_1 - x_2 + x_3 + 2x_5, 2x_1 - 2x_2 + 2x_4 + 2x_5, - x_1 + x_2 + 2x_3 - 3x_4 + x5, - 2x_1 + 2x_2 + 2x_3 - 3x_4 - x5)$: $R^5 \to R^4$

Exercise: Which of the following are matrix transformations? For the ones that are, find the corresponding matrices. For the ones that are not, provide the reason.

1. $T_1(x_1, x_2, x_3) = (2x_1 + 3x_2 - 7x_3, 10x_1 - 4x_2 - 8x_3, - 4x_1 + x_2)$

Ans. This is a matrix transformation.

$$\begin{bmatrix} 2 & 3 & -7 \\ 10 & -4 & -8 \\ -4 & 1 & 0 \end{bmatrix}$$

2. $T_2(x_1, x_2, x_3) = (2x_1 + 3x_2 - 7x_3 + 1, 10x_1 - 4x_2 - 8x_3, - 4x_1 + x_2 - 5)$

Ans. This is not a matrix/linear transformation because $T_2(0, 0, 0) = (1, 0, - 5)$, whereas any linear transformation must send zero vector to zero vector.

3. $T_3(x_1, x_2, x_3) = 2x_1 + 3x_2 - 7x_3$

Ans. This is a matrix transformation.

$$\begin{bmatrix} 2 & 3 & -7 \end{bmatrix}$$

4. $T_4(x_1, x_2, x_3) = 2x_1 + 3x_2 - 7x_3 + 1$

Ans. This is not a matrix/linear transformation because $T_4(0, 0, 0) = 1$, which is nonzero.

5. $T_5(x_1, x_2, x_3) = (2x_1, 3x_2, - 7x_3)$

Ans. This is a matrix transformation.

$$\begin{bmatrix} 2 & 0 & 0 \\ 0 & 3 & 0 \\ 0 & 0 & -7 \end{bmatrix}$$

6. $T_6(x_1, x_2, x_3) = (2x_2, 3x_3, - 7x_1)$

Ans. This is a matrix transformation.

$$\begin{bmatrix} 2 & 0 & 0 \\ 0 & 0 & 3 \\ -7 & 0 & 0 \end{bmatrix}$$

7. $T_7(x_1, x_2, x_3) = (3x_2, - 7x_3, 2x_1)$

Ans. This is a matrix transformation.

$$\begin{bmatrix} 0 & 3 & 0 \\ 0 & 0 & -7 \\ 2 & 0 & 0 \end{bmatrix}$$

8. $T_8(x) = (2x, 3x, -7x)$

Ans. This is a matrix transformation.

$$\begin{bmatrix} 2 \\ 3 \\ -7 \end{bmatrix}$$

9. $T_9(x, y, z) = (x - y, y - z, z - x)$

Ans. This is a matrix transformation.

$$\begin{bmatrix} 1 & -1 & 0 \\ 0 & 1 & -1 \\ -1 & 0 & 1 \end{bmatrix}$$

10. $T_{10}(x, y) = (x + y, xy)$

Ans. This is not a matrix/linear transformation because $T_{10}'(1, 0) = T_{10}(0, 1) = (1,0)$, $T_{10}(1, 1) = (2, 1) \rightarrow T_{10}(1, 0) + T_{10}(0, 1)$.

11. $T_{11}(x, y) = (x + y, 2^{1/2}x + 3^{1/2}y)$

Ans. This is a matrix transformation.

$$\begin{bmatrix} 1 & 1 \\ 2^{1/2} & 3^{1/2} \end{bmatrix}$$

12. $T_{12}(u, v) = (u + v, u^2 + v^2, u^3 + v^3)$

Ans. This is not a matrix/linear transformation because $T_{12}(1, 1) = (2, 2, 2)$, $T_{12}(2, 2) = (4, 8, 16)$, so that $T_{12}(2, 2)$ "" $2T_{12}(1, 1)$.

Exercise: Convince yourself that the following transformations are linear by drawing the picture and indicating how the parallelogram and scaling by 1/2 are preserved by the transformations. Then find the corresponding matrices.

1. *Reflection in diagonal line*: $R^2 \rightarrow R^2$

Ans. The following is the picture for the transformation.

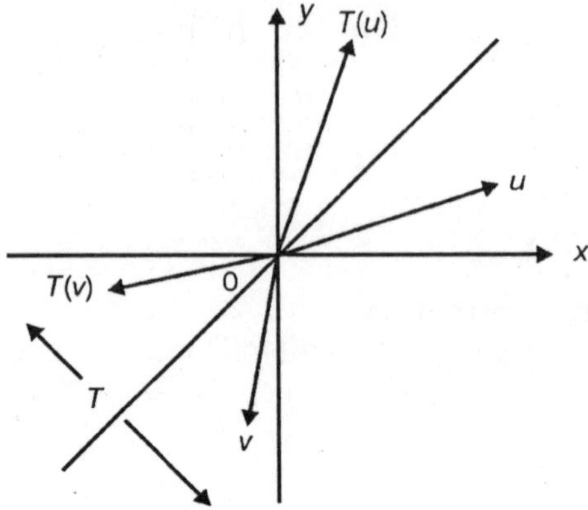

The following illustrates that the transformation is linear.

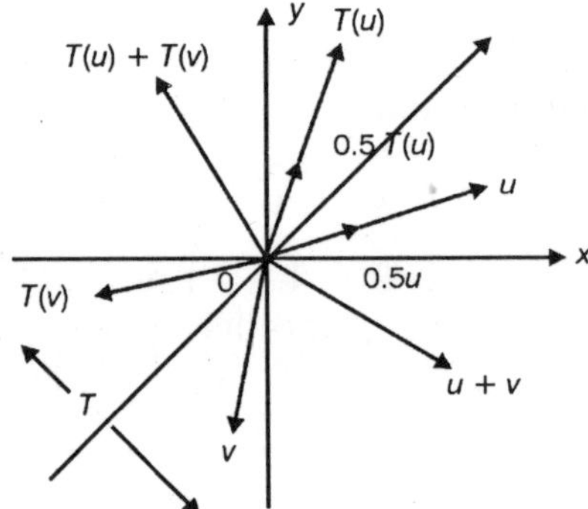

From this picture,

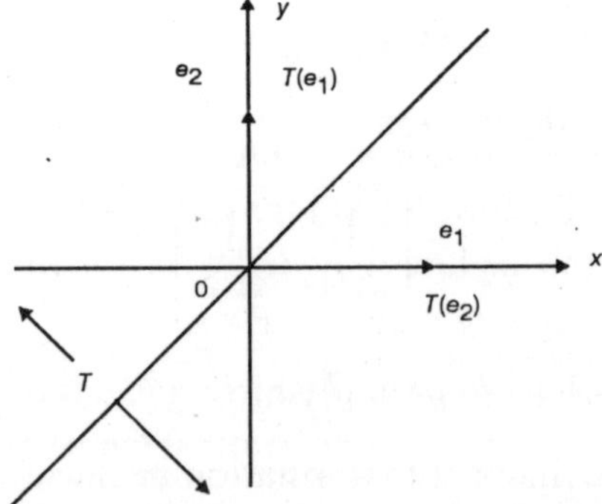

we see $T(e_1) = e_2 = (0, 1)$, $T(e_2) = e_1 = (1, 0)$. Therefore the matrix is

$$\begin{bmatrix} 0 & 1 \\ 1 & 0 \end{bmatrix}$$

2. *Projection to x_1–line*: $R^2 \rightarrow R^1$

Ans. The following is the picture for the transformation.

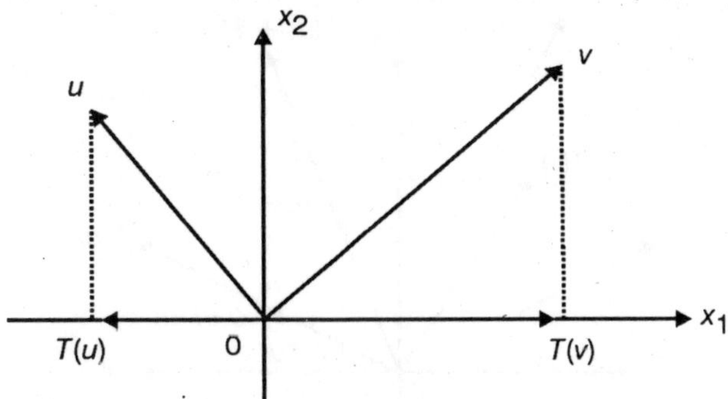

The following illustrates that the transformation is linear.

We clearly have $T(e_1) = e_1 = (1, 0)$, $T(e_2) = (0, 0)$. Therefore the matrix is

$$\begin{bmatrix} 1 & 0 \\ 0 & 0 \end{bmatrix}$$

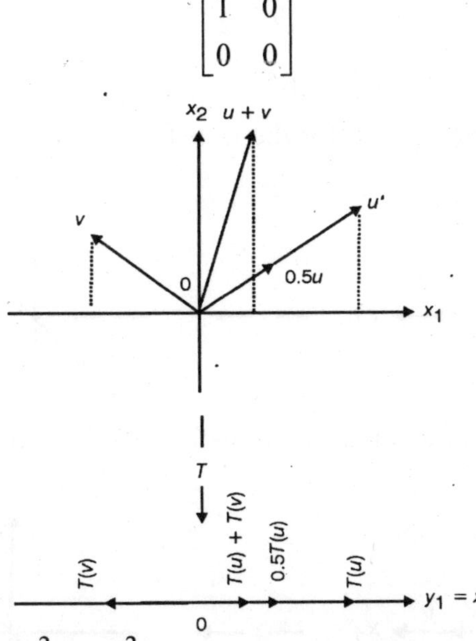

3. *Multiplying by* 2: $R^2 \to R^2$

Ans. The following is the picture for the transformation.

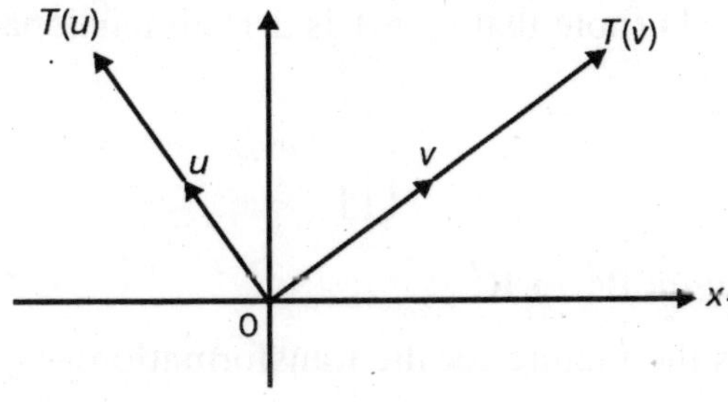

The following illustrates that the transformation is linear.

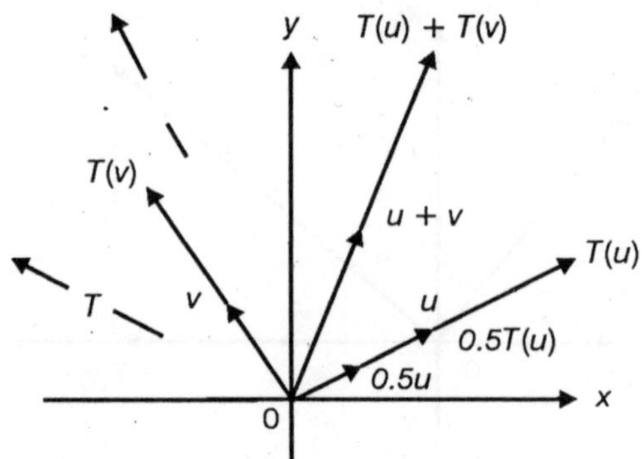

Since $T(e_1) = 2e_1 = (2, 0)$, $T(e_2) = 2e_2 = (0, 2)$, the matrix is

$$\begin{bmatrix} 2 & 0 \\ 0 & 2 \end{bmatrix}$$

4. *Embedding as diagonal*: $R^1 \to R^2$

Ans. The following is the picture for the transformation.

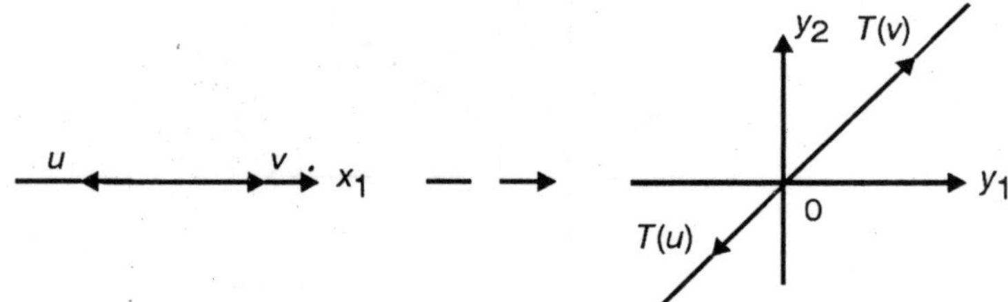

The following illustrates that the transformation is linear.

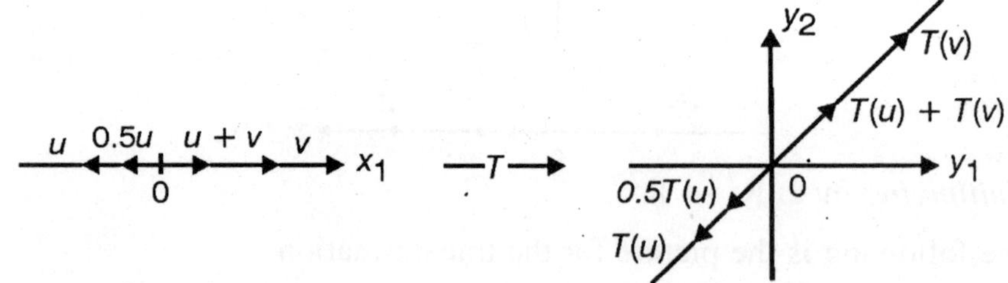

Since $T(e_1) = (1, 1)$ (note that $e_1 = 1$ is a 1–dimensional vector), the matrix is

$$\begin{bmatrix} 1 \\ 1 \end{bmatrix}$$

5. *Reflection in origin*: $R^2 \to R^2$

Ans. The following is the picture for the transformation.

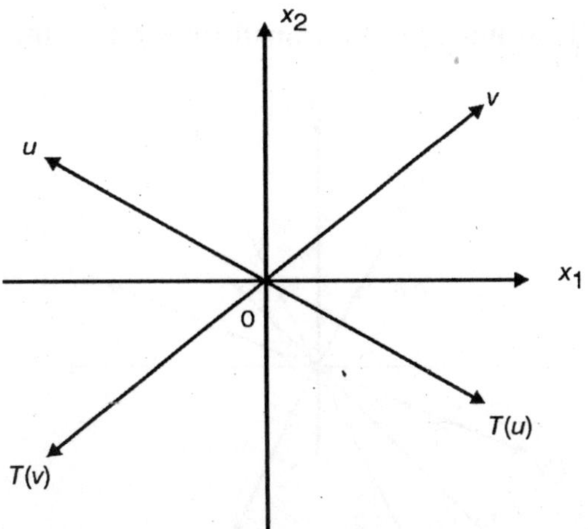

The following illustrates that the transformation is linear.

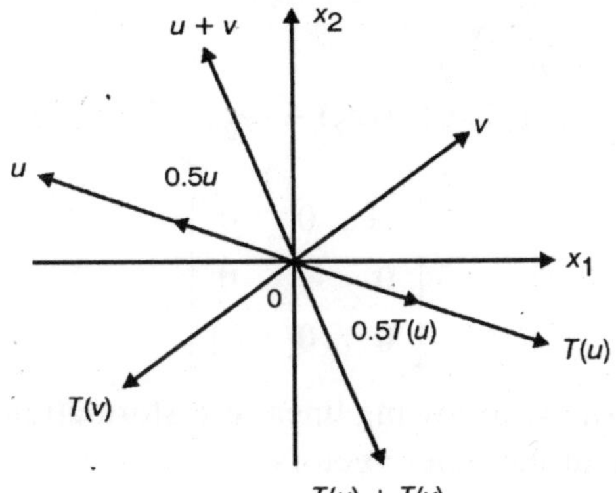

Since $T(e_1) = -e_1 = (-1, 0)$, $T(e_2) = -e_2 = (0, -1)$, the matrix is

$$\begin{bmatrix} -1 & 0 \\ 0 & -1 \end{bmatrix}$$

6. *Reflection in origin*: $R^3 \to R^3$

Ans. The following is the picture for the transformation.

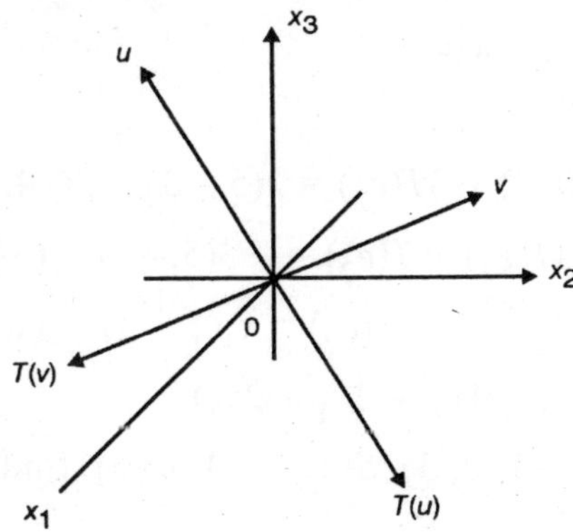

The following illustrates that the transformation is linear.

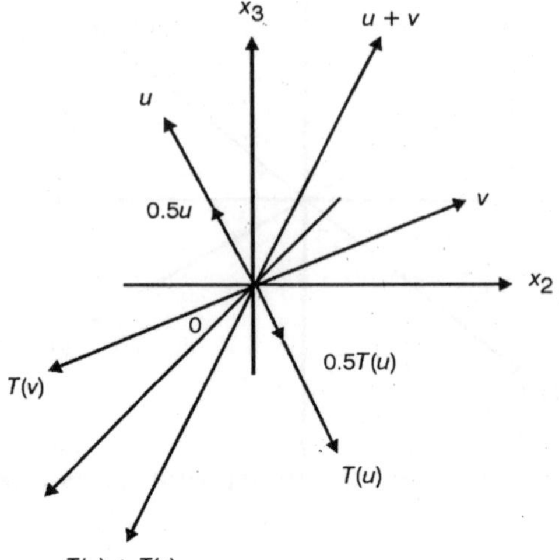

Since $T(e_1) = -e_1 = (-1, 0, 0)$, $T(e_2) = -e_2 = (0, -1, 0)$, $T(e_3) = -e_3 = (0, 0, -1)$, the matrix is

$$\begin{bmatrix} -1 & 0 & 0 \\ 0 & -1 & 0 \\ 0 & 0 & -1 \end{bmatrix}$$

Exercise: The values of the following linear transformations at the standard basis have been given. Find the value at the other vectors.

Ans.

1. T: $R^2 \to R^2$, $T(e_1) = (5, -3)$, $T(e_2) = (-4, 2)$, find $T(2, 3)$, $T(-3, 1)$, $T(x_1, x_2)$.

Ans. By

$$(2, 3) = 2e_1 + 3e_2$$
$$(-3, 1) = -3e_1 + e_2$$
$$(x_1, x_2) = x_1e_1 + x_2e_2$$

We have

$$T(2, 3) = 2T(e_1) + 3T(e_2) = 2(5, -3) + 3(-4, 2) = (-2, 0)$$
$$T(-3, 1) = -3T(e_1) + T(e_2) = -3(5, -3) + (-4, 2) = (-19, 9)$$
$$T(x_1, x_2) = x_1T(e_1) + x_2T(e_2) = x_1(5, -3) + x_2(-4, 2)$$
$$= (5x_1 - 4x_2, -3x_1 + 2x_2)$$

2. S: $R^2 \to R^3$, $S(e_1) = (1, 2, 3)$, $S(e_2) = (4, 5, 6)$, find $S(2, 3)$, $S(-3, 1)$, $S(x_1, x_2)$.

Ans. By

$$(2, 3) = 2e_1 + 3e_2$$
$$(-3, 1) = -3e_1 + e_2$$
$$(x_1, x_2) = x_1e_1 + x_2e_2$$

We have

$$S(2, 3) = 2S(e_1) + 3S(e_2) = 2(1, 2, 3) + 3(4, 5, 6) = (14, 19, 24)$$
$$S(-3, 1) = -3S(e_1) + S(e_2) = -3(1, 2, 3) + (4, 5, 6) = (1, -1, -3)$$
$$S(x_1, x_2) = x_1S(e_1) + x_2S(e_2) = x_1(1, 2, 3) + x_2(4, 5, 6)$$
$$= (x_1 + 4x_2, 2x_1 + 5x_2, 3x_1 + 6x_2)$$

3. R: $R^3 \to R^2$, $R(e_1) = (1, 2)$, $R(e_2) = (3, 4)$, $R(e3) = (5, 6)$, find $R(1, 0, 2)$, $R(-2, 1, 3)$, $R(x_1, x_2, x_3)$.

Ans. By

$$(1, 0, 2) = e_1 + 2e_3$$
$$(-2, 1, 3) = -2e_1 + e_2 + 3e_3$$
$$(x_1, x_2, x_3) = x_1e_1 + x_2e_2 + x_3e_3$$

We have

$$R(1, 0, 2) = R(e_1) + 2S(e_3) = (1, 2) + 2(5, 6) = (11, 14)$$
$$R(-2, 1, 3) = -2R(e_1) + R(e_2) + 3R(e_3)$$
$$= -2(1, 2) + (3, 4) + 3(5, 6) = (16, 18)$$
$$R(x_1, x_2, x_3) = x_1R(e_1) + x_2R(e_2) + x_3R(e3) = x_1(1, 2) + x_2(3, 4) + x_3(5, 6)$$
$$= (x_1 + 3x_2 + 5x_3, 2x_1 + 4x_2 + 6x_3)$$

Exercise: Let T: $R^2 \to R^3$ be a linear transformation. Assume we already know $T(1, 2) = (1, 2, 3)$, $T(3, 4) = (4, 5, 6)$.

- Use this exercise and the linearity of T to compute $T(4, 6)$, $T(1, 1)$, $T(1, 0)$, $T(0, 1)$.
- Use $T(1, 0)$, $T(0, 1)$ to compute $T(x_1, x_2)$ in general.
- Use $3(1, 1, 1) = (4, 5, 6) - (1, 2, 3)$ to find x_1, x_2 such that $T(x_1, x_2) = (1, 1, 1)$

Answer: From the exercise, we have

$(4, 6) = (1, 2) + (3, 4)$

$(1, 1) = -1/2\ (1, 2) + 1/2\ (3, 4)$

(1, 0) = –2(1, 2) + (3, 4)

(0, 1) = 3/2 (1,2) – 1/2 (3, 4)

Thus

T(4, 6) = ***T***(1, 2) + ***T***(3, 4) = (1, 2, 3) + (4, 5, 6) = (5, 7, 9)

T(1, 1) = –1/2 ***T***(1, 2) + 1/2 ***T***(3, 4) = –1/2 (1, 2, 3) + 1/2 (4, 5, 6) = (3/2, 3/2, 3/2)

T(1, 0) = –2***T***(1, 2) + ***T***(3, 4) = – 2(1, 2, 3) + (4, 5, 6) = (2, 1, 0)

T(0, 1) = 3/2 ***T***(1,2) – 1/2 ***T***(3, 4) = 3/2 (1, 2, 3) – 1/2 (4, 5, 6) = (–1/2, 1/2, 3/2)

Moreover,

$T(x_1, x_2) = x_1T(1, 0) + x_2T(0, 1) = x_1(2, 1, 0) + x_2(-1/2, 1/2, 3/2) = (2x_1 - 1/2x_2, x_1 + 1/2x_2, 3/2x_2)$.

Finally, from the equality 3(1, 1, 1) = (4, 5, 6) – (1, 2, 3) = *T*(3, 4) – *T*(1, 2) = *T*(2, 2) we have (1, 1, 1) = *T*(2/3, 2/3). Therefore for the last question, we may take $x_1 = x_2 = 2/3$.

Exercise: Let *T*: R^3 '! R^2 be a linear transformation. Assume we already know *T*(1, 1, 1) = (1, 2), *T*(1, 1, 0) = (3, 4), *T*(1, 0, 0) = (5, 6). Find the general formula (or the matrix) for *T*.

Ans: We need to find $T(x_1, x_2, x_3)$. We use the same idea as in the previous exercise. First we solve the three systems $c_1(1, 1, 1) + c_2(1, 1, 0) + c3(1, 0, 0) = e_i$ for $i = 1, 2, 3$ and find

$$e_1 = (1, 0, 0)$$

$$e_2 = (1, 1, 0) - (1, 0, 0)$$

$$e_3 = (1, 1, 1) - (1, 1, 0)$$

Then we use the linearity of ***T*** to get

$$T(e_1) = T(1, 0, 0) = (5, 6)$$

$$T(e_2) = T(1, 1, 0) - T(1, 0, 0) = (3, 4) - (5, 6) = (-2, -2)$$

$$T(e_3) = T(1, 1, 1) - T(1, 1, 0) = (1, 2) - (3, 4) = (-2, -2)$$

Finally we use $(x_1, x_2, x_3) = x_1e_1 + x_2e_2 + x_3e_3$ and the linearity of *T* to get

$T(x_1, x_2, x_3) = x_1T(e_1) + x_2T(e_2) + x_3T(e_3) = x_1(5, 6) + x_2(-2, -2) + x_3(-2, -2) = (5x_1 - 2x_2 - 2x_3, 6x_1 - 2x_2 - 2x_3)$.

Exercise: Solve the following three systems of linear equations.

$$x_1 + 2x_2 = 1 \quad x_1 + 2x_2 = 0$$

$$2x_1 + 3x_2 = 0 \quad 2x_1 + 3x_2 = 1$$

Then use your results and the linearity to find solutions of the following systems

$$x_1 + 2x_2 = 2 \; x_1 + 2x_2 = 13$$
$$2x_1 + 3x_2 = -3 \; 2x_1 + 3x_2 = 9$$

Finally use the same technique to find the solution of the system

$$x_1 + 2x_2 = b_1$$
$$2x_1 + 3x_2 = b_2$$

Ans: For the two systems $Ax = e_1$ and $Ax = e_2$, we combine the augmented matrices together.

$$[A \; e_1 \; e_2] = \begin{bmatrix} 1 & 2 & 1 & 0 \\ 2 & 3 & 0 & 1 \end{bmatrix}$$

Row operations give us

$$\begin{bmatrix} 1 & 0 & -3 & 2 \\ 0 & 1 & 2 & -1 \end{bmatrix}$$

Thus the solution for the system with $e_1 = (1, 0)$ on the right is $x = (-3, 2)$, and the solution with $e2 = (0, 1)$ on the right is $x' = (2, -1)$.

By $(2, 3) = 2(1, 0) + 3(0, 1)$, the solution of the system with $(2, 3)$ on the right is $2x + 3x' = 2(-3, 2) + 3(2, -1) = (0, 1)$.

By $(13, 9) = 13(1, 0) + 9(0, 1)$, the solution of the system with $(13, 9)$ on the right is $13x + 9x' = 13(-3, 2) + 9(2, -1) = (-21, 17)$.

By $(b_1, b_2) = b_1(1, 0) + b_2(0, 1)$, the solution of the system with (b_1, b_2) on the right is $b_1x + b_2x' = b_1(-3, 2) + b_2(2, -1) = (-3b_1 + 2b_2, 2b_1 - b_2)$.

Exercise: Solve the following three systems of linear equations.

$$x_1 + 2x_3 = 1 \qquad x_1 + 2x_3 = 0 \qquad x_1 + 2x_3 = 0$$
$$2x_1 - x_2 + 3x_2 = 0 \; 2x_1 - x_2 + 3x_2 = 1 \;\; 2x_1 - x_2 + 3x_2 = 0$$
$$4x_1 - x_2 + 8x_2 = 0 \; 4x_1 - x_2 + 8x_2 = 0 \;\; 4x_1 - x_2 + 8x_2 = 1$$

Then use your results and the linearity to find a solution of the system

$$x_1 + 2x_3 = b_1$$
$$2x_1 - x_2 + 3x_2 = b_2$$
$$4x_1 - x_2 + 8x_2 = b_3$$

Ans: For the three systems $Ax = e_1$, $Ax = e_2$, $Ax = e_3$, we combine the augmented matrices together.

$$[A\ e_1\ e_2\ e_3] = \begin{bmatrix} 1 & 0 & 2 & 1 & 0 & 0 \\ 2 & -1 & 3 & 0 & 1 & 0 \\ 4 & -1 & 8 & 0 & 0 & 1 \end{bmatrix}$$

Row operations give us

$$\begin{bmatrix} 1 & 0 & 0 & 5 & 2 & -2 \\ 0 & 1 & 0 & 4 & 0 & -1 \\ 0 & 0 & 1 & -2 & -1 & 1 \end{bmatrix}$$

Thus we conclude

$Ax = e_1$ has solution $x = (5, 4, -2)$

$Ax = e_2$ has solution $x' = (2, 0, -1)$

$Ax = e_3$ has solution $x'' = (-2, -1, 1)$

By $(b_1, b_2, b_3) = b_1e_1 + b_2e_2 + b_3e_3$, the combination

$b_1x + b_2x' + b_3x'' = b_1(5, 4, -2) + b_2(2, 0, -1) + b_3(-2, -1, 1) = (5b_1 + 2b_2 - 2b_3, 4b_1 - b_3, - 2b_1 - b_2 + b_3)$

is a solution of the system $Ax = b$.

Chapter 5

Matrix Multiplication

MATRIX MULTIPLICATION

In mathematics, matrix multiplication is the operation of multiplying a matrix with either a scalar or another matrix. This article gives an overview of the various ways to perform matrix multiplication.

ORDINARY MATRIX PRODUCT

This is the most often used and most important way to multiply matrices. It is defined between two matrices only if the number of columns of the first matrix is the same as the number of rows of the second matrix. If A is an *m*-by-*n* matrix (where m is the number of rows and n is the number of columns) and B is an *n*-by-*p* matrix, then their product is an *m*-by-*p* matrix denoted by AB (or sometimes A·B). If C = AB, and $c_{i,j}$ denotes the entry in C at position (*i*,*j*), then

$$c_{i,j} = \sum_{r=1}^{n} a_{i,r} b_{r,j} = a_{i,1} b_{1,j} + a_{i,2} b_{2,j} + \ldots + a_{i,n} b_{n,j}.$$

for each pair *i* and *j* with $1 \leq i \leq m$ and $1 \leq j \leq p$. The algebraic system of "matrix units" summarises the abstract properties of this kind of multiplication.

Calculating Directly from the Definition

The picture to the left shows how to calculate the (1,2) element and the (3,3) element of AB if A is a 4 X 2 matrix, and B is a 2 X 3 matrix. Elements from each matrix are paired off in the direction of the arrows; each pair is multiplied and the products are added. The location of the resulting number in AB corresponds to the row and column that were considered.

$$(AB)_{1,2} = \sum_{r=1}^{2} a_{1,r} b_{r,2} = a_{1,1} b_{1,2} + a_{1,2} b_{1,2} + \ldots + a_{1,2} b_{2,2}$$

$$(AB)_{3,3} = \sum_{r=1}^{2} a_{3,r} b_{r,3} = a_{3,1} b_{1,3} + a_{3,2} b_{2,3}$$

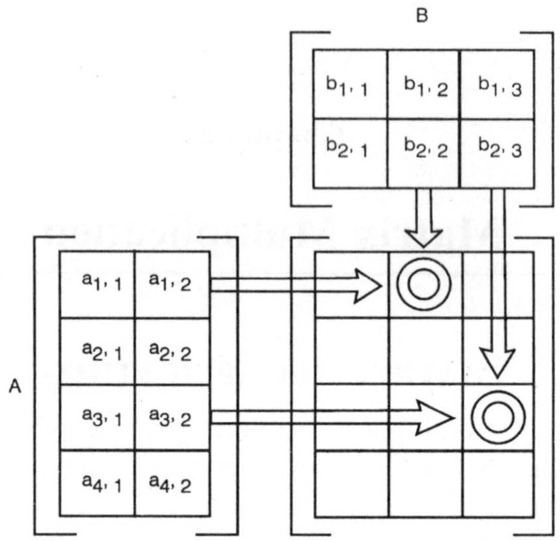

For example:

$$\begin{bmatrix} 1 & 0 & 2 \\ -1 & 3 & 1 \end{bmatrix} \cdot \begin{bmatrix} 3 & 1 \\ 2 & 1 \\ 1 & 0 \end{bmatrix} = \begin{bmatrix} 1\times 3+0\times 2+2\times 1 & 1\times 1+0\times 1+2\times 0 \\ -1\times 3+3\times 2+1\times 1 & -1\times 1\times 3+1+1\times 0 \end{bmatrix} = \begin{bmatrix} 5 & 1 \\ 4 & 2 \end{bmatrix}.$$

The Coefficients-vectors Method

This matrix multiplication can also be considered from a slightly different viewpoint: it adds vectors together after being multiplied by different coefficients. If A and B are matrices given by:

$$A = \begin{bmatrix} a_{1,1} & a_{1,2} & \cdots \\ a_{2,1} & a_{2,2} & \cdots \\ \vdots & \vdots & \ddots \end{bmatrix} \text{ and } B = \begin{bmatrix} b_{1,1} & b_{1,2} & \cdots \\ a_{2,1} & b_{2,2} & \cdots \\ \vdots & \vdots & \ddots \end{bmatrix} = \begin{bmatrix} B_1 \\ B_2 \\ \vdots \end{bmatrix}$$

then

$$AB = \begin{bmatrix} a_{1,1}B_1 + a_{1,2}B_2 + \ldots \\ a_{2,1}B_1 + a_{2,2}B_2 + \ldots \\ \vdots \end{bmatrix}$$

The example revisited:

$$\begin{bmatrix} 1 & 0 & 2 \\ -1 & 3 & 1 \end{bmatrix} \cdot \begin{bmatrix} 3 & 1 \\ 2 & 1 \\ 1 & 0 \end{bmatrix} = \begin{bmatrix} 1[3 \;\; 1] + 2[2 \;\; 1] + 2[1 \;\; 0] \\ -1[3 \;\; 1] + 3[2 \;\; 1] + 1[1 \;\; 0] \end{bmatrix}$$

$$= \begin{bmatrix} [3 \;\; 1]+[0 \;\; 0]+[2 \;\; 0] \\ [-3 \;\; -1]+[6 \;\; 3]+[1 \;\; 0] \end{bmatrix} = \begin{bmatrix} 5 & 1 \\ 4 & 2 \end{bmatrix}.$$

The rows in the matrix on the left are the list of coefficients. The matrix on the right is the list of vectors. In the example, the first row is [102], and thus we take 1 times the first vector, 0 times the second vector, and 2 times the third vector.

$$A = [A_1 \;\; A_2 \;\; \ldots] \Rightarrow AB = \sum_i A_i B_i$$

The equation can be simplified further by using outer products:

The terms of this sum are matrices of the same shape, each describing the effect of one column of A and one row of B on the result. The columns of A can be seen as a coordinate system of the transform, i.e. given a vector x we have $Ax = A_1x_1 + A_2x_2 + \ldots$. where x_i are coordinates along the A_i "axes". The terms A_iB_i are like A_ix_i, except that B_i contains the ith coordinate for each column vector of B, each of which is transformed independently in parallel.

The example revisited:

$$\begin{bmatrix} 1 & 0 & 2 \\ -1 & 3 & 1 \end{bmatrix} \cdot \begin{bmatrix} 3 & 1 \\ 2 & 1 \\ 1 & 0 \end{bmatrix} = \begin{bmatrix} 1 \\ -1 \end{bmatrix} [3 \;\; 1] + \begin{bmatrix} 0 \\ 3 \end{bmatrix} [2 \;\; 1] + \begin{bmatrix} 2 \\ 1 \end{bmatrix} [1 \;\; 0]$$

$$= \begin{bmatrix} 1 \cdot 3 & 1 \cdot 1 \\ -1 \cdot 3 & -1 \cdot 1 \end{bmatrix} + \begin{bmatrix} 0 \cdot 2 & 0 \cdot 1 \\ 3 \cdot 2 & 3 \cdot 1 \end{bmatrix} + \begin{bmatrix} 2 \cdot 1 & 2 \cdot 0 \\ 1 \cdot 1 & 1 \cdot 0 \end{bmatrix} = \begin{bmatrix} 5 & 1 \\ 4 & 2 \end{bmatrix}$$

The vectors $[3 \;\; 2 \;\; 1]^T$ and $[1 \;\; 1 \;\; 0]^T$ have been transformed to $[5 \;\; 4]^T$ and $[1 \;\; 2]^T$in parallel. One could also transform them one by one with the same steps:

$$\begin{bmatrix} 1 & 0 & 2 \\ -1 & 3 & 1 \end{bmatrix} \cdot \begin{bmatrix} 3 & 1 \\ 2 & 1 \\ 1 & 0 \end{bmatrix} = \begin{bmatrix} 1 \\ -1 \end{bmatrix} 3 + \begin{bmatrix} 0 \\ 3 \end{bmatrix} 2 + \begin{bmatrix} 2 \\ 1 \end{bmatrix} 1 = \begin{bmatrix} 1 \cdot 3 \\ -1 \cdot 3 \end{bmatrix} + \begin{bmatrix} 0 \cdot 2 \\ 3 \cdot 2 \end{bmatrix} + \begin{bmatrix} 2 \cdot 1 \\ 1 \cdot 1 \end{bmatrix} = \begin{bmatrix} 5 \\ 4 \end{bmatrix}$$

Vector-lists Method

The ordinary matrix product can be thought of as a dot product of a column-list of vectors and a row-list of vectors. If A and B are matrices given by:

$$A = \begin{bmatrix} a_{1,1} & a_{1,2} & a_{1,3} & \cdots \\ a_{2,1} & a_{2,2} & a_{2,3} & \cdots \\ a_{3,1} & a_{3,2} & a_{3,3} & \cdots \\ \vdots & \vdots & \vdots & \ddots \end{bmatrix} = \begin{bmatrix} A_1 \\ A_2 \\ A_3 \\ \cdots \end{bmatrix} \text{ and}$$

$$B = \begin{bmatrix} b_{1,1} & b_{1,2} & b_{1,3} & \cdots \\ b_{2,1} & b_{2,2} & b_{2,3} & \cdots \\ b_{3,1} & b_{3,2} & b_{3,3} & \cdots \\ \vdots & \vdots & \vdots & \ddots \end{bmatrix} = \begin{bmatrix} B_1 & B_2 & B_3 & \ldots \end{bmatrix}$$

where

A_1 is the row vector of all elements of the form $a_{1,x}$ A_2 is the row vector of all elements of the form $a_{2,x}$ etc, and B_1 is the column vector of all elements of the form $b_{x,1}$ B_2 is the column vector of all elements of the form $b_{x,2}$ etc,

then

$$AB = \begin{bmatrix} A_1 \\ A_2 \\ A_3 \\ \cdots \end{bmatrix} * \begin{bmatrix} B_1 & B_2 & B_3 & \ldots \end{bmatrix} = \begin{bmatrix} (A_1 \cdot B_1) & (A_1 \cdot B_2) & (A_1 \cdot B_3) & \cdots \\ (A_2 \cdot B_1) & (A_2 \cdot B_2) & (A_2 \cdot B_3) & \cdots \\ (A_3 \cdot B_1) & (A_3 \cdot B_2) & (A_3 \cdot B_3) & \cdots \\ \vdots & \vdots & \vdots & \ddots \end{bmatrix}$$

Properties of Matrix Multiplication

Matrix multiplication is not commutative (that is, AB ≠ BA), except in special cases. It is easy to see why: you cannot expect to switch the proportions with the vectors and get the same *result*. It is also easy to see how the order of the factors determines the result when one knows that the number of columns in the proportions matrix has to be the same as the number of rows in the vectors matrix: they have to represent the same number of vectors.

Although matrix multiplication is not commutative, the determinants of AB and BA are always equal (if A and B are square matrices of the same size). See the article on determinants for an explanation. However matrix multiplication is commutative when both matrices are diagonal and of the same dimension.

This notion of multiplication is important because if A and B are interpreted as linear transformations, then the matrix product AB corresponds to the composition of the two linear transformations, with B being applied first.

Additionally, all notions of matrix multiplication described here share a set of common properties described below.

Algorithms

The complexity of matrix multiplication, if carried out naively, is $O(n^3)$, but more efficient algorithms do exist. Strassen's algorithm, devised by Volker Strassen in 1969 and often referred to as "fast matrix multiplication", is based on a clever way of multiplying two 2×2 matrices which requires only 7 multiplications (instead of the usual 8), at the expense of several additional addition and subtraction operations. Applying this trick recursively gives an algorithm with a multiplicative cost of $O(n\log_2^7) \approx O(n^{2.809})$. Strassen's algorithm is awkward to implement, compared to the naive algorithm, and it lacks numerical stability. Nevertheless, it is beginning to appear in libraries such as BLAS, where it is computationally interesting for $n > 100$.

The algorithm with the lowest known exponent, which was presented by Don Coppersmith and Shmuel Winograd in 1990, has an asymptotic complexity of $O(n^{2.376})$. It is similar to Strassen's algorithm: a clever way is devised for multiplying two kXk matrices with fewer than k^3 multiplications, and this technique is applied recursively. It improves on the constant factor in Strassen's algorithm, reducing it to 4.537. However, the constant term implied in the $O(n^{2.376})$ result is so large that the Coppersmith–Winograd algorithm is only worthwhile for matrices that are too big to handle on present-day computers.

Since any algorithm for multiplying two nXn matrices has to process all $2\times n^2$ entries, there is an asymptotic lower bound of $\Omega(n^2)$ operations. Raz proves a lower bound of $\Omega(m^2\log m)$ for bounded coefficient arithmetic circuits over the real or complex numbers.

Cohn *et al.* put methods such as the Strassen and Coppersmith–Winograd algorithms in an entirely different, group-theoretic context.

They show that if families of wreath products of Abelian with symmetric groups satisfying certain conditions exists, matrix multiplication algorithms with essential quadratic complexity exist. Most researchers believe that this is indeed the case.

Relationship to Linear Transformations

Matrices offer a concise way of representing linear transformations between vector spaces, and (ordinary) matrix multiplication corresponds to the composition of linear transformations. This will be illustrated here by means of an example using three vector spaces of specific dimensions, but the correspondence applies equally to any other choice of dimensions.

Let X, Y, and Z be three vector spaces, with dimensions 4, 2, and 3, respectively, all over the same field, for example the real numbers. The coordinates of a point in X will be denoted as x_i, for $i = 1$ to 4, and analogously for the other two spaces.

Two linear transformations are given: one from Y to X, which can be expressed by the system of linear equations

$$x_1 = a_{1,1}y_1 + a_{1,2}y_2$$

$$x_2 = a_{2,1}y_1 + a_{2,2}y_2$$

$$x_1 = a_{3,1}y_1 + a_{3,2}y_2$$

$$x_1 = a_{4,1}y_1 + a_{4,2}y_2$$

and one from Z to Y, expressed by the system

$$y_1 = b_{1,1}z_1 + b_{1,2}z_2 + b_{1,3}z_3$$
$$y_2 = b_{2,1}z_1 + b_{2,2}z_2 + b_{2,3}z_3$$

These two transformations can be composed to obtain a transformation from Z to X. By substituting, in the first system, the right-hand sides of the equations of the second system for their corresponding left-hand sides, the x_i can be expressed in terms of the z_k:

$$\begin{aligned}
x_1 &= a_{1,1}(b_{1,1}z_1 + b_{1,2}z_2 + b_{1,3}z_3) + a_{2,1}(b_{2,1}z_1 + b_{2,2}z_2 + b_{2,3}z_3) \\
&= (a_{1,1}\, b_{1,1} + a_{1,2}b_{2,1})z_1 + (a_{1,1}b_{1,2} + a_{1,2}b_{2,2}z_2) + (a_{1,1}b_{1,3} + a_{1,2}b_{2,3})z_3 \\
x_2 &= a_{2,1}\,(b_{1,1} + z_{1,2}b_{2,1}z_2 + b_{1,3,\,z_3}) + a_{2,2}(b_{2,1}z_1 + b_{2,2}b_{2,2}z_2 + b_{2,3}z_3) \\
&= (a_{2,1}b_{1,1} + a_{2,2}b_{2,1})z_1 + (a_{2,1}b_{1,2} + a_{2,2}b_{2,2})z_1 + (a_{2,1}b_{1,3} + a_{2,2}b_{2,3})z_3 \\
x_3 &= a_{3,1}\,(b_{1,1}\; z_1 + b_{1,2}z_2 + b_{1,2}z_3) + a_{3,2}\,(b_{2,1}z_1 + b_{2,2}z_2 + b_{2,3}z_3) \\
&= (a_{3,1}b_{1,1} + a_{3,2}b_{2,1})z_1 + (a_{3,1}b_{1,2} + a_{3,2}b_{2,2})z_2 + (a_{3,1}b_{1,3} + a_{2,2}b_{2,3})z_3 \\
x_4 &= a_{4,2}(b_{1,1}z_1 + b_{1,2}z_2 + b_{1,3}z_3) + a_{4,2}(b_{2,1}z_1 + b_{2,2}z_2 + b_{2,3}z_3) \\
&= (a_{4,1}b_{1,1} + a_{4,2}b_{2,1})z_1 + (a_{4,1}b_{1,2} + a_{4,2}b_{2,2})z_2 + (a_{4,1}b_{1,3} + a_4 b_{2,2}b_{2,3})z_3
\end{aligned}$$

These three systems can be written equivalently in matrix–vector notation – thereby reducing each system to a single equation – as follows:

$$\begin{bmatrix} x_1 \\ x_2 \\ x_3 \\ x_4 \end{bmatrix} = \begin{bmatrix} a_{1,1} & a_{1,2} \\ a_{2,1} & a_{2,2} \\ a_{3,1} & a_{3,2} \\ a_{4,1} & a_{4,2} \end{bmatrix} \cdot \begin{bmatrix} y_1 \\ y_2 \end{bmatrix}.$$

$$\begin{bmatrix} y_1 \\ y_2 \end{bmatrix} = \begin{bmatrix} b_{1,1} & b_{1,2} & b_{1,3} \\ b_{2,1} & b_{2,2} & b_{2,3} \end{bmatrix} \cdot \begin{bmatrix} z_1 \\ z_2 \\ z_3 \end{bmatrix}$$

$$\begin{bmatrix} x_1 \\ x_2 \\ x_3 \\ x_4 \end{bmatrix} = \begin{bmatrix} a_{1,1}b_{1,1} + a_{1,2}b_{2,1} & a_{1,1}b_{1,2} + a_{1,2}b_{2,2} & a_{1,1}b_{1,3} + a_{1,2}b_{2,3} \\ a_{2,1}b_{1,1} + a_{2,2}b_{2,1} & a_{2,1}b_{1,2} + a_{2,2}b_{2,2} & a_{2,1}b_{1,3} + a_{2,2}b_{2,3} \\ a_{3,1}b_{1,1} + a_{3,2}b_{2,1} & a_{3,1}b_{1,1} + a_{3,2}b_{2,2} & a_{3,1}b_{1,3} + a_{3,2}b_{2,3} \\ a_{4,1}b_{1,1} + a_{4,2}b_{2,1} & a_{4,1}b_{1,2} + a4b_{2,2} & a_{4,1}b_{1,3} + a_{4,2}b_{2,3} \end{bmatrix} \cdot \begin{bmatrix} z_1 \\ z_2 \\ z_3 \end{bmatrix}$$

Representing these three equations symbolically and more concisely as

$$x = Ay$$
$$y = Bz$$
$$x = Cz$$

inspection of the entries of matrix C reveals that C = AB.

This can be used to formulate a more abstract definition of matrix multiplication, given the special case of matrix–vector multiplication: the product AB of matrices A and B is the matrix C such that for all vectors z of the appropriate shape Cz = A(Bz).

Scalar Multiplication

The scalar multiplication of a matrix A = (a_{ij}) and a scalar r gives a product rA of the same size as A. The entries of rA are given by

$$(r\mathrm{A})_{ij} = r \cdot a_{ij}.$$

For example, if

$$A = \begin{bmatrix} 1 & 2 \\ 3 & 4 \end{bmatrix}$$

then

$$7A = \begin{bmatrix} 7\cdot 1 & 7\cdot 2 \\ 7\cdot 3 & 7\cdot 4 \end{bmatrix} = \begin{bmatrix} 7 & 14 \\ 21 & 28 \end{bmatrix}$$

If we are concerned with matrices over a ring, then the above multiplication is sometimes called the *left multiplication* while the *right multiplication* is defined to be

$$(r\mathrm{A})_{ij} = a_{ij} \cdot r.$$

When the underlying ring is commutative, for example, the real or complex number field, the two multiplications are the same. However, if the ring is not commutative, such as the quaternions, they may be different. For example

$$i\begin{bmatrix} i & 0 \\ 0 & j \end{bmatrix} = \begin{bmatrix} -1 & 0 \\ 0 & k \end{bmatrix} = \begin{bmatrix} -1 & 0 \\ 0 & -k \end{bmatrix} = \begin{bmatrix} I & 0 \\ 0 & J \end{bmatrix}.$$

SCALAR AND MATRIX MULTIPLICATION

There are two types of multiplication for matrices: scalar multiplication and matrix multiplication. Scalar multiplication is easy. You just take a number (called a "scalar") and multiply it on every entry in the matrix. For example:

1. For the following matrix A, find $2A$ and $-1A$.

$$A = \begin{bmatrix} 1 & 2 \\ 3 & 4 \end{bmatrix}$$

To do the scalar multiplication, I just multiply a 2 on every entry in the matrix:

$$2A = 2 \cdot \begin{bmatrix} 1 & 2 \\ 3 & 4 \end{bmatrix} = \begin{bmatrix} 2 \cdot 1 & 2 \cdot 2 \\ 2 \cdot 3 & 2 \cdot 4 \end{bmatrix} = \begin{bmatrix} 2 & 4 \\ 6 & 8 \end{bmatrix}$$

The other scalar multiplication works the same way:

$$-1A = -1 \cdot \begin{bmatrix} 1 & 2 \\ 3 & 4 \end{bmatrix} = \begin{bmatrix} -1 \cdot 1 & -1 \cdot 2 \\ -1 \cdot 3 & -1 \cdot 4 \end{bmatrix} = \begin{bmatrix} -1 & -2 \\ -3 & -4 \end{bmatrix}$$

So the final answer is:

$$2A = \begin{bmatrix} 2 & 4 \\ 6 & 8 \end{bmatrix}$$

$$-1A = \begin{bmatrix} -1 & -2 \\ -3 & -4 \end{bmatrix}$$

Scalar multiplication is easy. Matrix multiplication, however, is quite another story: it's a royal pain. Your text probably gave you a complex formula for the process, and that formula probably didn't make any sense to you. That's okay. The process is messy, and that complicated formula is the best they can do for an explanation in a formal setting like a textbook. Here's how the process works:

1. Find the product AB for the following matrices:

$$A = \begin{bmatrix} 1 & 0 & -2 \\ 0 & 3 & -1 \end{bmatrix}$$

$$B = \begin{bmatrix} 0 & 3 \\ -2 & -1 \\ 0 & 4 \end{bmatrix}$$

To calculate AB, I write down A and B next to each other like this:

$$\begin{bmatrix} 1 & 0 & -2 \\ 0 & 3 & -1 \end{bmatrix} \begin{bmatrix} 0 & 3 \\ -2 & -1 \\ 0 & 4 \end{bmatrix}$$

Now I need to multiply the ROWS of A by the COLUMNS of B. By this I mean that I first take the first row of A and the first column of B, and I multiply the first entries, then the second entries, and then the third entries, and then I add the three products. The sum is one entry in the product matrix AB; in fact, being the product of row 1 and column 1, the result is the 1,1-entry of AB. Then I continue in like manner. For instance, the sum of the products from row 2 of A and column 1 of B is the 2,1-entry of AB.

$$\begin{bmatrix} 1 & 0 & -2 \\ 0 & 3 & -1 \end{bmatrix}\begin{bmatrix} 0 & 3 \\ -2 & -1 \\ 0 & 4 \end{bmatrix} = \begin{bmatrix} 1\cdot 0+0\cdot(-2)+(-2)\cdot 0 & 1\cdot 3+0\cdot(-1)+(-2)\cdot 4 \\ 0\cdot 0+3\cdot(-2)+(-1)\cdot 0 & 3\cdot 3+3\cdot(-1)+(-1)\cdot 4 \end{bmatrix}$$

$$= \begin{bmatrix} 0+0+0 & 3+0-8 \\ 0-3\cdot 6+0 & 3-3-4 \end{bmatrix} = \begin{bmatrix} 0 & -5 \\ -6 & -7 \end{bmatrix}$$

The final answer is:

$$AB = \begin{bmatrix} 0 & -5 \\ -6 & -7 \end{bmatrix}$$

As you saw on the above example, the general rule is that the product of the i-th row of A and the j-th column of B is the i,j-th entry of the product matrix AB. This general rule is, in large part, what that complicated formula in your text was all about.

For instance, when I, in the above example, multiplied the first row (of A) and the second column (of B), this gave me the first-row-second-column entry in the product matrix AB.

Just as with adding matrices, the sizes of the matrices matter when we are multiplying. For matrix multiplication to work, the columns of the second matrix have to have the same number of entries as the rows of the first matrix.

$$AB = \begin{bmatrix} 1 & 0 & -2 \\ 0 & 3 & -1 \end{bmatrix}\begin{bmatrix} 0 & 3 \\ -2 & -1 \\ 0 & 4 \end{bmatrix}$$

If, using the above matrices, B had had only two rows, its columns would have been too short to multiply against the rows of A. Then "AB" would not have existed. Likewise, if B had had, say, four rows, or alternatively if A had had two or four columns, then AB would not have existed, because A and B would not have been the right sizes.

In other words, for AB to exist (that is, for the very process of matrix multiplication to be able to function), A must have the same number of columns as B has rows; looking at the matrices, the rows of A must be the same length as the columns of B.

You can use this fact to check quickly whether a given multiplication is defined. Write the product in terms of the matrix dimensions. In the case of the above problem, *A* is 2X3 and *B* is 3 X 2, so *AB* is (2 X 3)(3 X 2). The middle values match:

$$\overbrace{(2\times 3)(3\times 2)}^{\text{Product is defined}}$$

...so the multiplication is defined. By the way, you will recall that *AB*, the product matrix, was 2 X 2. You can also see this on the dimensions:

$$\underbrace{(2\times 3)(3\times 2)}_{\text{Product is will be } 2\times 2}$$

Using this, you can see that *BA* ;must be a different matrix from *AB*, because:

$$\overbrace{\underbrace{(2\times 3)(3\times 2)}_{\text{Product is will be } 3\times 3}}^{\text{Product is defined}}$$

The product *BA* is defined (that is, we can do the multiplication), but the product, when the matrices are multiplied in this order, will be 3 X 3, not 2 X 2.

1. Given the following matrices, find the product *BA*.

$$A = \begin{bmatrix} 1 & 0 & -2 \\ 0 & 3 & -1 \end{bmatrix}$$

$$B = \begin{bmatrix} 0 & 3 \\ -2 & -1 \\ 0 & 4 \end{bmatrix}$$

The multiplication works the same way as the previous problem, going across the rows and down the columns. I won't try drawing my hands again, but you can see the computations in the colors:

$$A = \begin{bmatrix} 0 & 3 \\ -2 & -1 \\ 0 & 4 \end{bmatrix} \begin{bmatrix} 1 & 0 & -2 \\ 0 & 3 & -1 \end{bmatrix}$$

$$= \begin{bmatrix} 0\cdot 1 + 3\cdot 0 & 0\cdot 0 + 3\cdot 3 & 0\cdot(-2) + 3\cdot(-1) \\ -2\cdot 1 + (-1)\cdot 0 & -2\cdot 0 + (-1)\cdot 3 & -2\cdot(-2) + (-1)\cdot(-1) \\ 0\cdot 1 + 4\cdot 0 & 0\cdot 0 + 4\cdot 3 & 0\cdot(-2) + 4\cdot(-1) \end{bmatrix}$$

$$= \begin{bmatrix} 0+0 & 0+9 & 0-3 \\ -2+0 & 0-3 & 4+1 \\ 0+0 & 0+12 & 0-4 \end{bmatrix} = \begin{bmatrix} 0 & 9 & -3 \\ -2 & -3 & 5 \\ 0 & 12 & -4 \end{bmatrix}$$

Then the answer is:

$$BA = \begin{bmatrix} 0 & 9 & -3 \\ -2 & -3 & 5 \\ 0 & 12 & -4 \end{bmatrix}$$

Here are a couple more examples of matrix multiplication:

1. Find *CD* and *DC*, if they exist, given that *C* and *D*are the following matrices:

$$C = \begin{bmatrix} 2 & -1 \\ 0 & 3 \\ 1 & 0 \end{bmatrix} \quad D = \begin{bmatrix} 0 & 1 & 4 & -1 \\ -2 & 0 & 0 & 2 \end{bmatrix}$$

C is a 3 X 2 matrix and *D* is a 2X4 matrix, so first I'll look at the dimension product for *CD*:

Product is defined

$$(2 \times 3)(3 \times 4)$$

Product is will be 3 × 4

So the product *CD* is defined (that is, I can do the multiplication); also, I can tell that I'm going to get a 3X4 matrix for my answer. Here's the multiplication:

$$CD = \begin{bmatrix} 2 & -1 \\ 0 & 3 \\ 1 & 0 \end{bmatrix} \begin{bmatrix} 0 & 1 & 4 & -1 \\ -2 & 0 & 0 & 2 \end{bmatrix}$$

$$= \begin{bmatrix} 2 \cdot 0 + (-1)(-2) & 2 \cdot 1 + (-2) \cdot 0 & 2 \cdot 4 + (-1) \cdot 0 & 2 \cdot (-1) + (-1) \cdot 2 \\ 0 \cdot 0 + 3 \cdot (-2) & 0.1 + 3 \cdot 0 & 0 \cdot 4 + 3 \cdot 0 & 0 \cdot (-1) + 3 \cdot 2 \\ 1 \cdot 0 + 0 \cdot (-2) & 1 \cdot 1 + 0 \cdot 0 & 1 \cdot 4 + 0 \cdot 0 & 1 \cdot (-1) + 0 \cdot 2 \end{bmatrix}$$

$$= \begin{bmatrix} 0+2 & 2+0 & 8+0 & -2-2 \\ 0-6 & 0+0 & 0+0 & 0+6 \\ 0+0 & 1+0 & 4+0 & -1+0 \end{bmatrix} = \begin{bmatrix} 2 & 2 & 8 & -4 \\ -6 & 0 & 0 & 6 \\ 0 & 1 & 4 & -1 \end{bmatrix}$$

However, look at the dimension product for *DC*:

Product is not defined

$$(2 \times 4)(3 \times 2)$$

Since the inner dimensions don't match, I can't do the multiplication. (The columns of *C* aren't the same length as the rows of *D*; the columns of *C* are too short, or, if you prefer, the rows of *D* are too long.) Then the answer is:

$$CD = \begin{bmatrix} 2 & 2 & 8 & -4 \\ -6 & 0 & 0 & 6 \\ 0 & 1 & 4 & -1 \end{bmatrix}$$

DC is not defined.

1. Given the following matrices A and B, and defining C as $AB = C$, find the values of entries $c_{3,2}$ and $c_{2,3}$ in matrix C.

$$A = \begin{bmatrix} 4 & 3 & 6 & -1 \\ 0 & 2 & 1 & 4 \\ 3 & -2 & -2 & -2 \\ 0 & 0 & 1 & 3 \end{bmatrix} \quad B = \begin{bmatrix} 2 & 3 & 0 \\ 0 & 4 & -2 \\ 1 & 0 & -2 \\ 0 & -1 & 0 \end{bmatrix}$$

The dimension product of AB is (4 X 4)(4 X 3), so the multiplication will work, and C will be a 4 X 3 matrix. But to find $c_{3,2}$, I don't need to do the whole matrix multiplication. The 3,2-entry is the result of multiplying the third row of A against the second column of B, so I'll just do that:

$$c_{3,2} = (3)(3) + (-2)(4) + (-2)(0) + (-2)(-1) = 9 - 8 + 0 + 2 = 3$$

On the other hand, $c_{2,3}$ is the result of multiplying the second row of A against the third column of B, so:

$$c_{2,3} = (0)(0) + (2)(-2) + (1)(-2) + (4)(0) = 0 - 4 - 2 + 0 = -6$$

$$c_{3,2} = 3 \text{ and } c_{2,3} = -6$$

This type of problem serves as a reminder that, in general, to find $c_{i,j}$ you multiply row i of A against column j of B.

MULTIPLYING BY THE IDENTITY

Multiplying by the identity matrix I doesn't change anything, just like multiplying a number by 1 doesn't change anything. This property is why I and 1 are each called the "multiplicative identity". But while there is only one "multiplicative identity" for regular numbers (namely the number 1), there are lots of different identity matrices. Why? Because the identity matrix you need will depend upon the size of the matrix that it is being multiplied on! For instance, suppose you have the following matrix A:

$$A = \begin{bmatrix} 1 & 2 & 3 \\ 4 & 5 & 6 \end{bmatrix}$$

To multiply A on the right by the identity (that is, to do AI), you have to use I_3, the 3×3 identity, in order to have the right number of rows for the multiplication to work:

$$AI_3 = \begin{bmatrix} 1 & 2 & 3 \\ 4 & 5 & 6 \end{bmatrix} \begin{bmatrix} 1 & 0 & 0 \\ 0 & 1 & 0 \\ 0 & 0 & 1 \end{bmatrix}$$

$$= \begin{bmatrix} 1\cdot 1 + 2.0 + 3\cdot 0 & 1\cdot 0 + 2\cdot 1 + 3\cdot 0 & 1\cdot 0 + 2\cdot 0 + 3\cdot 1 \\ 4\cdot 1 + 5\cdot 0 + 6\cdot 0 & 4\cdot 0 + 5\cdot 1 + 6\cdot 0 & 4\cdot 0 + 5\cdot 0 + 6\cdot 1 \end{bmatrix}$$

$$= \begin{bmatrix} 1 & 2 & 3 \\ 4 & 5 & 6 \end{bmatrix} = A$$

On the other hand, to multiply A on the left by the identity, you have to use I_2, the 2 X 2 identity, in order to have the right number of columns:

$$I_2A = \begin{bmatrix} 1 & 0 \\ 0 & 1 \end{bmatrix} \begin{bmatrix} 1 & 2 & 3 \\ 4 & 5 & 6 \end{bmatrix}$$

$$= \begin{bmatrix} 1\cdot 1 + 2.0 + 3\cdot 0 & 1\cdot 0 + 2\cdot 1 + 3\cdot 0 & 1\cdot 0 + 2\cdot 0 + 3\cdot 1 \\ 4\cdot 1 + 5\cdot 0 + 6\cdot 0 & 4\cdot 0 + 5\cdot 1 + 6\cdot 0 & 4\cdot 0 + 5\cdot 0 + 6\cdot 1 \end{bmatrix}$$

$$= \begin{bmatrix} 1 & 2 & 3 \\ 4 & 5 & 6 \end{bmatrix} = A$$

That is, if you are dealing with a non-square matrix (such as A in the above example), the identity matrix you use will depend upon the side that you're multiplying on. This is just another example of matrix weirdness. Don't let it scare you. Matrices aren't bad; they're just different... really, really different.

Hadamard Product

For two matrices of the same dimensions, we have the Hadamard product, also known as the entrywise product and the Schur product. It can be generalized to hold not only for matrices but also for operators. The Hadamard product of two *m*-by-*n* matrices A and B, denoted by A • B, is an *m*-by-*n* matrix given by $(A \bullet B)_{ij} = a_{ij}b_{ij}$. For instance

$$\begin{bmatrix} 1 & 2 \\ 3 & 1 \end{bmatrix} \bullet \begin{bmatrix} 0 & 3 \\ 2 & 1 \end{bmatrix} = \begin{bmatrix} 1\cdot 0 & 2\cdot 3 \\ 3\cdot 2 & 1\cdot 1 \end{bmatrix} = \begin{bmatrix} 0 & 6 \\ 6 & 1 \end{bmatrix}.$$

COMPOSITION OF TRANSFORMATIONS

If the target of one transformation $S: X \to Y$ is the same as the source of another transformation $T: Y \to Z$, then we can combine the two transformations and form the composition $TS: X \to Z$, $TS(x) = T(S(x))$.

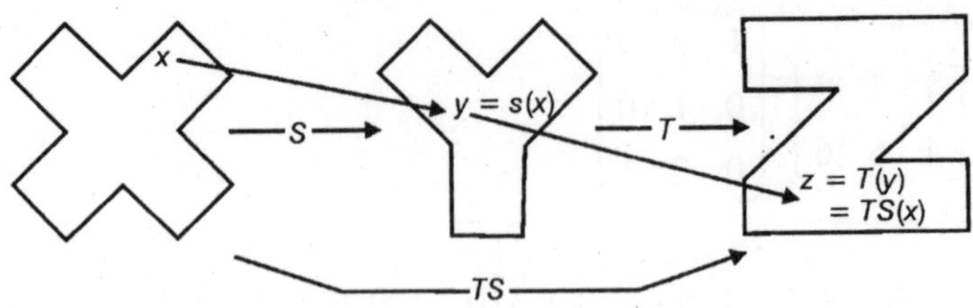

Example: The composition

$(ID_p)(Instructor)$: course → professor → number

is the same as

ID Number of the Instructor: course → number

The composition

(Population)(Capital City): country → city → number

is the same as

Population of the Capital City: country → number

Example: We are familiar with the following exercise: Suppose $y = x + 1$ and $z = y^2$. Then $z = (x + 1)^2 = x^2 + 2x + 1$. What we have done is actually composing two transformations

Add One: $R \to R, x \to x + 1$

Square: $R \to R, y \to y^2$

to form a transformation

Add One and then Take square: $R \to R, x \to (x + 1)^2$

For a multi–variable example, consider the transformations

$T: R^2 \to R^3, (y_1, y_2) \to (z_1, z_2, z_3) = (y_1^2 - 2y_2, y_1^3 - 3y_1y_2, y_1^4 - 4y_1^2y_2 - 2y_2^2)$

$S: R^2 \to R^2, (x_1, x_2) \to (y_1, y_2) = (x_1 + x_2, x_1x_2)$

The composition TS is computed by substituting the formulae of y_* (in terms of x_*) into the formulae of z_* (in terms of y_*).

$z_1 = (x_1 + x_2)^2 - 2x_1x_2 = x_1^2 + x_2^2$

$z_2 = (x_1 + x_2)^3 - 3(x_1 + x_2)x_1x_2 = x_1^3 + x_2^3$

$z_3 = (x_1 + x_2)^4 - 4(x_1 + x_2)^2x_1x_2 - 2(x_1x_2)^2 = x_1^4 + x_2^4$

Thus we have

$TS(x_1, x_2) = (x_1^2 + x_2^2, x_1^3 + x_2^3, x_1^4 + x_2^4)$.

If transformations are given by formulae, then the composition is simply the substitution.

Example: Let T and S be the reflection in x–axis and the rotation by 90 degrees on the plane R^2. The following picture illustrates the composition TS.

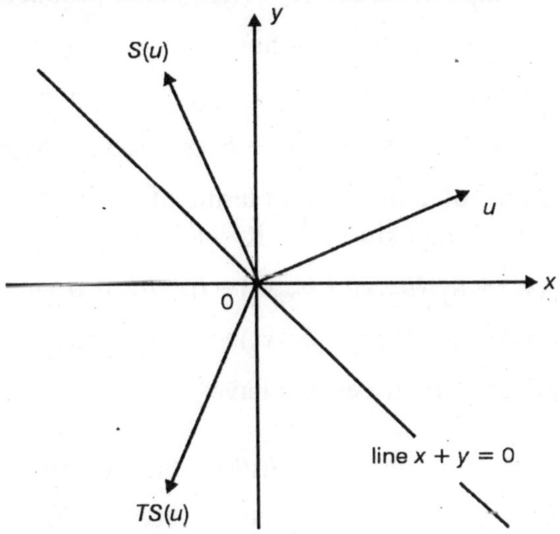

From the picture, we see that the composition TS is the reflection with respect to the line $x + y = 0$.

Similarly, the following picture illustrates that the composition ST is the reflection with respect to the diagonal line $x = y$.

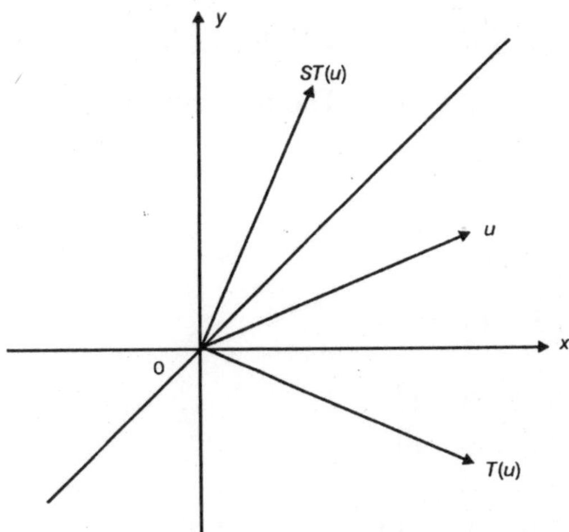

MULTIPLICATION OF MATRICES

The composition of two linear transformations is also linear. The geometrical reason is clear: if parallelograms and scalings are preserved by each transformation, then they are preserved by the combined transformation.

Sincc linear transformations between euclidean spaces are equivalent to matrices, the composition of linear transformations should have corresponding operation on matrices.

Let A and B be the matrices for linear transformations T: $R^n \to R^m$ and S: $R^k \to R^n$. Then the matrix for the composition TS: $R^k \to R^m$ is the product AB of A and B

Example Consider linear transformations

T: $R^2 \to R^2$, $(y_1, y_2) \to (z_1, z_2) = (a_{11}y_1 + a_{12}y_2, a_{21}y_1 + a_{22}y_2)$

S: $R^2 \to R^2$, $(x_1, x_2) \to (y_1, y_2) = (b_{11}x_1 + b_{12}x_2, b_{21}x_1 + b_{22}x_2)$

By substituting the formulae of y_1, y_2 in terms of x_1, x_2 into the formulae of z_1, z_2 in terms of y_1, y_2, we find the composition TS: $R^2 \to R^2$, $(x_1, x_2) \to (z_1, z_2)$ is given by

$$z_1 = a_{11}(b_{11}x_1 + b_{12}x_2) + a_{12}(b_{21}x_1 + b_{22}x_2) = (a_{11}b_{11} + a_{12}b_{21})x_1 + (a_{11}b_{12} + a_{12}b_{22})x_2$$

$$z_2 = a_{21}(b_{11}x_1 + b_{12}x_2) + a_{22}(b_{21}x_1 + b_{22}x_2) = (a_{21}b_{11} + a_{22}b_{21})x_1 + (a_{21}b_{12} + a_{22}b_{22})x_2$$

In terms of the product of matrices, we have

$$\begin{bmatrix} a_{11} & a_{12} \\ a_{21} & a_{22} \end{bmatrix} \begin{bmatrix} b_{11} & b_{12} \\ b_{21} & b_{22} \end{bmatrix} = \begin{bmatrix} a_{11}b_{11} + a_{12}b_{21} & a_{11}b_{12} + a_{12}b_{22} \\ a_{21}b_{11} + a_{22}b_{21} & a_{21}b_{12} + a_{22}b_{22} \end{bmatrix}$$

Note that in the example above, the first entry $a_{11}b_{11} + a_{12}b_{21}$ of the matrix AB is obtained by "multiplying" the first row $(a_{11}\ a_{12})$ of A with the first column $(b_{11}\ b_{21})$ of B. Similar observation can be made to the other entries. In fact, we have the following general rule for multiplying an m by n matrix A and an m by k matrix B together.

(i, j)-entry of AB = (i–th row of A) • (j-th row of B)

The result AB is an m by k matrix. The rigorous proof of the rule can be found in the next section.

Example: Let

$$A = \begin{bmatrix} 1 & 2 \\ 3 & 4 \end{bmatrix}$$

$$B = \begin{bmatrix} -1 & 0 & 2 \\ 0 & 1 & 3 \end{bmatrix}$$

$$C = \begin{bmatrix} 0 & 1 & -1 \\ 2 & 0 & 3 \\ -1 & 1 & 2 \end{bmatrix}$$

$$D = \begin{bmatrix} 1 & 2 \\ 2 & 3 \\ 3 & 4 \end{bmatrix}$$

Then

$$AB = \begin{bmatrix} 1\times(-1)+2\times 0 & 1\times 0+2\times 1 & 1\times 2+2\times 3 \\ 3\times(-1)+4\times 0 & 3\times 0+4\times 1 & 3\times 2+4\times 3 \end{bmatrix} = \begin{bmatrix} -1 & 2 & 8 \\ -3 & 4 & 18 \end{bmatrix}$$

$$A^2 = AA = \begin{bmatrix} 1\times 1+2\times 3 & 1\times 2+2\times 4 \\ 3\times 1+4\times 3 & 3\times 2+4\times 4 \end{bmatrix} = \begin{bmatrix} 7 & 10 \\ 15 & 22 \end{bmatrix}$$

$$DA = \begin{bmatrix} 1\times 1+2\times 3 & 1\times 2+2\times 4 \\ 2\times 1+3\times 3 & 2\times 2+3\times 4 \\ 3\times 1+4\times 3 & 3\times 2+4\times 4 \end{bmatrix} = \begin{bmatrix} 7 & 10 \\ 11 & 16 \\ 15 & 22 \end{bmatrix}$$

$$BC = \begin{bmatrix} (-1)\times 0+0\times 2+2\times(-1) & (-1)\times 1+0\times 0+2\times 1 & (-1)\times(-1)+0\times 3+2\times 2 \\ 0\times 0+1\times 2+3\times(-1) & 0\times 1+1\times 0+3\times 1 & 0\times(-1)+1\times 3+3\times 2 \end{bmatrix}$$

$$= \begin{bmatrix} -2 & 1 & 5 \\ -1 & 3 & 9 \end{bmatrix}$$

$$(AB)C = \begin{bmatrix} (-1)\times 0+2\times 2+8\times(-1) & (-1)\times 1+2\times 0+8\times 1 & (-1)\times(-1)+2\times 3+8\times 2 \\ (-3)\times 0+4\times 2+18\times(-1) & (-3)\times 1+4\times 0+18\times 1 & (-3)\times(-1)+4\times 3+18\times 2 \end{bmatrix}$$

$$= \begin{bmatrix} -4 & 7 & 23 \\ -10 & 15 & 51 \end{bmatrix}$$

$$A(BC) = \begin{bmatrix} 1\times(-2)+2\times(-1) & 1\times 1+2\times 3 & 1\times 5+2\times 9 \\ 3\times(-2)+4\times(-1) & 3\times 1+4\times 3 & 3\times 5+4\times 9 \end{bmatrix} = \begin{bmatrix} -4 & 7 & 23 \\ -10 & 15 & 51 \end{bmatrix}$$

On the other hand, the products such as AD, AC, BA, CB, B^2 are meaningless.

Example: According to an earlier example, the matrix for the rotation by angle α is

$$R_\alpha = \begin{bmatrix} \cos a & -\sin a \\ \sin a & \cos a \end{bmatrix}$$

Since rotation by β and then rotation by α is the same as rotation by $\alpha + \beta$, we must have $R_\alpha R_\beta = R_{\alpha+\beta}$, i.e.,

$$\begin{bmatrix} \cos a & -\sin a \\ \sin a & \cos a \end{bmatrix} \begin{bmatrix} \cos\beta & -\sin\beta \\ \sin\beta & \cos\beta \end{bmatrix} = \begin{bmatrix} \cos(a+\beta) & -\sin(a+\beta) \\ \sin(a+\beta) & \cos(a+\beta) \end{bmatrix}$$

If we compute matrix multiplication on the left side in the usual way, we get the following familiar formulae in trigonometry.

$$\cos\alpha\cos\beta - \sin\alpha\sin\beta = \cos(\alpha + \beta)$$

$$\sin\alpha\cos\beta + \cos\alpha\sin\beta = \sin(\alpha + \beta)$$

COMPOSITION OF LINEAR TRANSFORMATIONS IS LINEAR

Let T: $\mathrm{R}^m \rightarrow \mathrm{R}^n$ and S: $\mathrm{R}^k \rightarrow \mathrm{R}^n$ be linear transformations. Then by the definition, we have

- $T(u + v) = T(u) + T(v)$,
- $T(cu) = cT(u)$,
- $S(u + v) = S(u) + S(v)$,
- $S(cu) = cS(u)$.

To show the composition TS: $\mathrm{R}^k \rightarrow \mathrm{R}^m$ is also linear means the verification of the following equalities:

$$TS(u + v) = TS(u) + TS(v),$$

$$TS(cu) = cTS(u).$$

The first equality is verified as follows:

$$\begin{aligned} TS(u + v) &= T(S(u + v)) \\ &= T(S(u) + S(v)) \\ &= T(S(u)) + T(S(v)) \\ &= TS(u) + TS(v) \end{aligned}$$

The second equality is verified as follows:

$$\begin{aligned} TS(cu) &= T(S(cu)) \\ &= T(Cs(u)) \\ &= cT(S(u)) \\ &= CTS(u) \end{aligned}$$

This completes the proof that TS is a linear transformation.

PROPERTIES OF MATRIX MULTIPLICATION

The matrix multiplication is consistent with our notation for the left side of a system of linear equations. Specifically, by thinking of a vertical vector x as a 1 by n matrix, the matrix product

$$Ax = \begin{bmatrix} a_{11} & a_{12} & \cdots & a_{1n} \\ a_{21} & a_{22} & \cdots & a_{2n} \\ \vdots & \vdots & & \vdots \\ a_{m1} & a_{m2} & \cdots & a_{mn} \end{bmatrix}$$

is exactly the left of the system $Ax = b$. The remark also shows that the rule for matrix multiplication is equivalent to

$$B = [b_1\ b_2 \ldots\ b_k] \Rightarrow AB = [Ab_1\ Ab_2 \ldots\ Ab_k].$$

Now we give a rigorous proof of the rule for matrix multiplication by showing the above is true.

Proof Let A and B be matrices for linear transformations T: $\mathrm{R}^m \to \mathrm{R}^n$ and S: $\mathrm{R}^k \to \mathrm{R}^n$. Then by this formula and the additional reason provided below, we have

$$\begin{aligned} AB &= [TS(e_1)\ TS(e_2) \ldots\ TS(e_k)] \text{ (definition: } AB \text{ is matrix for } TS) \\ &= [T(S(e_1))\ T(S(e_2)) \ldots\ T(S(e_k))] \text{ (definition of composition } TS) \\ &= [T(b_1)\ T(b_2) \ldots\ T(b_k)] \text{ (} B \text{ is matrix for } S\text{: } B = [S(e_1)\ S(e_2) \ldots\ S(e_k)]) \\ &= [Ab_1\ Ab_2 \ldots\ Ab_k]. \text{ (} A \text{ is matrix for } T\text{: } T(x) = Ax) \end{aligned}$$

We have seen the *associativity* $(AB)C = A(BC)$ for the matrix multiplication in this example. In general, this is the consequence of the corresponding associativity $(TS)R = T(SR)$ for the compositions of transformations (which is clearly true). Moreover, composing with the identity does not change the transformations: $id\,T = T = T\,id$. Since the identity transformation corresponds to the identity matrix I, we conclude that the product with the identity matrix does not change the matrix.

associative

$$(AB)C = A(BC)$$

identity

$$IA = A = AI$$

Thus the matrix multiplication is similar to number multiplication. However, the following example shows that one should be careful in stretching the analogue.

Example: In an earlier example, we have shown

(Reflection in x–axis)(Rotation by 90 degrees) = *(Reflection in the line* $x + y = 0$*)*

(Rotation by 90 degrees)(Reflection in x–axis) = *(Reflection in the line* $x = y$*)*

In particular, we see that the compositions TS and ST are different in general. In terms of the matrix product, we have

$$\begin{bmatrix} 1 & 0 \\ 0 & -1 \end{bmatrix}\begin{bmatrix} 0 & -1 \\ 1 & 0 \end{bmatrix} = \begin{bmatrix} 0 & -1 \\ -1 & 0 \end{bmatrix}$$

$$\begin{bmatrix} 0 & -1 \\ 1 & 0 \end{bmatrix}\begin{bmatrix} 1 & 0 \\ 0 & -1 \end{bmatrix} = \begin{bmatrix} 0 & 1 \\ 1 & 0 \end{bmatrix}$$

which shows $AB \neq BA$ in general.

Example: Consider two transformations T = *Projection to x–axis* and S = *Projection to y–axis* from R^2 to itself.

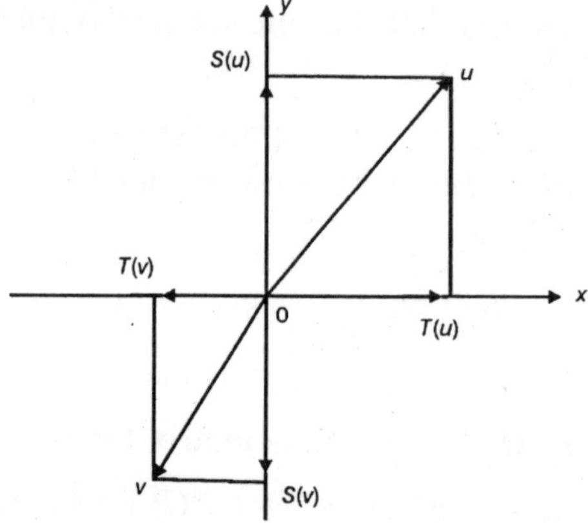

The composition TS is the zero transformation – taking everything to the zero vector 0.

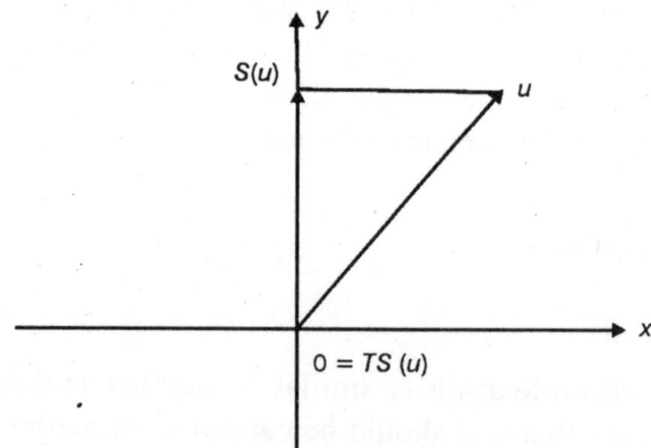

Correspondingly, we have two nonzero matrices with product being zero.

$$\begin{bmatrix} 1 & 0 \\ 0 & 0 \end{bmatrix}\begin{bmatrix} 0 & 0 \\ 0 & 1 \end{bmatrix} = \begin{bmatrix} 0 & 0 \\ 0 & 0 \end{bmatrix}$$

Finally, we denote

$$A^0 = I$$

$$A^1 = A$$
$$A^2 = AA$$
$$A^3 = AAA$$

....

In general, for positive k and square matrix A, we use Ak to denote $AA...A$, the product of k copies of A. The same notation is used for a transformation with the same same domain and codomain.

***Example*:** For a diagonal matrix,

$$A = \begin{bmatrix} a_1 & 0 & .. & 0 \\ 0 & a_2 & .. & 0 \\ \vdots & \vdots & & \vdots \\ 0 & 0 & .. & a_n \end{bmatrix}.$$

we have

$$AK = \begin{bmatrix} a_1^k & 0 & .. & 0 \\ 0 & a_2^k & .. & 0 \\ \vdots & \vdots & & \vdots \\ 0 & 0 & .. & a_n \end{bmatrix}$$

COMPOSITION/MULTIPLICATION MIXED WITH ADDITION AND SCALAR MULTIPLICATION

Consider the following equivalent columns.

Linear Transformation	*Matrix*
composition	1?
2?	addition
3?	scalar multiplication

Our answer to 1? is the matrix multiplication. What should be the answer to 2? and 3??

From the explicit formula for Ax, it is easy to see that

$(A + B)x = Ax + Bx$, $(cA)x = c(Ax)$.

Therefore the answer to 2? is the following definition,

For linear transformations T, S: $R^n \to R^m$, their addition $T + S$ is defined by the formula $(T + S)(x) = T(x) + S(x)$

and the answer to 3? is the following definition.

For a linear transformation $T{:}\cdot R^n \rightarrow R^m$, and a number r, the scalar multiplication rT is defined by the formula $(rT)(x) = r(T(x))$

Example: Let T and S be the reflection in x–axis and the rotation by 90 degrees on the plane R^2. The following picture illustrates the addition $T + S$.

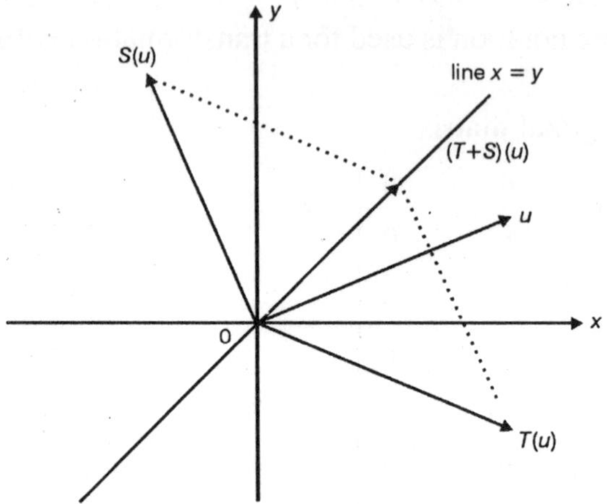

The formula for the addition is $(T + S)(x, y) = T(x, y) + S(x, y) = (x, -y) + (-y, x) = (x - y, x - y)$, and the corresponding matrix is

$$\begin{bmatrix} 1 & 0 \\ 0 & -1 \end{bmatrix} + \begin{bmatrix} 0 & -1 \\ 1 & 0 \end{bmatrix} = \begin{bmatrix} 1 & -1 \\ 1 & -1 \end{bmatrix}$$

For another example, consider two transformations T = *Projection to x–axis* and S = *Projection to y–axis* from R^2 to itself. The following picture shows their addition is the identity.

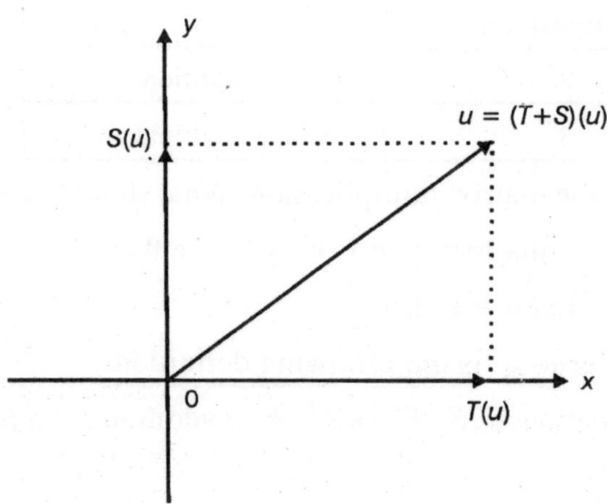

In terms of the corresponding matrix, we have

$$\begin{bmatrix} 1 & 0 \\ 0 & 0 \end{bmatrix} + \begin{bmatrix} 0 & 0 \\ 0 & 1 \end{bmatrix} = \begin{bmatrix} 1 & 0 \\ 0 & 1 \end{bmatrix}$$

When the multiplication is mixed with addition and scalar multiplication, we have the following properties.

distributive

$$(A + B)C = AC + BC$$

distributive

$$A(B + C) = AB + AC$$

associative

$$A(rB) = r(AB) = (rA)B$$

The properties are easy to prove in terms of linear transformations. For example, the following argument proves $(T + S)R = TR + SR$.

$$\begin{aligned} ((T + S)R)(x) &= (T + S)(R(x)) \text{ (definition of composition)} \\ &= T(R(x)) + S(R(x)) \text{ (definition of addition)} \\ &= (TR)(x) + (SR)(x) \text{ (definition of composition)} \\ &= (TR + SR)(x). \end{aligned}$$

The other properties can be proved in similar way.

Exercise: Compute the composition TS

1. $T(y) = 1 - y^2: \mathrm{R} \to \mathrm{R}$

   $S(x) = \sin x: \mathrm{R} \to \mathrm{R}$

Ans. $TS(x) = 1 - (\sin x)^2 = \cos^2 x: \mathrm{R} \to \mathrm{R}$

2. $T(x, y) = x^2 + y^2: \mathrm{R}^2 \to \mathrm{R}$

   $S(t) = (\cos t, \sin t): \mathrm{R} \to \mathrm{R}^2$

Ans. $TS(t) = (\cos t)^2 + (\sin t)^2 = 1: \mathrm{R} \to \mathrm{R}$

3. $T(x, y) = (x + y, x - y): \mathrm{R}^2 \to \mathrm{R}^2$

   $S(u, v) = (1/2(u + v), 1/2(u - v)): \mathrm{R}^2 \to \mathrm{R}^2$

Ans. $TS(u, v) = (1/2(u + v) + 1/2(u - v), 1/2(u + v) - 1/2(u - v)) = (u, v): \mathrm{R}^2 \to \mathrm{R}^2$. The composition is identity.

4. $T(x_1, x_2, x_3, x_4, x_5) = (x_1 - 2x_2 + 3x_3 - 4x_4 + 5x_5): \mathrm{R}^5 \to \mathrm{R}$

   $S(t) = (8t, 7t, 6t, 5t, 4t): \mathrm{R} \to \mathrm{R}^5$

Ans. $TS(t) = (8t - 2X7t + 3X6t - 4X5t + 5X4t) = 12t$: R '! R

5. $T(t) = (8t, 7t, 6t, 5t, 4t)$: $R \to R^5$

$S(x_1, x_2, x_3, x_4, x_5) = (x_1 - 2x_2 + 3x_3 - 4x_4 + 5x_5)$: $R^5 \to R$

Ans. $TS(x_1, x_2, x_3, x_4, x_5)$

$= (8(x_1 - 2x_2 + 3x_3 - 4x_4 + 5x_5), 7(x_1 - 2x_2 + 3x_3 - 4x_4 + 5x_5), 6(x_1 - 2x_2 + 3x_3 - 4x_4 + 5x_5), 5(x_1 - 2x_2 + 3x_3 - 4x_4 + 5x_5), 4(x_1 - 2x_2 + 3x_3 - 4x_4 + 5x_5))$

$= (8x_1 - 16x_2 + 24x_3 - 32x_4 + 40x_5, 7x_1 - 14x_2 + 21x_3 - 28x_4 + 35x_5, 6x_1 - 12x_2 + 18x_3 - 24x_4 + 30x_5, 5x_1 - 10x_2 + 15x_3 - 20x_4 + 25x_5, 4x_1 - 8x_2 + 12x_3 - 16x_4 + 20x_5)$: R^5 '! R^5

6. $T(x_1, x_2, x_3) = (x_1 - 2x_2 + 3x_3, - 2x_1 + 3x_2 - 4x_3)$: R^3 '! R^2

$S(y_1, y_2) = (- 2y_1 - y_2, y_2, y_1 + y_2)$: $R^2 \to R^3$

Ans. $TS(y_1, y_2) = ((- 2y_1 - y_2) - 2y_2 + 3(y_1 + y_2), - 2(- 2y_1 - y_2) + 3y_2 - 4(y_1 + y_2)) = (y_1, y_2)$: $R^2 \to R^2$. The composition is identity.

7. $T(y_1, y_2) = (- 2y_1 - y_2, y_2, y_1 + y_2)$: $R^2 \to R^3$

$S(x_1, x_2, x_3) = (x_1 - 2x_2 + 3x_3, - 2x_1 + 3x_2 - 4x_3)$: $R^3 \to R^2$

Ans. $TS(x_1, x_2, x_3)$

$= (- 2(x_1 - 2x_2 + 3x_3) - (- 2x_1 + 3x_2 - 4x_3), - 2x_1 + 3x_2 - 4x_3, (x_1 - 2x_2 + 3x_3) + (- 2x_1 + 3x_2 - 4x_3))$

$= (x_2 - 2x_3, - 2x_1 + 3x_2 - 4x_3, - x_1 + x_2 - x_3)$: $R^3 \to R^3$

Exercise: Draw the picture and identify the composition *TS*.

1. *T = Reflection in x–axis*: $R^2 \to R^2$

S = Reflection in y–axis: $R^2 \to R^2$

Ans. The composition *TS* is *Reflection in origin*: $R^2 \to R^2$

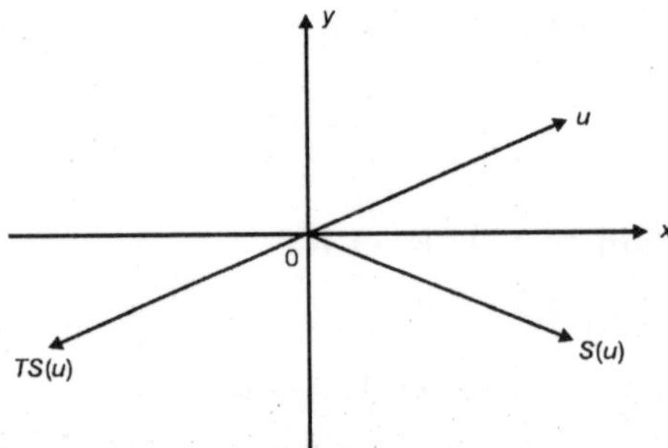

2. *T = Rotation by 90 degrees*: $R^2 \to R^2$

S = Reflection in origin: $R^2 \to R^2$

Ans. The composition TS is *Rotation by –90 degrees*: $R^2 \to R^2$

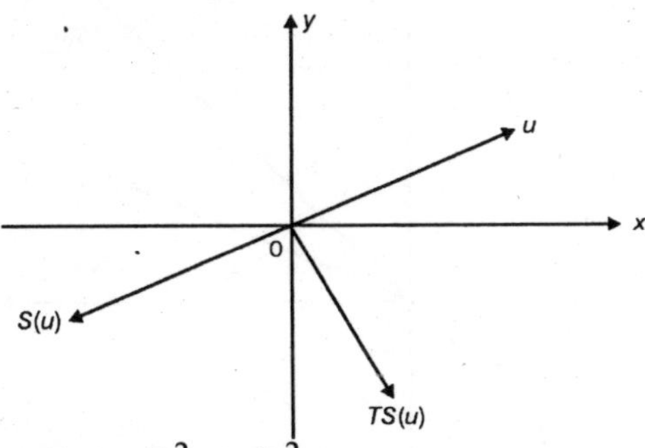

3. T = *Reflection in origin*: $R^2 \to R^2$

   S = *Rotation by 90 degrees*: $R^2 \to R^2$

Ans. The composition TS is *Rotation by –90 degrees*: $R^2 \to R^2$

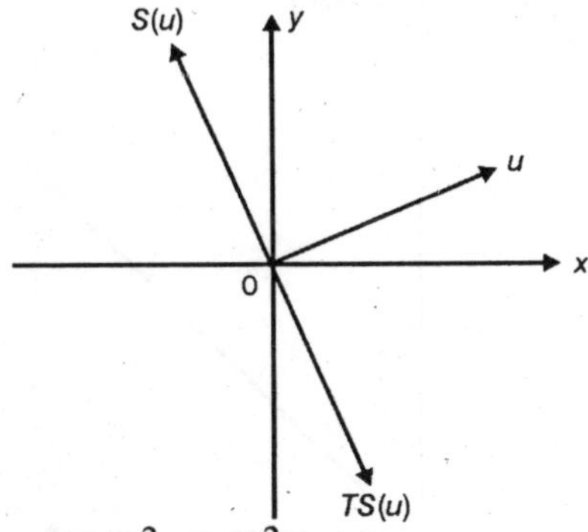

4. T = *Reflection in x–axis*: $R^2 \to R^2$

   S = *Embedding as diagonal*: $R \to R^2$

Ans. The composition TS is *Embedding as skew–diagonal*: $R^2 \to R^2$

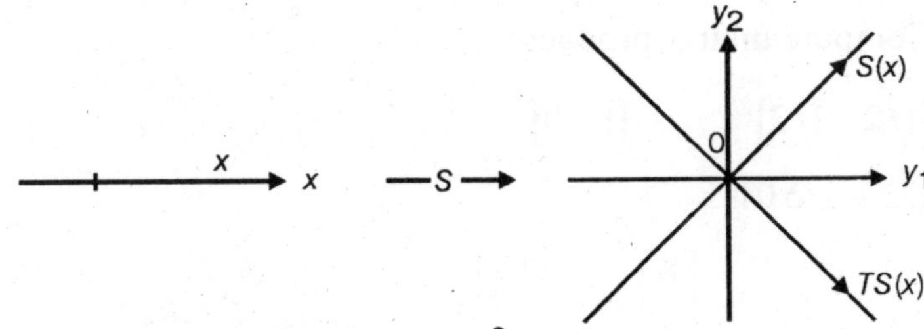

5. T = *Embedding as diagonal*: $R \to R^2$

   S = *Projection onto x–axis*: $R^2 \to R$

Ans. The composition TS is *Vertical intersection with diagonal*: $R^2 \to R^2$

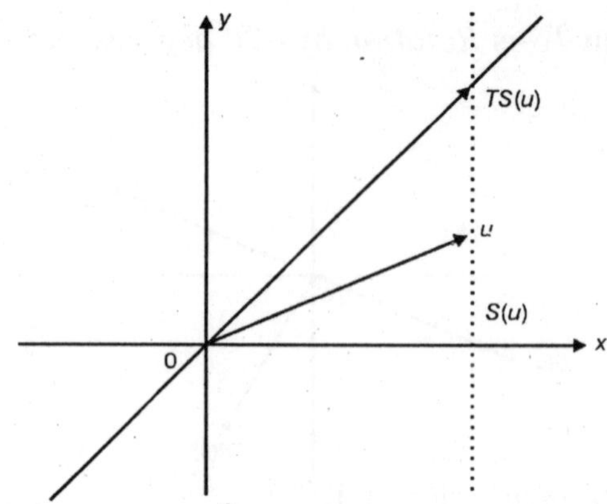

6. T = *Projection onto x–axis*: $R^2 \to R$

   S = *Embedding as diagonal*: $R \to R^2$

Ans. The composition TS is the identity.

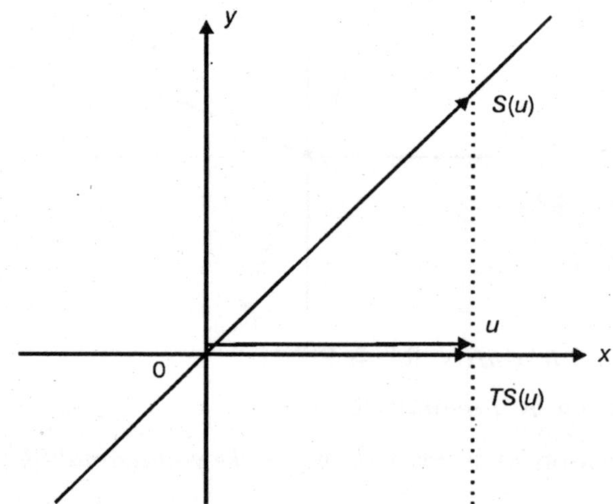

Exercise: Compute matrix products.

Ans. $\begin{bmatrix}1 & 1\end{bmatrix}\begin{bmatrix}1/2 & 1/2\end{bmatrix} = \begin{bmatrix}1 & 0\end{bmatrix}$

1 – 1 1/2 – 1/2 01

$$\begin{bmatrix}1 & -2 & 3 & -4 & 5\end{bmatrix}\begin{bmatrix}8\\7\\6\\5\\4\end{bmatrix} = [12]$$

$$\begin{bmatrix} 8 \\ 7 \\ 6 \\ 5 \\ 4 \end{bmatrix} \begin{bmatrix} 1 & -2 & 3 & -4 & 5 \end{bmatrix} = \begin{bmatrix} 8 & -16 & 24 & -32 & 40 \\ 7 & -14 & 21 & -28 & 35 \\ 6 & -12 & 18 & -24 & 30 \\ 5 & -10 & -15 & -20 & 25 \\ 4 & -8 & 12 & -16 & 20 \end{bmatrix}$$

$$\begin{bmatrix} 1 & -2 & 3 \\ -2 & 3 & -4 \end{bmatrix} \begin{bmatrix} -2 & -1 \\ 0 & 1 \\ 1 & 1 \end{bmatrix} = \begin{bmatrix} 1 & 0 \\ 0 & 1 \end{bmatrix}$$

$$\begin{bmatrix} -2 & -1 \\ 0 & 1 \\ 1 & 1 \end{bmatrix} \begin{bmatrix} 1 & -2 & 3 \\ -2 & 3 & -4 \end{bmatrix} = \begin{bmatrix} 0 & 1 & -2 \\ -2 & 3 & -4 \\ -1 & 1 & -1 \end{bmatrix}$$

$$\begin{bmatrix} -2 & -1 \\ 0 & 1 \end{bmatrix} \begin{bmatrix} 1 & -2 & 3 \\ -2 & 3 & -4 \end{bmatrix} = \begin{bmatrix} 0 & 1 & -2 \\ -2 & 3 & -4 \end{bmatrix}$$

$$\begin{bmatrix} 1 & 2 & 3 & 4 \\ 5 & 6 & 7 & 8 \\ 9 & 10 & 11 & 12 \end{bmatrix} \begin{bmatrix} 1 & 1 \\ -1 & -1 \\ 1 & -1 \\ -1 & 1 \end{bmatrix} = \begin{bmatrix} -2 & 0 \\ -2 & 0 \\ -2 & 0 \end{bmatrix}$$

$$\begin{bmatrix} 1 & -1 & 1 \\ -1 & 2 & -1 \\ 2 & 0 & -1 \end{bmatrix} \begin{bmatrix} 1 & 2 & 3 & 4 \\ 5 & 6 & 7 & 8 \\ 9 & 10 & 11 & 12 \end{bmatrix} = \begin{bmatrix} 5 & 6 & 7 & 8 \\ 0 & 0 & 0 & 0 \\ -7 & -6 & -5 & -4 \end{bmatrix}$$

The following exercises provide more examples that $AB = BA$ is generally not true.

Exercise: For each of the following matrix A, find all the matrices X satisfying $AX = XA$.

1. $A = [1 \quad 2]$

Ans. No X will satisfy $AX = XA$ because AX and XA have different sizes. Specifically, in order for AX to be meaningful, X must have two rows. In order for XA to be meaningful X must have one column. Then AX is a 1 by 1 matrix, and XA is a 2 by 2 matrix.

2. $A = \begin{bmatrix} 1 & 1 \\ 1 & -1 \end{bmatrix}$

Ans. Similar argument as in the first question shows that X must be a 2 by 2 matrix. Let

$$X = \begin{bmatrix} x & y \\ z & w \end{bmatrix}$$

Then

$$AX = \begin{bmatrix} x+z & y+w \\ x-z & y-w \end{bmatrix}, XA = \begin{bmatrix} x+y & x-y \\ z-w & z-w \end{bmatrix}$$

so that $AX = XA$ means

$$x + z = x + y, y + w = x - y, x - z = z + w, y - w = z - w.$$

The solution is $x = w + 2z$, $y = z$, and w, z are arbitrary. In summary, $AX = XA$ if and only if

$$X = \begin{bmatrix} x+2z & z \\ z & w \end{bmatrix} = w\begin{bmatrix} 1 & 0 \\ 0 & 1 \end{bmatrix} + z\begin{bmatrix} 2 & 1 \\ 1 & 0 \end{bmatrix}$$

3. $A = \begin{bmatrix} 1 & 0 \\ 0 & 2 \end{bmatrix}$

Ans. Again we assume

$$X = \begin{bmatrix} x & y \\ z & w \end{bmatrix}$$

and find the condition for $AX = XA$ to be $x = x$, $y = 2y$, $2z = z$, $2w = 2w$. The solution is $y = z = 0$, and x, w arbitrary. In other words, X is a diagonal matrix.

4. $A = \begin{bmatrix} 0 & 0 \\ 0 & 0 \end{bmatrix}$

Ans. For any 2 by 2 matrix X, we have $AX = O = XA$ (O is the 2 by 2 zero matrix). Therefore $AX = XA$ for any X.

5. $\begin{bmatrix} 2 & 0 \\ 0 & 2 \end{bmatrix}$

Ans. For any 2 by 2 matrix X, we have $AX = 2X = XA$. Therefore $AX = XA$ for any X.

6. $\begin{bmatrix} 2 & 0 & 0 \\ 0 & 2 & 0 \end{bmatrix}$

Ans. For AX and XA to be meaningful, X has to be a 3 by 2 matrix. Then AX is a 2 by 2 matrix, XA is a 3 by 3 matrix, and they cannot be equal.

Remark: Here are some further problems for you to think about.

1. If A is not a square matrix, then we can never have $AX = XA$ due to the contradiction in size.
2. In the 2nd and the 3rd questions, we have seen that two entries in X can be arbitrary. In the 4th and the 5th questions, all four entries in X can be arbitrary. Can you find examples of 2 by 2 matrix A, so that only one entry can be arbitrary? How about three arbitrary entries, or no arbitrary entries, in X?
3. What matrix commutes with all the other matrices? In other words, what A has the property that $AX = XA$ for all X? First try the 2 by 2 case, and then guess the general case.
4. What matrix commutes with all the diagonal matrices? First try the 2 by 2 diagonal matrices

$$X = \begin{bmatrix} x & 0 \\ 0 & w \end{bmatrix}$$

5. And then the general n by n matrices.
6. What matrix commutes with all the matrices of the form

$$X = \begin{bmatrix} w+2z & z \\ z & w \end{bmatrix} = w\begin{bmatrix} 1 & 0 \\ 0 & 1 \end{bmatrix} + z\begin{bmatrix} 2 & 1 \\ 1 & 0 \end{bmatrix}$$

7. We know the matrix A in the third problem is one such. The problem is to find all the others.

Remark: The following exercises show that, for the matrix product, $AB = O$ does not necessarily imply $A = O$ or $B = O$.

Exercise: For each of the following matrix A, find all the two column matrices X satisfying $AX = O$ (the zero matrix).

1. $A = [1 \;\; 2]$

Ans. The matrix X has to be 2 by 2. Let

$$X = \begin{bmatrix} x & y \\ z & w \end{bmatrix}$$

Then

$AX = [x + 2zy + 2w]$

so that $AX = O$ means

$x + 2z = 0,$

$y + 2w = 0.$

This further means $x = -2z$, $y = -2w$, with z and w arbitrary. We conclude that

$$X = \begin{bmatrix} -2z & -2w \\ z & w \end{bmatrix} = \begin{bmatrix} -2 & 0 \\ 1 & 0 \end{bmatrix} + \begin{bmatrix} 0 & -2 \\ 0 & 1 \end{bmatrix}.$$

2. $A = \begin{bmatrix} 1 \\ 2 \end{bmatrix}$

Ans. The matrix X has to be 1 by 2. Let $X = [x\ y]$. Then

$$AX = \begin{bmatrix} x & y \\ 2x & 2y \end{bmatrix}$$

so that $AX = O$ means

$x = 0, 2x = 0,$

$y = 0, 2y = 0.$

We conclude that $X = O = [0\ 0]$.

3. $A = \begin{bmatrix} 1 & 1 \\ 1 & -1 \end{bmatrix}$

Ans. The matrix X has to be 2 by 2. Let

$$X = \begin{bmatrix} x & y \\ z & w \end{bmatrix}$$

Then

$$AX = \begin{bmatrix} x + z & y + w \\ x - z & y - w \end{bmatrix}$$

so that $AX = O$ means

$x + z = 0, x - z = 0,$

$y + w = 0, y - w = 0.$

The first two equations tell us $x = z = 0$. The last two equations tell us $y = w = 0$. We conclude that $AX = O$ implies $X = O$.

4. $A = \begin{bmatrix} 2 & 4 \\ -1 & -2 \end{bmatrix}$

Ans. The matrix X has to be 2 by 2. Let X be as before. Then

$$AX = \begin{bmatrix} 2x + 4z & 2y + 4w \\ -x - 2z & -y - 2w \end{bmatrix}$$

so that $AX = O$ means

$2x + 4z = 0, - x - 2z = 0,$

$2y + 4w = 0, - y - 2w = 0.$

The first two equations tell us $x = - 2z$, z arbitrary. The last two equations tell us $y = - 2w$, w arbitrary. We conclude that

$$X = \begin{bmatrix} -2z & -2w \\ z & w \end{bmatrix} = z\begin{bmatrix} -2 & 0 \\ 1 & 0 \end{bmatrix} + w\begin{bmatrix} 0 & -2 \\ 0 & 1 \end{bmatrix}$$

5. $A = \begin{bmatrix} 0 & 0 \\ 0 & 0 \end{bmatrix}$

Ans. For any 2 by 2 matrix X, we always have $AX = O$.

6. $A = \begin{bmatrix} 1 & -2 & 3 \\ -2 & 3 & -4 \end{bmatrix}$

Ans. The matrix X has to be 3 by 2. Let

$$X = \begin{bmatrix} x & y \\ z & w \\ u & v \end{bmatrix}$$

Then $AX = O$ means

$x - 2z + 3u = 0, - 2x + 3z - 4u = 0,$

$y - 2w + 3v - 0, - 2y + 3w - 4v = 0.$

The first two equations tell us $x = u$, $z = 2u$, u arbitrary. The last two equations tell us $y = v$, $w = 2v$, v arbitrary. We conclude that

$$X = \begin{bmatrix} u & v \\ 2u & 2v \\ u & v \end{bmatrix} = u\begin{bmatrix} 1 & 0 \\ 2 & 0 \\ 1 & 0 \end{bmatrix} + v\begin{bmatrix} 0 & 1 \\ 0 & 2 \\ 0 & 1 \end{bmatrix}$$

7. $A = \begin{bmatrix} 1 & -2 \\ -2 & 3 \\ 3 & -4 \end{bmatrix}$

Ans. The matrix X has to be 2 by 2. Let

$$X = \begin{bmatrix} x & y \\ z & w \end{bmatrix}$$

Then $AX = O$ means

$x - 2z = 0, - 2x + 3z = 0, 3x - 4z = 0,$

$y - 2w = 0, - 2y + 3w = 0, 3y - 4w = 0.$

From the equations, we conclude that X has to be the zero matrix.

Remark: Here are some further problems for you to think about.

1. Let $X = [x\ y]$. Then $AX = [Ax\ Ay]$ by this formula. Thus $AX = O$ means solving $Ax = 0$ and $Ay = 0$. More precisely, we solve the homogeneous system $Ax = 0$ and put two independent solutions as the two columns of X. Check out the discussion with the concrete examples we worked out above.
2. When can we find nonzero X so that $AX = 0$? Answer the question in terms of properties of the systems $Ax = b$.
3. What if X has 3 columns? four columns?
4. For a given matrix A, find X such that $XA = O$. First try the problem with the matrices given in the exercise.

Exercise: For each of the following matrix A, find all the matrices X satisfying $AX = I$ (the identity matrix).

1. $A = [1 \quad 2]$

Ans. The matrix X has to be 2 by 1. Let

$$X = \begin{bmatrix} x \\ z \end{bmatrix}$$

Then $AX = [x + 2z]$, and $AX = I$ means $x + 2z = 1$. Thus we conclude that

$$X = \begin{bmatrix} 1-2z \\ z \end{bmatrix} = \begin{bmatrix} 1 \\ 0 \end{bmatrix} + z\begin{bmatrix} -2 \\ 1 \end{bmatrix}$$

where z is arbitrary.

2. $A = \begin{bmatrix} 1 \\ 2 \end{bmatrix}$

Ans. The matrix X has to be 1 by 2. Let $X = [x\ y]$. Then

$$AX = \begin{bmatrix} x & y \\ 2x & 2y \end{bmatrix}$$

so that $AX = I$ means $x = 1, y = 0, 2x = 0, 2y = 1$. The conditions are contradictory. Therefore there is no X such that $AX = I$.

3. $A = \begin{bmatrix} 1 & 1 \\ 1 & -1 \end{bmatrix}$

Ans. The matrix X has to be 2 by 2. Let

$$X = \begin{bmatrix} x & y \\ z & w \end{bmatrix}$$

Then

$$AX = \begin{bmatrix} x+z & y+w \\ x-z & y-w \end{bmatrix}$$

so that $AX = I$ means

$x + z = 1, x - z = 0,$

$y + w = 0, y - w = 1.$

The first two equations tell us $x = z = 1/2$. The last two equations tell us $y = 1/2$, $w = -1/2$. We conclude that $AX = I$ implies

$$X = \begin{bmatrix} 1/2 & 1/2 \\ 1/2 & -1/2 \end{bmatrix}$$

4. $A = \begin{bmatrix} 2 & 4 \\ -1 & -2 \end{bmatrix}$

Ans. The matrix X has to be 2 by 2. Let X be as before. Then

$$AX = \begin{bmatrix} 2x+4z & 2y+4w \\ -x-2z & -y-2w \end{bmatrix}$$

so that $AX = I$ means

$2x + 4z = 1, - x - 2z = 0,$

$2y + 4w = 0, - y - 2w = 1.$

The first two equations are contradictory. The last two equations are also contradictory. We conclude that there is no X such that $AX = I$.

5. $A = \begin{bmatrix} 0 & 0 \\ 0 & 0 \end{bmatrix}$

Ans. We have $AX = O$ for all X. Therefore AX can never be the identity matrix.

6. $A = \begin{bmatrix} 1 & -2 & 3 \\ -2 & 3 & -4 \end{bmatrix}$

Ans. The matrix X has to be 3 by 2. Let

$$X = \begin{bmatrix} x & y \\ z & w \\ u & v \end{bmatrix}$$

Then $AX = I$ means

$x - 2z + 3u = 1, - 2x + 3z - 4u = 0,$

$y - 2w + 3v = 0, - 2y + 3w - 4v = 1.$

The first two equations tell us $x = 1 + u$, $z = 2u$, u arbitrary. The last two equations tell us $y = - 2 + v$, $w = -1 + 2v$, v arbitrary. We conclude that

$$X = \begin{bmatrix} 1+u & -2+v \\ 2u & 2v \\ u & v \end{bmatrix}$$

where u and v are arbitrary.

7. $A = \begin{bmatrix} 1 & -2 \\ -2 & 3 \\ 3 & -4 \end{bmatrix}$

Ans. The matrix X has to be 2 by 3. Let

$$X = \begin{bmatrix} x & y & z \\ u & v & w \end{bmatrix}$$

Then $AX = I$ means

$x - 2u = 1, - 2x + 3u = 0, 3x - 4u = 0,$

$y - 2v = 0, - 2y + 3v = 1, 3y - 4v = 0,$

$z - 2w = 0, - 2z + 3w = 0, 3z - 4w = 1.$

Note that we actually have three systems of linear equations, with different variables ((x, u) for the first system, (y, v) for the second, (z, w) for the third), but with the same matrix A as the coefficient matrix. Since the system in (x, u) has no solutions, there is no X such that $AX = I$.

Exercise: Let

$$A = \begin{bmatrix} 2 & 0 & 2 & 1 \\ 1 & 1 & -1 & 0 \\ 0 & 1 & -2 & 1 \end{bmatrix}$$

For each of the following matrix B, find a matrix X such that $AX = B$.

1. $B = \begin{bmatrix} 1 \\ -1 \\ -2 \end{bmatrix}$

Ans. X must be a vertical vector of dimension 4. The equation $AX = B$ is the usual system of 3 linear equations in 4 variables. It is in fact this system. By the earlier computation, the augmented matrix $[A\ B]$ has the following row echelon form.

$$\begin{bmatrix} 1 & 1 & -1 & 0 & -1 \\ 0 & 1 & -2 & 1 & -2 \\ 0 & 0 & 0 & 3 & -1 \end{bmatrix}$$

From this we find one solution

$$X = \begin{bmatrix} -1 \\ -2 \\ 0 \\ -1/3 \end{bmatrix}$$

2. $B = \begin{bmatrix} 1 & 1 \\ -1 & 0 \\ -2 & 0 \end{bmatrix}$

Ans. Let

$$b_1 = \begin{bmatrix} 1 \\ -1 \\ -2 \end{bmatrix}, \quad b_2 = \begin{bmatrix} 1 \\ 0 \\ 1 \end{bmatrix}$$

be the two columns of B. Let x, y be the two columns of X. By $AX = A[x\ y] = [Ax\ Ay]$, $AX = B$ is equivalent to two separate systems $Ax = b_1$, $Ay = b_2$. The first system has been studied in the problem above and a solution has been obtained. The second system has an obvious solution

$$y = \begin{bmatrix} 0 \\ 0 \\ 0 \\ 1 \end{bmatrix}$$

Combining the solutions together, we get a solution

$$X = \begin{bmatrix} 1 & 0 \\ -2 & 0 \\ 0 & 0 \\ -1/3 & 1 \end{bmatrix}$$

of $AX = B$.

3. $B = \begin{bmatrix} 1 & 0 & 0 \\ 0 & 1 & 0 \\ 0 & 0 & 1 \end{bmatrix}$

Ans. The columns of B are the standard basis vectors e_1, e_2, e3. By a similar argument as before, the columns of X are the solutions of the systems $Ax = e_i$. Combining the augmented matrices of the three systems, we get

$$[A\ e_1\ e_2\ e_3] = [A\ I] = \begin{bmatrix} 2 & 0 & 2 & 1 & 1 & 0 & 0 \\ 1 & 1 & -1 & 0 & 0 & 1 & 0 \\ 0 & 1 & -2 & 1 & 0 & 0 & 1 \end{bmatrix}$$

Row operations (e.g., as in this exercise) change the matrix into

$$\begin{bmatrix} 1 & 1 & -1 & 0 & 0 & 1 & 0 \\ 0 & 1 & -2 & 1 & 0 & 0 & 1 \\ 0 & 0 & 0 & 3 & 1 & -2 & 2 \end{bmatrix}$$

By taking [col 1–5], we get a solution (1/3, –1/3, 0, 1/3) of $Ax = e_1$. By taking [col 1–4] and [col 6], we get a solution (1/3, 2/3, 0, –2/3) of $Ax = e_2$. By taking [col 1–4] and [col 7], we get a solution (–1/3, 1/3, 0, 2/3) of $Ax = e_3$. Putting the three columns together, we get a solution

$$X= \begin{bmatrix} 1/3 & 1/3 & -1/3 \\ -1/3 & 2/3 & 1/3 \\ 0 & 0 & 0 \\ 1/3 & -2/3 & 2/3 \end{bmatrix}$$

of $AX = B$.

Remark: From this sequence of exercises, you can see the relation between solving equations like $AX = B$ and our usual systems $Ax = b$ of linear equations. Apply your new insight to this exercise and this exercise. Moreover, study the existence/uniqueness/general solution problem for $AX = B$, just as we have done for systems of linear equations.

Exercise: Let

$$A = \begin{bmatrix} 2 & -2 & 0 & 2 & 2 \\ -1 & 1 & 2 & -3 & 1 \\ -2 & 2 & 1 & -3 & -1 \end{bmatrix}, B = \begin{bmatrix} 0 & 2 & 0 \\ 0 & 3 & 2 \\ 0 & 0 & 1 \end{bmatrix}$$

Can you find a matrix X such that $AX = B$? In general, what is the criterion for the existence of X such that $AX = B$? What about the uniqueness of X?

Answer: As we have seen in the previous exercise, the equation $AX = B$ is equivalent to solving three separate systems $Ax = b_1$, $Ax = b_2$, $Ax = b_3$, where b_1, b_2, b_3 are the three columns of b. In an earlier exercise, we have argued that the system $Ax = b_1$ has no solution. Therefore the equation $AX = B$ has no solution.

In general, $AX = B$ has solutions if and only if no column of B is a pivot column of the "augmented matrix" $[A\ B]$. The solution is unique if and only if all columns of A are pivot.

Exercise: Describe the first two rows and the last three columns of AB in terms of row and columns of A and B.

Answer: The first two rows of AB is the product of the first two rows of A with B. The last three columns of AB is the product of A with the last three columns of B.

Exercise: The trace trA of a square matrix is the sum of its diagonal entries. For example,

$$tr\begin{bmatrix} 1 & 2 \\ 3 & 4 \end{bmatrix} = 1 + 4 = 5$$

$$tr\begin{bmatrix} 0 & 3 & -4 \\ 2 & 5 & 7 \\ 0 & 4 & -3 \end{bmatrix} = 0 + 5 - 3 = 2$$

Prove the trace has the following properties

$$tr(A + B) = trA + trB,\ tr\ cA = c\ trA,\ trA^{T} = trA,\ trAB = trBA.$$

Answer: The first two properties mean that the trace is linear. The third property means the trace is not changed by the transpose. All are quite obvious.

In the property $trAB = trBA$, we do not need A and B to be square matrices. We will only assume A is an m by n matrix, and B is an n by m matrix.

The i-th diagonal entry of AB is $\Sigma_{1 \le j \le n} a_{ij} b_{ji} = a_{i1} b_{1i} + a_{i2} b_{2i} + \ldots + a_{in} b_{ni}$. Thus

$trAB = \Sigma_{1 \le i \le m}(a_{i1} b_{1i} + a_{i2} b_{2i} + \ldots + a_{in} b_{ni}) = \Sigma_{1 \le i \le m,\ 1 \le j \le n} a_{ij} b_{ji}$.

Exchanging B and A, we have

$trBA = \Sigma_{1 \le j \le n}(b_{j1} a_{1j} + b_{j2} a_{2j} + \ldots + b_{jm} a_{mj}) = \Sigma_{1 \le j \le n, 1 \le i \le m} b_{ij} a_{ji} = \Sigma_{1 \le i \le m, 1 \le j \le n} a_{ji} b_{ij} = trAB.$

Exercise: What does the multiplication of two upper triangular matrices look like? What about lower triangular matrices? diagonal matrices?

Answer: Given two upper triangular matrices,

$$A = \begin{bmatrix} a_1 & * & .. & * \\ 0 & a_2 & .. & * \\ \vdots & \vdots & & \vdots \\ 0 & 0 & .. & a_n \end{bmatrix},\ B = \begin{bmatrix} b_1 & * & .. & * \\ 0 & b_2 & .. & * \\ \vdots & \vdots & & \vdots \\ 0 & 0 & .. & b_n \end{bmatrix}.$$

their multiplication is still upper triangular, with the diagonal entries simply multiplied together (the other entries are more complicated).

$$AB = \begin{bmatrix} a_1b_1 & * & .. & * \\ 0 & a_2b_2 & .. & * \\ : & : & & : \\ 0 & 0 & .. & a_nb_n \end{bmatrix}$$

The similar statement is true for the multiplication of two lower triangular matrices.

For two diagonal matrices,

$$A = \begin{bmatrix} a_1 & 0 & .. & 0 \\ 0 & a_2 & .. & 0 \\ : & : & & : \\ 0 & 0 & .. & a_n \end{bmatrix}$$

their multiplication is still diagonal, with the diagonal entries simply multiplied together.

$$AB = \begin{bmatrix} a_1b_1 & 0 & .. & 0 \\ 0 & a_2b_2 & .. & 0 \\ : & : & & : \\ 0 & 0 & .. & a_nb_n \end{bmatrix}$$

Exercise: What are the matrices A satisfying $AX = XA$ for any X?

Answer: If A is an m by n matrix, then in order for both AX and XA to make sense, X has to be an n by m matrix. Then AX is an m by m matrix, and XA is an n by n matrix. Now for the two matrix to be equal, we must have $m = n$. In other words, A has to be a square matrix.

Now assume A is an n by n matrix. Let E_{ij} be the n by n matrix with 1 as the ij–entry and 0 as all the other entries. For example, in case $n = 3$ we have

$$E_{11} = \begin{bmatrix} 1 & 0 & 0 \\ 0 & 0 & 0 \\ 0 & 0 & 0 \end{bmatrix}, E_{23} = \begin{bmatrix} 0 & 0 & 0 \\ 0 & 0 & 1 \\ 0 & 0 & 0 \end{bmatrix}, E_{31} = \begin{bmatrix} 0 & 0 & 0 \\ 0 & 0 & 0 \\ 1 & 0 & 0 \end{bmatrix}$$

Then AE_{ij} is obtained by moving the i-th column of A to j-th column, and leaving all the other columns zero. For example,

$$AE_{23} = \begin{bmatrix} a_{11} & a_{12} & a_{13} \\ a_{21} & a_{22} & a_{23} \\ a_{31} & a_{32} & a_{33} \end{bmatrix} \begin{bmatrix} 0 & 0 & 0 \\ 0 & 0 & 1 \\ 0 & 0 & 0 \end{bmatrix} = \begin{bmatrix} 0 & 0 & a_{12} \\ 0 & 0 & a_{22} \\ 0 & 0 & a_{32} \end{bmatrix}.$$

And $E_{ij}A$ is obtained by moving the j-th row of A to i-th row, and leaving all the other rows zero. For example,

$$E_{23}A = \begin{bmatrix} 0 & 0 & 0 \\ 0 & 0 & 1 \\ 0 & 0 & 0 \end{bmatrix} \begin{bmatrix} a_{11} & a_{12} & a_{13} \\ a_{21} & a_{22} & a_{23} \\ a_{31} & a_{32} & a_{33} \end{bmatrix} = \begin{bmatrix} 0 & 0 & 0 \\ a_{31} & a_{32} & a_{23} \\ 0 & 0 & 0 \end{bmatrix}.$$

Thus from $AE_{ij} = E_{ij}A$, we have

1. Any entry of AE_{ij} not in the i-th row should vanish. This means in the i-th column of A, all but a_{ii} should vanish. In the E_{23} example above, we mean $a_{12} = a_{32} = 0$.
2. Any entry of $E_{ij}A$ not in the j-th column should vanish. This means in the j-th row of A, all but a_{jj} should vanish. In the E_{23} example above, we mean $a_{31} = a_{32} = 0$.
3. The ij–entry of AE_{ij} and $E_{ij}A$ are equal. This means that $a_{ii} = a_{jj}$.

   Since the above should hold for any i, j, we conclude that $a_{ij} = 0$ for i "" j, and $a_{ii} = a_{jj}$ for any i, j. This is the same as $A = aI$, which is a constant multiple of the identity.

SOLVED EXAMPLES

1. Prove that transformations T and S are onto $\Rightarrow$ The composition TS is onto.

Ans. For any b, by T onto, we can find y, such that $T(y) = b$. For this y, by S onto, we can find x, such that $S(x) = y$. Then we have

$TS(x) = T(S(x)) = T(y) = b$.

Thus any b is the image of some element under the transformation TS. This shows TS is onto.

2. What about the converse? In other words, if the composition TS is onto, are the transformations T and S onto?

Ans. If TS is onto, then for any b, we have x, such that $T(S(x)) = b$. In particular, b is the image of $S(x)$ under the transformation T. The observation tells us T must be onto.

On the other hand, S is not necessarily onto. An example is given by $S = ID_s$: student $\rightarrow$ integer, T = *Last–Digit*: integer $\rightarrow$ {0, 1, 2, 3, 4, 5, 6, 7, 8, 9}.

3. Rephrase your conclusions in the first and second part in terms of matrix products.

Ans. If $Ay = b$ and $Bx = c$ has solutions for any b and c, then $ABx = b$ has solution for any b.

Conversely, if $ABx = b$ has solutions for any b, then $Ay = b$ has solutions for any b. However, $Bx = c$ may not have solution for any c.

Chapter 6

Hyperbolic Geometry

Hyperbolic geometry was created in the first half of the nineteenth century in the midst of attempts to understand Euclid's axiomatic basis for geometry. It is one type of *non-Euclidean geometry*, that is, a geometry that discards one of Euclid's axioms. Einstein and Minkowski found in non-Euclidean geometry a geometric basis for the understanding of physical time and space. In the early part of the twentieth century every serious student of mathematics and physics studied non-Euclidean geometry. This has not been true of the mathematicians and physicists of our generation. Nevertheless with the passage of time it has become more and more apparent that the negatively curved geometries, of which hyperbolic non-Euclidean geometry is the prototype, are the generic forms of geometry. They have profound applications to the study of complex variables, to the topology of two- and three-dimensional manifolds, to the study of finitely presented infinite groups, to physics, and to other disparate fields of mathematics.

A working knowledge of hyperbolic geometry has become a prerequisite for workers in these fields. These notes are intended as a relatively quick introduction to hyperbolic geometry. They review the wonderful history of non-Euclidean geometry. They give five different analytic models for and several combinatorial approximations to non-Euclidean geometry by means of which the reader can develop an intuition for the behavior of this geometry. They develop a number of the properties of this geometry that are particularly important in topology and group theory. They indicate some of the fundamental problems being approached by means of non-Euclidean geometry in topology and group theory. Volumes have been written on non-Euclidean geometry, which the reader must consult for more exhaustive information. We recommend for starters, and for more advanced readers. In mathematics, hyperbolic geometry is a non-Euclidean geometry, meaning that the parallel postulate of Euclidean geometry is replaced.

The parallel postulate in Euclidean geometry states, in two dimensional space, for any given line l and point P not on l, there is exactly one line through P that does not intersect l; i.e., that is parallel to l. In hyperbolic geometry there are at least two distinct lines through P which do not intersect l, so the parallel postulate is false. Models have been constructed within Euclidean geometry that obey the axioms of hyperbolic geometry, thus proving that the parallel postulate is independent of the other postulates of Euclid. Since there is no precise hyperbolic analogue to Euclidean parallel lines, the hyperbolic use of *parallel* and related terms varies among writers. In this article, the two limiting lines are called *asymptotic* and lines sharing a common perpendicular are called *ultraparallel*; the simple word *parallel* may apply to both.

HYPERBOLIC GEOMETRY

It is an example of one of many non-euclidean geometries. It is often called *Bolyai-Lobachevskiian geometry* after two of its *discovers* Jαnos Bolyai and Nikolai Ivanovich Lobachevsky. Bolyai first announced his discoveries in a 26 page appendix to a book by his father, the *Tentamen,* in 1831. You can read about what happened to him and his work in the text. Another of the great mathematicians who seems to have preceeded Bolyai in his work is Carl Fredrich Gauss. He seems to have done some work in the area dating from 1792, but never published it. The first to publish a complete account of non-euclidean geometry was Lobachevsky in 1829. It was first published in Russian and was not widely read. In 1840 he published a treatise in German. We shall add one more axiom to the list of axioms for Neutral Geometry.

In hyperbolic geometry there exists a line l and a point P not on l such that at least two distinct lines parallel to l pass through P.

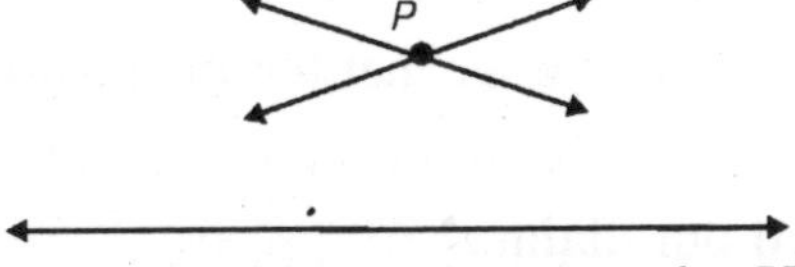

We shall denote the set of all points in the plane by H^2 and call this the *hyperbolic plane*.

Lemma: *There exists a triangle whose angle sum is less than* 180°

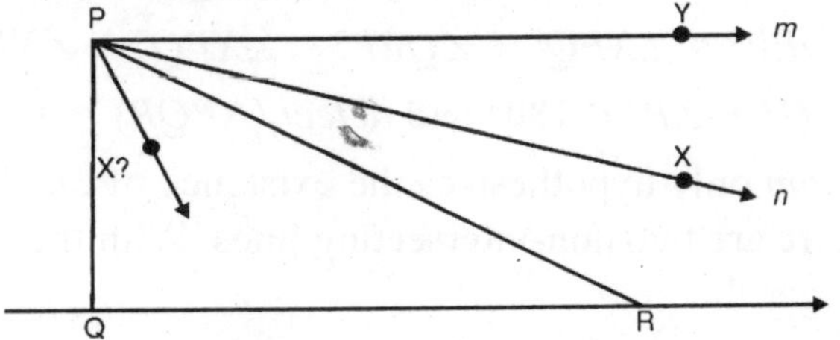

Proof: Let l be a line and P a point not on l such that two parallels to l pass through P. We can construct one of these parallels as previously done using perpendiculars. Let Q be the foot of the perpendicular to l through P. Let m be the perpendicular to $\overleftrightarrow{PQ}$ through P. Then m and l are non-intersecting. Let n be another line through P which does not intersect

l. This line exists by the *Hyperbolic Axiom*. Let $\overrightarrow{PX}$ be a ray of n lying between $\overleftrightarrow{PQ}$ and a ray $\overrightarrow{PY}$ of m.

Claim: There is a point $R \in l$ on the same side of $\overleftrightarrow{PQ}$ as X and Y so that $\angle QRP < \angle XPY$.

PROOF OF CLAIM. The idea is to construct a sequence of angles

$\angle QR_1\ P, \angle QR_2\ P, ..., \angle AR_n\ P, ...$

So that $\angle QR_{j+1}\ P° < \frac{1}{2} \angle QRj\ P°$ We will then apply Archimedes Axiom for real numbers to complete the proof.

There is a point $R_1 \in l$ so that $QR_1 \cong PQ$ Then ΔQR_1P is isosceles and $\angle QR_1P° \leq 45°$ Also, there is a point $R2 \in l$ so that $F * R_1 * R_2$ and $R_1R_2 \cong PR_1$ Then ΔPR_1R_2 is isosceles and $\angle R_1PR_2 \cong \angle QR_2P$ Since $\angle QR_1P$ is exterior to ΔPR_1R_2 it follows that.

$$\angle R_1PR°_2 + \angle QR_2P° \leq \angle QR_1P°,$$

So then $\angle QR_2P° \leq 22\frac{1}{2}°$ Continuing with this construction, we find a point $R_n \in l$ so that $Q * R_{n-1} * R_n$ and

$$\angle QR_nP° \leq \left(\frac{45}{2^n}\right)^{\circ}.$$

Applying the Archimedean axiom we see that for any positive real number, for example $\angle XPY°$ there is a point $R \in l$ so that R is on the same side of $\overleftrightarrow{PQ}$ as X and Y and $\angle QRP°$ $< \angle XPY°$ Thus, we have proved our claim.

Now, $\overrightarrow{PR}$ lies in the interior of $\angle QPX$ for if not then $\overrightarrow{PX}$ is in the interior of $\angle QRP$ By the *Crossbar Theorem* it follows that $\overrightarrow{PX} \cap l \neq \theta$ which implies that n and l are not non-intersecting --a contradiction. Thus, $\angle RPQ° < \angle XPQ°$ Then,

$$\angle RPQ° + \angle QRP° < \angle XPQ° + \angle QRP° < \angle XPQ° + \angle XPY° = 90°.$$

Therefore, $\angle P° + \angle Q° + \angle R° < 180°$ and *defect* $(\Delta PQR) > 0$.

The *Hyperbolic Axiom* only hypothesizes the existence of one line and one point not on that line for which there are two non-intersecting lines. With the above theorem we can now prove a much stronger theorem.

Universal Hyperbolic Theorem

Theorem: *In* H^2 *for every line l and for every point P not on l there pass through P at least two distinct lines, neither of which intersect l*

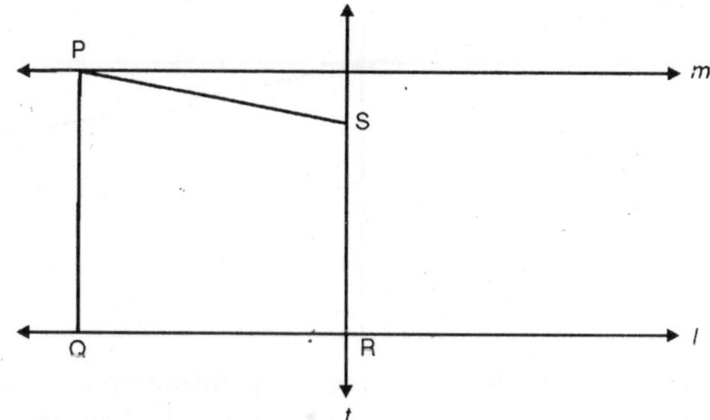

Proof: Drop a perpendicular $\overleftrightarrow{PQ}$ to l and construct a line m through P perpendicular to $\overleftrightarrow{PQ}$. Let R be any other point on l, and construct a perpendicular t to l through R. Now, let S be the foot of the perpendicular to t through P. Now, $\overleftrightarrow{PS}$ does not intersect l since both are perpendicular to t. At the same time $\overleftrightarrow{PS} \neq m$ Assume that $S \in m$ then $PQRS$ is a rectangle. By Theorem, if one rectangle exists all triangles have defect 0. We have a contradiction to Lemma. Thus, $\overleftrightarrow{PS} \neq m$, and we are done.

ANGLE SUMS

We have just proven the following theorem.

Theorem: *In* H^2 *rectangles do not exist and all triangles have angle sum less than* 180°.

This tells us that in hyperbolic geometry the defect of any triangle is a *positive* real number. We shall see that it is a very important quantity in hyperbolic geometry.

Corollary: *In* H^2 *all convex quadrilaterals have angle sum less than* 360°

Proof: Given any quadrilateral $ABCD$ Take the diagonal AC and consider triangles ΔABC and ΔACD By Theoremboth of these triangles have angle sum less than 180° The assumption that $\square ABCD$ is convex implies that $\angle BAC° + \angle CAD° = \angle BAD°$ and $\angle ACB° + \angle ACD° = \angle BCD°$ By adding all six angles we have that the angle sum of $\square ABCD$ is less that 360°

CLASSIFICATION OF PARALLELS

FAN ANGLES

A line l subdivides an angle if it passes through the vertex of the angle and intersects the interior of the angle. Let P be an arbitrary point in the hyperbolic plane. The pencil of lines through P, denoted P (P) is the set of all lines coincident with P.

Theorem: *If* $P \notin l$ *then there exists an angle with vertex at P whose interior contains* l *and whose bisector is perpendicular to* l.

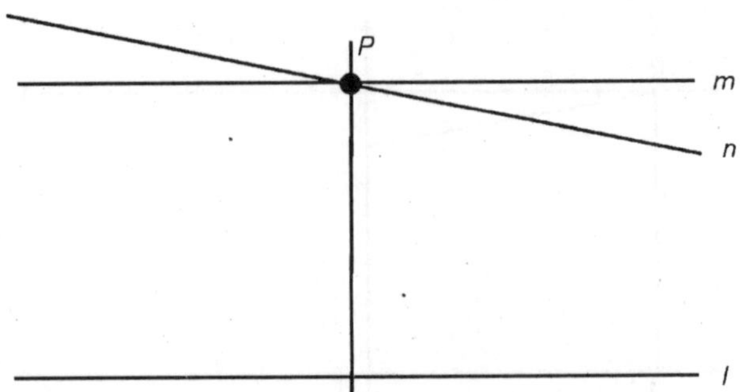

Proof: In the pencil P (P) let p be the perpendicular to l and let $m \in P$ (p) be perpendicular to p at P. Thus, we know that $m \cap l = \theta$ By the *Universal Hyperbolic Theorem* there is another line $n \in p(P)$ which does not intersect l. Since n is distinct from m, n is not perpendicular to p. Thus, one pair of the vertical angles formed by p and n is a pair of acute angles. Thus, there exists a ray $\overrightarrow{PA} \subset n$ so that

where Q is the foot of P in l.

Now, by *Congruence Axiom 4* there is a ray $\overrightarrow{PB}$ on the opposite side of p from A so that $\angle QPB \cong \angle QPA$. The line $\overrightarrow{PB}$ does not intersect l. If it did, then let R be the point of intersection. Choose R' on the same side of p as A so that $QR \cong QR'$ Then by *SAS* $\Delta PRQ \cong \Delta PR'Q$ Thus, $\angle QPR \cong \angle QPR \cong \angle QPA$ Thus, by *Congruence Axiom 4* $R' \in n$ This implies that $n \cap l \neq \theta$ which is a contradiction. Therefore, $\overrightarrow{PB} \cap l = \theta$.

Clearly, $\overrightarrow{PQ}$ bisects $\angle BPA$ We only need to show that l is contained in the interior of $\angle BPA$ or that every point on l lies in the interior of $\angle BPA$.

Let $X \in l$ be on the same side of p as A, and let Y be the unique point on l on the opposite side of p as X so that $QX \cong QY$ Now, we know that the ray $\overrightarrow{PQ}$ lies in the interior of $\angle APB$ and, in fact, using an abuse of notation, $\angle BPA = \angle QPB \cup \angle QPA$.

Claim: $\overrightarrow{QX}$ lies in the interior of $\angle QPA$ Likewise, $\overrightarrow{QY}$ lies in the interior of $\angle QPB$.

To prove this claim, first note that since X is on the same side of $P = \overrightarrow{PQ}$ as A, every point of $\overrightarrow{QX}$ lies on the same side of p as A. Thus, we are left only to show that if $R \in \overrightarrow{QX}$ then R is on the same side of $n = \overrightarrow{PA}$ as Q. This must be the case, for if not we would have that $n \cap l \neq \theta$, a contradiction. Thus, the above claim is true.

We then have that

$$l = \overrightarrow{QY} \cup \{Q\} \cup \overrightarrow{QX},$$

which from above must lie in the interior of $\angle APB$

Corollary: *Every line in* $p(P)$ *that intersects* l *subdivides the angle* $\angle APB$.

We have almost proven the following result.

Theorem: *If $P \notin l$ then there is a unique angle with the following properties:*

1. The angle contains l in its interior;
2. The bisector of this angle is perpendicular to l;
3. Every line passing through P that subdivides the angle intersects l

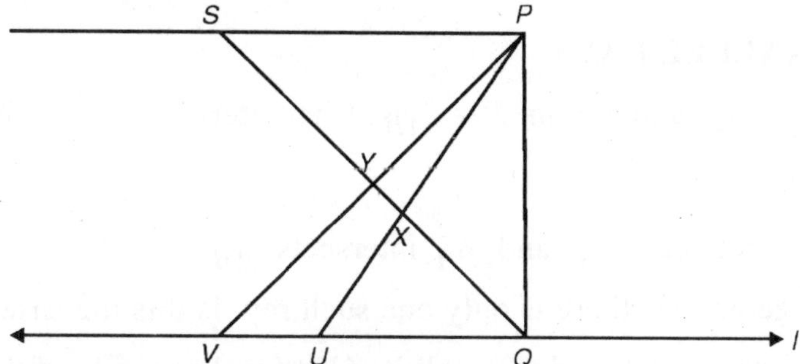

Proof: To prove that this angle exists, we need to construct the sides of the angle. We already know that there are angles with vertex at *P* that contain the line *l*. We want to prove much more in this theorem. In some sense we are proving that there is a smallest such angle.

Let *S* be a point on the line *m* which is the perpendicular at *P* to the perpendicular through *P* to *l*. Consider the line $\overleftrightarrow{SQ}$ Let Σ_1 be the set of all points *T* on *SQ*, such that $\overrightarrow{PT}$ intersects *l* together with all points on the ray opposite to $\overrightarrow{QS}$ Let $\Sigma_2 = \overleftrightarrow{SQ} \setminus \Sigma_1$ By the *Crossbar Theorem* if $T \in SQ$ and $T \in \Sigma_1$ then the entire subsegment $TQ \subset \Sigma_1$ Hence, (Σ_1, Σ_2) is a Dedekind cut. By Dedekind's Axiom there is a unique point *X* on $\overleftrightarrow{SQ}$ such that for $P_1, P_2 \in \overleftrightarrow{SQ}$, $P_1 * X * P_2$ if and only if $X \neq P_1$ and $X \neq P_2$, $P_1 \in \Sigma_1$ and $P_2 \in \Sigma_2$.

By the definition of Σ_1 and Σ_2 rays below $\overrightarrow{PX}$ all intersect *l* and rays above $\overrightarrow{PX}$ do not. Now, we wish to prove that $\overrightarrow{PX} \cap l = \theta$.

Assume on the contrary that $\overrightarrow{PX} \cap l = \{U\}$ Choose a point *V* on *l* so that $V * U * Q$ Since *V* and *U* are on the same side of $\overleftrightarrow{SQ}$, *V* and *P* are on opposite sides, so that *VP* meets *SQ* in a point *Y*. By Propositionwe have that $Y * X * Q$. Thus, $Y \in \Sigma_2$ This is impossible, for if $Y \in \Sigma_2$, then $\overrightarrow{PY} \cap l = \theta$ Thus, $\overrightarrow{PX}$ does not meet

Doing the same construction on the opposite side of $\overrightarrow{PQ}$ at *P* gives us a similar ray $\overrightarrow{PX'}$ with similar properties. We need to show that $\angle QPX \cong \angle QPX'$ Assume not, then we may assume that $\angle QPX < \angle QPX'$ By Congruence Axiom 4 there is a ray $\overrightarrow{PR'}$ between $\overrightarrow{PX'}$ and $\overrightarrow{PQ}$ such that $\angle QPR' \cong \angle QPX$. By the properties of $\overrightarrow{PX'}$, $\overrightarrow{PR'}$ must intersect *l*. Let *R* be the point on *l* on the opposite of $\overleftrightarrow{PQ}$ from *R*' so that $QR \simeq QR'$ Then by *SAS* $\Delta QPR' \cong \Delta QPR$ This implies that $\angle QPR \cong \angle QPR' \cong \angle QPX$. By Congruence Axiom 4 we have to have that $R \in \overrightarrow{PX}$ or $\overrightarrow{PX} \in l \neq \theta$, a contradiction. Thus, $\angle QPX \cong \angle QPX'$.

Checking the properties, clearly $\overrightarrow{PQ}$ bisects the angle, the angle contains l and by the definition of the ray $\overrightarrow{PX}$ any line that subdivides the angle intersects l.

For $P \notin l$ this angle at P is called the fan angle of P and l, and will be denoted by $\angle(P, l)$.

LIMITING PARALLEL RAYS

Consider a ray $\overrightarrow{AB}$ and a point $P \notin \overleftrightarrow{AB}$. Characterize all rays $\overrightarrow{PX}$ satisfying:

1. $\overleftrightarrow{PX} \cap \overleftrightarrow{AB} = \theta$,
2. Every ray between $\overrightarrow{PX}$ and $\overrightarrow{PA}$ intersects $\overrightarrow{AB}$.

In euclidean geometry, there is only one such ray. Is this the case in H^2 ?

There is a fan angle of P and $\overleftrightarrow{AB}$ call it $\angle XPY$ and one side of this angle lies on the same side of $\overleftrightarrow{AP}$ as B. Let us say that it is $\overrightarrow{PX}$ Now, $\overleftrightarrow{PX} \cap \overleftrightarrow{AB} = \theta$. Also, if $\overrightarrow{PR}$ lies between $\overrightarrow{PX}$ and $\overrightarrow{PA}$ then $\overrightarrow{PR}$ subdivides *angle XPY*. Thus, $\overrightarrow{PR} \cap \overleftrightarrow{AB} \neq \theta$, but due to its relative position with respect to $\overrightarrow{PA}$ and $\overrightarrow{PX}$ it must, in fact, intersect $\overrightarrow{AB}$ Thus, the side of the fan angle satisfies the two aforementioned conditions. Do any others?

Let $\overrightarrow{PT}$ be different from $\overrightarrow{PX}$.

If $\overrightarrow{PT}$ is on the line $\overleftrightarrow{PA}$ then $\overrightarrow{PT}$ nintersects $\overleftrightarrow{AB}$ and the ray $\overrightarrow{PT}$ does not have property

If $\overrightarrow{PT}$ lies on the same side of $\overleftrightarrow{PA}$ as Y, let $\overrightarrow{PU}$ be between $\overrightarrow{PA}$ and $\overrightarrow{PT}$ Then the line $\overleftrightarrow{PA}$ separates $\overrightarrow{PT}$ and $\overrightarrow{PU}$ from $\overrightarrow{AB}$ which means that $\overrightarrow{PU} \cap \overrightarrow{AB} = \theta$ so that $\overrightarrow{PT}$ does not satisfy property

If $\overrightarrow{PT}$ lies on the same side of $\overleftrightarrow{PA}$ as X, then one of $\overrightarrow{PT}$ and $\overrightarrow{PX}$ is between the other and $\overrightarrow{PA}$

1. If $\overrightarrow{PT}$ is between $\overrightarrow{PX}$ and $\overrightarrow{PA}$ then $\overrightarrow{PT}$ must intersect $\overleftrightarrow{AB}$ by the properties of the fan angle. Thus, $\overrightarrow{PT}$ would not satisfy property (1).
2. If $\overrightarrow{PX}$ is between $\overrightarrow{PA}$ and $\overrightarrow{PT}$ then $\overrightarrow{PT}$ cannot satisfy property (2), for $\overleftrightarrow{PX} \cap \overleftrightarrow{AB} = \theta$.

Thus, $\overrightarrow{PX}$ is the only ray at P with these two properties. This gives rise to the following definition.

Definition: A ray $\overrightarrow{PQ}$ is parallel to a ray $\overrightarrow{AB}$ if

1. $\overleftrightarrow{PQ} \cap \overleftrightarrow{AB} = \theta$,
2. Every ray between $\overrightarrow{PQ}$ and $\overrightarrow{PA}$ intersects $\overrightarrow{AB}$.

We have just shown that this ray is unique. In this case the rays $\overrightarrow{PQ}$ and $\overrightarrow{AB}$ are called limiting parallel rays and we say that $\overrightarrow{PQ}$ is *limiting parallel* to $\overrightarrow{AB}$.

QUESTION: If $\overrightarrow{PQ}$ is limiting parallel to $\overrightarrow{AB}$ is $\overrightarrow{AB}$ is limiting parallel to $\overrightarrow{PQ}$?

We need to following lemma for the proof of the main theorem. I leave the proof to you as an exercise.

Lemma: *If $\angle PAB$ is acute, then the foot of P in the line $\overleftrightarrow{AB}$ lies in the ray $\overrightarrow{AB}$ and is different from A.*

I offer the following theorems without proof.

Theorem: *If $\overrightarrow{PQ}$ is limiting parallel to $\overrightarrow{AB}$ then $\overrightarrow{AB}$ lies in the interior of the angle $\angle APQ$.*

Theorem: *Let $P \notin l$ and let $A * B * C$ in l. Then $\overrightarrow{PQ}$ is limiting parallel to $\overrightarrow{AB}$ if and only if $\overrightarrow{PQ}$ is limiting parallel to $\overrightarrow{BC}$.*

Theorem: *Let $A * B\ C$ on l $\overrightarrow{AB}$ is limiting parallel to $\overrightarrow{DE}$ if and only if $\overrightarrow{BC}$ is limiting parallel to $\overrightarrow{DE}$.*

Theorem: *If $\overrightarrow{PQ}$ is limiting parallel to $\overrightarrow{AB}$ then $\overrightarrow{AB}$ is limiting parallel to $\overrightarrow{PQ}$.*

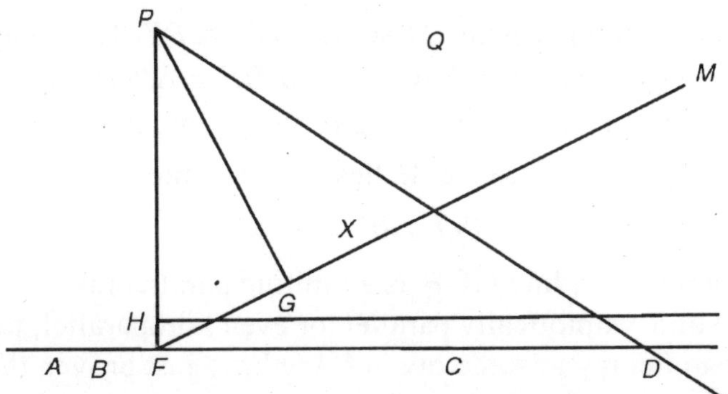

Proof: Let F be the foot of P in $\overleftrightarrow{AB} = l$ Let $C \in l$ so that $\overrightarrow{FC}$ is like directed to $\overrightarrow{AB}$ This means that either $\overrightarrow{AB} \subset \overrightarrow{FC}$ or $\overrightarrow{FC} \subset \overrightarrow{AB}$ It follows from Theorem that $\overrightarrow{PQ}$ is limiting parallel to $\overrightarrow{FC}$ so that C lies on the same side of $\overleftrightarrow{PF}$ as does Q and $\overrightarrow{FC} \cap \overrightarrow{PQ} = \theta$ Let $\overrightarrow{FX}$ lie in the interior of $\angle PFC$

Claim: $\overrightarrow{FX} \cap \overrightarrow{PQ} \neq \theta$.

From this it follows that $\overrightarrow{FC}$ is limiting parallel to $\overrightarrow{PQ}$ Applying Theorem again, we have that $\overrightarrow{AB}$ is limiting parallel to $\overrightarrow{PQ}$ Thus, we need to establish this claim to finish the proof.

Since $\angle PFX^\circ < \angle PFC^\circ = 90^\circ$, $\angle PFX$ is acute. Thus, by Lemmathe foot of P in $\overleftrightarrow{FX}$ must lie in the ray $\overrightarrow{FX}$ Label this point G. It follows that G lies on the same side of $\overleftrightarrow{PF}$ as C.

If $G \in \overleftrightarrow{PQ}$ then it must lie in $\overrightarrow{PQ}$. It follows that $\overrightarrow{FX} \cap \overrightarrow{PQ} \neq \theta$, and we are done.

If G lies on the opposite side of $\overleftrightarrow{PQ}$ from F, then again $\overrightarrow{FX}$ intersects $\overrightarrow{PQ}$ and we are done.

Assume then, as the final case, that G lies on the same side of $\overleftrightarrow{PQ}$ as F. Then, this puts G in the interior of $\angle QPF$ Let $\alpha^\circ = \angle Q^\circ PG$ We have that $\alpha^\circ < \angle QPF^\circ$ By Congruence Axiom 4 there is a unique ray $\overrightarrow{PZ}$ in the interior of $\angle QPF$ so that

$$\angle FPZ^\circ = \alpha^\circ = \angle QPG^\circ$$

Since $\overrightarrow{PQ}$ is limiting parallel to $\overrightarrow{FC}$, $\overrightarrow{PZ}$ must intersect in some point D. From Lemma $G \neq F$ so that $PG<PF$ by an Exercise (the hypotenuse of a right triangle is longer than either leg). Thus, there is a point $H \in PF$ so that $PH \cong PG$.

Let $\overleftrightarrow{HY}$ be perpendicular to $\overleftrightarrow{PF}$ at H. Since $\overleftrightarrow{AB}$ is perpendicular to $\overleftrightarrow{PF}$ it follows that $\overleftrightarrow{AB} \cap \overleftrightarrow{HY} = \theta$ Now, $\overleftrightarrow{HY}$ does intersect one side of the triangle ΔPFD By *Pasch's Theorem* it must intersect a second side. That side must be PD. Let the point of intersection be E. Now, $E \neq P$ and $E \neq D$.

On $\overrightarrow{PQ}$ there is a unique point M so that $PM \cong PE$ Recalling that $PH \cong PG$ and $\angle HPE = \angle FPZ \cong \angle QPG$ we then have that $\Delta PHE \cong \Delta PGM$ by *SAS*. Thus, $\angle PGM \cong \angle PHE$ and is a right angle. Thus, $\overleftrightarrow{GM} = \overleftrightarrow{FX}$ since both are perpendicular to $\overleftrightarrow{PG}$ at G. Therefore, $\overleftrightarrow{FX} \cap \overrightarrow{PQ} = \{M\}$ Since M lies on the same side of $\overleftrightarrow{FC}$ as Q, it follows, finally, that $M \in \overrightarrow{FX}$ and $\overrightarrow{FX} \in \overrightarrow{PQ} \neq \theta$.

A ray $\overrightarrow{X}$ is parallel to a line l if $\overrightarrow{R}$ is a limiting parallel ray to some ray in l. A line k is limiting parallel or asymptotically parallel, or even horoparallel, to a line l. if some ray in k is a limiting parallel ray to some ray in l. We have just proven that these parallelisms are symmetric and we may denote them by

$$\overrightarrow{X} \, \| l \quad l \| \overrightarrow{X} \quad l \| k \quad k \| l.$$

If $k = \overleftrightarrow{AB}$, $l = \overleftrightarrow{CD}$ and $\overrightarrow{AB} \,\|\, \overrightarrow{CD}$, then $\overleftrightarrow{AB}$ is said to be parallel to l in the direction of $\overrightarrow{CD}$ on l. Furthermore, k and are said to be parallel in the direction of $\overrightarrow{AB}$ on k and in the direction of $\overrightarrow{CD}$ on l.

Theorem: *If $P \notin l$ then there are exactly two lines through P that are limiting parallel to l. Each contains an arm of the fan angle $\angle(P, l)$ and they are limiting parallel to l in opposite directions.*

As we have mentioned several times already, there is no simple transitivity of parallelism in hyperbolic geometry. There is a weak form of transitivity.

Weak Transitivity of Parallels

Theorem: *Two lines parallel to a third in the same direction on the third are parallel to each other.*

HYPERPARALLEL LINES

Let k and l be lines in H^2 and let A_n $n = 1, 2, ...$ be points on k. Let A_n' be the foot of A_n in l. We say that the points $\{A_n\}$ are equidistant from l. if $A_iA_i' \cong A_jA_j'$ for all choices of i and j.

Theorem: *If l and l' are non-intersecting lines in H^2, then any set of points on l equidistant from l' has at most two elements.*

Proof: Assume not; *i.e.*, assume that $A, B, C \in l$ are equidistant from l' Thus, the quadrilaterals $AA'\ B'B$, $AA'CC$, and are all Saccheri quadrilaterals. Thus, $\angle A'\ AB \cong \angle B'\ BA$, $\angle AAC \cong \angle C'CA$, and $\angle B'BC \cong C'CB$. Stringing these together, we have

$$\angle BBA \cong \angle AAB \cong \angle C'CA \cong \angle B'BC.$$

Since $\angle B'BA$ and $\angle B'BC$ are supplementary, they are right angles. Thus all three quadrilaterals are rectangles, a contradiction.

This theorem states that at most two points at a time on l can be equidistant from l' It does not put a limit on the number of pairs of such points, though. We may have pairs (A,B) and (C,D) so that $AA' \cong BB'$ and $CC' \cong DD'$ but it is not possible that $AA' \cong CC'$ or any of the other possibilities. It may also occur that the number of pairs of such points will be zero. There is no guarantee that such a pair will exist. This will help us distinguish types of parallel lines.

Theorem: *Let l and l' be non-intersecting lines for which there is a pair of points A, $B \in l$ which are equidistant from l'. Then, l and l' have a common perpendicular segment which is the shortest segment between l and l'.*

Proof: $AA'\ B'B$ is a Saccheri quadrilateral. Its altitude is perpendicular to both l and l' by Theorem. Also, the quadrilateral formed by this altitude and any other segment from a point $X \in l$ and its foot $X' \in l'$ is a Lambert quadrilateral. By Theorem the altitude is less than XX' and hence is the shortest segment between l and l'.

Theorem: *In H^2 if lines l and l' have a common perpendicular segment MM', then they are non-intersecting and MM' is unique. Moreover, if A, $B \in l$ so that M is the midpoint of AB then A and B are equidistant from l'.*

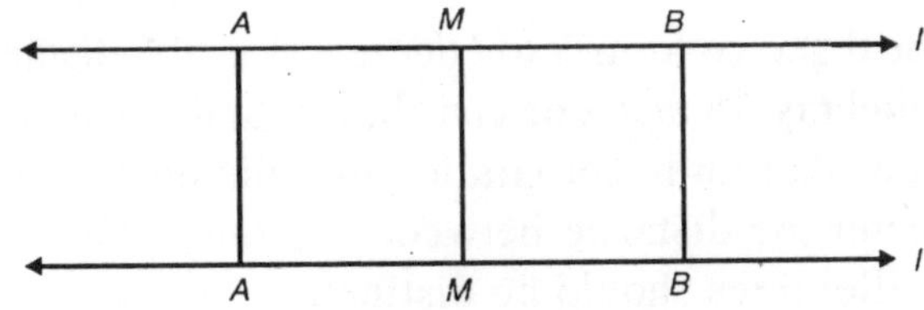

Proof: We already know that the lines do not intersect from Corollary to the *Alternate Interior Angles Theorem*. If *PP'* were another common perpendicular then □*MM' P'P* would be a rectangle, which cannot exist. Thus, *MM'* is unique.

Let $A, B \in l$ be so that M is the midpoint of AB. We have that $AM \cong BM$ and $\angle AMM' \cong \angle BMM'$ so by *SAS* $\Delta AMM' \cong \Delta BMM'$ Thus, $AM' \cong BM'$ and $\angle AMM \cong \angle BM'M$ By *Angle Subtraction* $\angle A'M'A \cong \angle B'M'B$. Then by *AAS* $\Delta AA'M' \cong \Delta BB'M'$ and $AA' \cong BB'$.

Lemma: *Let MM' be the common perpendicular to l andl' Let* $A, B \in l$ *so that* $M * A * B$, *then* $AA' < BB'$.

If *k* and *l* are non-intersecting lines which admit a common perpendicular then they are said to be hyperparallel. This is denoted by *k*)(*l*. They are sometimes called *divergently parallel lines.*

Recall that for $P \notin l$ there is a unique fan angle $\angle(P, l)$ for *P* and *l*. Let *F* be the foot of *P* in *l* and let *X* and *Y* be on opposite arms of the fan angle. $\angle FPX \cong \angle FPY$ The angle is called the angle of parallelism at *P* with respect to *l* and its angle measure is denoted by $\Pi(PF)°$ We know that its measure must be less than 90°.

Theorem: *If l is any line in* H^2 *and* $0 < \alpha < 90$ *then there is a point* $P \notin l$ *so that* $\Pi(PF)° = \alpha°$.

CLASSIFICATION OF PARALLELS

We have defined two types of parallel lines--hyperparallel and horoparallel. Are these the only types of parallel lines in hyperbolic geometry? Are there lines that are both hyperparallel and limiting parallel? These are the questions that we wish to answer in this section.

Theorem: *Let* $\overrightarrow{PX}$ *be the limiting parallel ray to l through P and let Q and X' be the feet of P and X, respectively, in l. Then* $PQ > XX'$.

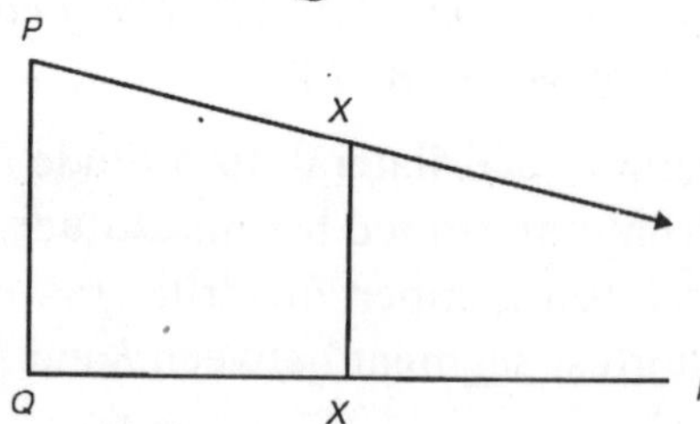

Proof: If $P * X * Y$ then $\overrightarrow{XY} \| l$ by Theorem. Thus, $\angle X'XY$ is the angle of parallelism for *X* and *l*. Thus, it is acute. Therefore, $\angle X'XP$ is obtuse. This implies that $\angle QPX < \angle X'XP$ and therefore $PQ > XX'$.

This shows that the distance form *X* to *l* decreases as the distance from *P* to *X* increases along the limiting parallel ray. In fact, one can show that the distance from *X* to *l* approaches 0. Note that we have seen that for hyperparallel lines the further one gets from the common perpendicular, the greater the distance between the lines. This gives one the impression that these types of parallel lines should be distinct.

Corollary: *If $l||k$ in a given direction, then l and k have no common perpendicular.*

Proof: Let $P \in l$ so that $\overrightarrow{PX} \| k$ and assume that there is a point $M \in l$ so that

- *$P * M * X$, and*
- *MM' is a common perpendicular to l and k.*

From Theorem $PQ > MM' > XX'$. By Theorem $MM' < XX'$, a contradiction. Thus, l and k have no common perpendicular.

This tells us that if two lines are limiting parallel, then they are not hyperparallel. We now need to show that if two lines are non-intersecting and are not limiting parallel, then they must be hyperparallel. Having done this we will then know that there are only two types of parallel lines.

Theorem: *Given lines m and n so that*

- *$m \cap n = \theta$, and*
- *m does not contain a limiting parallel ray to n in either direction.*

Then $m)(n$.

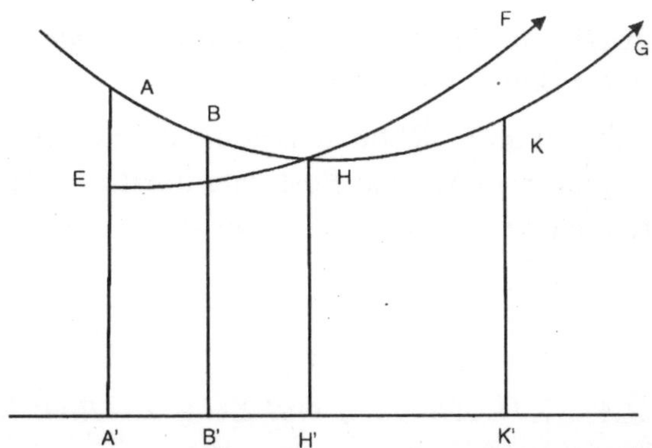

Proof: We need to find two points $H, K \in n$ that are equidistant from m; *i.e.*, $HH' \cong KK'$ If this is true, then by Theorem the perpendicular bisector of HK is the common perpendicular to m and n. By Theorem this common perpendicular is unique. It follows then that $m)(n$.

(i) The search for H and K.

Let $A, B \in n$ with feet $A', B' \in m$ If $AA' \cong BB'$ we are done. Assume that these segments are not congruent. Then, we may assume that $AA' > BB'$, since one must be greater than the other. There is a point $E \in AA'$ so that $A'E \cong BB'$.

There is a unique ray $\overrightarrow{EF}$ on the same side of $\overleftrightarrow{AA'}$ as B so that $\angle A'EF \cong \angle B'BG$ with $A * B * G$.

Claim: $\overrightarrow{EF} \cap \overrightarrow{AG} \neq \theta$ (Proof will follow.)

Call this point of intersection H. The there is a unique point $K \in \overrightarrow{BG}$ so that $BK \cong EH$ Then, by *SAS* $\Delta EHA' \cong BKB'$ Thus, $A'HB'K'$ We also have that $\angle EA'H \cong \angle BB'K$ so that by *Angle Subtraction* it follows that $\angle HA'H' \cong \angle KB'K'$ Since HH' and KK' are perpendicular to m, we have by *Hypotenuse-Angle* that $\Delta HA'H' \cong \Delta KB'K'$ Therefore, $HH' \cong KK'$ and H and K are equidistant from m.

Thus, we are finished once we have proved the claim.

It turns out that the claim is the hardest item to prove in the whole theorem. It takes us through some strange territory.

SIMILAR TRIANGLES

In euclidean geometry we are used to having two triangles similar if their angles are congruent. It is obvious that we can construct two non-congruent, yet similar, triangles. In fact John Wallis attempted to prove the Parallel Postulate of Euclid by adding another postulate.

WALLIS' POSTULATE

Given any triangle ΔABC and given any segment DE. There exists a triangle ΔDEF having DE as one of its sides that is similar to ΔABC.

However, it is proven that Wallis' Postulate is equivalent to Euclid's Parallel Postulate. Thus, we know that the negation of Wallis' Postulate must hold in hyperbolic geometry. That is, under certain circumstances similar triangles do not exist. We can prove a much stronger statement.

AAA Criterion

Theorem: *In* H^2 *if* $\angle A \cong \angle D$, $\angle B \cong \angle E$ *and* $\angle C \cong \angle F$ *then* $\Delta ABC \cong \Delta DEF$ *That is, if two triangles are similar, then they are congruent.*

Proof: Assume that $\Delta ABC \not\cong \Delta DEF$ Then there can be no corresponding sides which are congruent, else by *ASA* the triangles would be congruent. Consider the triples $\{AB, BC, AC\}$ and $\{DE, EF, DF\}$ One of these triples contains at least two segments that are larger than the corresponding segments of the other. Without loss of generality, we can assume that

$$AB > DE \text{ and } AC > DF.$$

There exist points $B' \in AB$ and $C' \in AC$ such that $AB' \cong DE$ and $AC' \cong DF$ Then by *SAS* $\Delta AB'C' \cong \Delta DEF$. Then $\angle AB'C' \cong \angle E \cong \angle B$ and $\angle AC'B' \cong \angle F \cong \angle C$ Thus, by the *Alterate Interior Angles Theorem* we can show that $\overleftrightarrow{B'C'}$ does not intersect $\overleftrightarrow{BC}$ Thus, it follows that $\square BB'C'C$ is convex. Using the fact that $\angle AB'C'$ and $\angle BB'C'$ are supplementary, as are $\angle AC'B'$ and $\angle CC'B'$ the angle sum of $\square BB'CC'$ is 360°, contradicting Corollary.

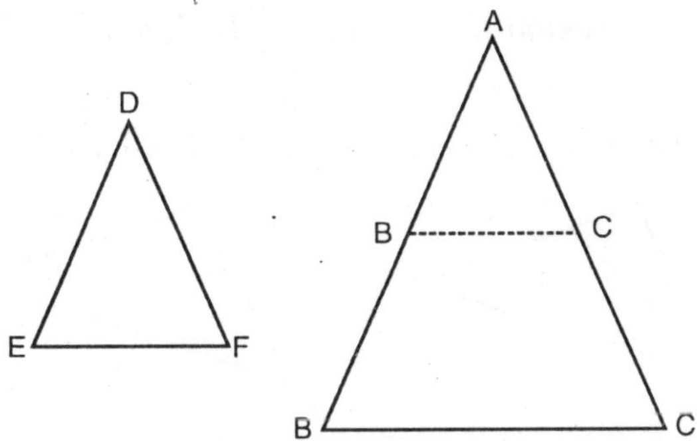

As a consequence of Theorem we shall see that in hyperbolic geometry a segment can be determined with the aid of an angle. For example, an angle of an equilateral triangle determines the length of a side uniquely. Thus in hyperbolic geometry there is an *absolute unit of length* as there is in elliptic geometry.

PROOF OF CLAIM AND ASYMPTOTIC TRIANGLES

If $\overrightarrow{PB}$ and $\overrightarrow{QD}$ are limiting parallel rays, the convention is to *pretend* that they pass through an ideal point Ω. We then denote the rays by $P\Omega$ and $Q\Omega$ The figure consisting of these rays and the segment PQ is called a singly asymptotic triangle. You can have doubly asymptotic triangles, with 2 ideal points, and trebly asymptotic triangles, with 3 ideal points.

Exterior Angle Theorem

Lemma: *If $\Delta PQ\Omega$ is a singly asymptotic triangle, the exterior angles at P and Q are greater than their respective opposite interior angles.*

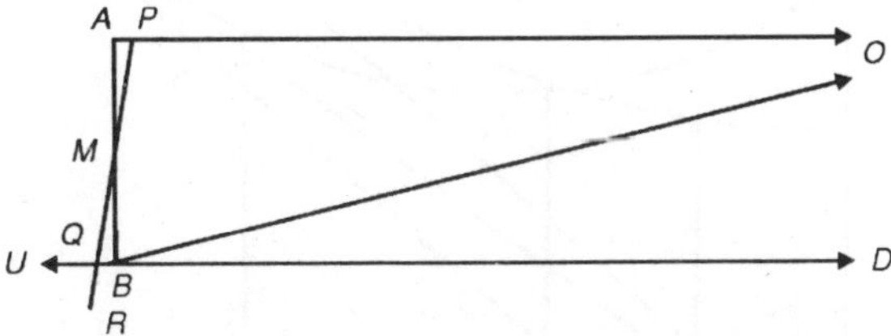

Proof: Choose $R \in \overrightarrow{PQ}$ so that $P * Q * R$ We need to show that $\angle RQ\Omega > \angle QP\Omega$.

There is a unique ray $\overrightarrow{PD}$ lying on the same side of $\overleftrightarrow{PQ}$ as Ω so that $\angle RQD \cong \angle QP\Omega$. If $U \in \overleftrightarrow{QD}$ is so that $U * Q * D$ then $\angle UQP \cong \angle RQD \cong \angle QP\Omega$ Thus, we have that the alternate interior angles are congruent. Let M be the midpoint of PQ and let A and B be the feet of M in $\overleftrightarrow{Q\Omega}$ and $\overleftrightarrow{QD}$ respectively. By AAS $\Delta APM \cong \Delta BQM$ so that $AM \cong BM$ and $\angle AMP \cong \angle BMQ$ Thus, A, M, and B are collinear. Thus, $\overleftrightarrow{Q\Omega}$ and $\overleftrightarrow{QD}$ have a common perpendicular. Thus, $\overleftrightarrow{Q\Omega})(\overleftrightarrow{QD}$ Thus $\overrightarrow{QD}$ lies between $\overrightarrow{Q\Omega}$ and $\overrightarrow{QR}$ from which it follows that $\angle RQ\Omega > \angle QP\Omega$.

Lemma: *If in the singly asymptotic triangles* $\Delta AB\Omega$ *and* $\Delta CD\Theta$ *we have* $\angle BA\Omega \cong \angle DC\Theta$ *then*

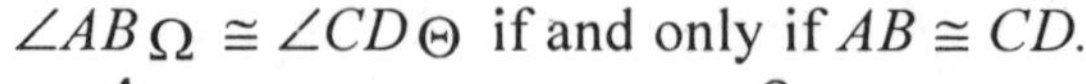

$$\angle AB\Omega \cong \angle CD\Theta \text{ if and only if } AB \cong CD.$$

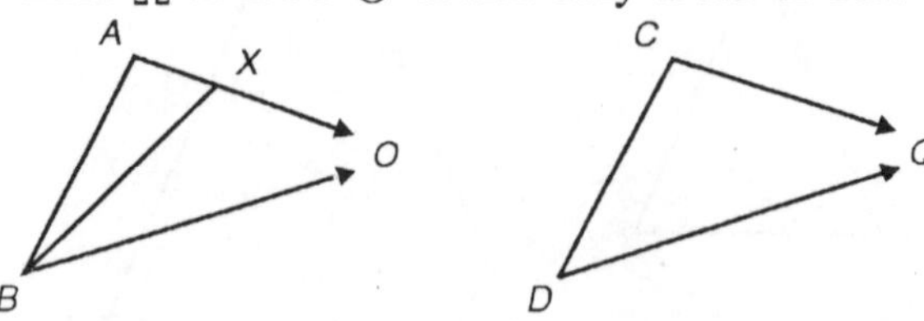

Proof: First, let us assume that $AB \cong CD$ but $\angle AB\Omega \not\cong \angle CD\Theta$ We may suppose that $\angle AB\Omega > \angle CD\Theta$ Then, there is a unique ray $\overrightarrow{BW}$ between $\overrightarrow{B\Omega}$ and $\overrightarrow{BA}$ so that $\angle ABW \cong \angle CD\Theta$ Since $\overrightarrow{A\Omega}$ and $\overrightarrow{B\Omega}$ are limiting parallel, $\overrightarrow{BW}$ must intersect $\overrightarrow{A\Omega}$ in a point X. There is a point $Y \in \overrightarrow{C\Theta}$ so that $AX \cong CY$ Then by SAS $\Delta ABX \cong \Delta CDY$ Therefore, $\angle CDY \cong \angle ABX \cong \angle CD\Theta$ Then, $\overrightarrow{C\Theta}$ and $\overrightarrow{D\Theta}$ are not limiting parallel, a contradiction.

Now assume that $\angle AB\Omega \cong \angle CD\Theta$ and $AB \not\cong CD$ We may then assume that $AB>CD$. There is a unique point $P \in AB$ so that $BP \cong CD$ Let $\overrightarrow{P\Omega}$ be the limiting parallel ray to $\overrightarrow{A\Omega}$ from P. Thus, $\overrightarrow{P\Omega}$ is also parallel to $\overrightarrow{B\Omega}$ The first part of this proof now implies $\angle BP\Omega \cong \angle DC\Theta \cong \angle BA\Omega$ But $\angle BP\Omega$ is exterior to $\Delta AP\Omega$ by the *Exterior Angle Theorem* $\angle BP\Omega > \angle BA\Omega$ a contradiction.

Corollary: *$PQ \cong RS$ if and only if $\prod(PQ)° = \prod(RS)°$.*

We are now in a position to show that $\overrightarrow{EF} \cap \overrightarrow{AG} \neq \theta$.

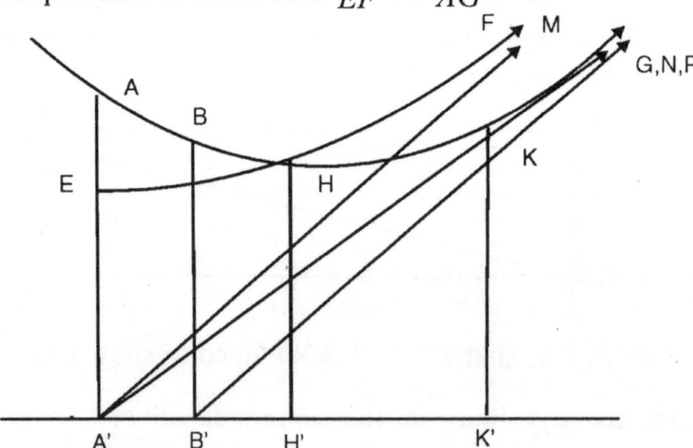

Let $\overrightarrow{A'M} \parallel \overrightarrow{EF}$, $\overrightarrow{A'N} \parallel \overrightarrow{AG}$, and $\overrightarrow{B'P} \parallel \overrightarrow{BG}$ Recall that $EA' \cong BB'$ and $\angle A'EF \cong \angle B'BG$ By Lemma $\angle EA'M \cong \angle BB'P$ Choose a point $L \in m$ so that $A' * B' * L$.

Since m and n were not limiting parallel lines, we have that $\overrightarrow{B'L} \neq \overrightarrow{B'P}$ and $\overrightarrow{A'L} \neq \overrightarrow{A'N}$ Now, $\angle MA'L \cong \angle PB'L$ by *Angle Subtraction*. Since $\overrightarrow{A'N} \parallel \overrightarrow{AG}$, $\overrightarrow{A'N} \parallel \overrightarrow{BG}$ and $\overrightarrow{B'P} \parallel \overrightarrow{BG}$ we must have that $\overrightarrow{A'N} \parallel \overrightarrow{B'P}$ Thus, $\angle PB'L$ is exterior to the singly asymptotic triangle $\Delta A'B'\Omega$ Therefore, $\angle NA'L < \angle PB'L$ This means that $\overrightarrow{A'N}$ lies between $\overrightarrow{A'M}$

and $\overrightarrow{A'L}$ which implies that $\overrightarrow{A'M}$ lies between $\overrightarrow{A'N}$ and $\overrightarrow{A'A}$ But since $\overrightarrow{A'N} \parallel \overrightarrow{AG}$, we have that

$$\overrightarrow{A'M} \cap \overrightarrow{AG} = \{J\}.$$

Since $J \in \overrightarrow{A'M}$ it lies on the same side of $\overleftrightarrow{EF}$ as does *A*'. Thus, *A* and *J* are on opposite sides of $\overleftrightarrow{EF}$ Thus, $AJ \cap \overleftrightarrow{EF} = \{H\} \neq \theta$ But, $H \in \overleftrightarrow{EF}$ since *H* lies on the same side of $\overleftrightarrow{AA'}$ as does *J*. It is possible to construct the limiting parallel ray to a line through a given point with a compass and straightedge. Let $P \notin l$ and let *Q* be the foot of *P* in *l*. Let *m* be perpendicular to $\overleftrightarrow{PQ}$ at *P* so that *m*)(*l* Let $R \in l$, $R \neq Q$ and let *S* be the foot of *R* in *m*.

Then *PQRS* is a Lambert quadrilateral and *PS*<*QR*. *QR*<*PR* since *PR* is a hypotenuse. Take a circle centered at *P* of radius *QR*. *S* is inside this circle and *R* is outside this circle. There is a unique point $X \in RS$ where *RS* intersects the circle. One can show that the ray $\overrightarrow{PX}$ is limiting parallel to *l*!

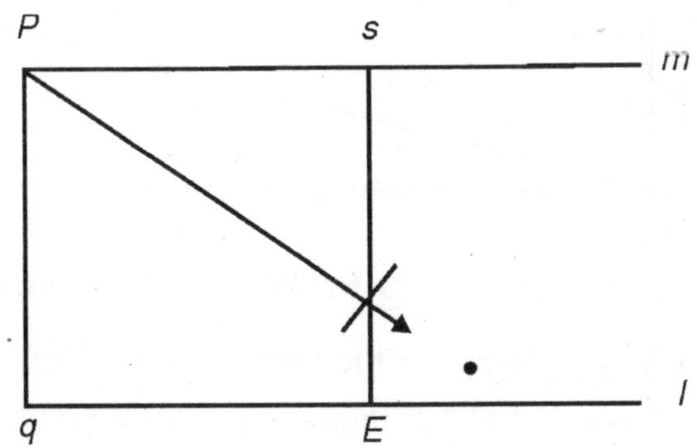

STRANGE NEW TRIANGLES

If $\overrightarrow{PB}$ and $\overrightarrow{QD}$ are limiting parallel to each other, the convention is to pretend that they go through an *ideal point* Ω and denote them simply by the *segments* $P\Omega$ and $Q\Omega$ The figure consisting of these rays and the segment *PQ* is then called a trilateral.

Note that here we do not assume that *PQ* is necessarily perpendicular to $Q\Omega$ Trilaterals are also called *singly asymptotic triangles*. The angles of a trilateral are the angles $\angle \Omega PQ$ and $\angle \Omega QP$ and these are called its *interior angles*. Just as we have defined a trilateral--singly asymptotic triangle--we can define a DAT--doubly asymptotic triangle. Given a point *P* and two non-opposite rays through *P*-- $\overrightarrow{PX}$ and $\overrightarrow{PY}$ --let *l* be the line that is limiting parallel to $\overrightarrow{PX}$ and limiting parallel to $\overrightarrow{PY}$ in the opposite direction.

Let Ω denote the ideal point in the direction of $\overrightarrow{PX}$ and let Σ denote the ideal point in the direction of $\overrightarrow{PY}$ Then we can identify the line *l* with the *segment* $\Omega\Sigma$ and the *segments* $P\Omega$, $P\Sigma$ and $\Omega\Sigma$ form a doubly asymptotic triangle. The angle $\angle \Omega P\Sigma$ is the fan angle of *l* and *P*.

Many of the properties of trilaterals are analogous to those of triangles.

Crossbar Theorem for Trilaterals

Lemma: *A line which subdivides an angle of a trilateral intersects the opposite side.*

Pasch's Theorem for Trilaterals

Lemma: *A line which intersects a side of the trilateral* $\Delta AB\Omega$ *but does not pass through a vertex will intersect another side, provided the line is not limiting parallel to either* $A\Omega$ *or* $B\Omega$ *at* Ω.

Lemma: *The sum of the interior angles of a trilateral is less than* 180°.

Exterior Angle Theorem

Lemma: *If* $\Delta PQ\Omega$ *is a trilateral, the exterior angles at P and Q are greater than their respective opposite interior angles.*

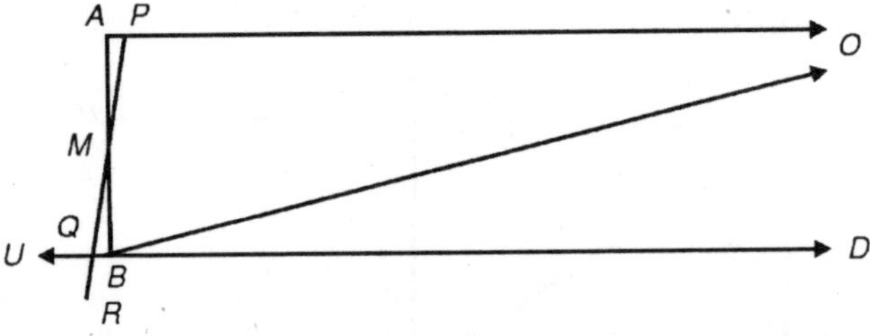

Proof: Choose $R \in \overleftrightarrow{PQ}$ so that $P * Q * R$ We need to show that $\angle RQ\Omega > \angle QP\Omega$.

There is a unique ray $\overrightarrow{PD}$ lying on the same side of $\overleftrightarrow{PQ}$ as Ω so that $\angle RQD \cong \angle QP\Omega$. If $U \in \overleftrightarrow{QD}$ is so that $U * Q * D$ then $\angle UQP \cong \angle RQD \cong \angle QP\Omega$ Thus, we have that the alternate interior angles are congruent. Let M be the midpoint of PQ and let A and B be the feet of M in $\overrightarrow{Q\Omega}$ and $\overrightarrow{QD}$ respectively. By AAS $\Delta APM \cong \Delta BQM$ so that $AM \cong BM$ and $\angle AMP \cong \angle BMQ$ Thus, A, M, and B are collinear. Thus, $\overleftrightarrow{Q\Omega}$ and $\overleftrightarrow{QD}$ have a common perpendicular. Thus, $\overrightarrow{Q\Omega})(\overrightarrow{QD}$ Thus $\overrightarrow{QD}$ lies between $\overrightarrow{Q\Omega}$ and $\overrightarrow{QR}$ from which it follows that $\angle RQ\Omega > \angle QP\Omega$.

Lemma: *If in the singly asymptotic triangles* $\Delta AB\Omega$ *and* $\Delta CD\Theta$ *we have* $\angle BA\Omega \cong \angle DC\Theta$ *then*

$$\angle AB\Omega \cong \angle CD\Theta \text{ if and only if } AB \cong CD.$$

Proof: First, let us assume that $AB \cong CD$ but $\angle AB\Omega \not\cong \angle CD\Theta$ We may suppose that $\angle AB\Omega > \angle CD\Theta$ Then, there is a unique ray $\overrightarrow{BW}$ between $\overrightarrow{B\Omega}$ and $\overrightarrow{BA}$ so that

$\angle ABW \cong \angle CD\Theta$ Since $\overrightarrow{A\Omega}$ and $\overrightarrow{B\Omega}$ are limiting parallel, $\overrightarrow{BW}$ must intersect $\overrightarrow{A\Omega}$ in a point X. There is a point $Y \in \overrightarrow{C\Theta}$ so that $AX \cong CY$ Then by *SAS* $\Delta ABX \cong \Delta CDY$ Therefore, $\angle CDY \cong \angle ABX \cong \angle CD\Theta$ Then, $\overrightarrow{C\Theta}$ and $\overrightarrow{D\Theta}$ are not limiting parallel, a contradiction.

Now assume that $\angle AB\Omega \cong \angle CD\Theta$ and $AB \not\cong CD$ We may then assume that $AB>CD$. There is a unique point $P \in AB$ so that $BP \cong CD$ Let $\overrightarrow{P\Omega}$ be the limiting parallel ray to $\overrightarrow{A\Omega}$ from P. Thus, $\overrightarrow{P\Omega}$ is also parallel to $\overrightarrow{B\Omega}$ The first part of this proof now implies $\angle BP\Omega \cong \angle DC\Theta \cong \angle BA\Omega$ But $\angle BP\Omega$ is exterior to $\Delta AP\Omega$ by the *Exterior Angle Theorem* $\angle BP\Omega > \angle BA\Omega$ a contradiction.

INVERSION IN EUCLIDEAN CIRCLES

In order to define congruence in the Poincarı model and verify the axioms of congruence, we must study the operation of inversion in a Euclidean circle. This operation is the process of reflecting across a line in the Poincarı disk model of the hyperbolic plane.

Let γ be a circle of radius r and center O. For any point $P \neq O$ the inverse P' of P with respect to the circle γ is the unique point $P' \in \overrightarrow{OP}$ so that $OP.\ OP' = r^2$ In this case γ is called the circle of inversion.

These are Euclidean geometry theorems about circles, so we may use the facts that we know about Euclidean geometry.

Proposition: *For P, P′ and γ as above,*

1. $P = P'$ if and only if $P \in \gamma$.
2. If $P \in$ int γ then $P' \in$ int γ and vice versa.
3. $(P')' = P$.

Proposition: *Let P* $\in$ int γ *and let TU be the chord through P perpendicular to* $\overleftrightarrow{OP}$ *Then P'= P(TU), the pole of TU, i.e., the intersection of the tangents to γ at T and U.*

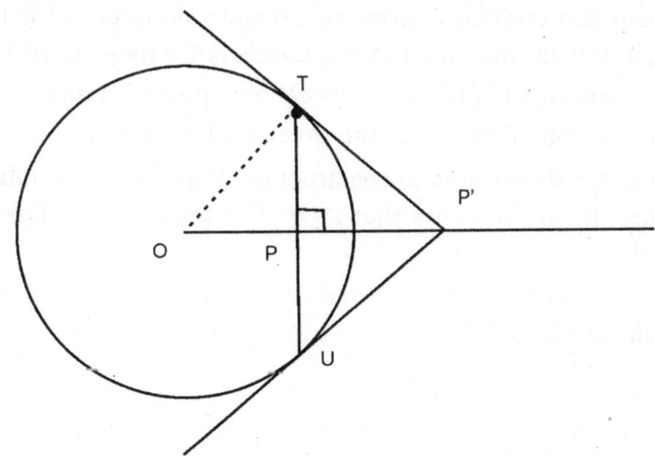

Proof: Suppose the tangent to γ at T cuts $\overrightarrow{OP}$ at the point P'. The right triangle ΔOPT is similar to right triangle OTP' (since they have $\angle TOP$ in common and the angle sum is 180°). Hence corresponding sides are proportional. Since $l(OT) = r$ we get that

$$\frac{l(OP)}{r} \cdot = \frac{r}{l(OP')}$$

which shows that P' is the inverse to P. Reflect across the line $\overleftrightarrow{OP}$ and we see that the tangent to γ at U also passes through P', so that P' is the pole of TU.

Proposition: *If P is outside of γ let Q be the midpoint of OP. Let be the circle of radius QP centered at Q. Then*

1. $\sigma \cap \gamma = \{T, U\}$.
2. $\overrightarrow{PT}$ and $\overrightarrow{PU}$ are tangent to γ.
3. $P' = TU \cap OP$.

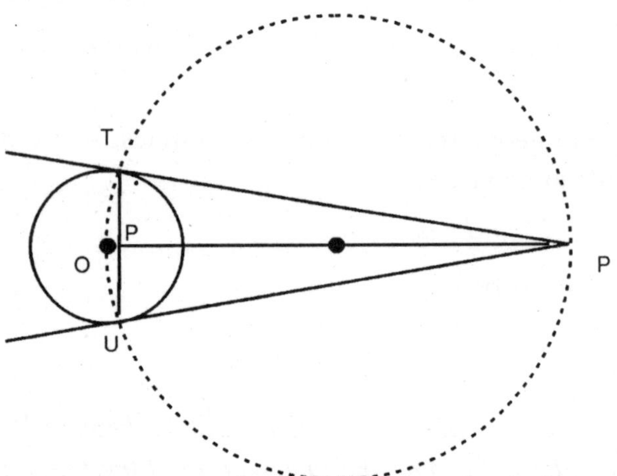

Proof: By the circular continuity principle, σ and γ do intersect in two points T and U. Since $\angle OTP$ and $\angle OUP$ are inscribed in semicircles of σ they are right angles. Therefore $\overleftrightarrow{PT}$ and $\overleftrightarrow{PU}$ are tangent to γ If TU intersects OP in a point P', then P is the inverse of P', by the previous proposition. Thus, P' is the inverse of P in γ.

The next proposition shows how to construct the Poincarı line joining two ideal points--the line of enclosure. Its proof shows that $\angle OP'T$ in the previous figure is indeed a right angle, as we needed.

Proposition: *Let T and U be points on γ that are not contained on a the same diameter, and let P be the pole of TU. Then*

1. $PT \cong PU$,
2. $\angle PTU \cong \angle PUT$,

3. $\overleftrightarrow{OP} \perp \overleftrightarrow{TU}$,
4. The circle έ with center P and radius PT cuts γ orthogonally at T and U.

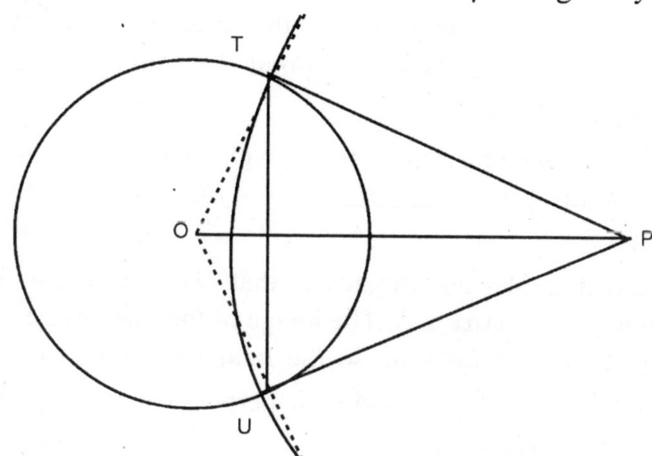

Proof: By definition of pole, $\angle OTP$ and $\angle OUP$ are right angles, so by the Hypotenuse-Leg criterion, $\Delta OTP \cong \Delta OUP$ Therefore, $PT \cong PU$ and $\angle OPT \cong \angle OPU$ The base angles $\angle PTU$ and $\angle PUT$ of the isosceles triangle ΔTPU are thus congruent, and the angle bisector $\overrightarrow{PQ}$ is perpendicular to the base TU. The circle έ is then well-defined because $PU \cong PT$ and έ intersects γ orthogonally by the hypothesis that $\overrightarrow{PT}$ and $\overrightarrow{PU}$ are tangent to γ.

Let P be a point in the plane and γ a circle with center O. The power of P with respect to γ is defined to be

$$Pw\ (P) = (l(OP))^2 - r^2.$$

Proposition: *Assume $P \notin \gamma$ and assume that two lines through P intersect γ in points $\{R_1, R_2\}$ and $\{S_1, S_2\}$. Then*

1. $(PR_1).\ (PR_2) = (PS_1).\ (PS_2) = Pw\ (P)$.
2. *If one of these lines through P is tangent to γ at T, then $Pw(P) = PT^2$.*

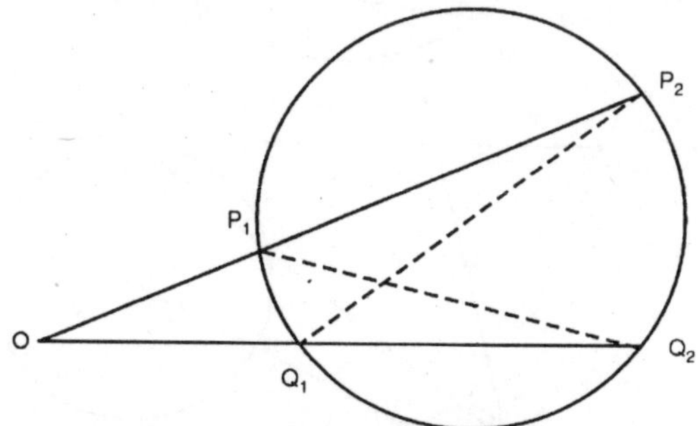

Proposition: *Let $P \notin \gamma$ and $P \neq O$ the center of γ Let έ be a circle through P. δ is orthogonal to γ if and only if έ passes through P', the inverse of P with respect to γ.*

Proposition: *Let γ have radius r, δ have radius t and let P be the center of δ. δ is orthogonal to γ if and only if $Pw(P) = t^2$ where the power is computed with respect to γ.*

Let O be a point and $k > 0$. The dilation with center O and ratio k is a mapping of the Euclidean plane that fixes O and maps every point $P \neq O$ to a unique $P^* \in \overrightarrow{OP}$ such that $OP^* = kOP$ Call this map Do,k.

Proposition: *Let δ be a circle with center $C \neq O$ and radius s. Do, α maps δ to a circle δ with center $C^* = Do,\alpha(C)$ and radius αs If $Q \in \delta$ the tangent to δ^* at Q* is parallel to the tangent to δ at Q.*

Proof: Choose rectangular coordinates so that O is the origin. Then the dilation is given by the mapping $(x, y) \mapsto (\alpha x, \alpha y)$. The image of the line have equation $ax + by = c$ is the line having equation $\alpha x + by = \alpha c$ so the image is parallel to the original line. In particular, $\overleftrightarrow{CQ}$ is parallel to $\overleftrightarrow{C^*Q^*}$ and their perpendiculars at Q and Q^* respectively, are also parallel. If δ has equation $(x - h)^2 + (y - k)^2 = s^2$ then έ* has equation $(x - \alpha h)^2 + (y - \alpha k)^2 = (\alpha s)^2$.

Proposition: *Let γ be the circle of radius r centered at O and let έ be the circle of radius s centered at C. Assume O lies outside δ; let $p = Pw(O)$ with respect to δ and let $\alpha = r^2/p$ The image δ' of δ under inversion in γ is the circle of radius αs whose center is $C^* = Do,_{\alpha}(C)$. If $P \in \delta$ and P' is the inverse of P with respect to γ then the tangent t' to δ at P' is the reflection across the perpendicular bisector of PP' of* the tangent to δ at P.

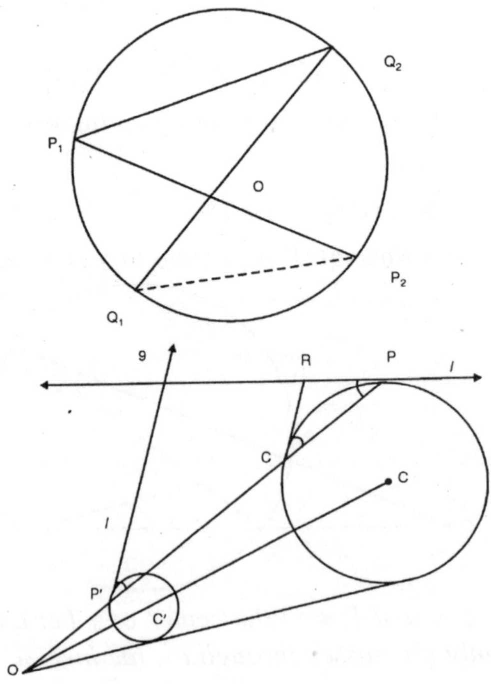

Corollary: *δ is orthogonal to γ if and only if ἑ is mapped to itself by inversion in γ.*

Proposition: *Let l be a line so that O ∉ l The image of l under inversion by γ is a punctured circle with missing point O. The diameter through O of the completed circle δ is perpendicular to l.*

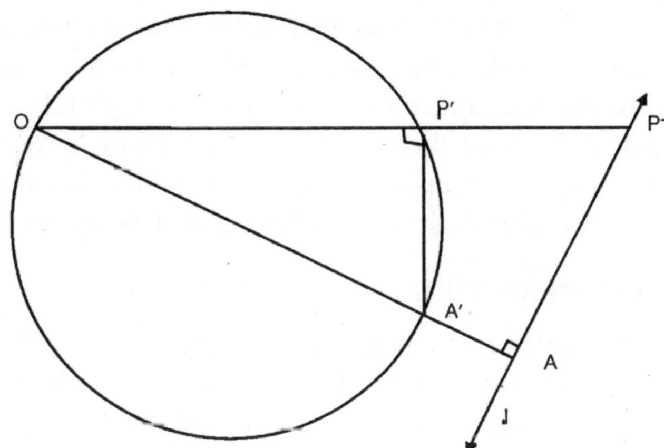

Proposition: *Let δ be a circle passing through the center O of γ The image of δ minus O under inversion in γ is a line l so that O ∉ l and l is parallel to the tangent to δ at O.*

Proposition: *A directed angle of intersection of two circles is preserved in magnitude by an inversion. The same applies to the angle of intersection of a circle and a line or the intersection of two lines.*

Proposition: *Let ἑ be orthogonal to Γ Inversion in δ maps Γ onto Γ and the interior of Γ onto itself. Inversion in δ preserves incidence, betweenness and congruence in the sense of the Poincarı disk model.*

CONSISTENCY OF HYPERBOLIC GEOMETRY

Is this system of hyperbolic geometry a consistent mathematical system? Does it lead to a contradiction? This is a question of *metamathematics*; *i.e.*, a question outside of a mathematical system about the system itself. The question is not about lines and points, but about the *whole system* itself. If it is inconsistent, then an ordinary mathematical argument would derive a contradiction. This is what Saccheri and Lambert attempted to do, and failed.

Theorem: *If Euclidean geometry is consistent, so is hyperbolic geometry.*

Corollary: *If Euclidean geometry is consistent, then no proof or disproof of the parallel postulate from the rest of Hilbert's postulates will ever be found, i.e., the parallel postulate is independent of the other postulates.*

Proof: Assume not, then there is a proof of the parallel postulate. This implies that hyperbolic geometry is inconsistent since the *Hyperbolic Axiom* contradicts a proven result.

Our above theorem asserts that hyperbolic geometry is as consistent as Euclidean geometry. Thus, Euclidean geometry is inconsistent. This contradiction implies that no proof o the parallel postulate exists.

Note also that the hypothesis that euclidean geometry is consistent implies that do disproof exists either. You should recognize that had mathematicians succeeded in proving *Euclid's Fifth Postulate* from the other axioms with the intention of making Euclidean geometry more secure and elegant, they would have completely destroyed Euclidean geometry as a consistent body of thought. What we will do now, in order to *see* that hyperbolic geometry is as consistent as Euclidean geometry, is to construct a model of hyperbolic geometry. We will look at three different models for H^2

THE BELTRAMI-KLEIN MODEL

This model is often referred to as the *Klein model*, because of the extensive work done by the German mathematician Felix Klein in geometry with this model. Fix, once and for all, a circle Γ in E^2 with center O and radius OR. Let

$$\text{int}\,\Gamma = \{X \in E^2 \mid OX < OR\}.$$

Identify the points of H^2 with the points in int Γ.

A chord of Γ is the Euclidean segment AB joining two points $A, B \in \Gamma$. $AB \cap \text{int}\,\Gamma = (A, B)$ is an open chord. These open chords represent the lines of the hyperbolic plane. To say that P lies on (A,B) means $P \in \overleftrightarrow{AB}$, the Euclidean line, and $A * P * B$.

It is easy to see that this model satisfies the *Hyperbolic Axiom*.

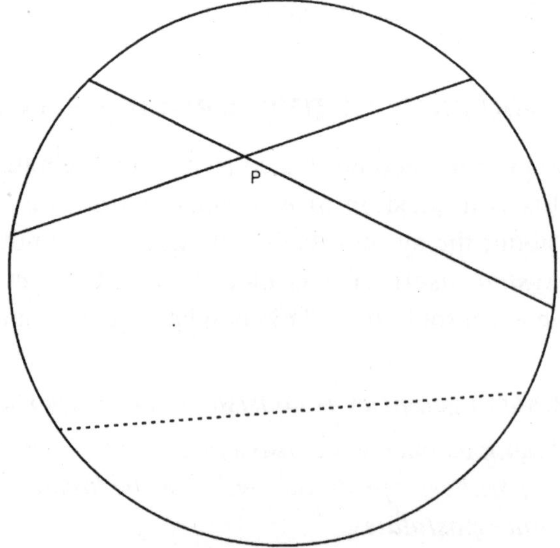

Fig. Parallel Lines in the Klein Model

Here, $k \cap \Gamma = \{A, B\}$ The lines $\overrightarrow{PA}$ and $\overrightarrow{PB}$ are two lines that do not intersect k in H^2

as is the line l The difference is that $\overrightarrow{PA}$ and $\overrightarrow{PB}$ are limiting parallel to k in different directions, while l and k are hyperparallel, admitting a common perpendicular. Recall that the points of Γ do not belong to the hyperbolic plane. They are called ideal points of H^2 The points outside of Γ are called ultraideal points.

By saying that the lines k and l admit a common perpendicular raises the question about how one defines congruence of segments and angles. It is not obvious, for it would seem that lines must be of finite extent, no line measuring more than twice the diameter of Γ. If this were the case, we would not have a model for hyperbolic geometry, as the congruence axioms, Archimedes' axiom, and Dedekind's axiom definitely would not hold. We must define a different method for measuring the length of segments and the measure of angles in this model.

I will at this point only mention the method for measuring the length of segments and the definition of right angles. The remaining measurement will follow from work that we do later, and will be noted then.

Let $A, B \in H^2$ and let $P, Q \in \Gamma$ denote the endpoints of the chord through A and B. Let $\overline{AB}$ denote the Euclidean distance from A to B, or the length of the segment AB. Define the *Klein distance*

$$dk\,(A, B) = \frac{1}{2}|\log\,(AB, PQ)|$$

$$= \frac{1}{2}\left|\log\left(\frac{\overline{AP}.\,\overline{BQ}}{\overline{BP}.\,\overline{AQ}}\right)\right|$$

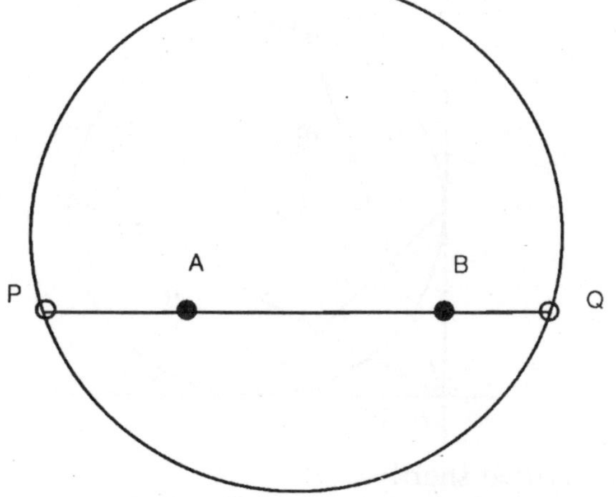

Fig. Length in the Klein Model

Note then that as B approaches Q, $\overline{BQ}$ approaches 0, thus

$$\lim_{B \to Q} d_k(A, B) = +\infty$$

This allows us to see then that we can find a segment of any length on any ray (Axiom C-1). Also, it will follow that Dedekind's and Archimedes' axioms are valid.

Our technique for measuring angles will be introduced at a later time. It depends on the model for hyperbolic geometry due to Poincarı. It is not the way in which angles are measured in Euclidean geometry. For this reason the Klein model is not a conformal model of geometry. We can talk about right angles, without too much difficulty. Let l and m be two lines in the Klein model of the hyperbolic plane, or *K-lines*. $l \perp m$ in the Klein sense if,

- l or m is a diameter and l and m are perpendicular in the Euclidean sense, or
- If neither l or m is a diameter, do the following: Let t_1 and t_2 be the tangents to Γ at the ends of l. Since l is not a diameter, $t_1 \cap t_2 \neq \theta$ Put $P(l) = t_1 \cap t_2$ $P(l)$ is called the pole of l. m is perpendicular to J in the Klein sense if and only if the Euclidean line extending m passes through $P(l)$.

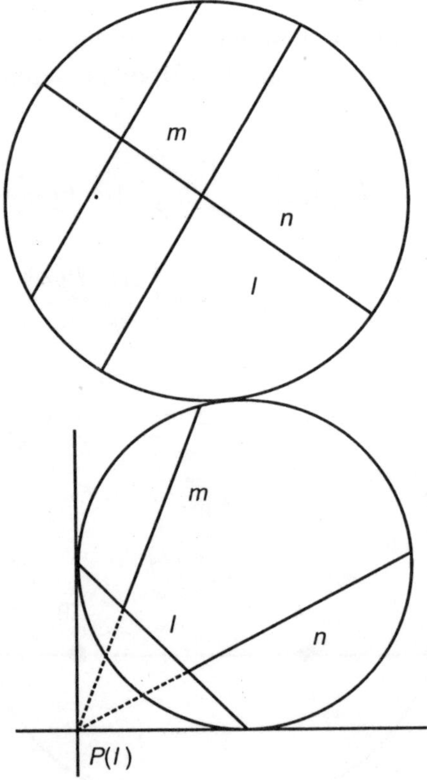

All of this will be verified shortly.

THE POINCARI HALF-PLANE MODEL

In this model take H^2 to be the set of all points in the coordinate Euclidean plane that satisfy $H^2 = \{(x, y) \in E^2 \mid y > 0\}$

We interpret lines in this model to be one of the following:

- The intersection of a Euclidean vertical line with H^2 or
- The intersection of a Euclidean circle, whose center lies on the x-axis, with H^2.

Once again, our concept of *congruence* must change. As before, we must define a new measure for segments. However, the manner of measuring angles is exactly the method that is used in Euclidean geometry. To measure the angle formed by two of these lines, we simply measure the Euclidean angle formed by the tangent lines to the two Poincarı lines.

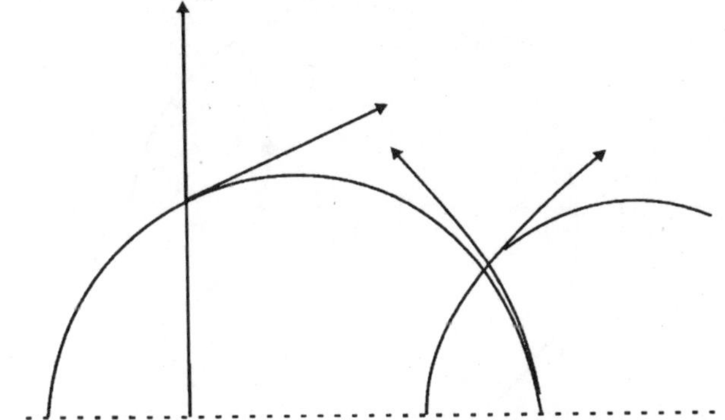

Fig. Measuring angles in the Poincarı Half-Plane Model

The method of measuring segments is more easily described in the next model, which is derived from this model.

THE POINCARI DISK MODEL

Let γ denote the circle of radius 1 centered at the point (0,1) in the real plane. Let N denote the point (0,2). Then N and each point on the circle γ determine a line which has non-zero slope. It must then intersect the x-axis in some point. On the other hand, given any point on the x-axis, it and the point N determine a line which must intersect the circle γ in some point. Thus, there is a one-to-one correspondence between the points of the real line (here, the x-axis) and the points of the circle γ, less N. The mappings are

$$f : \mathbb{R} \rightarrow \gamma \setminus N \text{ given by } f(r) = \left(\frac{-4r}{r^2+4}, \frac{2r^2}{r^2+4} \right);$$

And

$$g : \gamma \setminus N \rightarrow \mathbb{R} \text{ given by } g(x, y) = \frac{2r^2}{2-y}.$$

Not only can you see that the mappings are continuous, but are differentiable as well.

What this means to us is that we may add a *point at infinity* to the real line and we get a circle, in some sense. Think of taking the Poincarı half-plane model and adding a point at infinity to the real axis. We will then have the interior of a circle sitting in the real plane.

The vertical lines will correspond to diameters of the circle and the circles in the half-plane model will correspond to circles which are *orthogonal* to the boundary circle.

To make the setting more precise, let H^2 = int Γ. A *p*-line in this model is either

1. An open diameter of Γ, or
2. An open arc of a circle orthogonal to Γ. A circle is orthogonal to Γ if at each point of intersection of έ and Γ the radii of Γ and έ through that point are perpendicular. έ ∩ int Γ = *m* an open arc which represents a *p*-line in H^2.

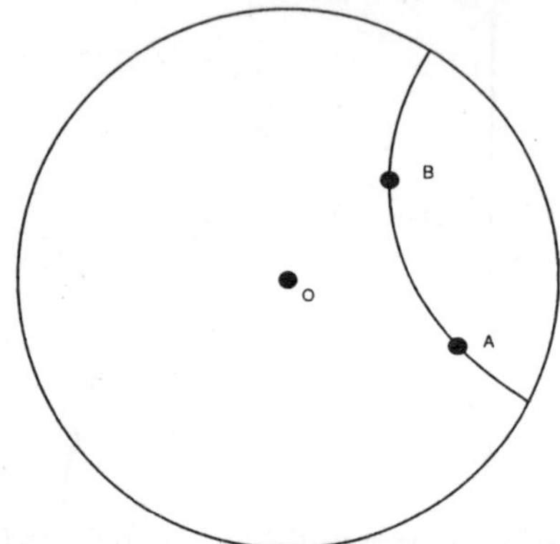

Fig. A *p*-line in the Poincarι disk model

A point in H^2 lies on a *p*-line if and only if it lies on it in the Euclidean sense. Betweenness has the same interpretation, though if *A*, *B*, and *C* are on an open arc from έ with center *P* then *A* * *B* * *C* if and only if $\overrightarrow{PB}$ is between $\overrightarrow{PA}$ and $\overrightarrow{PC}$.

If $A, B \in H^2$ and if *P* and *Q* are the ends of the *p*-line through *A* and *B*, then we define the Poincarι distance from *A* to *B* to be

$$d_p(A, B) = |\log(AB, PQ)| = \left|\log \frac{\overline{AP}.\overline{BQ}}{\overline{AQ}.\overline{BP}}\right|$$

As with the above model, this Poincarι model is conformal. If two directed circular arcs meet at *A* then the measure of the angle they form is the measure of the angle between their tangent rays at *A*.

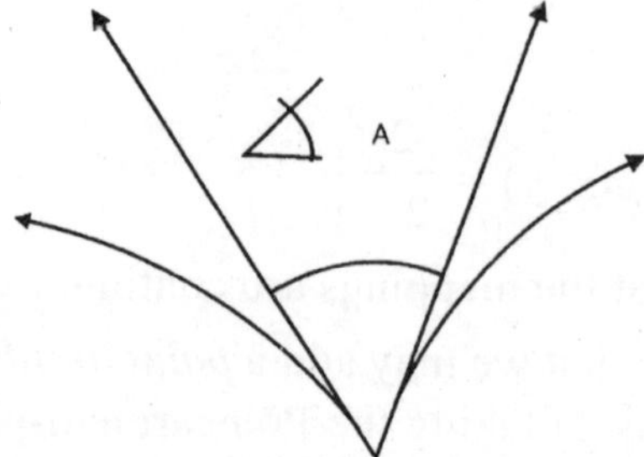

Fig. Angle Measure in the Poincarι Disk Model

Question: IF $A, B \in H^2$ and A and b do not lie on a diameter of Γ, how do you construct the p-line through A and B?

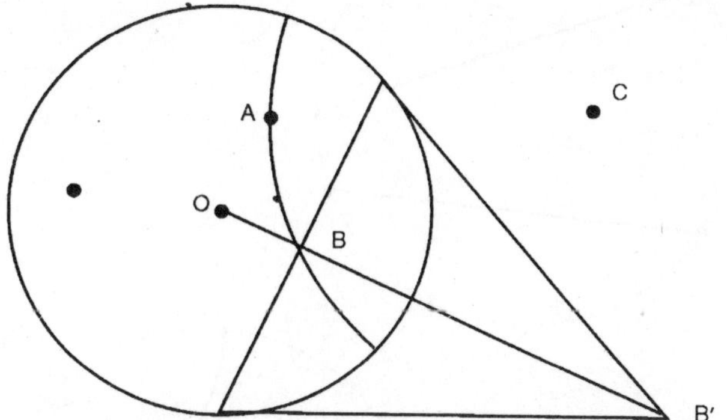

Fig. Construction of the p-line through A and B

We will verify these steps later.

- Construct the perpendicular to $\overrightarrow{OA}$ at A, intersecting Γ in P and Q.
- Construct the tangents to Γ at P and Q. These tangents intersect at a point A'.
- Construct the circle, δ containing A, A', and B.

δ is orthogonal to Γ and $m = \delta \cap int\ \Gamma$ is the p-line through A and B.

Let us compare the two disk models by looking at certain familiar objects in each of them.

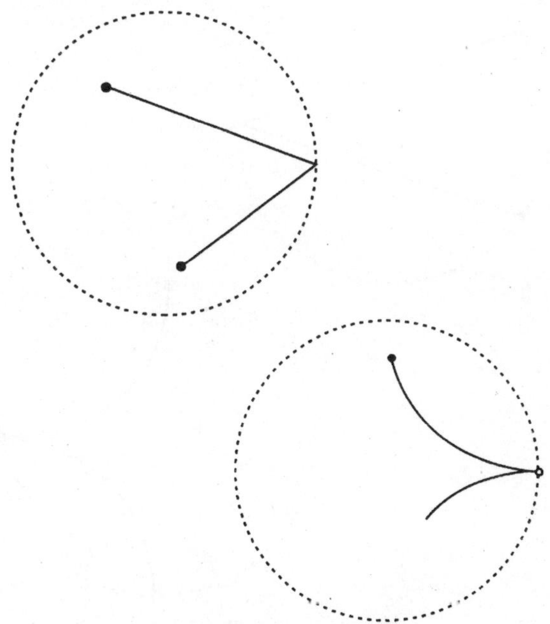

Fig. Limiting Parallel Rays

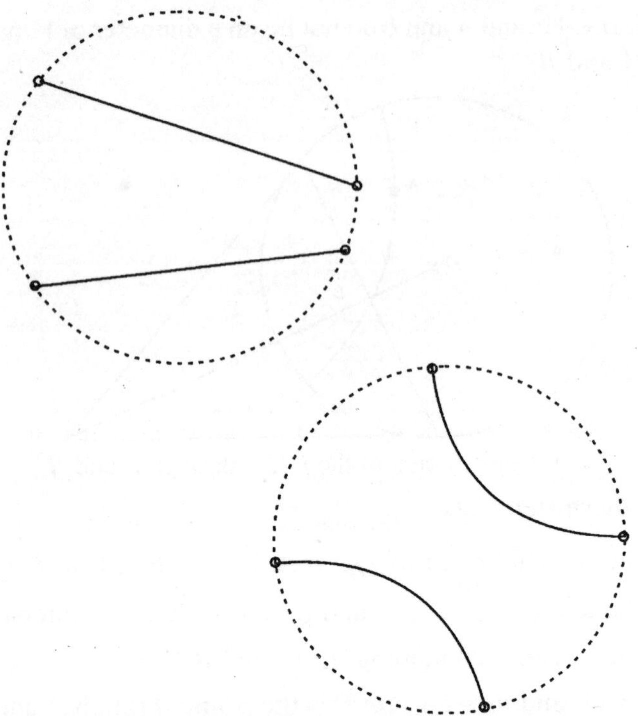

Fig. Hyperparallel Lines

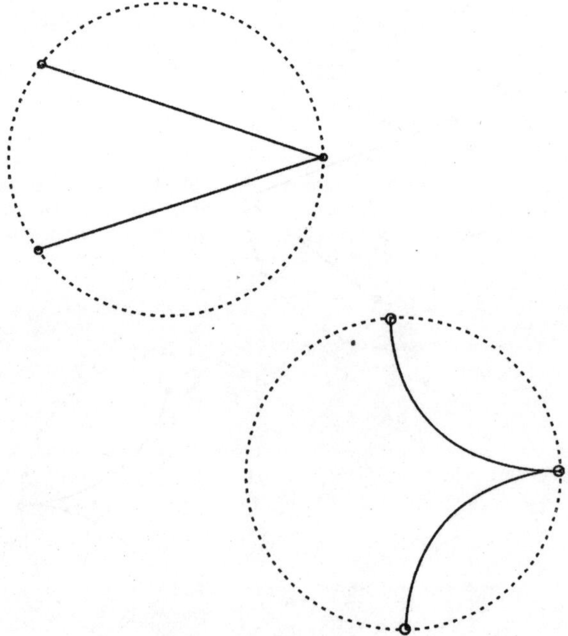

Fig. Horoparallel Lines

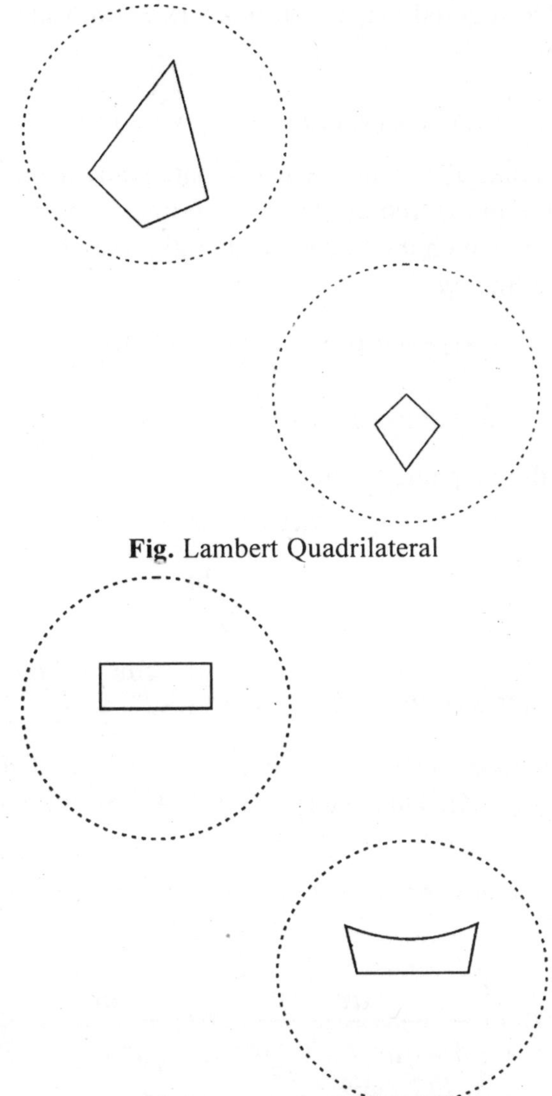

Fig. Lambert Quadrilateral

Fig. Saccheri Quadrilateral

ISOMORPHISM OF MODELS

You will, doubtless, be surprised to learn that these two models are *isomorphic*;, *i.e.*, there is a one-to-one correspondence between the points and lines of the two models such that the relations of incidence, betweenness, and congruence are preserved.

Let S^2 be a sphere of the same radius as the radius of Γ, tangent to *int* Γ at the origin. Use *orthogonal projection* of the Klein model in Γ upward onto the lower hemisphere of the sphere and then stereographic projection from the north pole back to the plane. The image will be the Poincarι model.

Let $(x, y) \in int\ \Gamma$ Orthogonal projection maps (x,y) onto the bottom hemisphere of $x^2 + y^2 + (z - a)^2 = a^2$ by

$$(x, y) \mapsto (x, y, a - \sqrt{a^2 - x^2 - y^2}).$$

Stereographic projection of S^2 onto the coordinate plane is the analogous map to that described for mapping the punctured circle onto the line. A point $(x, y, z) \in S^2$ is sent to the point on the xy-plane at which the line through $(0,0,2a)$ and (x,y,z) intersects the plane. We can parameterize the line by:

$$l(t) = \langle 0, 0, 2a \rangle + t\langle x, y, z - 2a \rangle$$

$$= \langle tx, ty, 2a + t(z - 2a) \rangle$$

$l(t)$ passes through the xy-plane when

$$2a + t(z - 2a) = 0$$

$$t = \frac{2a}{2a - z}$$

Thus, stereographic projection sends (x,y,z) to $\left(\frac{2ax}{2a - z}, \frac{2ay}{2a - z}\right)$.

We need a shrinking map to shrink the image back into Γ. The equator of the sphere (x,y,a) maps to $(2x,2y) = (x',y')$. Thus, $(x')^2 + (y')^2 = 4r^2$ Shrinking back into Γ it is only necessary to divide by 2.

Combining the three maps, we have a mapping from the Klein model to the Poincarı model given by

$$(x, y) \mapsto \left(\frac{ax}{a + \sqrt{a^2 - x^2 - y^2}}, \frac{ay}{a + \sqrt{a^2 - x^2 - y^2}}\right).$$

It is easy to check that this is a continuous and differentiable map from $int\ \Gamma$ to itself. It sends k-lines to p-lines and is an isomorphism of geometries.

CONSTRUCTIONS IN THE POINCARI MODEL

To construct the Poincarı line joining two ideal points, or the so-called line of enclosure, we have the following result.

Lamma: *Let T, $U \in \Gamma$, not on a diameter, and Let $P = P(TU)$, the pole of TU. Then*

1. $PT \cong PU$,
2. $\angle PTU \cong \angle PUT$,

3. $\overrightarrow{OP} \perp \overrightarrow{TU}$,
4. The circle δ with center P and radius $PU \cong PT$ is orthogonal to γ at T and U.

The previous description for the construction of the unique line between two ordinary points follows from the fifth item above. We have already discussed the measure of segments, as well.

How do you copy a given angle at a given point, A, on a given ray, $\overrightarrow{AB}$?

1. If A is the center of Γ then the ray $\overrightarrow{AB}$ is a diameter of Γ and the angle is copied as in Euclidean geometry, using another diameter of Γ.
2. If $A \neq O$ the center of Γ, then we must find the unique circle δ through A orthogonal to Γ and tangent to the given Euclidean line l which forms the Euclidean angle with the tangent to $\overrightarrow{AB}$ at A.
3. For δ orthogonal to Γ, by the above facts, δ must pass through A'_Γ the inverse of A with respect to Γ. The center, C, of δ lies on the perpendicular bisector of AA'. Let m be this bisector.
4. For δ to be tangent to l at A, we must have $\overrightarrow{CA}$ perpendicular to l Thus, C lies on the unique perpendicular to l at A, say $n.\{C\} = m \cap n$.

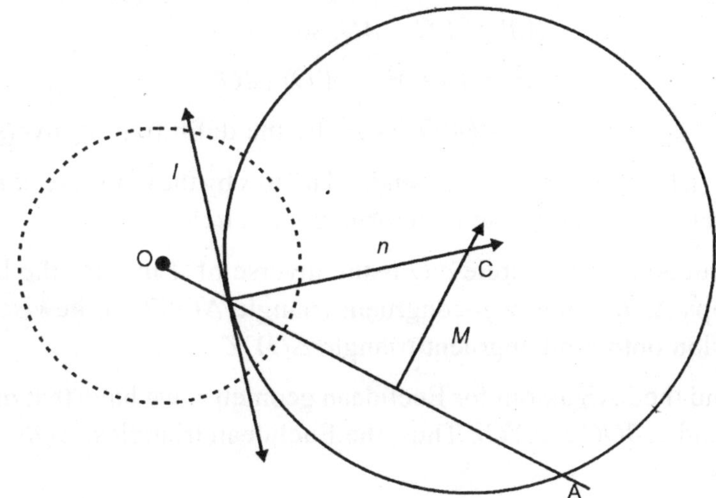

Lemma: *If* $d_p(O, B) = x$ *then*

$$d(O, B) = r\frac{e^x - 1}{e^x + 1},$$

Where r is the radius of Γ *and e is the number so that* log e = 1.

Proof: If P and Q are the ends of the diameter through OB then

$$x = \log (OB, PQ)$$

$$e^x = \frac{OP.BQ}{OQ.BP}$$

$$= \frac{BQ}{BP} = \frac{r+OB}{r-OB}$$

$$OB = r\frac{e^x+1}{e^x-1}$$

Which is what was to be proven.

This will give us enough information to prove that all of the axioms of neutral geometry are satisfied, except for the *SAS* axiom. It requires a bit of work. Assume that we are given two *p*-triangles ΔABC and ΔXYZ inside Γ so that $\angle A \cong \angle X$, $AC \cong XY$ and $AB \cong XY$ We need to show that these triangles are congruent in the Poincarı model.

First, we show that we may assume that $A = X = O$, where O is the center of Γ. Let δ be the circle through A and B orthogonal to Γ and let be the circle through A and C orthogonal to Γ. Then, since both circles are orthogonal to Γ it follows that $\delta \cap \sigma = \{A, A'\}$.

Let $s^2 = (AA')(A'O)$ and let be the circle of radius s centered at A'. Then

$$AA' = A'O - AO, \text{ so}$$

$$s^2 = (A'O)^2 - (A'O)(AO)$$

$$= (A'O)^2 - r^2, \text{ by the definition of inverse}$$

Thus, $r^2 + s^2 = (A'O)^2$ so that r is "perpendicular" to s by the converse of the Pythagorean theorem. This means then that Γ and c are orthogonal circles.

By the construction of this circle c O is the inverse of A in c By the last item above, inversion in c maps ΔABC onto a *p*-congruent triangle $\Delta OB'C'$ Likewise, ΔXYZ can be mapped by inversion onto a *p*-congruent triangle $\Delta OY'Z'$.

By Lemma and the SAS axiom for Euclidean geometry, we have that $d(O,B) = d(O,Y)$, $d(O,C) = d(O,Z)$ and $\angle BOC \cong \angle YOZ$ Thus, the Euclidean triangles, ΔOBC and ΔOYZ are congruent by *SAS*.

Therefore, there is a rotation about the center of Γ and possibly a reflection through a diameter which sends ΔOBC to ΔOYZ This will map Γ onto itself and the orthogonal circle through B and C to the orthogonal circle through Y and Z. This preserves the *p*-length, and hence the triangles are *p*-congruent.

Theorem: *Two triangles in the Poincarı model of hyperbolic geometry are p-congruent if and only if they can be mapped onto each other by a succession of inversions in circles orthogonal to* Γ *and/or reflections in diameters of* Γ.

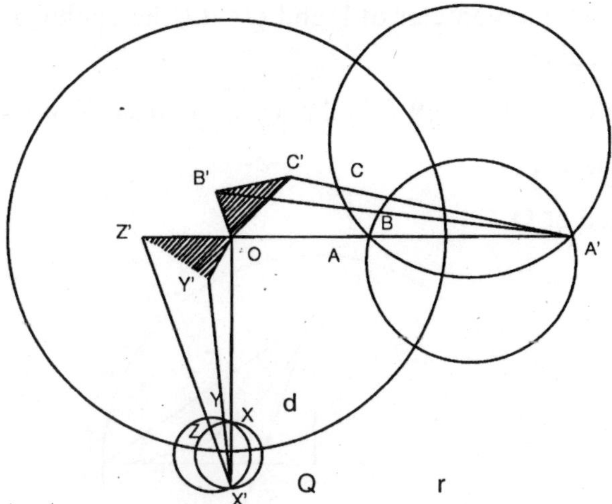

Fig. SAS in the Poincarı Model

At this point, I would rather think of measuring angles in radian measure, not degree measure. In this case we know that the sum of the angles of any triangle in hyperbolic geometry is less than π. Also, let $\prod(d)$ denote the radian measure of the angle of parallelism corresponding to the hyperbolic distance d. We can define the standard trigonometric functions, not as before--using right triangles--but in a standard way. Define

$$\sin x = \sum_{n=0}^{\infty}(-1)^n \frac{x^{2n+1}}{(2n+1)!}$$

$$\cos x = \sum_{n=0}^{\infty}(-1)^n \frac{x^{2n}}{(2n)!}$$

$$\text{ten } x = \frac{\sin x}{\cos x}.$$

In this way we have avoided the problem of the lack of similarity in triangles, the premise upon which all of real Euclidean trigonometry is based. What we have done is to define these functions *analytically*, in terms of a power series expansion. These functions are defined for all real numbers x and satisfy the usual properties of the trigonometric functions.

Theorem: *In the Poincarı model of hyperbolic geometry the angle of parallelism satisfies the equation*

$$e^{-d} = \tan\left(\frac{\Pi(d)}{2}\right).$$

Proof: By the definition of the angle of parallelism, $d = d_p(P, Q)$ for some point P to its foot Q in some p-line l Now, $\prod(d)$ is half of the radian measure of the fan angle at P, or is the radian measure of $\Pi QP\Omega$ where $\overrightarrow{P\Omega}$ is the limiting parallel ray to l through P.

We may choose l to be a diameter of Γ and $Q = O$, the center of Γ, so that P lies on a diameter of Γ perpendicular to l.

The limiting parallel ray through P is the arc of a circle δ so that

1. δ is orthogonal to Γ,
2. l is tangent to δ at Ω.

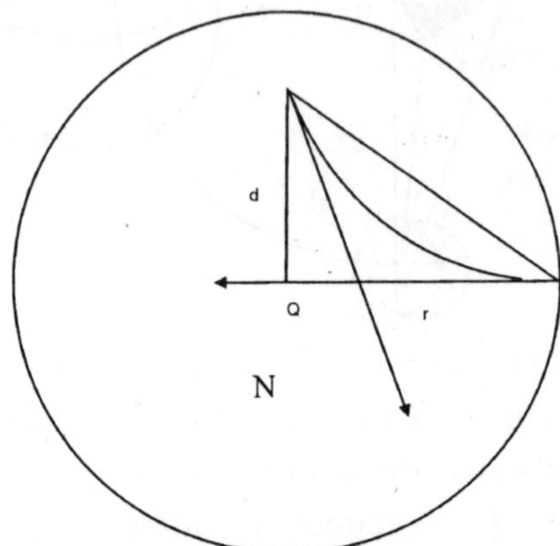

Fig. Angle of Parallelism

The tangent line to δ at P must meet l at a point R inside Γ. This point R is the pole of $P\Omega$ from δ By Lemma$\angle QP\Omega = \angle Q\Omega P = \beta$ radians. Let us denote $\Pi(d) = \alpha$ Then in $\Delta PQ\Omega$,

$\alpha + 2\beta = \frac{\pi}{2}$ or

$$\beta = \frac{\pi}{4} - \frac{\alpha}{2}.$$

Now, $d(P, Q) = r \tan \beta = r$ ten $\left(\frac{\pi}{4} - \frac{\alpha}{2}\right)$ Applying Lemma we have

$$e^d = \frac{r + d(P,Q)}{r - d(P,Q)} = \frac{1 + \tan\beta}{1 - \tan\beta}.$$

Using the identity tan $\left(\frac{\pi}{4} - \frac{\alpha}{2}\right) = \frac{1 + \tan\alpha/2}{1 - \tan\alpha/2}$ it follows that

Simplifying this it becomes $e^d = \frac{1}{1 - \tan\alpha/2}$.

$$= e^{-d} = \tan\left(\frac{\Pi(d)}{2}\right).$$

Also, we can write this as $\angle(d) = 2 \arctan (e^{-d})$

RETURN TO THE KLEIN MODEL

Now, we have that the Poincarı model is a model for hyperbolic geometry. Since the Klein model is isomorphic to the Poincarı model, it too is a model of the hyperbolic plane, but that does not explain why we do the things the way that we do.

We shall use another isomorphism to study the Klein model. Let Σ denote the unit sphere in Euclidean three-space,

$$\Sigma = \{(x, y, z) \mid x^2 + y^2 + z^2 = 1\}$$

And let σ denote the equatorial circle

$$\sigma = \{(x, y, 0) \mid x^2 + y^2 = 1\}$$

We shall consider the interior of σ for both the Klein and Poincarı models. Let N =(0,0,1) and map the interior of onto the southern hemisphere by projection from N.

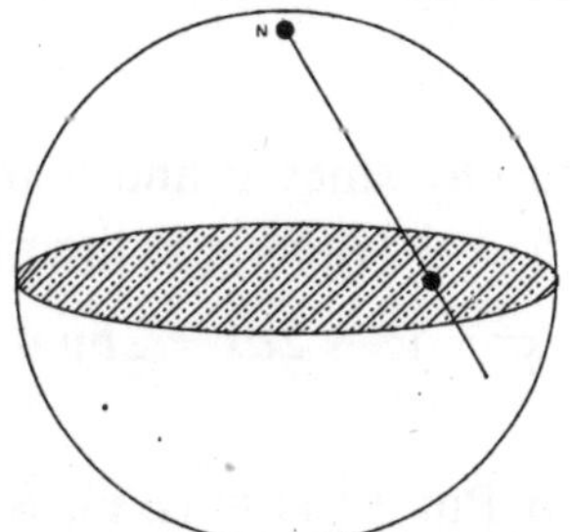

Fig. Central Projection from N

$$f(x, y, 0) = \left\langle \frac{2x}{1+x^2+y^2}, \frac{2y}{1+x^2+y^2}, \frac{x^2+y^2-1}{1+x^2+y^2} \right\rangle.$$

Now, apply the projection up to the xy-plane,

$$\pi(x, y, z) = (x, y, 0).$$

The composition of these two maps is the map from the interior of σ to itself given by:

$$F(x, y, 0) = \left\langle \frac{2x}{1+x^2+y^2}, \frac{2y}{1+x^2+y^2}, 0 \right\rangle.$$

A common way of studying the circle is to recall that the Euclidean plane can represent the complex line by the point (a,b) represents the complex number $a + bi$. The circle is then the set $\sigma = \{z \in \mathbb{C} \mid z\bar{z} = 1\}$ Defining the modulus of z by $|z| = z\bar{z}$ we can rewrite the above function F by

$$F(z) = \frac{2z}{1+|z|^2}.$$

Clearly, F maps a diameter of onto itself, but fixes only the origin and the endpoints of the diameter. In fact, as F maps the interior of onto itself, it fixes only the origin and the points on σ

Consider a p-line in σ say δ --a circle orthogonal to σ. $\delta \cap \sigma = \{P, Q\}$ What is $F(\delta)$?

Claim: If $A \in \delta \cap$ int σ and PQ is the chord of δ in *int* σ then $F(A) = PQ \cap \overrightarrow{OA}$.

Proof: Let $C = (a,b)$ be the center of δ. Then we found P and Q by taking the circle with CO as diameter and intersecting it with σ. It s center is $\left(\frac{\alpha}{2}, \frac{b}{2}\right)$ and its radius is $\frac{1}{2}\sqrt{a^2 + b^2}$. Applying standard analytic geometry we find that its equation is $x^2 - \alpha^x + y^2 - by = 0$ Intersecting this with σ $(x^2 + y^2 = 1)$ gives

$$ax + by = 1$$

for the line joining P and Q.

Now CQ is perpendicular to OQ since σ and δ are orthogonal circles. Thus, $OC^2 = OQ^2 + CQ^2$ or $r^2 = CQ^2 = \alpha^2 + b^2 - 1$ where r is the radius of δ Then δ has equation

$$x^2 + y^2 = 2ax + 2by - 1.$$

Let $A = (\alpha, \beta) \in \delta \cap$ *int* σ Put $F(A) = (u,v)$. Solving we find $u = \frac{\alpha}{a\alpha + b\beta}$ and $u = \frac{\beta}{a\alpha + b\beta}$ Thus, $au + bv = 1$ which implies that $(u, v) \in PQ$.

We can now justify the definitions we made earlier in the Klein model.

- $AB \cong_k CD$ if and only if $F^1(AB) \cong_p F^{-1}(CD)$.
- $\angle ABC \cong_k \angle DEF$ if and only if $F{-}1(\angle ABC) \cong F^{-1}(\angle DEF)$.

PERPENDICULARITY

Now, we see that $l \perp m$ if and only if $F^{-1}(l) \perp F^{-1}(m)$ in the Poincarı model.

1. If l and m are both diameters, then $F^{-1}(l)$ and $F^{-1}(m)$ are both diameters. In fact, $F^{-1}(l) = l$ and $F^{-1}(m) = m$ Hence, the usual Euclidean meaning applies.
2. Assume that only l is a diameter. Then $F^{-1}(l) = l$. $F^{-1}(m)$ is a circular arc perpendicular to l Thus, the tangent to $F^{-1}(m)$ is perpendicular to l and l passes through the center of $F^{-1}(m)$. m is the chord of $F{-}1(m)$. Since l passes through this center, l is the perpendicular bisector of m in the Euclidean sense. Therefore, l is perpendicular to m in our previous definition. Conversely, if $l \perp_k m$ then l is the perpendicular bisector of m. Thus, l passes through the center of the circle and l is perpendicular to the circular arc $F^{-1}(m)$.

3. Assume that neither l nor m is a diameter. F^{-1} (l) and F^{-1} (m) are arcs of circles λ and μ orthogonal to σ.
4. Suppose $\lambda \perp \mu$ in the Poincarı model. By one of the facts, the center of λ is $P(l)$ and that of μ is $P(m)$. We need to show that m passes through $P(l)$ then we will have that $m \perp_k l$.
   Let m H $\sigma = \{P, Q\}$ Sinçe λ and μ are orthogonal inversion in λ interchanges P and Q, since λ and μ are also orthogonal. Thus, $P = Q$' in μ and the Euclidean line through P and Q--namely m--passes through the center of λ which is $P(l)$.
5. Suppose m passes through $P(l)$ Then $P = Q$' in λ so that λ and μ are orthogonal.

HYPERCYCLES AND HOROCYCLES

There is a curve peculiar to hyperbolic geometry, called the *horocycle*.Consider two limiting parallel lines, l and m, with a common direction, say Ω Let P be a point on one of these lines $P \in l$. If there exists a point $Q \in m$ such that the *trilateral*, $\Delta PQ\Omega$, has the property that

$$\angle PQ\Omega \cong \angle QP\Omega$$

Then we say that Q *corresponds* to P. If the trilateral $\Delta PQ\Omega$ has the above property we shall say that it is *equiangular*. Note that it is obvious from the definition that if Q corresponds to P, then P corresponds to Q. The points P and Q are called a pair of *corresponding points*.

Theorem: *If points P and Q lie on two limiting parallel lines in the direction of the ideal point, Ω, they are corresponding points on these lines if and only if the perpendicular bisector of PQ is limiting parallel to the lines in the direction of Ω.*

Theorem: *Given any two limiting parallel lines, there exists a line each of whose points is equidistant from them. The line is limiting parallel to them in their common direction.*

Proof: Let l and m be limiting parallel lines with common direction Ω. Let $A \in l$ and $B \in m$ The bisector of $\angle BA\Omega$ in the trilateral $\Delta AB\Omega$ meets side $B\Omega$ in a point X and the bisector of $\Delta AB\Omega$ meets side AX of the triangle ΔABX in a point C. Thus the bisectors of the angles of the trilateral $\Delta AB\Omega$ meet in a point C. Drop perpendiculars from C to each of l and m, say P and Q, respectively. By *Hypothesis-Angle* $\Delta CAP \cong \Delta CAM$ (M is the midpoint of AB) and $\Delta CBQ \cong \Delta CBM$ Thus, $CP \cong CM \cong CQ$ Thus, by *SAS* for trilaterals, we have that

$\Delta CP\Omega \cong \Delta CQ\Omega$

And thus the angles at C are congruent. Now, consider the line $\overleftrightarrow{C\Omega}$ and let F be any point on it other than C. By *SAS* we have $\Delta CPF \cong \Delta CQF$ If S and T are the feet of F l in and m, then we get that $\Delta PSF \cong \Delta QTF$ and $FS \cong \Delta FT$. Thus, every point on the line $\overleftrightarrow{C\Omega}$ is equidistant from l and m.

This line is called the *equidistant line*.

Theorem: *Given any point on one of two limiting parallel lines, there is a unique point on the other which corresponds to it.*

Theorem: *If three points P, Q, and R lie on three parallels in the same direction so that P and Q are corresponding points on their parallels and Q and R are corresponding points on theirs, then P, Q, and R are noncollinear.*

Theorem: *If three points P, Q, and R lie on three parallels in the same direction so that P and Q are corresponding points on their parallels and Q and R are corresponding points on theirs, then P and R are corresponding points on their parallels.*

Consider any line l,any point $P \in l$ and an ideal point in one direction of l, say Ω. On each line parallel to l in the direction Ω there is a unique point Q that corresponds to P. The set consisting of P and all such points Q is called a horocycle, or, more precisely, the horocycle determined by l, P, and Ω The lines parallel to l in the direction Ω together with l, are called the *radii* of the horocycle. Since l may be denoted by $\overleftrightarrow{P\Omega}$ we may regard the horocycle as determined simply by P and Ω, and hence call it the *horocycle through P with direction* Ω, or in symbols, the horocycle (P, Ω).

All the points of this horocycle are mutually corresponding points by Theorem, so the horocycle is equally well determined by any one of them and Ω. In other words if Q is any point of horocycle (P, Ω) other than P, then horocycle (Q, Ω) is the same as horocycle (P, Ω) If, however, P' is any point of l other than P, then horocycle (P', Ω) is different from horocycle (P, Ω) even though they have the same direction and the same radii. Such horocycles, having the same direction and the same radii, are called *codirectional horocycles*.

There are analogies between horocycles and circles. We will mention a few.

Lemma: *There is a unique horocycle with a given direction which passes through a given point.* (*There is a unique circle with a given center which passes through a given point.*)

Lemma: *Two codirectional horocycles have no common point.* (*Two concentric circles have no common point.*)

Lemma: *A unique radius is associated with each point of a horocycle.* (*A unique radius is associated with each point of a circle.*)

A *tangent to a horocycle* at a point on the horocycle is defined to be the line through the point which is perpendicular to the radius associated with the point.

No line can meet a horocycle in more than two points. This is a consequence of the fact that no three points of a horocycle are collinear inasmuch as it is a set of mutually corresponding points.

Theorem: *The tangent at any point A of a horocycle meets the horocycle only in A. Every other line through A except the radius meets the horocycle in one further point B. If α is the acute angle between this line and the radius, then d(A,B) is twice the segment which corresponds to α as angle of parallelism.*

Proof: Let t be the tangent to the horocycle at A and let Ω be the direction of the horocycle. If t met the horocycle in another point B, we would have a trilateral with two right angles, since A and B are corresponding points. In fact the entire horocycle, except for A, lies on the same side of t, namely, the side containing the ray $\overrightarrow{A\Omega}$.

Let k be any line through A other than the tangent or radius. We need to show that k meets the horocycle in some other point.

Let α be the acute angle between k and the ray $\overrightarrow{A\Omega}$ Let C be the point of k, on the side of t containing the horocycle, such that AC is a segment corresponding to as angle of parallelism. (RECALL: $e^{-d} = \tan(\alpha/2)$). The line perpendicular to k at C is then parallel to $\overleftrightarrow{A\Omega}$ in the direction Ω. Let B be the point of k such that C is the midpoint of AB. The trilaterals $\Delta AC\Omega$ and $\Delta BC\Omega$ are congruent. Hence $\angle CB\Omega = \alpha$ B corresponds to A, and $B \in (A, \Omega)$.

A *chord* of a horocycle is a segment joining two points of the horocycle.

Theorem: *The line which bisects a chord of a horocycle at right angles is a radius of the horocycle.*

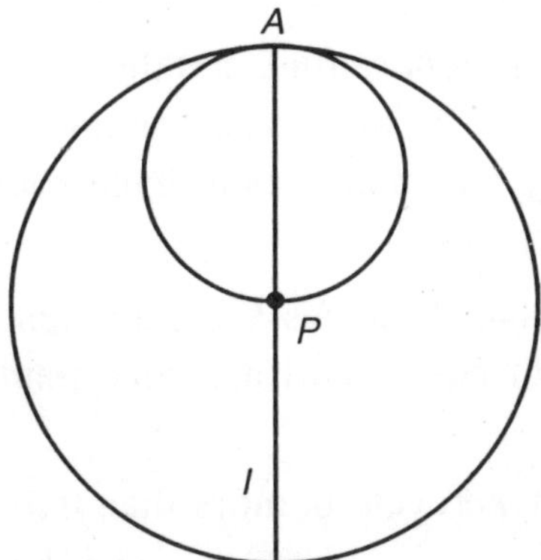

Fig. A Horocycle in the Poincarı Model

We can visualize a horocycle in the Poincarı model as follows.

Let l be the diameter of the euclidean circle Γ whose interior represents the hyperbolic plane, and let O be the center of Γ It is a fact that the *hyperbolic circle* with hyperbolic center P is represented by a euclidean circle whose euclidean center R lies between P and A.

As P recedes from A towards the ideal point Ω, R is pulled up to the euclidean midpoint of ΩA, so that the horocycle (A, Ω) is a euclidean circle tangent to Γ at Ω and tangent to l at A. It can be shown that all horocycles are represented in the Poincarı model by euclidean circles inside Γ and tangent to Γ. Moreover, all the Poincarι lines passing through the ideal point Ω are orthogonal to the horocycle (A, Ω).

In the Poincarι model two horocycles tangent to Γ at Ω are said to be *concentric*. It is by studying the ratio of corresponding arcs on concentric horocycles that the Bolyai-Lobachevsky formula can be derived.

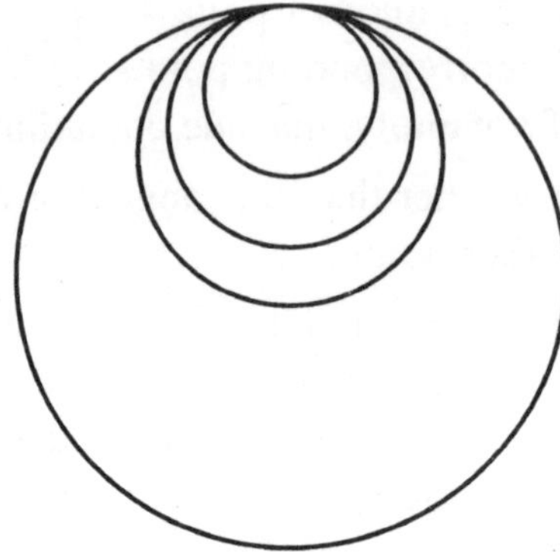

Fig. Concentric Horocycles in the Poincarι Model

Another curve found specifically in the hyperbolic plane and nowhere else is the *equidistant curve,* or *hypercycle*. Given a line l and a point P not on l consider the set of all points Q on one side of l so that the perpendicular distance from Q to l is the same as the perpendicular distance from P to l

The line l is called the *axis*, or *base line*, and the common length of the perpendicular segments is called the *distance*. The perpendicular segments defining the hypercycle are called its *radii*. The following statements about hypercycles are analogous to statements about regular euclidean circles.

1. Hypercycles with equal distances are congruent, those with unequal distances are not. (Circles with equal radii are congruent, those with unequal radii are not.)
2. A line cannot cut a hypercycle in more than two points.
3. If a line cuts a hypercycle in one point, it will cut it in a second unless it is tangent to the curve or parallel to it base line.
4. A tangent line to a hypercycle is defined to be the line perpendicular to the radius at that point. Since the tangent line and the base line have a common perpendicular, they must be hyperparallel. This perpendicular segment is the shortest distance between the two lines. Thus, each point on the *tangent line* must be at a greater perpendicular distance from the base line than the corresponding point on the hypercycle. Thus, the hypercycle can intersect the hypercycle in only one point.

5. A line perpendicular to a chord of a hypercycle at its midpoint is a radius and it bisects the arc subtended by the chord.
6. Two hypercycles intersect in at most two points.
7. No three points of a hypercycle are collinear.

In the Poincarı model let P and Q be the ideal end points of l It can be shown that the hypercycle to l through P is represented by the arc of the euclidean circle passing through A, B, and P. This curve is orthogonal to all Poincarı lines perpendicular to the line l

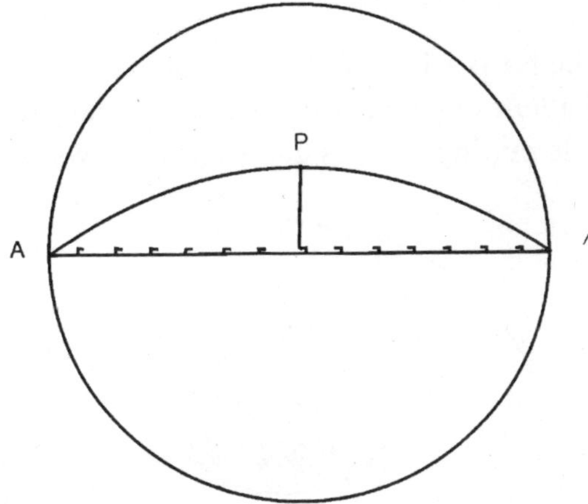

Fig. A hypercycle in the Poincarı Model

In the Poincarı model a euclidean circle represents:

1. A *hyperbolic circle* if it is entirely inside Γ;
2. A *horocycle* if it is inside Γ except for one point where it is tangent to Γ;
3. An *hypercycle* if it cuts Γ nonorthogonally in two points;
4. A *hyperbolic line* if it cuts Γ orthogonally.

It follows that in the hyperbolic plane three non-collinear points lie either on a circle, a horocycle, or a hypercycle accordingly, as the perpendicular bisectors of the triangle are *concurrent* in an ordinary point, an ideal point, or an ultraideal point.

THE PSEUDOSPHERE

Some of the difficulties of the different plane models for the hyperbolic plane are readily apparent. The Klein model is not conformal--angles are not preserved--we cannot use a Euclidean protractor--and it distorts distances. The Poincarı disk model, while conformal, distorts distances--we cannot use a Euclidean ruler. People have searched for another model which is conformal and that represents hyperbolic lengths faithfully by Euclidean lengths. Such a model would be called *isometric*. We are able to construct a model of elliptic geometry on the surface of the sphere in 3-space. Is there a surface in 3-

space on which we can isometrically model hyperbolic geometry? If so, the lines of the hyperbolic plane would then be represented by *geodesics∠* on the surface. Since there seems to be some feeling that hyperbolic lines cannot be straight (*i.e.*, modelled on Euclidean lines), we might expect this surface to be curved in Euclidean 3-space.

A theorem first proved by David Hilbert states that it is impossible to embed the entire hyperbolic plane isometrically as a surface in Euclidean space. On the other hand it *is* possible to embed the Euclidean plane isometrically in hyperbolic space, as the surface of the horosphere.

It is possible to embed a portion of the hyperbolic plane isometrically in Euclidean space, the portion called a *horocyclic sector*, bounded by an arc of a horocycle and the two diameters of the horocycle cutting off this arc. Figure shows such a sector in the Poincarι model.

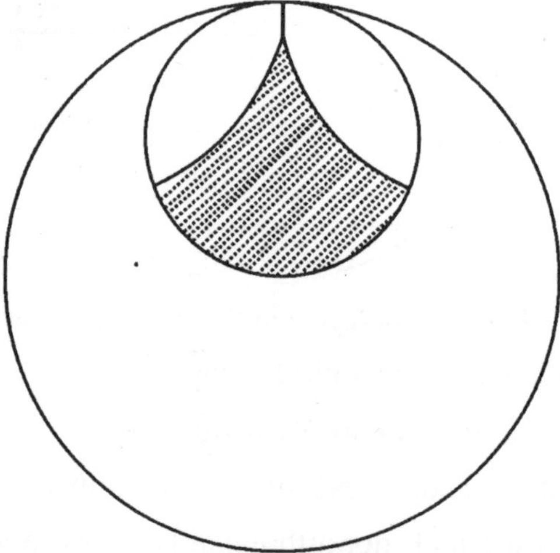

Fig. Horocyclic Sector in the Poincarι Model

We must embed this sector isometrically onto a surface in Euclidean 3-space. The surface that represents this region is called a *pseudosphere*. It is obtained by rotating a curved known as the *tractrix* around its asymptote. The tractrix is the curve characterized by the condition that the length of the segment of the tangent line to the curve from the curve to the y-axis is constant. It has the following equation for a given constant a:

$$x = a \ln \left(\frac{a + \sqrt{a^2 - y^2}}{y} \right) - \sqrt{a^2 - y^2}.$$

The representation on the pseudosphere gives us the opportunity to give some geometric meaning to the fundamental constant k in the theorem on area. Gauss discovered a way in which we can measure the *curvature* of a surface. Intuitively, it is a number K which measures how much a surface bends. The sphere of radius R, for example, has constant curvature $1/R^2$ This makes sense as you think about, because the larger a sphere is the less it bends. In general the curvature will change from point to point, or region to region, on a surface. It should be zero at points where the surface is flat and it should be large where the surface bends sharply. Surfaces where the curvature is the same at all points are called *surfaces of constant curvature*. The sphere is such an example.

For a surface of constant curvature K, Gauss found a fundamental formula which relates the curvature, area, and angular measure. Take a geodesic triangle, ΔABC --a triangle where the sides are geodesic segments. Let $(\angle A)'$ denote the radian measure of angle A. Gauss showed that

$$K \times \text{arca } (\Delta ABC) = (\angle A)' + (\angle B)' + (\angle C)' - \pi.$$

What does this mean? There are three possible cases.

Case 1. Let K be positive. This means that both sides of the above equation are positive. This means that the angle sum in radians of a geodesic triangle is greater than π(or in degrees, greater than 180°), and the area of the triangle is proportional to the *excess*. On a sphere of radius R, then we know that the area of a geodesic triangle is

$$\text{area } (\Delta ABC) = \frac{1}{K} \times ((\angle A)' + (\angle B)' + (\angle C)' - \pi)$$

$$= R2 \times ((\angle A)' + (\angle B)' + (\angle C)' - \pi)$$

If we have a geodesic triangle on the surface of the sphere with three right angles, then the excess of this triangle is $\frac{\pi}{2}$. Thus, the area of this geodesic triangle is

$$\text{area } (\Delta ABC) = \frac{1}{2} \pi R^2.$$

A theorem by Liebmann, Hopf, and Rinow proves that spheres are the only complete surfaces of constant positive curvature in Euclidean three-space. It follows from this information that elliptic geometry cannot be embedded isometrically in Euclidean three-space.

Case 2. $K = 0$. Gauss' formula then says that the angle sum of a geodesic triangle, in radians, is exactly π An example is the Euclidean plane. Another example is an infinitely long cylinder.

Case 3. $K<0$. Now Gauss' formula shows that the angle sum in radians is less than π and the area is proportional to the *defect*.

The pseudosphere is a surface of constant negative curvature. Since the pseudosphere represents a portion of the hyperbolic plane isometrically, we can compare Gauss' formula with the formula for the area in the hyperbolic plane. The comparison gives $K = 1/k^2$ Thus, $-1/k^2$ *is the curvature of the hyperbolic plane*.

Of interest, let $r = ik$ where $i = \sqrt{-1}$ Then $K = 1/r^2$ so that the hyperbolic plane can be described as a *sphere of imaginary radius* $r = ik$, as was noted by the mathematician Lambert. This description is very reasonable, for all the formulas of hyperbolic trigonometry can be obtained from the formulas known for spherical trigonometry by replacing r with ik.

Finally, notice that as k gets very large, the curvature K approaches zero, and the geometry of the surface resembles more and more the geometry of the Euclidean plane. It is in this sense that Euclidean geometry is a *limiting case* of hyperbolic geometry.

HYPERBOLIC TRIGONOMETRY

Trigonometry is the study of the relationships among sides and angles of a triangle. In Euclidean geometry we use similar triangles to define the trigonometric functions—but the theory of similar triangles in not valid in hyperbolic geometry. Thus, we have defined them in terms of their power series expansion for any real number.

Define two hyperbolic trigonometric functions

$$\sin h\, x = \frac{e^x - e^{-x}}{2}$$

$$\cos h\, x = \frac{e^x + e^{-x}}{2}$$

Since $e^x = \sum_{n=0}^{\infty} \frac{x^n}{n!}$ converges for all $x \in \mathbb{R}$ the power series expansions of the hyperbolic trigonometric functions are

$$\sin h\, x = \sum_{n=0}^{\infty} \frac{x^{2n+1}}{(2n+1)!}$$

$$\cos h\, x = \sum_{n=0}^{\infty} \frac{x^{2n}}{(2n)!}$$

And converge for all real x. In fact, using complex analysis and letting $i = \sqrt{-2}$, we can easily see that

$$\sinh x = -i \sin(ix) = i \sin\left(\frac{x}{i}\right)$$

$$\cosh x = \cos(ix) = \cos\left(\frac{x}{i}\right)$$

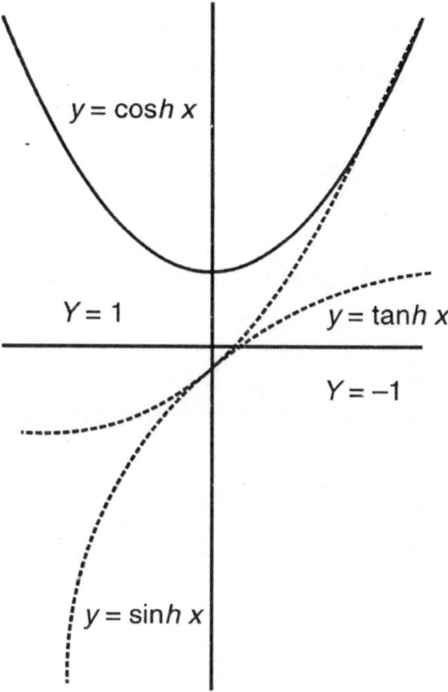

Fig. Graphs of Hyperbolic Functions

We get to choose our unit of measurement in H^2 -- not a particular model now. We would like to choose a unit so that $k=1$. It can be shown that this is analogous to the choice of a unit of angle measurement so that a right angle has *radian* measure $\frac{\pi}{2}$. By measuring in radians the theorem on area becomes

$\alpha(\Delta ABC) = defect\ (\Delta ABC)$,

And the Fundamental Formula becomes

$\prod(x) = 2 \arctan\ ^{e-x}$.

A straightforward calculation using double angle formulas for the circular functions gives the following formulas:

$\sin \prod (x) = \text{sech}\ x$

$\cos \prod (x) = \text{tenh}\ x$

$\tan \prod (x) = \text{esch}\ x$

For example, to derive the first equation:

$$\sin \prod (x) = \sin (2 \arctan e^{-x})$$

$$= 2 \sin (\text{arctane}\ e{-}x) \cos (\arctan e{-}x)$$

$$= 2\left(\frac{e^{-x}}{1+e^{-2x}}\right) = 2\left(\frac{1}{e^{x}+e^{-x}}\right)$$

$$= \frac{1}{\cosh x} = \text{sech } x.$$

This function $\Pi\left(0, \frac{\pi}{2}\right) \to H^2$ provides a connection between the hyperbolic and circular functions.

Given ΔABC, let $a = d_p(B, C)$, $b = d_p(A, C)$, and $c = d_p(A, B)$ First, we will derive some formulζ of hyperbolic geometry from the Poincarι model. Choose our circle Γ to have radius 1.

Let O denote the center of Γ and let $B \in H^2$ Let $x = dp(O, B)$ be the Poincarι distance from O to B and let $t = d(O,B)$ be the Euclidean distance from O to B. From Lemma we have

$$ex = \frac{1+t}{1-t}.$$

We then get that

$$\sinh x = \frac{2t}{1-t^2}$$

$$\cosh x = \frac{1+t^2}{1-t^2}$$

$$\tanh x = \frac{2t}{1+t^2} = F(t)$$

Where F is the isomorphism of the Poincarι model onto the Klein model described in the previous chapter.

Theorem: *Given any right triangle ΔABC in the hyperbolic plane and k=1 with the right angle (having measure π/2 then*

$$\sin A = \frac{\sinh a}{\sinh c} \quad \cos A = \frac{\tanh b}{\tanh c}$$

$$\cosh C = \cosh a \cosh = \cot A \cot B$$

$$\cosh a = \frac{\cos A}{\sin B}$$

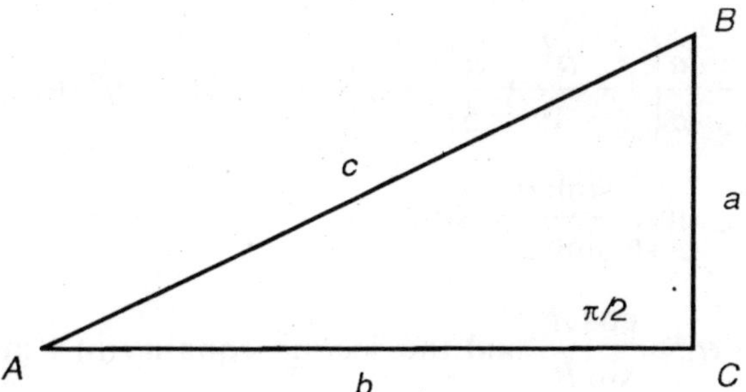

Before we prove these equations, compare them with the formulζ for a right triangle in Euclidean geometry.

1. Equation $\cosh a = \dfrac{\cos A}{\sin B}$ is the hyperbolic analogue of the Pythagorean theorem.

   $\cosh c = \cosh a \cosh b$

$$1+\frac{c^2}{2!}+\frac{c^4}{4!}+\ldots = \left(1+\frac{a^2}{2!}+\frac{a^4}{4!}+\ldots\right)\cdot\left(1+\frac{b^2}{2!}+\frac{b^4}{4!}+\ldots\right)$$

$$= 1+\left(\frac{a^2}{2!}+\frac{b^4}{2!}\right)+\ldots$$

   Thus, if ΔABC is sufficiently small so that higher powers of a, b, and c are negligible, then

$$1+\frac{c^2}{2} \approx 1+\frac{1}{2}(a^2+b^2)$$

$$c^2 \approx a^2+b^2$$

2. Equation $\cosh a = \dfrac{\cos A}{\sin B}$ says that for ΔABC sufficiently small $\sin A \approx \dfrac{a}{c}$ and $\sin B \approx \dfrac{b}{c}$. How close are these approximations? Consider right triangles $\angle A$ with fixed and let c approach 0. Since $a<c$ $a \to 0$.

$$\frac{1}{\sinh x} \approx \frac{1}{c+\dfrac{c^3}{3!}} = \frac{1}{c}\frac{1}{1+u} = \frac{1}{c}(1-u+u^2-u^3+\ldots)$$

   where $\lim_{c\to 0} u = 0$. Thus,

Thus, $\dfrac{\sinh a}{\sinh c} = \dfrac{a}{c}\left(1+\dfrac{a^2}{3!}+\dfrac{a^4}{5!}+...\right)(1-u+u^2-u^3+...)$

$$\lim_{c\to 0}\frac{a}{c} \approx \lim_{c\to 0}\frac{\sinh a}{\sinh c} = \sin A.$$

3. Equation $\cosh a = \dfrac{\cos A}{\sin B}$ and the second equality in Equation $\cosh a = \dfrac{\cos A}{\sin B}$ have no euclidean parallels for there the angles do not determine the lengths of the sides.

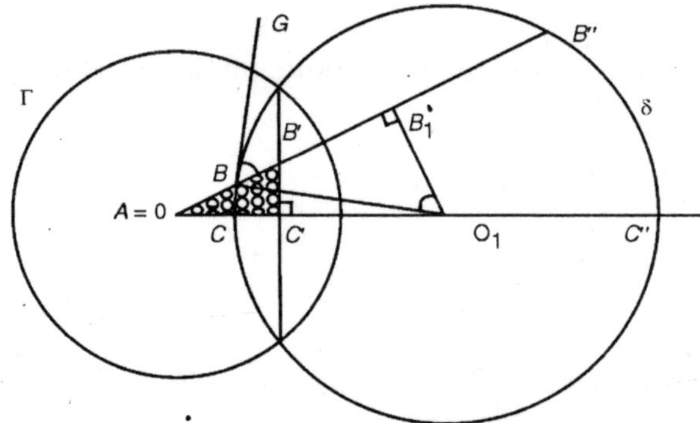

Proof: We may assume that $A = O$ is the center of Γ Let $B' = F(B)$ and $C'=F(C)$. Then

$$\cos A = \frac{AC'}{AB'}$$

From our above observation we have

$$\tanh b = F(d_p(A, C)) = AC'$$

$$\tanh c = F(d_p(A, B)) = AB'$$

The best way to see this is to think of t as $t = t+0i$, so that t sits on the x-axis and is sent to a point on the x-axis. The point to which it is sent must be at a distance $F(t)$ from the origin.

It is then easy to see that

$$\cos A = \frac{\tanh b}{\tanh c}.$$

Now, let $B'' = \overrightarrow{OB} \cap \acute{\varepsilon}$, $B'' \neq B$ By our previous results on the inversion of circles, B'' is the inverse of B in Γ Thus,

$$BB'' = OB'' - OB = \frac{1}{t} - t = \frac{1-t^2}{t} = \frac{2}{\sinh c}.$$

Likewise, $CC'' = \dfrac{2}{\sinh b}$.

Let O_1 be the center of έ Let B_1 be the foot of O_1 on $\overleftrightarrow{BB''}.O_1B \cong O_1B''$ so that B_1 is the midpoint of BB''. Thus, if $\overrightarrow{BG}$ is tangent to έ at B, $\angle GBO_1 = \angle BB_1\,O_1 = \pi/2$ Since they are opposite angles, $\angle B \cong \angle GBB_1$ Thus, $\angle B = \angle GBO_1 - \angle B_1\,BO_1 = \dfrac{\pi}{2} - \angle B_1\,BO_1 = \angle BO_1\,B_1$ Then

$$\sin B = \frac{BB_1}{O_1B} = \frac{BB''}{CC''} = \frac{\sinh b}{\sinh c}.$$

Since $\angle B$ is an arbitrary acute angle in a right triangle we can relabel to get that

$$\sin A = \frac{\sinh a}{\sinh c}.$$

To prove $\cosh a = \dfrac{\cos A}{\sin B}$

$$\sin 2\,A + \cos 2\,A = 1$$

$$\frac{\sinh^2 a}{\sinh^2 c} + \frac{\tanh^2 b}{\tanh^2 c} = 1$$

$$1 + \sinh^2 a + \cosh^2 c \tanh^2 b = \sinh^2 c$$

$$1 + \sinh^2 a + \cosh^2 c\left(\frac{\sinh^2 b}{\cosh^2 b}\right) = 1 + \sinh^2 c$$

$$\cosh^2 a\cosh^2 b + \cosh^2 c \sinh^2 b = \cosh^2 b \cosh^2 c$$

$$\cosh^2 c(\cosh^2 b - \sin^2 b) = \cosh^2 a \cosh^2 b$$

$$\cosh^2 c = \cosh^2 a \cosh^2 b$$

$$\cosh c = \cosh a \cosh b$$

Then,

$$\frac{\cos A}{\sin B} = \frac{\tanh b}{\tanh c}.\frac{\sin c}{\sin b} = \frac{\cosh c}{\cosh b} = \cosh a.$$

$$\cot A \cot B = \frac{\cos A \cos B}{\sin B \sin A} = \cosh a \cosh b.$$

follows from what we have done above.

Theorem: *For any triangle ΔABC in the hyperbolic plane, with $k = 1$*

$$\cosh c = \cosh a \cosh b - \sinh a \sinh b \cos C$$

Hyperbolic Law of Cosines

$$\frac{\sin A}{\sinh a} = \frac{\sin B}{\sinh b} = \frac{\sin C}{\sinh c}$$

Hyperbolic Law of Cosines

$$\cosh c = \frac{\cos A \cos B + \cos C}{\sin A \sin B}$$

Let us take a look at a specific example. Consider equilateral triangles. In Euclidean geometry all are similar, since they all must have angles measuring 60°. If this were true in hyperbolic geometry, they would have to be congruent by *AAA*. What then are the angles in an equilateral triangle of differing sides?

Table: Hyperbolic Equilateral Triangles

Sides	*Radians*	*Degrees*
10	0.0135	0.77
5	0.1633	9.35
2.5	0.5359	30.71
1.5	0.7930	45.43
1	0.9188	52.64
.5	1.0122	57.99
.1	1.0458	59.92

Other interesting examples would be how the angles in a right isosceles triangle vary with the sides and what triangle is analogous to the standard 30-60-90 triangle in Euclidean geometry.

CIRCUMFERENCE AND AREA OF A CIRCLE

Theorem: *The circumference C of a circle of radius r is*

$$C = 2\pi \sinh r.$$

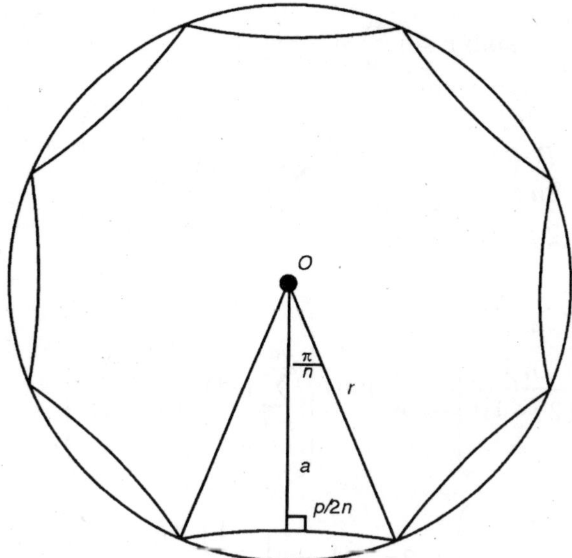

Fig. Hyperbolic Circle

In Euclidean geometry $C = \lim_{n\to\infty} P_n$ where P^n is the perimeter of the regular n-gon inscribed in the circle.

$$Pn = 2nr \sin \frac{\pi}{n} = 2nr\left[\frac{\pi}{n} - \frac{1}{3!}\left(\frac{\pi}{n}\right)^3 + \frac{1}{5!}\left(\frac{\pi}{n}\right)^5 - \cdots\right]$$

$$= 2\pi r - 2\pi r\left[\frac{1}{3!}\left(\frac{\pi}{n}\right)^2 + \frac{1}{5!}\left(\frac{\pi}{n}\right)^4 - \cdots\right]$$

Thus, $C = \lim_{n\to\infty} P_n$.

In hyperbolic geometry we can still compute the perimeter and compute the limit, but we will use Theorem to compute the perimeter.

The proof is nothing but the following computation.

$$\sin \frac{\pi}{2n} = \frac{\sinh \frac{p}{2n}}{\sinh r}$$

$$\sinh \frac{p}{2n} = \sinh r \sin \frac{\pi}{n}$$

$$\sum_{j=0}^{\infty} \frac{\left(\frac{p}{2n}\right)^{2j+1}}{(2j+1)!} = \sinh r \left(\sum_{j=0}^{\infty} (-1)^j \frac{\left(\frac{\pi}{n}\right)^{2j+1}}{(2j+1)!} \right)$$

$$\frac{p}{2n}\left(\sum_{j=0}^{\infty} \frac{\left(\frac{p}{2n}\right)^{2j}}{(2j+1)!} \right) = \frac{\pi}{n}\sinh r \left(\sum_{j=0}^{\infty} (-1)^j \frac{\left(\frac{\pi}{n}\right)^{2j}}{(2j+1)!} \right)$$

$$p\left(1+\frac{1}{3!}\left(\frac{p}{2n}\right)^2 + \cdots\right) = 2\pi \sinh r \left(1-\frac{1}{3!}\left(\frac{\pi}{n}\right)^2 + \cdots\right)$$

$$C = \lim_{n\to\infty} P = 2\pi \sinh r.$$

Let K be the area of ΔABC so that according to our choices, with k=1 and angles measured in radians, $K = \pi - \angle A - \angle B - \angle C$ Let ΔABC have a right angle at C, then $K = \frac{\pi}{2} - A - B$.

Theorem: *tan K/2 = tanh a/2 tanh b/2*

Once again the proof is a computation.

$$\tanh^2 \frac{a}{2} \tanh^2 \frac{b}{2} = \frac{\cosh a - 1}{\cos a + 1} \cdot \frac{\cosh b - 1}{\cos b + 1}$$

$$= \frac{\cos A - \sin B}{\cos A + \sin B} \cdot \frac{\cos B - \sin A}{\cos B + \sin A}$$

$$= \frac{1 - \sin(A+B)\cos(A-B)}{1 + \sin(A+B)\cos(A-B)}$$

$$= \frac{1-\cos K}{1+\cos K} = \tan^2 \frac{K}{2}.$$

Using this and our limiting approach we can now compute the area of a circle.

Theorem: *The area, A, of a circle of radius r is*

$$A = 4\pi \sinh^2 \frac{\pi}{2}$$

Proof: We do this just as before. If K_n is the area of the inscribed regular n-gon, then $A = \lim_{n\to\infty} K_n$. In the right triangle in Figure let K, a, and p denote K_n, a_n and P_n The area of the right triangle is $K_n/2_n$.

$$\tan\frac{K}{4n} = \tanh\frac{a}{2}\ \tanh\frac{p}{4n}\ 1$$

$$4n\tanh\frac{K}{4n} = (\tanh\frac{a}{2})\ 4n\tanh\frac{p}{4n}$$

Now,

$$4n\tan\frac{K}{4n} = 4n\left[\frac{K}{4n}+\frac{1}{3}\left(\frac{K}{4n}\right)^3+\frac{2}{15}\left(\frac{K}{4n}\right)^5+\cdots\right]$$

$$= K\left[1+\frac{1}{3}\left(\frac{K}{4n}\right)^2+\frac{2}{15}\left(\frac{K}{4n}\right)^4+\cdots\right]$$

and

$$4n\tanh\frac{p}{4n} = 4n\left[\frac{p}{4n}-\frac{1}{3}\left(\frac{p}{4n}\right)^3+\frac{2}{15}\left(\frac{p}{4n}\right)^5+\cdots\right]$$

$$= p\left[1+\frac{1}{3}\left(\frac{p}{4n}\right)^2-\frac{2}{15}\left(\frac{p}{4n}\right)^4+\cdots\right]$$

Thus, we find that

$$\lim_{n\to\infty} 4n\tanh\frac{p}{4n} = \lim_{n\to\infty} K = A$$

$$\lim_{n\to\infty} 4n\tanh\frac{p}{4n} = \lim_{n\to\infty} P = C$$

$$\lim_{n\to\infty} a = r$$

Putting all of this together we have that

$$A = C\tanh\frac{r}{2}$$

$$= 2\pi\sinh r\tanh\frac{r}{2}$$

$$= 2\pi \sinh r \frac{\cosh r - 1}{\sinh r}$$

$$= 2\pi (\cosh r - 1)$$

$$= 4\pi \sinh^2 \frac{r}{2}$$

Thus, we have computed the area of a circle.

HYPERBOLIC ANALYTIC GEOMETRY

MORE ON QUADRILATERALS

Now we need to consider a Saccheri quadrilateral which has base b, sides each with length a, and summit with length c. We showed that $c>a$, but we would like to know

- How much bigger?
- How are the relative sizes related to the lengths of the sides?

Theorem: For a Saccheri quadrilateral

$$\sinh \frac{c}{2} = (\cosh a) \cdot \left(\sinh \frac{b}{2} \right)$$

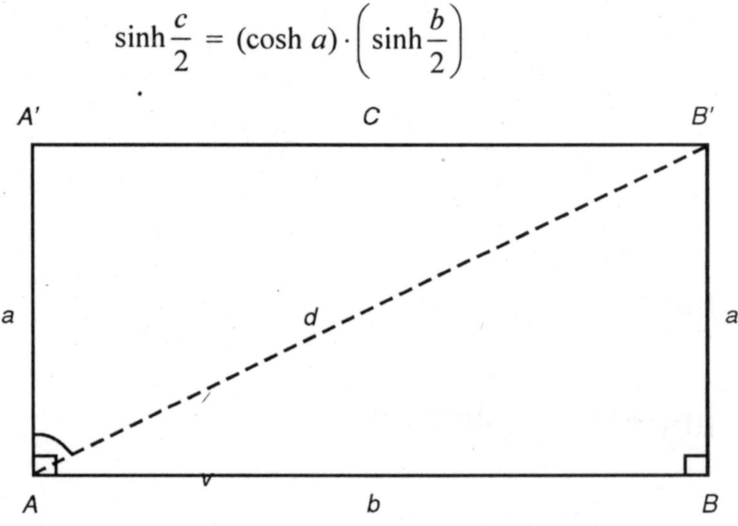

Fig. Saccheri Quadrilateral

Proof: Applying the Hyperbolic Law of Cosines from Theorem, we have

From Theoremwe know that

$$\cos (\theta) = \sin\left(\frac{\pi}{2} - \theta \right) = \frac{\sinh a}{\sinh d}$$

$$\cosh d = \cosh a \cosh b$$

Using these in Equationwe eliminate the variable d and have

$$\cosh c = \cosh^2 a \cosh b - \sinh^2 a$$
$$= \cosh^2 a(\cosh b - 1) + 1$$

Now, we need to apply the identity

$$2\sinh^2\left(\frac{x}{2}\right) = \cosh x - 1,$$

and we have the formula.

Corollary: *Given a Lambert quadrilateral, if c is the length of a side adjacent to the acute angle, a is the length of the other side adjacent to the acute angle, and b is the length of the opposite side, then*

Two segments are said to be complementary segments if their lengths x and x^* are related by the equation

$$\Pi(x) + \Pi(x^*) = \frac{x}{2}$$

The geometric meaning of this equation is shown in the following figure. These lengths then are complementary if the angles of parallelism associated to the segments are complementary angles. This is then an "ideal Lambert quadrilateral" with the fourth vertex an ideal point Ω.

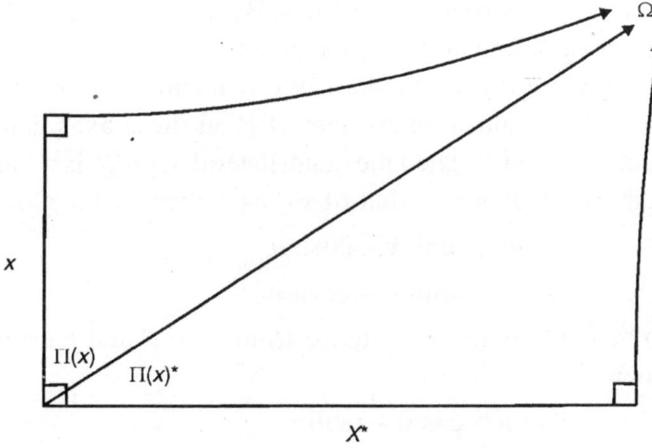

Fig. Complementary Segments

If we apply the earlier formulas for the angle of parallelism to these segments, we get

$$\sinh x^* = \operatorname{csch} x$$
$$\cosh x^* = \coth x$$

$$\tanh x^* = \text{scch } x$$

$$\tanh \frac{x^*}{2} = e^{-x}.$$

Engel's Theorem

Theorem: There is a right triangle with sides and angles as shown in Figure below if and only if there is a Lambert quadrilateral with sides as shown in Figure. Note that PQ is a complementary segment to the segment whose angle of parallelism is $\angle A$.

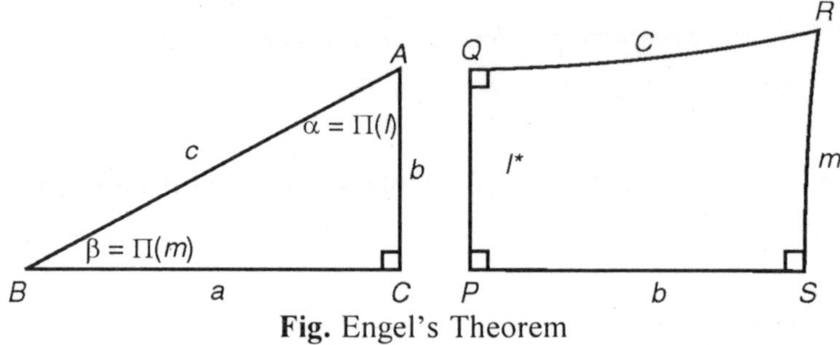

Fig. Engel's Theorem

COORDINATE GEOMETRY IN THE HYPERBOLIC PLANE

In the hyperbolic plane choose a point O for the origin and choose two perpendicular lines through $O - \overleftrightarrow{OX}$ and $\overleftrightarrow{OY}$. In our models—both the Klein and Poincarι—we will use the Euclidean center of our defining circle for this point O. We need to fix coordinate systems on each of these two perpendicular lines. By this we need to choose a positive and a negative direction on each line and a unit segment for each. There are other coordinate systems that can be used, but this is standard. We will call these the u-axis and the v-axis. For any point let $P \in H^2$ U and V be the feet of P on these axes, and let u and v be the respective coordinates of U and V. Then the quadrilateral $UOVP$ is a Lambert quadrilateral. If we label the length of UP as w and that of VP as z, then by the Corollary we have

$$\tanh w = \tanh v - \cosh u$$

$$\tanh z = \tanh u - \cosh v$$

Let $r = d_n(OP)$ be the hyperbolic distance from O to P and let θ be a real number so tha $-\pi < \theta < \pi$. Then

$$\tanh u = \cos \theta - \tanh r$$

$$\tanh v = \tanh \theta - \tanh r$$

We also set

$$x = \tanh u, \quad y = \tanh v$$

$$T = \cosh u, \cosh w \quad X \qquad = xT, \ Y = yT.$$

The ordered pair $\{\overleftrightarrow{OX}, \overleftrightarrow{OY}\}$ is called a frame with axes $\overleftrightarrow{OX}$ and $\overleftrightarrow{OY}$. With respect to this frame, we say the point P has

1. Axial coordinates (u,v),
2. Polar coordinates (r, θ),
3. Lobachevsky coordinates (u,w),
4. Beltrami coordinates (x,y),
5. Weierstrauss coordinates (T,X,Y).

If a point has Beltrami coordinates (x,y) and $t = 1 + \sqrt{1 - x^2 - y^2}$ put

$$p = x/t \quad q = y/t$$

then (p,q) are the Poincarı coordinates of the point.

Every point has a unique ordered pair of Lobachevsky coordinates, and, conversely, every ordered pair of real numbers is tha pair of Lobachevsky coordinates for somc unique point. In Lobachevsky coordinates

1. For $a \neq 0$, $u = a$ is the equation of a line;
2. For $a \neq 0$, $w = a$ is the equation of a hypercycle;
3. $e^{-u} = \tanh w$ is an equation of the line in the first quadrant that is horoparallel to both axes.
4. $e^{u} = \cosh w$ is an equation of the horocycle with radius

Thus, a line does not have a linear equation in Lobachevsky coordinates, and a linear equation does not necessarily describe a line.

Every point has a unique ordered pair of axial coordinates. However, not every ordered pair of real numbers is a pair of axial coordinates.

Let U and V be points on the axes with $V \neq 0$. Now the perpendiculars at U and V do not have to intersect. It is easy to see that they might be horoparallel or hyperparallel, especially by looking in the Poincarı model. If the two lines are limiting parallel (horoparallel) then that would make the segments OU and OV complementary segments. It can be shown then that these perpendiculars to the axes at U and V will intersect if and only if $| u | < | v |$*. It then can be shown that (u,v) are the axial coordinates of a point if and only if $\tanh^2 u + \tanh^2 v < 1$.

Lemma: *With respect to a given frame*

1. Every point has a unique ordered pair of Beltrami coordinates, and (x, y) is an ordered pair of Beltrami coordinates if and only if $x^2 + y^2 < 1$

2. If the point P_1 has Beltrami coordinates (x_1, y_2) and point has P_2 Beltrami coordinates (x_2, y_2) then the distance $d_h(P_1, P_2) = P_1P_2$ is given by the following formulae:

$$\cosh P_1 P_2 = \frac{1 - x_1x_2 - y_1y_2}{\sqrt{1 - x_1^2 - y_1^2}\sqrt{1 - x_2^2 - y_2^2}}$$

$$\tanh P_1 P_2 = \frac{\sqrt{(x_2 - x_1)^2 + (y_2 - y_1)^2 - (x_1y_2 - x_2y_1)^2}}{1 - x_1x_2 - y_1y_2}$$

3. $Ax + By + C = 0$ is an equation of a line in Beltrami coordinates if and only if $A^2 + B^2 > C^2$, and every line has such an equation.
4. Given an angle $\angle PQR$ and given that the Beltrami coordinates of P are (x_1, y_1) of Q are (x_2, y_2) and of R are (x_3, y_3), then the cosine of this angle is given by

$$\cos(\angle PQR) =$$

$$\frac{(x_2 - x_1)(x_3 - x_1) + (y_2 - y_1)(y_3 - y_1) - (x_2y_1 - x_1y_2)(x_3y_1 - x_1y_3)}{\sqrt{(x_2 - x_1)^2 + (y_2 - y_1)^2 - (x_1y_2 - x_2y_1)^2}\sqrt{(x_3 - x_1)^2 + (y_3 - y_1)^2 - (x_1y_3 - x_3y_1)^2}}$$

5. If $Ax + By + C = 0$ and $Dx + Ey + F = 0$ are equations of two intersecting line in Beltrami coordinates and is the angle formed by their intersection, then

$$\cos\theta = \pm\frac{AD + BE - CE}{\sqrt{A^2 + B^2 - C^2}\sqrt{D^2 + E^2 - F^2}}.$$

In particular the lines are perpendicular if and only if $AD + BE = CF$.

6. If (x_1, y_1) and (x_2, y_2) are the Beltrami coordinates of two distinct points, let $t_1 = \sqrt{1 - x_1^2 - y_1^2}$ and $t_2 = \sqrt{1 - x_2^2 - y_2^2}$. Then the midpoint of the segment joining the two points has Beltrami coordinates

$$\left(\frac{x_1t_2 + x_2t_1}{t_1 + t_2}, \frac{y_1t_2 + y_2t_1}{t_1 + t_2}\right)$$

and the perpendicular bisector of the two points has an equation

$(x_1 t_2 - x_2 t_1)x + (y_1 t_2 - y_2 t_1)y + (t_1 - t_2) = 0.$

7. If $A_1x + B_1y + C_1 = 0$ and $A_2x + B_2y + C_2 = 0$ are equations of lines in Beltrami coordinates and if $A_1B_2 = A_2B_1$, then the two lines are hyperparallel.

8. Every cycle has an equation in Beltrami coordinates that is of the form

$$\sqrt{1 - x^2 - y^2} = ax + by + c.$$

– The cycle is a circle if and only if $-1 < a^2 + b^2 - a^2 < 0$ and $c > 0$.

– The cycle is a horocycle if and only if $a^2 + b^2 - c^2$ and $c > 0$.

– The cycle is a hypercycle if and only if $a^2 + b^2 - c^2 < 0$.

In Poincarι coordinates (p,q)

$C((p^2 + q^2) + 2Ap + 2Bq + C = 0$

is an equation of a line if and only if $A^2 + B^2 > C^2$, and every line has such an equation.

AREA IN HYPERBOLIC GEOMETRY

A *polygonal region* is a plane figure which can be expressed as the union of a finite number of triangular regions, in such a way that if two of the triangular regions intersect, their intersection is an edge or a vertex of each of them.

Let R be a polygonal region. A *triangulation* of R is a finite collection,

$K = \{T_1, T_2, ..., T_n\}$

Of triangular regions T_i such that

1. The T_i 's intersect only at edges and vertices, and
2. Their union is R.

Let R_1 and R_2 be polygonal regions. Suppose that they have triangulations

$$K_1 = \{T_1, T_2, ..., T_n\},$$

$$K_2 = \{T'_1 T'_2, ..., T'_n\},$$

Such that for each i we have $T_i \cong T'_i$ Then we say that R_1 and R_2 are *equivalent by finite decomposition*, and we write $R_1 \equiv R_2$.

Let $\acute{\varepsilon}T$ = *defect* $(\Delta(T))$ for any triangular region T.

Theorem: *If* K_1 *and* K_2 *are triangulations of the same polygonal region R, then* $\acute{\varepsilon}K1 = \acute{\varepsilon}K_2$

Theorem: *If two Saccheri quadrilaterals have the congruent summits and equal defects, then their summit angles are congruent, in which case, the two Saccheri quadrilaterals are congruent.*

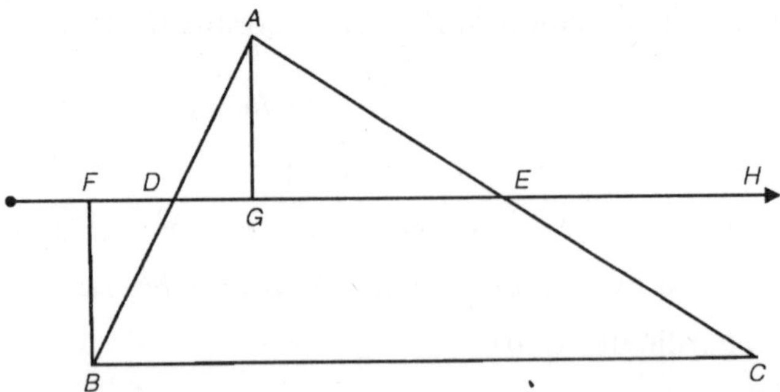

Fig. The Saccheri Quadrilateral *Associated* with ΔABC

Given ΔABC with BC considered as the base. Let D and E be the midpoints of AB and AC; let F, G, and H be the feet of the perpendiculars from B, A, and C, respectively, to $l = \overleftrightarrow{DE}$ As you have proven in the homework, $HCBF$ is a Saccheri quadrilateral. It is known as the quadrilateral *associated* withΔABC It depends on the choice of the base, but it should be clear which base we mean.

Theorem: *Every triangular region is equivalent by finite decomposition to its associated quadrilateral region.*

Theorem: *Every triangular region has the same defect as its associated quadrilateral region.*

Theorem: *If ΔABC and ΔDEF have the same defect and a pair of congruent sides, then the two triangular regions are equivalent by finite decomposition.*

Bolyai's Theorem in the Hyperbolic Plane

Theorem: *If T_1 and T_2 are triangular regions, and $\acute{\varepsilon}T_1 = \acute{\varepsilon}T_2$, then $T_1 \equiv T_2$.*

Theorem: *If $\acute{\varepsilon}\Delta ABC > \acute{\varepsilon}\Delta DEF$, then there is a point P between A and C such that $\acute{\varepsilon}\Delta ABP = \acute{\varepsilon}\Delta DEF$.*

REQUIREMENTS FOR AN AREA FUNCTION

Now, how do we define the area of a polygonal region in the hyperbolic plane? Should we define the area in the same way that we do in the Euclidean plane? If so, what are the minimum requirements for an area function of the Euclidean plane? Minimally it should satisfy the following: Let R be the set of all polygonal regions in H^2. An area function should be a function

$\alpha : R \to R$

Such that

1. $\alpha R > 0$ for every R;
2. If R_1 and R_2 intersect only in edges and vertices, then

$$\alpha(R_1 \cup R_2) = \alpha R_1 + \alpha R_2;$$

3. If T_1 and T_2 are triangular regions with the same base and altitude, then $\alpha T_1 = \alpha T_2$.

   If there is such a function α, then by (2) and (3) it will satisfy

4. If $R_1 \equiv R_2$, then $\alpha R_1 = \alpha R_2$ because congruent triangles have the same bases and altitudes.

   In Euclidean geometry we can show that this area function is unique and it must satisfy the formula $\alpha T = \frac{1}{2}bh$ for each triangular region, T, where b is the length of the base and h is the length of the altitude.

Theorem: *There is no such function* $\alpha : R \to \mathbb{R}$ *satisfying* (1), (2), (3), *and, hence,* (4).

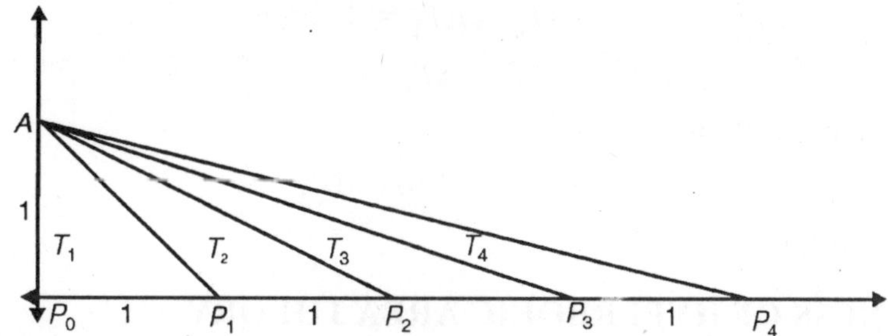

Proof: Consider the right angle $\angle AP_0P_1$, with $AP_0 = PP_0 = 1$ For each n, let P_n be the point of $\overrightarrow{P_0P_1}$ such that $P_0P_n = n$. This gives a sequence of triangles

$$\Delta AP_0P_1, \Delta AP_1P_2, ...,$$

And a corresponding sequence of triangular regions

$$T_1, T_2, ...$$

By condition (3) all the regions T_i have the same "area" $\alpha T_i = A$.

Now consider the defects of these triangles and let $d_i = \acute{\varepsilon}T_i$. For each n

$d_1 + d_2 + ... + d_n = \delta \Delta AP_0P_n < 180°$

Since the partial sums $d_1 + d_2 + ... + d_n$ are bounded, we have that the infinite series,

$$\sum_{n=0}^{\infty} d_n$$

Is convergent. Therefore,

$$\lim_{n\to\infty} d_n = 0$$

Hence $d_n < d_1$ for some n.

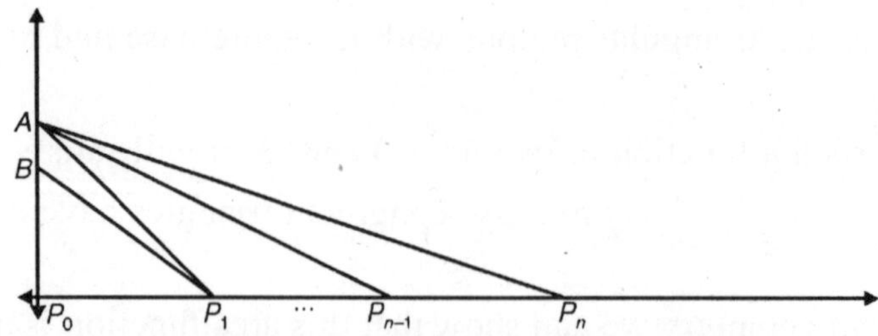

By Theorem there is a point, B, between A and P_0 such that

$$\delta\Delta BP_0P_1 = \delta\Delta AP_{n-1}\,P_n = d_n$$

Therefore by Bolyai's Theorem the regions T and T_n determined by these triangles are equivalent by finite decomposition. By condition (4) this means that $\alpha T = \alpha T_n$. But

$$\alpha T_n = \alpha T_1 = A.$$

Therefore, $$\alpha T = \alpha T_1.$$

Because $\alpha\Delta ABP_1 > 0$ and

$$\alpha T_1 = \alpha T + \alpha\Delta ABP_1$$

This must be impossible.

THE UNIQUENESS OF HYPERBOLIC AREA THEORY

Any reasonable area function α should have the following properties:

1. $\alpha R > 0$ for every R;
2. If R_1 and R_2 intersect only in edges and vertices, then
$$\alpha(R_1 \cup R_2) = \alpha R_1 + \alpha R_2;$$
3. If $R_1 \equiv R_2$ then $\alpha R_1 = \alpha R_2$.

There is such a function--the defect. Recall *defect* (ΔABC) = $\delta\Delta ABC$. We know that satisfies (1), (2) and (3). In fact, we have the following theorem:

Theorem: *Let*

$$\alpha : R \to \mathbb{R}$$

Be an area function satisfying (1), (2), *and* (3). *Then there is a positive constant* $k>0$ *such that*

$$\alpha\, R = \frac{\pi}{180} K^2 \delta R$$

for every $R \in R$..

We are not able to prove this at this time. The proof is within your grasp, but we do not have time to do it.

Corollary: *In* H^2 *the area of any triangle is at most* $\pi\, k^2$,

There is no finite triangle whose area equals the maximal value πk^2, although you can approach this area as closely as you wish (and achieve it with a trebly asymptotic triangle). J. Bolyai proved that you can construct a *circle* of area πk^2 and a regular 4-sided polygon with a 45° angle that also has this area.

ANGLE OF PARALLELISM

In the Poincarı model of hyperbolic geometry, we found a formula relating the angle measurement of the *angle of parallelism* and the distance to which it is associated. This formula is a special case of the general formula discovered independently by Bolyai and Lobachevsky.

Formula of Bolyai-Lobachevsky

Theorem:

$$\tan \frac{\alpha}{2} = e^{-d/k},$$

Where k is the proportionality constant from Theorem.

We have proven this formula in the Poincarı model where $k = 1$.

CARTESIAN COORDINATES IN THE PLANE

In cartesian coordinates (or rectangular coordinates), the ``address" of a point P is given by two real numbers indicating the positions of the perpendicular projections from the point to two fixed, perpendicular, graduated lines, called the axes. If one coordinate is denoted x and the other y, the axes are called the x-axis and the y-axis, and we write $P = (x,y)$. Usually the x-axis is drawn horizontal, with x increasing to the right, and the y-axis is drawn vertical, with y increasing going up. The point $x = 0$, $y = 0$ is the origin, where the axes intersect.

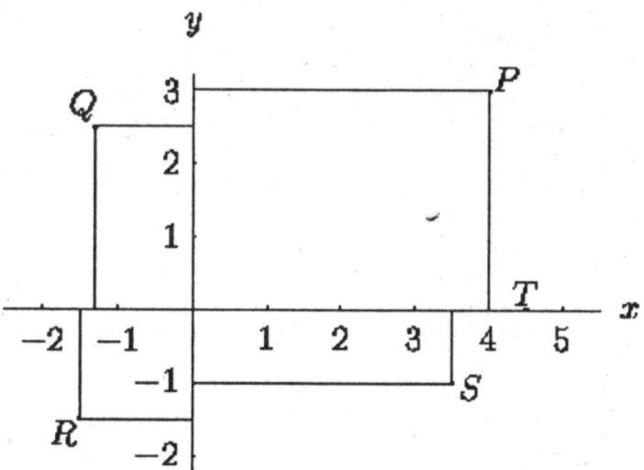

Fig. In Cartesian Coordinates, P =(4,3), Q=(-1.3,2.5), R =(-1.5,-1.5), S=(3.5,-1), and T=(4.5,0). The axes Divide the Plane into four Quadrants: P is in the First Quadrant, Q in the Second, R in the Third, and S in the Fourth. T is on the Positive x-axis.

POLAR COORDINATES IN THE PLANE

In polar coordinates a point P is also characterized by two numbers: the distance $r \geq 0$ to a fixed pole or origin O, and the angle the ray OP makes with a fixed ray originating at O, which is generally drawn pointing to the right (this is called the initial ray). The angle q is only defined up to a multiple of 360° or 2 p. In addition, it is sometimes convenient to relax the condition $r \geq 0$ and allow r to be a signed distance, so (r, θ) and $(-r, \theta+180°)$ represent the same point.

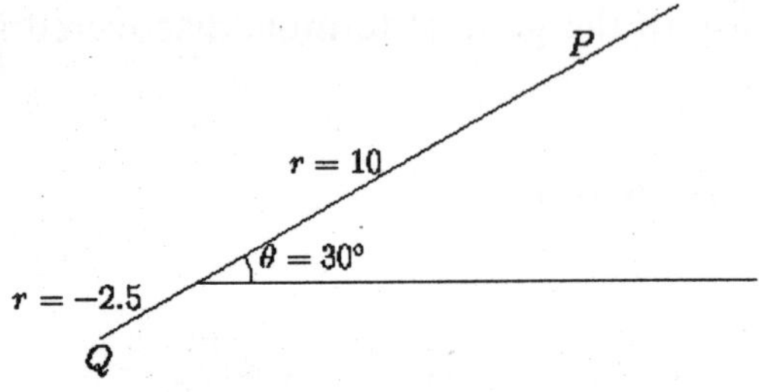

Fig. Among the Possible sets of Polar Coordinates for P are: (10, 30°), (10, 390°) and (10,-330°). Among the Sets of Polar Coordinates for Q are: (2.5, 210°) and (-2.5, 30°).

Chapter 7

Quasi-Periodic Sequence

The optical transmission properties of a quasiperiodic sequence of two types of directional (delta beta) couplers are studied. Under certain resonant conditions, the spatial evolution of the field amplitude is found to be fractal. We also show that the transmission coefficient as a function of the phase mismatch delta beta displays scaling with respect to the number of directional couplers in the sequence.

Consider the series defined by the recurrence

$$A(j)A(j-5) = A(j-1)A(j-4) + A(j-2) + A(j-3)$$

with the initial values A(0) = 1, A(1) = 1, A(2) = 1, A(3) = 2, and A(4) = 3. The values of this sequence up to A(22) are tabulated below:

j	*A(j)*	*j*	*A(j)*
0	1	12	22833
1	1	13	165713
2	1	14	1249441
3	2	15	9434290
4	3	16	68570323
5	5	17	1013908933
6	11	18	11548470571
7	37	19	142844426789
8	83	20	2279343327171
9	274	21	57760865728994
10	1217	22	979023970244321
11	6161		

The recurrence can also be applied in reverse, and since it is formally symmetrical in both directions, we have A(-j) = A(j). Also, the terms of the sequence clearly increase at a rate that is roughly proportional to g^{j^2} for some constant q. Indeed it's easy to verify that

$$A(j) = Cq^{j^2+aj+b}$$

is a solution of the recurrence (1) for any coefficients C, a, and b, provided q is any root of

$$q^{12} - q^4 - 1 = 0$$

This is just Perrin's polynomial in q^4, and the positive real root is $q^4 = 1.324717957...$, so one possible value of q is 1.0728298... To match the initial values of the sequence we set a = b = 0, so a rough estimate for A(j) is $(1.0728298)^{j^2}$. With j = 20 this gives $(1.63)10^{12}$, which is the same order of magnitude as the actual value of $(2.27)10^{12}$.

Unfortunately, the non-linearity of the recurrence prevents us from linearly combining solutions, i.e., we can't simply add quadradic powers of the characteristic roots to give a better fit. An alternative approach is to notice that the A terms also satisfies the relation

$$A(j-2)A(j+2) - pA(j-1)A(j+1) + A(j)^2 = 0$$

where p = 2 if j is even and p = 3 if j is odd. This suggests that a more uniform sequence would be given if we could modify the alternate terms of A slightly so that a single recurrence applies to all the terms. Suppose we define the sequence

$$a(j) = \begin{cases} A(j) & \text{if j is even} \\ QA(j) & \text{if j is odd} \end{cases}$$

where Q is some fixed constant. Then we can substitute a or a/Q into equation as appropriate for even j to give

$$A(j-2)A(j+2) - (2/Q^2)A(j-1)A(j+1) + A(j)^2 = 0$$

Likewise we can substitute into appropriately for odd j, and then multiply through by Q^2, to give

$$A(j-2)A(j+2) - (3Q^2)A(j-1)A(j+1) + A(j)^2 = 0$$

The two relations are identical if we choose the constant Q such that $2/Q^2 = 3Q^2$, which implies that $Q^4 = 2/3$. With this Q the initial values of the a sequence are

$$a(0)=1 \quad a(1)=Q \quad a(2)=1 \quad a(3)=2Q \quad a(4)=3$$

and the sequence satisfies the recurrence

$$a(j-2)a(j+2)-\sqrt{6}a(j-1)a(j-1)+a(j)^2=0$$

Again we note that a(-j) = a(j), just as with the A sequence itself. Now in order to arrive at a less rapidly growing sequence, we consider the sequence defined by

$$b(j)=\frac{a(j-1)a(j+1)}{a(j)^2}$$

with the initial values

$$b(0)=Q^2 \quad b(1)=\frac{1}{Q^2} \quad b(2)=2Q^2 \quad b(3)=\frac{3}{4}\frac{1}{Q^2}$$

Dividing equation by $a(j)^2$, we see that it can be expressed in terms of the b( ) sequence elements as follows

$$b(j-1)b(j)^2 b(j+1)-\sqrt{6}\,b(j)+1=0$$

Incidentally, although the b sequence satisfies this third-order recurrence with the irrational coefficient, it's not difficult to show that it also satisfies the following fourth-order recurrence with purely rational coefficients

$$b(j)b(j+1)b(j+2)b(j+3)=1+\frac{1}{b(j+1)b(j+2)}$$

This same recurrence arises when considering the analogous sequences without the Q factors, because each term contains both odd and even indices. We discuss this further at the end of this article.

The terms of the a sequence can be expressed in terms of the b sequence by means of a "telescoping" identity. For example

$$a(5)=\frac{a(1)^5}{a(0)^4}\left(\frac{a(5)a(3)}{a(4)^2}\right)^1\left(\frac{a(4)a(2)}{a(3)^2}\right)^2\left(\frac{a(3)a(1)}{a(2)^2}\right)^3\left(\frac{a(2)a(0)}{a(1)^2}\right)^4$$

Since a(0) = 1 and a(1) = Q, this can be written as

$$a(5)=Q^5 b(4)^1 b(3)^2 b(2)^3 b(1)^4$$

In general we have

$$a(n) = Q^n \prod_{j=1}^{n-1} b(n-j)^j$$

For numerical purposes it is sometimes preferable to take the natural log of both sides, giving the summation

$$\ln[a(n)] = n \ln(Q) + \sum_{j=1}^{n-1} j \ln[b(n-j)]$$

Letting B denote the asymptotic geometric mean of the terms of the b sequence, we can numerically determine the value B = 1.154022794... Using this value we can define the normalized sequence

$$c(j) = \frac{1}{B} b(j)$$

This sequence is normalized in the sense that the asymptotic geometric mean of the c terms is 1. We can now expressing the elements of the *a* sequence as follows:

$$a(n) = Q^n B^{n(n-1)/2} f(n)$$

where

$$f(n) = \prod_{j=1}^{n-1} c(n-j)^j$$

Since the values of Q and B are known, we need only compute the function f(n), which is a product of c terms, to find the value of a(n). Examining the terms of the c sequence, we find that they are cyclical, with a period of roughly 7/2 = 3.5. This is apparent if we plot every 7th point, giving a very low-frequency curve. By plotting every kth point for increasing values of k, we find that an even better "resonance" occurs when we plot every 123rd point, from which we infer that 123/35=3.514 is a better rational approximation to the period.

Going further, we find that by plotting every 1237th value of q(i) we get an extremely low-frequency signal, so we infer that 1237/352=3.514204... is an even better rational approximation of the period. In this way we arrive at an estimate of T = 3.5142040524... for the actual period. A plot of these terms versus the indices modulo T is shown below.

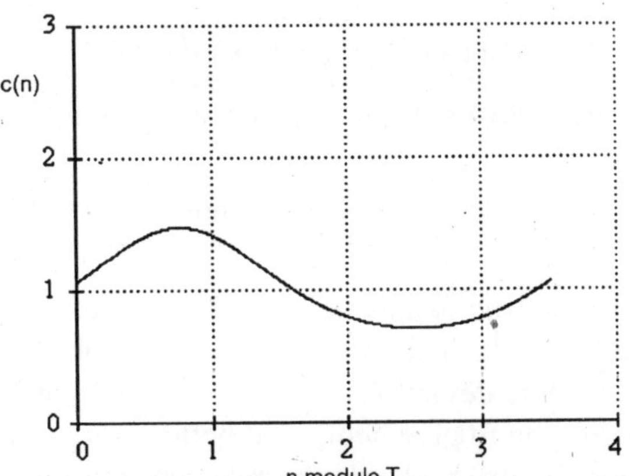

It might seem that the function f(n) should be bounded, since the asymptotic geometric mean of the c sequence terms is unity. However, the c terms do not occur uniformly in any given product f(n), so the asymptotic geometric mean does not necessarily apply. Also, we are not taking the nth root of the cumulative product. As a result, we find that it's necessary to normalize f(n) by a factor of the form R^n, where R is found numerically to equal approximately R = 1.18885806.... A plot of $f(n)/R^n$ versus n (mod T) is shown below.

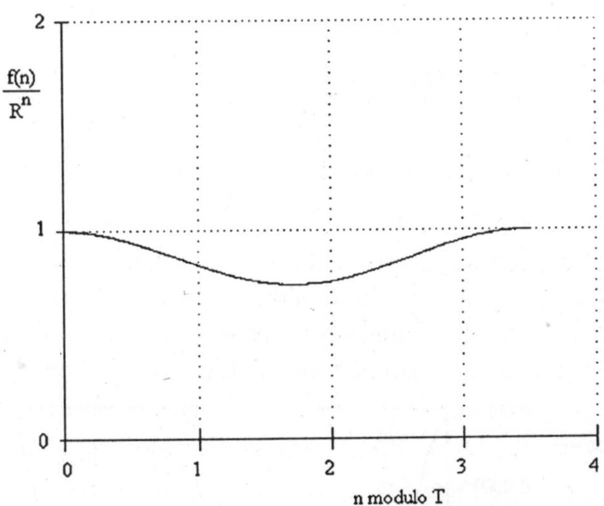

Remarkably, this conforms almost exactly to a pure (shifted and scaled) cosine function, with the frequency $\omega = 2\pi/T = 1.78793...$ Thus, letting δ denote the scale factor, we can represent the function very closely by

$$\frac{f(n)}{R^n} \approx 1 - \delta\left(1 - \cos(\omega n)\right)$$

Substituting this expression for *f*(*n*) into, letting σ denote $B^{1/2}$, and noting that $QR = \sigma$, we arrive at

$$a(n) \approx \sigma^{n^2}\left[1-\delta\left(1-\cos(\omega n)\right)\right]$$

This automatically gives a(0) = 1, and in order to have a(1) = Q we must set

$$\delta = \frac{1-Q/\sigma}{1-\cos(\omega)} = 0.13069879...$$

Although, it matches the true values very closely. For example, the true value of a is 83, and equation gives 83.00157... The true value is $(2.27934)10^{12}$, and the value given by equation is $(2.27928)10^{12}$. The deviations can be examined, and higher-order correction terms can be added to equation to give better and better representations. It's natural to try substituting the expression for a(n) back into the recurrence to see how closely it approaches being an exact solution. This yields the approximate equality

$$\sigma^{2n^2}\left[C_2(n)\sigma^8 - \sqrt{6}C_1(n)\sigma^2 + C_0(n)\right] \approx 0$$

where

$$C_j(n) = \left[1-\bar{\delta}\sin\left(\bar{\omega}(n-j)\right)^2\right]\left[1-\bar{\delta}\sin\left(\bar{\omega}(n+j)\right)^2\right]$$

and we have set $\bar{\delta} = 2\delta$ and $\bar{\omega} = \omega/2$. The expression would be an exact solution if σ was a root of

$$C_j(n) = x^8 - \sqrt{6}\,C_1(n)x^2 + C_0(n) = 0$$

for all indices n. At first this does not look very promising, because the individual coefficients $C_j(n)$ are not constants, nor are they in constant proportion to each other, making it unlikely that *any* single value of x satisfies this equation for all n. However, remarkably, it is very nearly the case that this equation is satisfied by x = s for all n. The figure below shows the relevant root as a function of n (mod T).

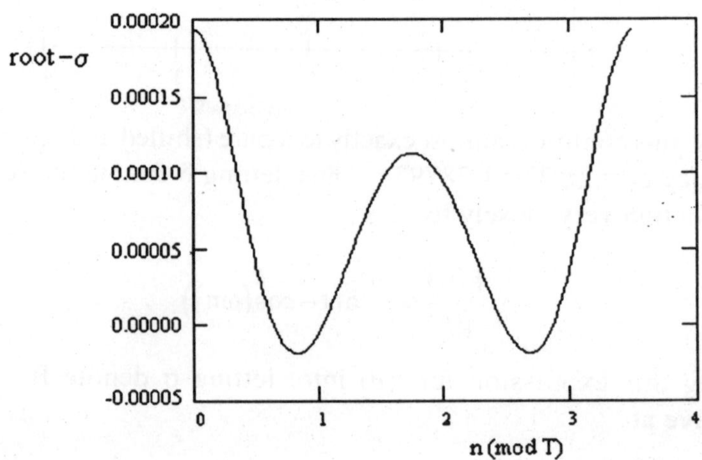

A more conventional approach to solving recurrence is to fit the terms to an infinite series of the form

$$a(n) = C\alpha^{n^2} \sum_{j=-\infty}^{\infty} u^{jn} \theta^{j^2}$$

for constants C, α, u, and q. This can be seen as a generalization. To match the given initial values of a(n) we must have

C = 0.95768989959... α = 0.788912868537...

u = 0.114194204160... θ = 0.02208942811...

Combining the summation terms with indices +j and -j, this can also be written as

$$a(n) = C\alpha^{n^2}\left[1 + 2\theta^1 \cosh(\rho n) + 2\theta^4 \cosh(2\rho n) + 2\theta^9 \cosh(3\rho n) + \ldots\right]$$

where ρ = ln(u). To give a good value we must include up to the 7th order term in this sequence.

In the preceding discussion we focused on the b sequence, which has irrational terms due to the alternating factors of Q. This led to a recurrence with the irrational coefficient P. An alternative approach would be to confine ourselves entirely to rational numbers, and focus on the sequence

$$g(j) = \frac{A(j)A(j+3)}{A(j+1)A(j+2)}$$

with the initial values g(0) = 2 and g(1) = 3/2. This sequence satisfies the recurrence

$$g(j)g(j+1)g(j+2) - g(j+1) - 1 = 0$$

In terms of this sequence we can express the ratios of consecutive terms of the A sequence as follows

$$\frac{A(2j+1)}{A(2j)} = g(0)g(2)g(4)\cdots g(2j-2)$$

$$\frac{A(2j+2)}{A(2j+1)} = g(1)g(3)g(5)\cdots g(2j-1)$$

Therefore, if all the values of g( ) were equal to a single fixed value g, we would have A(j) asymptotic to $g^{j^2/2}$. Consequently, the asymptotic geometric mean of the values of the g( ) sequence is the same as B^2 from the preceding discussion. We can use this fact as an alternate way of evaluating B.

Now, the values of the g( ) sequence oscillate around a fixed cycle but with an irrational period, so they never actually repeat, but we can determine the orbit of the cycle. To visualize the cycle, note that for any given pair [g(j),g(j+1)] the recurrence (8) enables us to compute the "next" pair, i.e., [g(j+1),g(j+2)], and we can regard these pairs as coordinates [x,y] of points on the plane. The figure below shows the orbit of all such points.

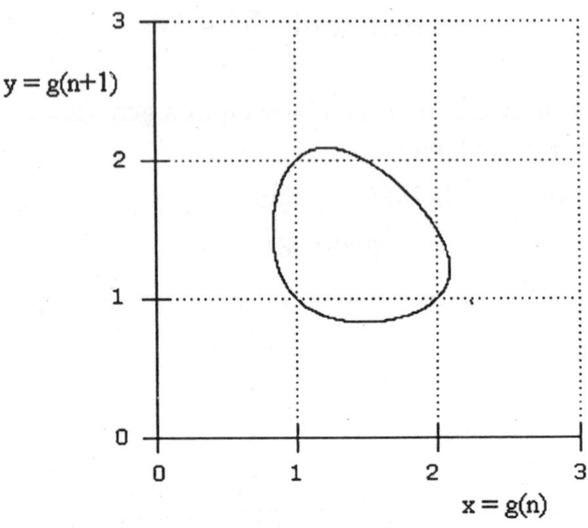

Naturally this locus includes the points (1,1), (1,2), and (2,1), and is symmetrical about the x = y line, because the recurrence is symmetrical forwards and backwards. We can fit the first several points and find that the locus is a simple closed curve given exactly by the equation

$$x^2y + xy^2 - 5xy + x + y + 1 = 0$$

which can also be written in the form

$$(a + xy)(5 - x - y) = 6$$

This can be studied as an elliptic curve, and the g( ) sequence terms are rational points on this curve. Asymptotically, the iterates are uniformly distributed over this curve, so in principle we could evaluate the geometric mean of the points of this curve. For example, we could integrate the logarithm of x (or y) over this curve. Numerically we find the geometric mean is approximately g = 1.331768502...., which as expected is equal to B^2.

Solving equation for y as a function of x gives a quadratic with the discriminant

$$\begin{aligned} D(x) &= x^4 - 10x^3 + 23x^2 - 14x + 1 \\ &= x(x-1)(x-2)(x-7) + 1 \end{aligned}$$

Therefore, D(g(j)) must be a rational square for all j. In fact, if g(j) = u/v for integers u and v, then there is an integer m such that

$$u^4 - 10u^3v + 23u^2v^2 - 14uv^3 + v^4 = m^2$$

Hence the recurrence generates infinitely many solutions of this Diophantine equation. Beginning with g(0) = 2/1 and g(1) = 3/2 we get the solutions

u	*v*	*m*
2	1	1
3	2	7
5	6	1
22	15	391
111	55	2729
415	407	175111
⋮	⋮	⋮

Another sequence is generated by reversing the sign on the right hand side. This gives the recurrence

$$F(j)F(j-5)=F(j-1)F(j-4)-F(j-2)F(j-3)$$

Taking the initial values 1, 1, 1, 2, 1 gives the integer sequence

1,	1,	1,	2,	1,	–1,	–3,	–5,	–4,
–11,	–13,	–7,	23,	86,	87,	199,	415,	799,
–152,	–4159,	–63,	–17047,	–38727,	...			

Incidentally, the A(n) sequence, formed by the convolution of successive terms, is quite similar to the sequence of coefficients for the power series solution of the equation

$$xx + ax^2 = b$$

ASYMPTOTIC APPROACH TO 2D ARRAYS

How would one go about determining the value of the general term of the array

1	1	1	1	1	1	1	1	...
1	3	5	7	9	11	13	15	...
1	5	13	25	41	61	85	113	...
1	7	25	63	129	231	377	575	...

etc.

where A(n,m) = A(n-1,m) + A(n-1,m-1) + A(n,m-1)? There is a nice approach based on generating functions, but it's also interesting to note that this array of numbers, like the binomial coefficients, has an asymptotic relation to the normal distribution. In fact, the Central Limit Theorem can be used to show that just about ANY symmetrical 2-D recurrence formula with arbitrary boundary values gives a distribution of numbers that asymptotically approaches normality.

In the example mentioned above, we can write the numbers sideways in a manner similar to Pascal's triangle

1							1							1
2						1		1						2
3					1		3		1					5
4				1		5		5		1				12
5			1		7		13		7		1			29
6		1		9		25		25		9		1		70
7	1		11		41		63		41		11		1	169

The numbers in the right hand column are the sums of the individual rows, and satisfy the linear recurrence s(n)=2s(n-1)+s(n-2). It follows that

$$s(n) = k\left(1+\sqrt{2}\right)^n - k\left(1-\sqrt{2}\right)^n$$

where $$k = \frac{1}{2\sqrt{2}}.$$

Knowing the total sum of each row, it only remains to determine how the values are distributed in a given row. If we are just interested in the asymptotic behavior, the Central Limit Theorem can be used to estimate the values of A(n,m), m=0 to n, because they approach a "histrogram" of a normal distribution as n increases.

For example, take the 11th row of the preceding array:

1 19 145 575 1289 1683 1289 575 145 19 1

The sum of this row is s(11)=5741, so this is the overall scale factor for the density. To find the scale factor for the "x axis", observe that the integral over the central strip is 1683/5741 = 0.293154, so we know the central strip extends from -0.3761 sigma to +0.3761 sigma, and all the strips are 0.7522 sigma wide. On this basis we can use the normal distribution to estimate the values in this row as follows [rounded to integers]:

2 22 148 572 1285 1683 1285 572 148 22 2

This demontrates that even at n=11 the distribution is already quite close to normal. The nice thing about this approach is that it works on a very wide class of arrays – just about any array generated by a symmetrical recurrence formula. It doesn't even have to be a linear recurrence. For example, the Eulerian numbers are generated by a recurrence of the form E(n,m)=nE(n-1,m) + mE(n,m-1), and they also asymptotically approach a normal distribution.

CONTINUOUS FROM DISCRETE TRANSFER FUNCTIONS

Second order filters are usually implemented digitally by means of recurrence relations of the form

$$y_n = c_1 y_{n-2} + c_2 y_{n-1} + c_3 x_{n-2} + c_4 x_{n-1} + c_5 x_n$$

where the coefficients c_j are constants, and the values of x_j and y_j are the jth discrete samples of the input and output functions respectively. The computations are performed once per Δt seconds. Assuming the input and output functions are scaled equally, the coefficients must satisfy the identity

$$c_1 + c_2 + c_3 + c_4 + c_5 = 1$$

It is often of interest to know how much lag (or lead) such a filter introduces. One way of characterizing this lag is in terms of the asymptotic response to a ramp input. Since the input and output both have constant and equal rates (ultimately), the second-order aspects of the filter drop out, and we effectively have just a first-order lead-lag. To determine the amount of lag, let us assign ramp values to the input and output as follows:

$$\begin{aligned} x_{n-2} &= 0 & y_{n-2} &= y_{n-2} \\ x_{n-1} &= 1 & y_{n-1} &= y_{n-2} + 1 \\ x_n &= 2 & y_n &= y_{n-2} + 2 \end{aligned}$$

Inserting these into equation we can solve for y_{n-2}, and since $x_{n-2} = 0$, the value of y_{n-2} is the amount by which y differs from x at constant t. Furthermore, since the rate of change of the ramp is 1 unit per iteration, the value of y_{n-2} also equals the number of iterations that y leads or lags the ramp input x. Hence if we multiply this by Δt we get the duration of the lag (if the quantity is positive) or lead (if the quantity is negative). Thus we have

$$\text{Lag} = \frac{2 - c_2 - c_4 - 2c_5}{1 - c_1 - c_2} \Delta t$$

For example, suppose we have a second-order digital filter of the form with the coefficients

$$c_1 = -0.8872 \qquad c_2 = 1.8800 \qquad c_3 = 0.0036 \qquad c_4 = 0.0018$$

(Note that these coefficients sum to 1, as required.) Assume the iteration rate of the filter is $\Delta t = 50$ msec. Inserting these values into the preceding equation, we find that the filter has a ramp lag of 0.783 seconds.

A previous note on Lead-Lag algorithms described how to determine the recurrence coefficients c_j to best represent a continuous second-order transfer function of the form

$$x \rightarrow \boxed{\frac{a_2 s^2 + b_2 s + 1}{a_1 s^2 + b_1 s + 1}} \rightarrow y$$

If, instead, we are given the recurrence coefficients and we wish to determine the coefficients of the corresponding continuous transfer function, we immediately have, for the parameters γ and δ from the previous note, $\gamma = -c_1$ and $\delta = c_2$. Also, we can compute the denominator coefficients as

$$a_1 \frac{\Delta t^2}{\text{In}(\sigma_1)\,\text{In}(\sigma_2)} \qquad b_1 \frac{-\Delta t\,\text{In}(-c_1)}{\text{In}(\sigma_1)\,\text{In}(\sigma_2)}$$

where

$$\sigma_{1,2} = \frac{c_2 \pm \sqrt{c_2{}^2 + 4c_1}}{2}$$

To determine the numerator coefficients, we solve the equations for the optimum second-order recurrence coefficients for the parameters α and β to give

$$\alpha = -\frac{4(1+c_1)(c_1+c_3)+(1+3c_1+c_2)(c_2+c_4)}{2(1-c_1-c_2)^2}$$

$$\beta = \frac{2c_1+c_2+2c_3+c_4}{(1-c_1-c_2)}$$

Then we have

$$a_2 = a_1 - \alpha\Delta t^2 - b_1\beta\Delta t \qquad b_2 = b_1 - \beta\Delta t$$

Using the coefficients and iteration interval from the previous example, we get the continuous transfer function shown below.

$$x \rightarrow \boxed{\frac{0.0004s^2 - 0.0005s + 1}{0.327s^2 + 0.783s + 1}} \rightarrow y$$

Thus the filter has essentially no numerator dynamics, and we can confirm from the first-order coefficient in the denominator that this continuous transfer function has a ramp lag of 0.783 seconds, in agreement with what we found earlier for the corresponding discrete digital filter.

The characteristic roots of the denominator dynamics can be expressed in terms of the coefficients of the continuous transfer function or of the original recurrence relations as

$$\rho_{1,2} = \frac{-b_1}{2a_1} \pm \sqrt{\left(\frac{b_1}{2a_1}\right)^2 - \frac{1}{a_1}} = \frac{1}{\Delta t}\,\text{In}\left(\frac{c_2 \pm \sqrt{c_2{}^2 + 4c_1}}{2}\right)$$

It's convenient to define the parameters

$$\omega_n = \sqrt{\frac{1}{a_1}} \qquad \xi = \frac{b_1}{2}\sqrt{\frac{1}{a_1}}$$

which are called the natural frequency and the damping ratio respectively. In terms of these parameters the characteristic roots can be written as

$$\rho_{1.2} = -\omega_n\zeta \pm \omega_n\sqrt{\xi^2 - 1}$$

Therefore, the solution y(t) for any fixed input x is of the form

$$y(t) = A + Be^{\left(-\omega_n\xi + \omega_n\sqrt{\xi^2 - 1}\right)t} + Ce^{\left(-\omega_n\xi - \omega_n\sqrt{\xi^2 - 1}\right)t}$$

$$= A + e^{-\omega_n\xi t}\left(Be^{\left(\omega_n\sqrt{\xi^2 - 1}\right)t} + Ce^{-\left(\omega_n\sqrt{\xi^2 - 1}\right)t}\right)$$

where A, B, C are constants that can be determined from the initial conditions. This shows that the critical value of the damping ratio ξ is 1. For damping ratios greater than 1, the two real exponents are both negative (noting that the square root of $\xi^2 - 1$ is necessarily smaller than ξ). Consequently the function y(t) monotonically approaches the fixed value of x. On the other hand, if the damping ratio is less than 1, then for a real solution we must have B = C, and we can express y(t) in terms of the cosine functions as

$$y(t) = A + 2Be^{-\omega_n\xi t}\cos\left(\omega_{n,d^t}\right)$$

where

$$\omega_{n,d} = \omega_n\sqrt{1 - \xi^2}$$

is the damped natural frequency. (We've assumed t is defined to give a zero phase angle.) For the numerical example discussed previously, the characteristic roots are

$$\rho_{1,2} = -1.197 \pm 1.275i$$

and the solution for any fixed input x is of the form

$$y(t) = A + 2Be^{-1.195t}\cos(1.275i)$$

The exponential factor in the second term has a negative coefficient, so the term approaches zero as time increases. The cosine factor shows that one overshoot cycle lasts for $2\pi/1.275 = 4.9$ units of time.

We can also assess the frequency response of this filter by analyzing the corresponding differential equation

$$a_1 \frac{d^2y}{dt^2} + b_1 \frac{dy}{dt} + y = x$$

Inserting y(t) = Acos(ωt) and x(t) = Bcos(ωt + ϕ) into this equation, and expanding the latter cosine by means of the angle-addition formula, we get

$$\begin{aligned} &-a_1A\omega^2 \cos(\omega t) - b_1A\omega \sin(\omega t) + A\cos(\omega t) \\ &= B\cos(\omega t + \phi) \\ &= B\left[\cos(\omega t)\cos(\phi) - \sin(\omega t)\sin(\phi)\right] \end{aligned}$$

This implies the two relations

$$-a_1A\omega^2 + A = B\cos(\phi)$$

$$b_1A\omega = B\sin(\phi)$$

Dividing the second by the first gives the tangent of the phase shift

$$\tan(\phi) = \frac{b_1\omega}{1 - a_1\omega^2} = \frac{2\xi(\omega/\omega_n)}{1 - (\omega/\omega_n)^2}$$

Also, solving for f and inserting into the second of the preceding equations, we can solve for the amplitude ratio A/B, giving

$$\frac{A}{B} = \frac{1}{\sqrt{\left(1 - a_1\omega^2\right)^2 + \left(b_1\omega\right)^2}} = \frac{1}{\sqrt{\left(1 - (\omega/\omega_n)^2\right)^2 + \left(2\xi\omega/\omega_n\right)^2}}$$

Hence if b_1 is non-zero (meaning that the system has damping) the amplitude ratio is never infinite. For a critically damped system (i.e., with ξ = 1) the amplitude ratio is a Cauchy distribution

$$\left(\frac{A}{B}\right)_{\zeta=1} = \frac{1}{1 + (\omega/\omega_n)^2}$$

On the other hand, for a totally undamped system (b_1 = ξ = 0), we get an infinite amplitude ratio when

$$\omega = \omega_n = \sqrt{\frac{1}{a_1}}$$

which, as we've seen, is the system's undamped "natural frequency". For a damped system, i.e., a system with non-zero b_1, the maximum amplitude ratio (for a non-zero frequency) can be found by differentiating the general expression for A/B with respect to w and setting the result to zero. This gives

$$\omega_{\max A/B} = \sqrt{\frac{1}{a_1} - \frac{1}{2}\left(\frac{b_1}{a_2}\right)^2} + \omega_n\sqrt{1 - \frac{1}{2}\frac{b_1^2}{a_1}} = \omega_n\sqrt{1 - 2\xi^2}$$

as the most responsive frequency. If the damping ratio exceeds $1/\sqrt{2}$ the responsiveness drops monotonically as the frequency increases, so there is no real-valued maximum (other than the one at $\omega = 0$).

The responsiveness drops monotonically as the frequency increases, so there is no real-valued maximum (other thatn the one at $\omega = 0$)

FROM THE GEOMETRIC SERIES TO STIRLING NUMBERS

The geometric series

$$\sum_{n-0}^{\infty} x^n = \frac{1}{1-x}$$

with $|x| < 1$ can be written in the form

$$\sum_{n-0}^{\infty} (1-x)x^n = 1$$

Differentiating with respect to x gives

$$\sum_{n-0}^{\infty} \left(\frac{(1-x)n - x}{x}\right) x^n = 0$$

The numerator inside the parentheses is a polynomial of degree 1 in n with coefficients that are functions of x. Letting $p_1(n,x)$ denote this polynomial, the above identity can be written as

$$\sum_{n\ 0}^{\infty} \frac{p_1(n,x)}{x^1} x^n = 0$$

In general, given a relation of the form

$$\sum_{n-0}^{\infty} \frac{p_k(n,x)}{x^k} x^n = 0$$

where k is a positive integer and $p_k(n,x)$ is a polynomial of degree k in n with coefficients that are functions of x, we can differentiate with respect to x to give

$$\sum_{n-0}^{\infty} \frac{(n-k)p_k + x\dfrac{dp_k}{dx}}{x^{k+1}} x^n = 0$$

The numerator of the coefficient of x^n in this summation is a polynomial of degree k+1 in n, so we will denote it by p_{k+1}. Thus we can define an infinite family of polynomials by means of the recurrence

$$p_{k+1} = (n-k)p_k + x\frac{dp_k}{dx}$$

The first several polynomials of this form are

$$\begin{aligned}
p_1(n,x) &= n(1)-(n+1)x \\
p_2(n,x) &= n(n-1)-n(n+1)x \\
p_3(n,x) &= n(n^2-3n+2)-n(n+1)(n-1)x \\
p_4(n,x) &= n(n^3-6n^2+1\,\text{In}-6)-n(n+1)(n^2-3n+2)x \\
&\vdots
\end{aligned}$$

Notice that these expressions contain the sub-polynomials

$$\begin{aligned}
(n-1) &= (n-1) \\
(n^2-3n+2) &= (n-1)(n-2) \\
(n^3-6n^2+1\,\text{In}-6) &= (n-1)(n-2)(n-3) \\
&\vdots
\end{aligned}$$

We will denote the kth such polynomial by $s_k(n)$, the roots of which are the integers 1 through k. Thus for all k > 1 we see that the first term of p_k vanishes for n equal to 0, 1,.., k-1, whereas the second term vanishes for n equal to 0, 1,.., k-2, and for n = k-1 the second term equals -(k!)x. This leads us to split the summation in equation (1) into two parts, as shown below

$$\sum_{n-0}^{k-1} \frac{p_k(n,x)}{x^k} x^n + \sum_{n-k}^{\infty} \frac{p_k(n,x)}{x^k} = 0$$

The first summation is simply –k!, independent of x, so we have

$$\sum_{n-k}^{\infty} p_k(n,x) x^{n-k} = k!$$

If we add k to each appearance of n in the summand, we can change the indices so they range from 0 to infinity. Thus we have

$$\sum_{n-k}^{\infty} p_k(n+k+x) x^n = k!$$

Letting $P_k(n,x)$ denote $p_k(n+k,x)$, we can write

$$\sum_{n-k}^{\infty} p_k(n,x) x^n = k!$$

where

$$P_1(n,x) = (n+1) - (n+2)x$$
$$P_2(n,x) = (n^2+3n+2) - (n^2+5n+6)x$$
$$P_3(n,x) = (n^3+6n^2+11n+6) - (n^3+9n^2+26n+24)x$$
$$p_4(n,x) = (n^4+10n^3+35n^2+50n+24) - (n^4+14n^3+71n^2+154n+120)x$$
$$\vdots$$

The coefficients of x in these expressions are actually the same polynomials as the "constant" coefficients, except n is replaced by n+1. Therefore, in terms of the polynomials

$$r_k(x) = (x+1)(x+2)\ldots(x+k)$$

we can write the general expression for $P_k(n,x)$ in the form

$$P_k(n,x) = r_k(n) - r_k(n+1)x$$

Obviously we have $r_k(n) = (-1)^k s_k(-n)$. The coefficients of the first several polynomials s_k are shown below.

k	*0*	*1*	*2*	*3*	*4*	*5*	*6*	*7*
0	1							
1	1	–1						
2	1	–3	2					
3	1	–6	11	–6				
4	1	–10	35	–50	24			
5	1	–15	85	–225	274	–120		
6	1	–21	175	–735	1624	–1764	720	
7	1	–28	322	–1960	6769	–13132	13068	–5040

Letting $S_{k,j}$ denote the jth number in the kth row, the magnitude of $S_{k,j}$ is the sum of all products of j of the first k positive integers (with the convention that the null product is 1). Obviously we have $S_{k,k} = k!$, and we have the generating function

$$(x-1)(x-2)...(x-k)=\sum_{j-0}^{k} S_{kj}x^{k-j}$$

It follows that the numbers $S_{k,j}$ satisfy the recurrence relation

$$s(k,j)=S(k-1,j)-kS(k-1,j-1)$$

which corresponds to the fact that the s_k polynomials satisfy the relation $s_k(n+1) = ns_{k-1}(n)$. The numbers $S_{k,j}$ are essentially Stirling numbers of the first kind. These numbers were named for James Stirling (1692-1770), a Scotsman who is also remembered for the approximate formula for factorials

$$n!\approx\left(\frac{n}{e}\right)^{n}\sqrt{2\pi n}$$

that appeared in his book *Methodus differentialis* on series and interpolation. Stirling was a follower of Newton, and elaborated Newton's scheme for classifying the cubics. He also worked on the problem of determining the earth's shape, agreeing with Newton that the earth bulges at the equator (contrary to Cassini's claim). Later, Stirling became manager of the Scottish Mining Company, a demanding job which made it difficult for him to continue with the study of mathematics and physics.

For example, he felt compelled to let lapse a correspondence with Euler, since his work at the mining company prevented him from keeping up with the discussion of Euler's "many ingenious results", "for after deliberations have been interrupted... patience is required before the mind can be brought to think about the same things once again". In terms of the usual nomenclature and indexing conventions for the Stirling number denoted by $S_k^{(j)}$ we have

$$S_k{}^{(j)} = S_{k-1j-1}$$

Speaking of Euler, there is a clear relation between Stirling numbers and the Eulerian numbers. As explained in the note on Eulerian Numbers, we have the identity

$$\sum_{n-1}^{\infty} n^k x^n = \frac{x}{(1+x)^{k+1}} \sum_{j-0}^{k-1} E_{k,j} x^j$$

where $E_{k,j}$ are the Eulerian Numbers. Thus we have

$$\sum_{n-1}^{\infty} x^n = \frac{x}{(1+x)}$$

$$\sum_{n-1}^{\infty} nx^n = \frac{x}{(1+x)^2}$$

$$\sum_{n-1}^{\infty} n^2 x^n = \frac{x}{(1-x)^3}(1+x)$$

$$\sum_{n-1}^{\infty} n^3 x^n = \frac{x}{(1-x)^4}\left(1+4x+x^2\right)$$

$$\vdots$$

These identities can be used to determine the coefficients of a polynomial $p_k(n)$ of degree k such that

$$\sum_{n-k}^{\infty} p_k(n) x^n = k!\left(\frac{x}{1-x}\right)^{k+1}$$

For example, with $k = 2$ we seek constant coefficients A, B, C such that

$$\sum_{n-k}^{\infty} \left(An^2 + Bn + C\right) x^n = 2\left(\frac{x}{1-x}\right)^3$$

Splitting up the summation by powers of n, and making use of the first three Eulerian identities, we have

$$\left[\frac{x}{(1-x)^3}(1+x)\right]A + \left[\frac{x}{(1-x)^2}\right] + B\left[\frac{x}{(1-x)}\right]C = 2\left(\frac{x}{1-x}\right)^3$$

Simplifying this relation, we find that it is satisfied if and only if

$$(C-2)x^2 + (A-B-2C)x + (A+B+C) = 0$$

Since we seek a solution that is valid for arbitrary values of x, each of the quantities in parentheses must vanish, so we have thee equations in three unknowns, which can be solved to give

$$\begin{bmatrix} A \\ B \\ C \end{bmatrix} = \begin{bmatrix} 1 & 1 & 1 \\ 1 & -1 & -2 \\ 0 & 0 & 1 \end{bmatrix}^{-1} \begin{bmatrix} 0 \\ 0 \\ 2 \end{bmatrix} = \begin{bmatrix} 1 \\ -3 \\ 2 \end{bmatrix}$$

In the same way we can determine that the polynomials $p_k(n)$ for any degree k are just the Stirling polynomials $s_k(n)$ discussed previously, i.e., we have

$$\sum_{n-1}^{\infty} s_k(n)x^n = k!\left(\frac{x}{1-x}\right)^{k+1}$$

This is a nice generalization of the geometric series. To illustrate, with k = 4 we have the identity

$$\sum_{n-1}^{\infty}\left(n^4 - 10n^3 + 35n^2 - 50n + 24\right)x^n = 24\left(\frac{x}{1-x}\right)^5$$

Our derivation was explicitly based on the case k = 2, but it's interesting to examine the system of equations for higher degrees. With k = 3 the system of equations is

$$\begin{bmatrix} 1 & 1 & 1 & 1 \\ 4 & 0 & -2 & -3 \\ 1 & -1 & 1 & 3 \\ 0 & 0 & 0 & -1 \end{bmatrix}^{-1} \begin{bmatrix} 0 \\ 0 \\ 0 \\ 6 \end{bmatrix} = \begin{bmatrix} 1 \\ -6 \\ 11 \\ -6 \end{bmatrix}$$

With k = 4 the system of equations is

$$\begin{bmatrix} 1 & 1 & 1 & 1 & 1 \\ 11 & 3 & -1 & -3 & -4 \\ 11 & -3 & -1 & 3 & 6 \\ 11 & -1 & 1 & -1 & -4 \\ 0 & 0 & 0 & 0 & 1 \end{bmatrix}^{-1} \begin{bmatrix} 0 \\ 0 \\ 0 \\ 0 \\ 24 \end{bmatrix} = \begin{bmatrix} 1 \\ -10 \\ 35 \\ -50 \\ 24 \end{bmatrix}$$

As a last example, we show the system of equations with k = 5

$$\begin{bmatrix} 1 & 1 & 1 & 1 & 1 & 1 \\ 26 & 10 & 2 & -2 & -2 & -5 \\ 66 & 0 & -6 & 0 & 6 & 10 \\ 26 & -10 & 2 & 2 & -4 & -10 \\ 1 & -1 & 1 & -1 & 1 & 5 \\ 0 & 0 & 0 & 0 & 0 & -1 \end{bmatrix}^{1} \begin{bmatrix} 0 \\ 0 \\ 0 \\ 0 \\ 0 \\ 120 \end{bmatrix} = \begin{bmatrix} - \\ -15 \\ 85 \\ -225 \\ 274 \\ -120 \end{bmatrix}$$

The left-most column of the coefficient matrix is a row of the Eulerian numbers, the two right-most columns of the coefficient matrix are rows of the binomial coefficients, and the right-hand column vector is a row of Stirling numbers (of the first kind). Notice that the rows of each of these kinds of numbers sum to factorials if we take their absolute values. On the other hand, with alternating signs each row of binomial or Stirling numbers sums to zero.

It's interesting to review the basic recurrences defining the binomial coefficients, the two kinds of Stirling numbers, and the Eulerian numbers. These are summarized below, with indices signifying left and right diagonal slices beginning with A(1,1) = 1.

Binomials

```
         1
      1     1
   1     2     1
1     3     3     1
```

$$A(m,n) = A(m,n-1) + A(m-1,n)$$

Stirling (2nd Kind)

```
         1
      1     1
   1     3     1
1     7     6     1
```

$$A(m,n) = A(m,n-1) + nA(m-1,n)$$

Eulerians

```
          1
       1     1
    1     4     1
1     11    11     1
```

$$A(m,n) = mA(m,n-1) + nA(m-1,n)$$

Stirling (1st)

```
          1
       1     1
    1     3     2
1     6     11     6
```

$$A(m,n) = A(m,n-1) + (m+n-1)A(m-1,n)$$

We also have a series closely related to equation, namely

$$\sum_{n-1}^{\infty} r_k(n)x^n = k!\left(\frac{1}{1-x}\right)^{k+1}$$

where, as before, we have $r_k(n) = (n+1)(n+2)...(n+k)$. For example, with k = 4 we have the identity

$$\sum_{n-1}^{\infty}\left(n^4+10n^3+35n^2+50n+24\right)x^n = 24\left(\frac{1}{1-x}\right)^5$$

Combining this with the corresponding identities based on equation (4), we can deduce identities with coefficients involving just the alternate terms of the Stirling polynomials.

THE EULER-MACLAURIN FORMULA

One of the fundamental concepts of calculus is the correspondence between sums and integrals. Given any sufficiently well-behaved continuous function f(x), consider the sum the integral defined by

$$s(n)=\sum_{k=0}^{n} f(k) \qquad \sigma(n)=\int_{x=0}^{n} f(x)dx$$

These two functions obviously have some similarities, but they also have significant differences. For example, the summation is taken over the values of f(k) at discrete integer arguments k = 0, 1, 2, ..., n, whereas the integral is taken over the values of f(x) for arguments x varying continuously from 0 to n. Strictly speaking, the function s(n) is defined only for integer arguments, while σ(n) is defined for any non-negative real value of n. Nevertheless, we can derive an interesting and useful relation between these two functions.

First, notice that σ(n) can be written in the form

$$\sigma(n)=\int_{x=0}^{1}\left[f(0+x)+f(1+x)+f(2+x)+\ldots+f(n-1+x)\right]dx$$

The Taylor series expansion of each of the individual terms of the integrand is of the form

$$f(k+x)=\frac{f^{(0)}(k)}{0!}+\frac{f^{(1)}(k)}{1!}x+\frac{f^{(2)}(k)}{2!}x^2+\frac{f^{(3)}(k)}{3!}x^3+\ldots$$

where $f^{(j)}$ denotes the jth derivative of f. The integral of this term from x = 0 to 1 is

$$\int_{x=0}^{1} f(k+x)=\frac{f^{(0)}(k)}{0!}+\frac{f^{(1)}(k)}{1!}x+\frac{f^{(2)}(k)}{2!}x^2+\frac{f^{(3)}(k)}{3!}x^3+\ldots$$

Inserting all these expressions into the original integral, we get

$$\sigma(n)=\frac{s^{(0)}}{1!}+\frac{s^{(1)}}{2!}+\frac{s^{(2)}}{3!}+\frac{s^{(3)}}{4!}+\ldots$$

where

$$S^{(j)} = \sum_{k=0}^{n-1} f^{(j)}(k)$$

Now, if we had started with the derivative of f in place of f, we would have arrived at the same expression (1), except that each $f^{(j)}$ would be replaced with $f^{(j+1)}$, and hence each $S^{(j)}$ would be replaced with $S^{(j+1)}$, and s(n) would be replaced with $s^{(1)}(n)$ where

$$\sigma^{(j)}(n) = \int_{x=0}^{n} f^{(j)}(x)\,dx = f^{(j-1)}(n) - f^{(j-1)}(0)$$

Thus beginning with successively higher derivatives of f, we get the infinite sequence of relations

$$\sigma^{(0)}(n) = \frac{S^{(1)}}{1!} + \frac{S^{(2)}}{2!} + \frac{S^{(3)}}{3!} + \frac{S^{(4)}}{4!} + \ldots$$

$$\sigma^{(1)}(n) = \frac{S^{(2)}}{1!} + \frac{S^{(3)}}{2!} + \frac{S^{(4)}}{3!} + \frac{S^{(5)}}{4!} + \ldots$$

$$\sigma^{(2)}(n) = \frac{S^{(3)}}{1!} + \frac{S^{(4)}}{2!} + \frac{S^{(5)}}{3!} + \frac{S^{(6)}}{4!} + \ldots$$

etc.

Each of these equations has (in general) infinitely many terms, but we can solve the system of equations for any finite numbers of terms by taking just a restricted portion of these equations. For example, taking just the terms up to $S^{(4)}$ from the first four equations, we have (in matrix form)

$$\begin{bmatrix} \frac{1}{1!} & \frac{1}{2!} & \frac{1}{3!} & \frac{1}{4!} & \frac{1}{5!} & \frac{1}{6!} \\ 0 & \frac{1}{1!} & \frac{1}{2!} & \frac{1}{3!} & \frac{1}{4!} & \frac{1}{5!} \\ 0 & 0 & \frac{1}{1!} & \frac{1}{2!} & \frac{1}{3!} & \frac{1}{4!} \\ 0 & 0 & 0 & \frac{1}{1!} & \frac{1}{2!} & \frac{1}{3!} \\ 0 & 0 & 0 & 0 & \frac{1}{1!} & \frac{1}{2!} \\ 0 & 0 & 0 & 0 & 0 & \frac{1}{1!} \end{bmatrix} \begin{bmatrix} S^{(1)} \\ S^{(2)} \\ S^{(3)} \\ S^{(4)} \\ S^{(5)} \\ S^{(6)} \end{bmatrix} = \begin{bmatrix} \sigma^{(0)} \\ \sigma^{(1)} \\ \sigma^{(2)} \\ \sigma^{(3)} \\ \sigma^{(4)} \\ \sigma^{(5)} \end{bmatrix}$$

Solving this system of equations gives

$$\begin{bmatrix} S^{(1)} \\ S^{(2)} \\ S^{(3)} \\ S^{(4)} \\ S^{(5)} \\ S^{(6)} \end{bmatrix} = \begin{bmatrix} \frac{B_0}{0!} & \frac{B_1}{1!} & \frac{B_2}{2!} & \frac{B_3}{3!} & \frac{B_4}{4!} & \frac{B_5}{5!} \\ 0 & \frac{B_0}{0!} & \frac{B_1}{1!} & \frac{B_2}{2!} & \frac{B_3}{3!} & \frac{B_4}{4!} \\ 0 & 0 & \frac{B_0}{0!} & \frac{B_1}{1!} & \frac{B_2}{2!} & \frac{B_3}{3!} \\ 0 & 0 & 0 & \frac{B_0}{0!} & \frac{B_1}{1!} & \frac{B_2}{2!} \\ 0 & 0 & 0 & 0 & \frac{B_0}{0!} & \frac{B_1}{1!} \\ 0 & 0 & 0 & 0 & 0 & \frac{B_0}{0!} \end{bmatrix} \begin{bmatrix} \sigma^{(0)} \\ \sigma^{(1)} \\ \sigma^{(2)} \\ \sigma^{(3)} \\ \sigma^{(4)} \\ \sigma^{(5)} \end{bmatrix}$$

where

$$B_0 = 1 \qquad B_1 = -\frac{1}{2} \qquad B_2 = \frac{1}{6} \qquad B_3 = 0 \qquad B_4 = -\frac{1}{30} \qquad B_5 = 0 \qquad B_6 = \frac{1}{42}$$

and so on are the Bernoulli numbers. Thus we have

$$S^{(1)} = \frac{B_0}{0!}\sigma^{(0)} + \frac{B_1}{1!}\sigma^{(1)} + \frac{B_2}{2!}\sigma^{(2)} + \frac{B_3}{3!}\sigma^{(3)} + \ldots$$

Inserting the expressions for each $\sigma^{(j)}$ from equation, this gives

$$S^{(0)} = \frac{B_0}{0!}\int_{x=0}^{n} f(x)dx + \frac{B_1}{1!}\left[f(n) - f(0)\right] + \frac{B_2}{2!}\left[f^{(1)}(n) - f^{(1)}(0)\right] + \ldots$$

Recall that $S^{(0)}$ is defined as the sum of f(k) for k = 0, 1, 2, ..., n-1. Therefore, if we wish to express this in terms of a summation from k = 0 to n, we must add f(n) to both sides. Noting that B_1 = -1/2, this leads to the result

$$\sum_{x=0}^{n} f(x) = \int_{x=0}^{n} f(x)dx + \frac{1}{2}\left[f(n) - f(0)\right] + \frac{B_2}{2!}\left[f^{(1)}(n) - f^{(1)}(0)\right] + \ldots$$

This is called the Euler-Maclaurin formula, a generalization of Bernoulli's formula for the sum of powers of the first n integers. We should note that this series is divergent for most applications, but the error is less than the first neglected term, so the Euler-Maclaurin formula is often a useful method of approximation, relating the series summation to the continuous integral of a function.

Incidentally, by examining the inversion of the system of equations in detail, we can gain some insight into the structure of the Bernoulli numbers, as well as the more general Bernoulli polynomials. First, we re-write equations in the form

$$S^{(1)} = \sigma^{(0)} - \frac{S^{(2)}}{2!} - \frac{S^{(3)}}{3!} - \frac{S^{(4)}}{4!} - \dots$$

$$S^{(2)} = \sigma^{(1)} - \frac{S^{(3)}}{2!} - \frac{S^{(4)}}{3!} - \frac{S^{(5)}}{4!} - \dots$$

$$S^{(3)} = \sigma^{(2)} - \frac{S^{(4)}}{2!} - \frac{S^{(5)}}{3!} - \frac{S^{(6)}}{4!} - \dots$$

etc.

where we have omitted the argument "(n)" from the $\sigma^{(k)}$ terms for convenience. These equations show the beginnings of how to express each $S^{(j)}$ as a function of the $\sigma^{(k)}(n)$ terms, and we must now repeatedly substitute from the left-hand side into the right side to eliminate higher and higher derivatives of S from the equation for $S^{(1)}$. Our objective now is to determine the coefficients of the $\sigma^{(j)}$ terms in the expression for $S^{(1)}$ once all the higher S derivatives have been eliminated. Clearly the coefficient of $\sigma^{(0)}$ will be 1, because none of the remaining terms contain any occurrences of that quantity. Also, we can see that the coefficient of $\sigma^{(1)}$ will be -1/2, which arises from substituting the expression for $S^{(2)}$ into the first equation. No other substitutions will give any more occurrences of that quantity.

In general, the coefficient of $\sigma^{(j)}$ will be the sum of products of the reciprocal factorials involved in the sequence of substitutions leading to the occurrence of that term in the first equation. For example, the term $\sigma^{(2)}$ reaches the first equation in two ways. First, it appears directly when we substitute $S^{(3)}$ into the first equation, and this appearance has a coefficient of -1/3!.

Second, it appears when we substitute $S^{(3)}$ into the second equation, where it is given a coefficient of -1/2!, and then this $S^{(2)}$ is substituted into the first equation, where it is given another factor of -1/2!. Consequently, the overall coefficient of $\sigma^{(2)}$ in the final expression for $S^{(1)}$ will be $-1/3! + (1/2!)^2 = 1/12$. Likewise the coefficient of $\sigma^{(k)}$ for any k consists of the sum of 2^{k-1} products of negative reciprocal factorials, corresponding to the distinct partitions of k (including order).

For example, the coefficient of $\sigma^{(4)}$ is given by the sum of the eight products shown below along with the corresponding partitions of 4:

4	(–5/5!)	–1/12
1+3	(–1/2!)(–1/4!)	1/48
2+2	(–1/3!)(–1/3!)	1/36
1+1+2	(–1/2!)(–1/2!)(–1/3!)	–1/24
3+1	(–1/4!)(–1/2!)	1/48
1+2+1	(–1/2!)(–1/3!)(–1/2!)	–1/24
2+1+1	(–1/3!)(–1/2!)(–1/2!)	–1/24
1+1+1+1	(–1/2!)(–1/2!)(–1/2!)(–1/2!)	1/16
		sum = –1/720

The sum in this case is, of course, simply $B_4/4!$. The fact that there are 2^{k-1} of these partitions follows because there are k-1 possible locations for slicing the interval from 0 to k, and each of these slices can either be present or absent. We can now express the nth Bernoulli number as an explicit sum of products of (signed) reciprocal factorials, although the number of summands doubles for each successive Bernoulli number, so this formula isn't of much practical value.

However, notice that the partitions on each level can be generated very simply from those on the previous level. We can generate the first half of the partitions on the next level by simply incrementing the last components of each partition on the current level. Then we can general the other half of the partitions on the next level by simply appending a "1" to each partition on the current level. Thus the sixteen partitions involved in determining the coefficient of $\sigma^{(5)}$ are as shown below.

5	4+1
1+4	1+3+1
2+3	2+2+1
1+1+3	1+1+2+1
3+2	3+1+1
1+2+2	1+2+1+1
2+1+2	2+1+1+1
1+1+1+2	1+1+1+1+1+1

It's also easy to adjust the product corresponding to each of these partitions, and the same adjustment factor applies to each partition ending in the same number. So, it will be convenient to represent each Bernoulli number B_k as the sum

$$\frac{B_k}{k!} = b_{k,1} + b_{k,2} + \ldots + b_{k,k}$$

where $b_{k,j}$ denotes the sum of the products in the expression for B_k whose corresponding partitions end with j. If we let $b_{k,0}$ denote B_k, then we have the simple relations

$$b_{k,1} = -\frac{1}{2}\sum_{j=1}^{k-1} b_{k-1,j} = -\frac{1}{2} b_{k-1,0} \qquad b_{k,j} = \frac{1}{j+1} b_{k-1,j-1}$$

Using these relations we can generate the table below.

		j					
		0	1	2	3	4	5
k	0	$\frac{1}{1\cdot 0!}$					
	1	$\frac{-1/2}{1\cdot 1!}$	$\frac{-1}{2\cdot 1!}$				
	2	$\frac{1/6}{1\cdot 2!}$	$\frac{1}{2\cdot 2!}$	$\frac{-1}{3\cdot 2!}$			
	3	$\frac{0}{1\cdot 3!}$	$\frac{-1/2}{2\cdot 3!}$	$\frac{3/2}{3\cdot 3!}$	$\frac{-1}{4\cdot 3!}$		
	4	$\frac{-1/30}{1\cdot 4!}$	$\frac{0}{2\cdot 4!}$	$\frac{-1}{3\cdot 4!}$	$\frac{2}{4\cdot 4!}$	$\frac{-1}{5\cdot 4!}$	
	5	$\frac{0}{1\cdot 5!}$	$\frac{-1/6}{2\cdot 5!}$	$\frac{0}{3\cdot 5!}$	$\frac{-5/3}{4\cdot 5!}$	$\frac{5/2}{5\cdot 5!}$	$\frac{-1}{6\cdot 5!}$

The values in this table can also be presented with suitably chosen scale factors as shown below, revealing that not only are the $b_{k,0}$ terms (which are the sums of all the other terms in the respective rows), but the $b_{k,j}$ terms are scaled version of the coefficients of the Bernoulli polynomials.

Since each number in the j = 0 column is the sum of the other elements in that row, and since those other elements can be expressed by reciprocal factorials, it follows that there is a recursive relation involving just the elements in the j = 0 column (i.e., the standard Bernoulli numbers). From the previous recurrence relations we have

$$b_k = -\frac{1}{(J+1)} b_{k-j,0}$$

Therefore we have

$$b_{k,0} = b_{k,1} - b_{k,2} + b_{k,3} + \ldots + b_{k,k}$$
$$= -\frac{b_{k-1,0}}{2!} - \frac{b_{k-2,0}}{3!} - \frac{b_{k-3,0}}{4!} - \ldots - \frac{b_{0,0}}{(k+1)!}$$

Since the terms in the j = 0 column are the standard Bernoulli numbers scaled according to $B_k/k! = b_{k,0}$, this represents the equation

$$\frac{B_k}{k!} + \frac{B_{k-1}}{(k-1)!2!} + \frac{B_{k-2}}{(k-2)!3!} + \frac{B_{k-3}}{(k-3)!4!} + \ldots + \frac{B_0}{0!(k+1)!} = 0$$

Multiplying through by (k+1)!, we get the familiar form of the recurrence

$$\binom{k+1}{1}B_k + \binom{k+1}{2}B_{k-1} + \binom{k+1}{3}B_{k-2} + \binom{k+1}{4}B_{k-3} + \ldots + \binom{k+1}{k+1}B_0 = 0$$

FOURIER TRANSFORMS AND UNCERTAINTY RELATIONS

The function $\exp(-x^2)$ has no simple closed-form indefinite integral, but the related function $x \exp(-x^2)$ does have a simple integral, namely,

$$\int xe^{-x^2} = -\frac{1}{2}e^{-x^2}$$

This identity can be used to evaluate the definite integral of $\exp(-x^2)$ from $x = -\infty$ to $+\infty$. Letting Q denote the value of this definite integral, we can write

$$Q^2 = \left(\int_{-\infty}^{\infty} e^{-x^2} dx\right)\left(\int_{-\infty}^{\infty} e^{-y^2} dy\right) = \int_{-\infty}^{\infty}\int_{-\infty}^{\infty} e^{-}\left(x^2 + y^2\right)dxdy$$

In terms of polar coordinates on the x,y plane we have $x = r\cos(\theta)$ and $y = r\sin(\theta)$, and therefore $x^2 + y^2 = r^2$. The Jacobian of the transformation is

$$\begin{vmatrix} \partial x/\partial r & \partial x/\partial\theta \\ \partial y/\partial r & \partial y/\partial\theta \end{vmatrix} = \begin{vmatrix} \cos(\theta) & -r\sin(\theta) \\ \sin(\theta) & r\cos(\theta) \end{vmatrix} = r$$

so the incremental area element is

$$dA = dx\,dy = r\,dr\,d\theta$$

Hence the double integral expressed in terms of r,θ coordinates is

$$Q^2 = \int_0^{2\pi}\int_0^{\infty} e^{-r^2} r\,dr\,d\theta$$

Making use of equation to evaluate the interior definite integral, we have

$$Q^2 = \int_0^{2\pi}\left(\int_0^{\infty} e^{-r^2}\, r\, dr\right)d\theta = \int_0^{2\pi}\frac{1}{2}d\theta = \pi$$

Therefore the definite integral of $\exp(-x^2)$ from $-\infty$ to ∞ is

$$Q = \sqrt{\pi} = \int_{-\infty}^{\infty} e^{-x^2}\,dx$$

More generally, if we replace the exponent $-x^2$ with $-ax^2 + bx - c$, we can define the parameter

$$y = x\sqrt{a} - \frac{b}{2\sqrt{a}}$$

in terms of which we can write

$$\int_{-\infty}^{\infty} e^{-ax^2+bx-c}\,dx = \frac{e^{(b^2-4ac)/(4a)}}{\sqrt{a}}\int_{-\infty}^{\infty} e^{-y^2}dy = \sqrt{\frac{\pi}{a}}e^{(b^2-4ac)/(4a)}$$

This definite integral is particularly useful when considering the Fourier transform of a normal density distribution. Recall that for any function f(x) we can define another function F(y) that satisfies the relations

$$F(y) = \frac{1}{\sqrt{2\pi}}\int_{-\infty}^{\infty} f(x)e^{ixy}dx$$

$$f(x) = \frac{1}{\sqrt{2\pi}}\int_{-\infty}^{\infty} F(x)e^{-ixy}dy$$

These two functions are a *Fourier transform pair*, i.e., each of them is the Fourier transform of the other. Now, if f(x) is the normal probability density function

$$f(x) = \frac{e^{-(x-\mu)^2/2\sigma^2}}{\sigma\sqrt{2\pi}}$$

then the Fourier transform is

$$F(y) = \frac{1}{2\pi\sigma}\int_{\infty}^{\infty} e^{-(x-\mu)^2/2\sigma^2+ixy}dx$$

so the exponent in the integral is of the form $-ax^2 + bx + c$ with $a = 1/2\sigma^2$, $b = iy + \mu/\sigma^2$, and $c = -\mu^2/(2\sigma^2)$. Hence the Fourier transform of the normal density distribution is

$$F(y) = e^{\mu^2/(2\sigma^2)} \frac{e^{-\left(y - i\mu/\sigma^2\right)^2/\left(2/\sigma^2\right)}}{\frac{1}{\sigma}\sqrt{2\pi}}$$

Choosing our scales so that the mean of f is zero, i.e., so that $\mu = 0$, the above expression reduces to

$$F(y) = \frac{e^{-y^2/\left(2/\sigma^2\right)}}{\frac{1}{\sigma}\sqrt{2\pi}}$$

In other words, the Fourier transform of the normal distribution with mean zero and standard deviation s is also a normal distribution with mean zero, but with standard deviation $1/\sigma$. This shows that the variances of the f and F distributions satisfy the "uncertainty relation" var(f) var(F) = 1. This equality is the limiting case of a general inequality on the product of variances of Fourier transform pairs. In general, if f is an arbitrary probability density distribution and F is its Fourier transform, then

$$\mathrm{var}(f)\,\mathrm{var}(F) \geq 1$$

Notice that f and F are just two different ways of characterizing the same distribution, one in the amplitude domain and the other in the frequency domain. Given either of these distributions, the other is completely determined.

The relation between conjugate variables (such as position and momentum) in quantum mechanics can be expressed in terms of the relation between Fourier transform pairs. Consider a physical system with just one degree of freedom, represented by the operator q, and let p denote the operator for the corresponding momentum (i.e., the generalized momentum of q in the usual Hamiltonian formulation). The basic commutation relation between these operators is

$$qp - pq = i\hbar$$

Notice the symmetry between the p and q operators if we replace i with -i. The usual Schrodinger representation of this system takes the observable q as a "diagonal" operator, i.e., with the eigenvalues specified explicitly, and then the corresponding momentum operator is defined as

$$p = -i\hbar\frac{d}{dq}$$

However, we could (in theory) just as well take p as a diagonal operator, and then q would be given by

$$p = -i\hbar \frac{d}{dp}$$

Dirac referred to this as the momentum representation of the system. This again shows the symmetry between q and p under an exchange of i and -i. Indeed if we let <q|S> and <p|S> for any given state S denote the probability amplitudes that measurements corresponding to the operators q and p will return the eigenvalues q and p respectively, then we find

$$\langle p/S \rangle = \frac{1}{\hbar} \int_{-\infty}^{\infty} \langle q \mid S \rangle e^{-iqp/\hbar} dq$$

$$\langle q/S \rangle = \frac{1}{\hbar} \int_{-\infty}^{\infty} \langle p \mid S \rangle e^{-iqp/\hbar} dq$$

In other words, the probability amplitude distributions of two conjugate variables are simply the (suitably scaled) Fourier transforms of each other. We saw previously that the dispersions (variances) of two density distributions that comprise a Fourier transform pair satisfy the inequality, so the variances of the probability amplitude distributions of conjugate observables in quantum mechanics satisfy such an inequality. Thus Heisenberg's uncertainty principle for conjugate pairs of observables follows directly from the fact that those observables are essentially the Fourier transforms of each other.

Of course, this attribute of Fourier transform pairs is purely mathematical, and has no *a priori* applicability to pairs of observables such as position and momentum, or time and energy. The physical content of quantum mechanics is based on the two relations

$$E = \hbar\omega \qquad\qquad p = \hbar k$$

where E is energy, p is momentum (in one dimension), $\hbar$ is Planck's (reduced) constant, w is the frequency with units second^{-1}, and k is the wave number with units meter^{-1}. These relations were introduced in the early 1900's by Planck, Einstein, and deBroglie to account for non-classical phenomena such as cavity radiation and the photo-electric effect, both of which depend on the particle-like behavior of entities that had previously been modeled as waves, as well as phenomena involving wave-like behavior of material particles. These are the relations that associate the familiar observables of energy, momentum, space, and time, with the frequency domain. Indeed in terms of the characteristic time $\tau = 1/\omega$ and distance $\lambda\!\!\!^{-} = 1/k$ the above relations can be written as

$$\tau E = \lambda\!\!\!^{-} p = \hbar$$

which already clearly reveals the conjugacy of time and energy, and of distance and momentum. In view of this, it isn't surprising to find that the product of the dispersions of two conjugate observables (such as position and momentum) cannot be less than one quanta of action, represented by $\hbar$.

In a sense, there is also a conjugacy between space and time – two observable that had been regarded as disjoint and independent prior to the early 1900s. In special relativity the inertial space and time intervals dx and dt between two events are components of a single invariant spacetime interval ds between those events. These intervals are related according to the Minkowski metric, which can be written in the form

$$\left(\frac{dx}{dt}+\frac{ds}{dt}\right)\left(\frac{dx}{dt}+\frac{ds}{dt}\right)=\frac{1}{c^2}$$

This can be regarded as an "uncertainty relation" for space and time. In general, physics was based, prior to 1900, on the premise that $\hbar$ and $1/c^2$ were both zero. With the advent of quantum mechanics and special relativity, it was realized that they both have non-zero values, although they are extremely small in terms of ordinary units.

TILINGS BASED ON PERRIN'S SEQUENCE

Perrin's sequence consists of the integers 3, 0, 2, 3, 2, 5, 5, 7, 10,... and so on, satisfying the recurrence relation $s_n = s_{n-2} + s_{n-3}$. We wish to construct a non-periodic tiling of the plane with the corresponding characteristic polynomial $p(x) = x^3 - x - 1$. One way of doing this is by an infinite geometric sequence of equilateral triangles in a spiral arrangement called the golden pentagon. However, if we want a tiling consisting of just a finite number of distinct tile types, we need to take a different approach. Following the method described in a previous note on non-periodic tilings, we seek a system of area equations with the characteristic cubic polynomial Q such that $Q(x^2) = -p(-x)p(x)$. This leads to

$$Q(\lambda) = \lambda^3 - 2\lambda^2 + \lambda - 1$$

for which we have

$$Q\left(x^2\right) = \left(x^3 - x - 1\right)\left(x^3 - x + 1\right)$$

From this we see that the trace of the system of area equations is 2, and since the area coefficients must be non-negative integers the only possibilities are 2+0+0 and 1+1+0. The first of these has no overall solution in non-negative integers, so the trace partition must be 1+1+0. Therefore, letting B, G, and Y denote the areas of the three distinct tile types, the system of area equations is of the form

$$\begin{bmatrix} 1 & c_{12} & c_{13} \\ c_{21} & 1 & c_{23} \\ c_{31} & c_{32} & 0 \end{bmatrix}\begin{bmatrix} B \\ G \\ Y \end{bmatrix} = \lambda \begin{bmatrix} B \\ G \\ Y \end{bmatrix}$$

In terms of the unknown coefficients the characteristic polynomial is

$$.\lambda^3 - 2\lambda^3 + (1 - c_{12}c_{21} - c_{23}c_{32} - c_{31}c_{13})\lambda - (c_{12}c_{23}c_{31} + c_{21}c_{32}c_{13} - c_{23}c_{32} - c_{13}c_{31})$$

Identifying this polynomial with Q(λ), we have the conditions

$$c_{12}c_{21} + c_{23}c_{32} + c_{31}c_{13} = 0$$
$$c_{12}c_{23}c_{31} + c_{21}c_{32}c_{13} - c_{23}c_{32} - c_{13}c_{31} = 1$$

The products of three coefficients in the second condition cannot both vanish, because each coefficient must be a non-negative integer, so the last two terms in the second condition could not equal 1. Consequently, the only solution (up to symmetry) is given by $c_{12} = c_{23} = c_{31} = 0$ and $c_{21} = c_{32} = c_{13} = 1$. Also, the dominant eigenvalue λ of the area equations is the square of the scale factor σ, so the area equations are

$$\begin{bmatrix} 1 & 0 & 1 \\ 1 & 1 & 0 \\ 0 & 1 & 0 \end{bmatrix} \begin{bmatrix} B \\ G \\ Y \end{bmatrix} = \sigma \begin{bmatrix} B \\ G \\ Y \end{bmatrix}$$

where σ = 1.3247179572447... is the root of $p(\sigma) = \sigma^3 - \sigma - 1$. This implies that the inflated one coloured tile is composed of one coloured and second coloured tile. Also, the inflated Green tile is composed of one Blue and one Green tile. Lastly, the inflated Yellow tile is identical with the Blue tile.

The G and Y tiles have the same shape, and we know that when G is appended to B it yields this same shape. In addition we know that when Y is appended to B it yields another figure similar to B. If these shapes are all triangles, they must all be similar right triangles, and we can arrange them schematically as shown below.

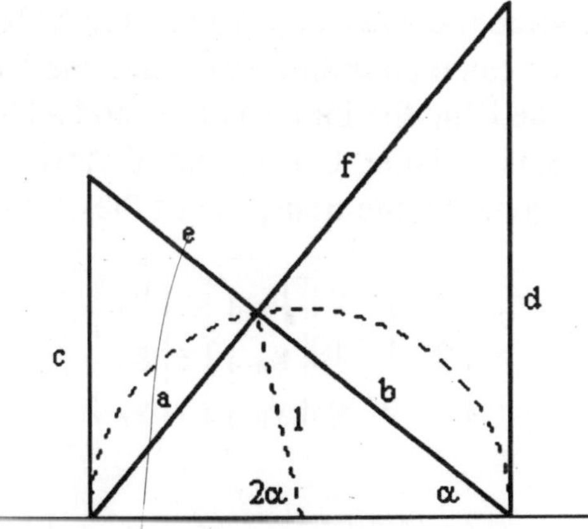

The necessary and sufficient condition for these shapes to combine in the required way is for the angle α to be such that b/2 = c/d. We have

$$b = 2\cos(\alpha) \qquad c = 2\tan(\alpha) \qquad d = 2\tan\left(\frac{\pi}{2} \div \alpha\right) = \frac{2}{\tan(\alpha)}$$

and therefore

$$\cos(\alpha) = \tan(\alpha)^2$$

Substituting $[1-\cos(\alpha)^2]/\cos(a)^2$ for $\tan(\alpha)^2$ and clearing the denominator, we get

$$\cos(\alpha)^3 = 1 - \cos(\alpha)^2$$

This shows that the secant (the reciprocal of the cosine) of a equals the scale factor s, because it is the positive real root of the characteristic polynomial $x^3 - x - 1$. The numerical value is a = 0.71532875... radians (about 40.985... degrees). Thus we arrive at the tiles given by three similar right triangles with the angles and edge lengths shown in the figure below.

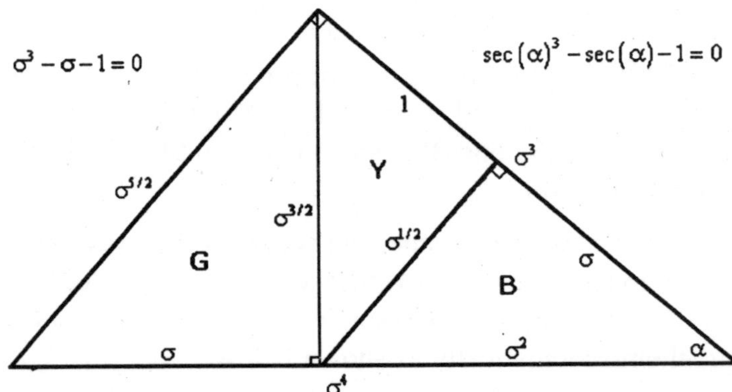

Beginning with single Blue tile, we can inflate by the factor s and sub-divide into one Blue and one Yellow tile. We can then inflate again, and the Yellow tile becomes a Green tile, and we can sub-divide the Blue tile into a Yellow and a Blue. And so on. If we let b_n, g_n, and y_n denote the numbers of Blue, Green, and Yellow tiles respectively on the nth stage, we have recurrences given by the transpose of the area equations, i.e.,

$$\begin{bmatrix} 1 & 1 & 0 \\ 0 & 1 & 1 \\ 1 & 0 & 0 \end{bmatrix} \begin{bmatrix} b_{n-1} \\ g_{n-1} \\ y_{n-1} \end{bmatrix} = \begin{bmatrix} b_n \\ g_n \\ y_n \end{bmatrix}$$

The numbers of tiles of the three types on the first several levels are shown below.

n	b_n	g_n	y_n
0	1	0	0
1	1	0	1
2	1	1	1
3	2	2	1
4	4	3	2
5	7	5	4
6	12	9	7
7	21	16	12
8	37	28	21
9	65	49	37
10	114	86	65
11	200	151	114

Each sequence individually satisfies the third order recurrence $s_n = 2s_{n-1} - s_{n-2} + s_{n-3}$, as well as the fourth order recurrence $s_n = s_{n-1} + s_{n-2} + s_{n-4}$, due to the factorization of the quartic

$$x^4 - x^3 - x^2 - 1 = (x+1)\left(x^3 - 2x^2 + 1 - 1\right)$$

Asymptotically the terms increase by a factor of $\lambda = \sigma^2$ on each step, and we can see that $y_n = b_{n-1}$ and $g_n = b_{n-1} + b_{n-3}$. Therefore, the fractions of tiles of each type as n goes to infinity approach

$$\frac{b_n}{T_n} \to \frac{\lambda^3}{\lambda^3 + 2\lambda^2 + 1} \qquad \frac{g_n}{T_n} \to \frac{\lambda^2 + 1}{\lambda^3 + 2\lambda^2 + 1} \qquad \frac{g_n}{T_n} \to \frac{\lambda^2}{\lambda^3 + 2\lambda^2 + 1}$$

where $T_n = b_n + g_n + y_n$. Substituting σ^2 for λ and making use of the identity $\sigma^3 = \sigma + 1$, these asymptotic ratios reduce to

$$\frac{b_n}{T_n} \to \frac{1}{\sigma^3} \qquad \frac{b_n}{T_n} \to \frac{1}{\sigma^4} \qquad \frac{b_n}{T_n} \to \frac{1}{\sigma^5}$$

These have the values 0.430159..., 0.324717..., and 0.245122... respectively. Equating the sum of these fractions to 1 gives the quartic noted above.

Incidentally, the cumulative sums of the b_n sequence give the g_n sequence, and the cumulative sums of the g_n sequence give values that are one less than the b_n sequence, so these two sequences are nearly the transform-duals of each other.

The figure below shows the tiling produced at the 11th stage, which consists of 200 one coloured tiles, 151 second coloured tiles, and 114 third coloured tiles, in accord with the table given previously.

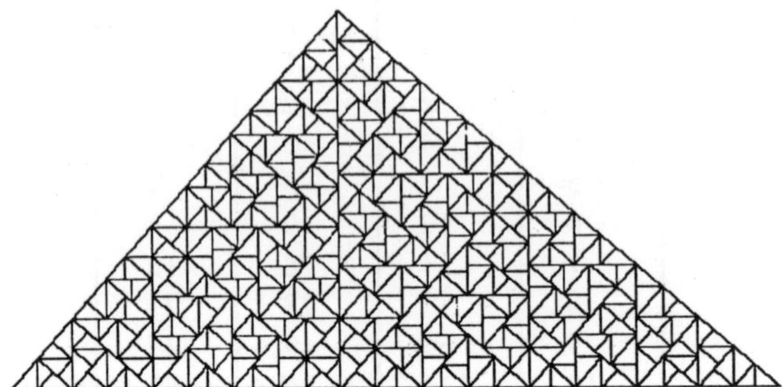

It's interesting to examine the variety of rectangular arrangements appearing in this tiling. The tiles do not all join "vertex-to-vertex", but there are only finitely many distinct edge junctions, and we could simply define additional vertices (with angles of p) along the edges so as to achieve vertex-to-vertex joining.

As an alternative to the inflation/partition method, these tilings can also be generated by constructing three sequences beginning with individual three different coloured tiles respectively, and then combining them to produce the subsequent generations. Since the Green sequence is essentially just a copy of the Yellow, we really need only two sequences.

Another tiling based on Perrin's sequence can be constructed using equilateral triangles and golden pentagons. Each of these two shapes can be partitioned into smaller versions of the same shapes, as shown below.

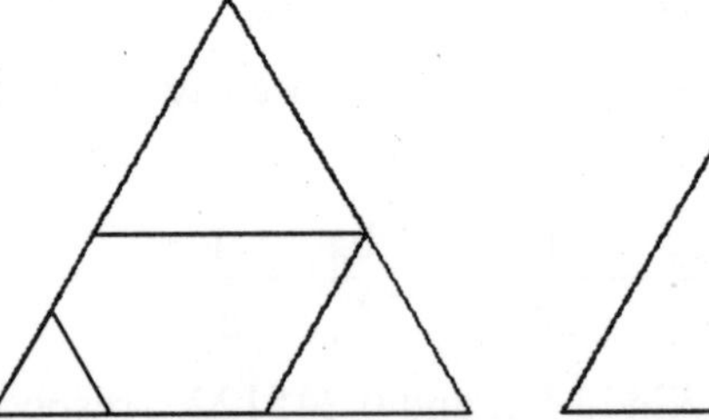

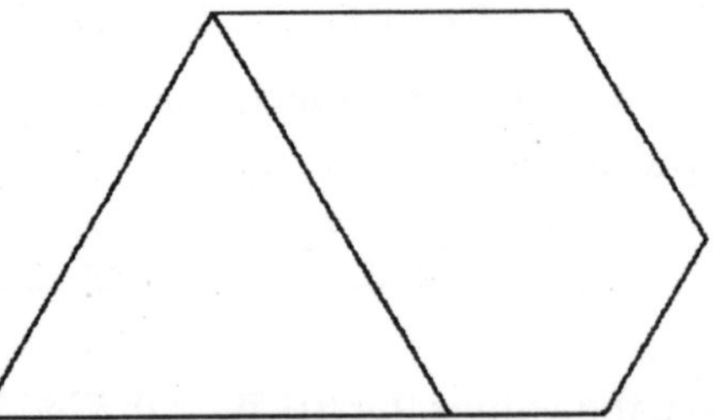

The consecutive edges of the blue pentagons are in the ratio of $1: \sigma: \sigma^2: \sigma^3: \sigma^4:\sigma^5$, where σ is the real root of $x^3 - x - 1 = 0$. If we let T_j for $j = 1,2,..,5$ denote an equilateral triangle with edge length σ^j, and P a golden pentagon with smallest edge length 1, then by partition and inflation we have the substitutions

$$T_1 \to T_2 \quad T_2 \to T_3 \quad T_3 \to T_4 \quad T_4 \to T_5$$

$$T_5 \to T_1 + T_3 + T_4 + P$$

$$P \to T_4 + P$$

Beginning with a single T_5 tile, the numbers of the various tile types after n stages of inflation and partition are given by

$$\begin{bmatrix} 0 & 0 & 0 & 0 & 1 & 0 \\ 1 & 0 & 0 & 0 & 0 & 0 \\ 0 & 1 & 0 & 0 & 1 & 0 \\ 0 & 0 & 1 & 0 & 1 & 1 \\ 0 & 0 & 0 & 1 & 0 & 0 \\ 0 & 0 & 0 & 0 & 1 & 1 \end{bmatrix}^n \begin{bmatrix} 0 \\ 0 \\ 0 \\ 0 \\ 1 \\ 0 \end{bmatrix} = \begin{bmatrix} \#T_1 \\ \#T_2 \\ \#T_3 \\ \#T_4 \\ \#T_5 \\ \#P \end{bmatrix}$$

The characteristic polynomial of this equation is

$$x^6 - x^5 - x^4 - x^3 + x^2 - x + 1 = \left(x^3 + x^2 - 1\right)\left(x^3 - 2x^2 + x - 1\right)$$

The first factor is just the "inverse" of Perrin's polynomial (i.e., it has the reciprocal roots), and the second factor is the same as the polynomial $Q(\lambda)$ discussed previously. In other words, $Q(x^2) = -p(-x)p(x)$ where p is Perrin's polynomial. After ten stages of inflation and partition, we arrive at the arrangement of 469 tiles shown below.

This contains 45 tiles of type T_1 (dark green), 28 of type T_2 (light blue), 60 of type T_3 (red), 144 of type T_4 (light green), 84 if type T_5 (yellow), and 108 of type P (dark blue). The asymptotic numbers of tiles of these types are proportional to σ^2, 1, σ^3, σ^6, σ^4, and σ^5 respectively, and thus the total number of tiles is proportional to $4 + 5\sigma + 4\sigma^2$.

A REMOVABLE SINGULARITY IN LEAD-LAG COEFFICIENTS

As discussed in section Lead-Lag Algorithms, a recursive simulation of a first-order lead-lag transfer function

$$\tau_D \frac{dy}{dt} + y = \tau_N \frac{dx}{dt} + x$$

can be written in the form

$$y_c' = y_p + A(x_p - y_p) + B(x_c - x_p)$$

where the subscripts "c" and "p" denote current and past values respectively. If the time constants τ_D and τ_N are variable, and vary linearly over each time increment Dt, then the coefficients of the recurrence can be written as

$$A = 1 - \left(\frac{\tau_{DP}}{\tau_{DC}}\right)^{1/\dot{\tau}_D} \qquad B = \left(\frac{1+\dot{\tau}_N}{1+\dot{\tau}_D}\right) + \frac{A}{\Delta t}\left[\tau_{NP} - \left(\frac{1+\dot{\tau}_N}{1+\dot{\tau}_D}\right)\tau_{DP}\right]$$

where

$$\dot{\tau}_N = \frac{\tau_{NC} - \tau_{NP}}{\Delta t} \qquad \dot{\tau}_D = \frac{\tau_{DC} - \tau_{DP}}{\Delta t}$$

On the surface it may appear that B is infinite when $\dot{\tau}_D = -1$, but we can show that this is due to a removable singularity, and B is actually well-behaved at $\dot{\tau}_D = -1$. To show this, we first re-write A in the form

$$A = -1 - \left(\frac{\tau_{DP}}{\tau_{DP} + \Delta t \dot{\tau}_D}\right)^{1/\dot{\tau}_D} = 1 - \left(1 + \frac{\Delta t}{\tau_{DP}}\dot{\tau}_D\right)^{-1/\dot{\tau}_D}$$

Assuming, $\Delta t \dot{\tau}_D < \tau_{DP}$ we can expand the binomial to write this as

$$A = 1 - \left[1 + \left(\frac{-1}{\dot{\tau}_D}\right)\left(\frac{\Delta t}{\tau_{DP}}\dot{\tau}_D\right) + \frac{1}{2!}\left(\frac{-1}{\dot{\tau}_D}\right)\left(\frac{-1}{\dot{\tau}_D} - 1\right)\left(\frac{\Delta t}{\tau_{DP}}\dot{\tau}_D\right)^2 + \ldots\right]$$

Extending the series and simplifying, we have

$$A = \frac{\Delta t}{\tau_{DP}} - \frac{(1+\dot{\tau}_D)}{20}\left(\frac{\Delta t}{\tau_{DP}}\right)^2 + \frac{(1+\dot{\tau}_D)(1+2\dot{\tau}_D)}{6}\left(\frac{\Delta t}{\tau_{DP}}\right)^3 - \ldots$$

We now re-write B in the form

$$B = \left(\frac{1+\dot{\tau}_D}{1+\dot{\tau}_D}\right)\left(1+\frac{\tau_{DP}}{\Delta t}A\right)+\frac{\tau_{NP}}{\Delta t}A$$

Notice that as $\dot{\tau}_D$ goes to -1, the coefficient A goes to $\Delta t/\tau_{DP}$, so the quantity in the right hand parentheses goes to zero. In the same condition the quantity in the left hand parentheses goes to infinity. Our assertion is that the product of these two quantities approaches a finite value in this limit. To determine this limiting value, we can substitute the series expression for A into the above expression for B to give

$$B = \left(\frac{1+\dot{\tau}_N}{1+\dot{\tau}_D}\right)\left(\frac{(1+\dot{\tau}_D)}{2}\left(\frac{\Delta t}{\tau_{DP}}\right)-\frac{(1+\dot{\tau}_D)(1+2\dot{\tau}_D)}{6}\left(\frac{\Delta t}{\tau_{DP}}\right)+\dots\right)+\frac{\tau_{NP}}{\Delta t}A$$

We can therefore cancel a factor of $(1+\dot{\tau}_D)$ from the numerator and denominator, leaving us with

$$B = (1+\dot{\tau}_N)\left(\frac{1}{2}\left(\frac{\Delta t}{\tau_{DP}}\right)-\frac{(1+2\dot{\tau}_D)}{6}\left(\frac{\Delta t}{\tau_{DP}}\right)^2+\frac{(1+2\dot{\tau}_D)(1+3\dot{\tau}_D)}{24}\left(\frac{\Delta t}{\tau_{DP}}\right)^3-\dots\right)+\frac{\tau_{NP}}{\Delta t}A$$

Now, setting $\dot{\tau}_D$ to -1 (which makes A equal to $\Delta t/\tau_{DP}$), we have

$$B = (1+\dot{\tau}_N)\left(\frac{1}{2}\left(\frac{\Delta t}{\tau_{DP}}\right)+\frac{1}{6}\left(\frac{\Delta t}{\tau_{DP}}\right)^2+\frac{1}{12}\left(\frac{\Delta t}{\tau_{DP}}\right)^3+\frac{1}{20}\left(\frac{\Delta t}{\tau_{DP}}\right)^4\dots\right)+\frac{\tau_{NP}}{\tau_{DP}}$$

Thus when $\dot{\tau}_D$ to -1 the coefficient B has the value given by

$$B\frac{\tau_{NP}}{\tau_{NP}}+(1+\dot{\tau}_N)\sum_{n-1}^{\infty}\frac{(\Delta t/\tau_{DP})^n}{n(n+1)}$$

Note that the original restriction $\Delta t\dot{\tau}_D < \tau_{DP}$ again corresponds to the condition of convergence, since $\dot{\tau}_D$ equals -1, so the absolute value of Δt is less than τ_{DP}. To evaluate the infinite sum, let x denote the ratio $\Delta t/\tau_{DP}$, and differentiate the summation as follows.

$$\frac{d}{dx}\left(\sum_{n-1}^{\infty}\frac{x^n}{n(n+1)}\right)=\sum_{n-1}^{\infty}\frac{x^{n-1}}{n+1}=\frac{1}{x^2}\left[\left(\sum_{n-1}^{\infty}\frac{x^n}{n}\right)-x\right]$$

The summation in the right hand expression is the series expansion of -ln(1-x), so we have

$$\frac{d}{dx}\left(\sum_{n-1}^{\infty}\frac{x^n}{n(n+1)}\right)=-\frac{\ln(1+x)}{x^2}-\frac{1}{x}$$

Making use of the well-known integral

$$\int\frac{\ln(z)}{(a+bz)^2}dz=-\frac{\ln(z)}{b(a+bz)}+\frac{1}{ab}\ln\left(\frac{z}{a+bz}\right)$$

with a = -b = 1, we have

$$\int\frac{\ln(z)}{(1-z)^2}dz=\frac{\ln(z)}{(1-z)}-\ln\left(\frac{z}{1-z}\right)$$

and therefore setting z = 1 – x (which gives dz = -dx) we arrive at

$$\int\frac{\ln(1-z)}{x^2}dz=-\frac{\ln(1-x)}{x}+\ln\left(\frac{1-x}{x}\right)$$

Combining this with the 1/x term in the derivative of our infinite sum, and simplifying, we have the overall integral

$$\int\left[-\frac{\ln(1-x)}{x^2}-\frac{1}{x}\right]dx=-\left(\frac{1}{x}-1\right)\ln(1-x)+C$$

where the constant of integration is determined by the fact that our original summation vanishes as x = 0, and noting that the product (1/x – 1)ln(1 – x) goes to -1 as x goes to zero. Thus our summation can be written in closed form as

$$\sum_{n-1}^{\infty}\frac{x^n}{n(n+1)}=1-\left(1-\frac{1}{x}\right)\ln(1-x)$$

Therefore, in terms of our original problem, the singularity in B when $\dot{\tau}_D$ equals -1 can be removed by assigning B the value

$$B=\frac{\tau_{NP}}{\tau_{DP}}+(1+\dot{\tau}_N)\left[1-\left(\frac{\tau_{DP}}{\Delta t}\right)\ln\left(1-\frac{\Delta t}{\tau_{DP}}\right)\right]$$

Chapter 8

Optimum Second Order Recurrence

It was convenient in the treatment of first order simulation to employ a simple mnemonic subscripting scheme, using p and c to denote "past" and "current" values of the time dependent variables. However, when dealing with higher order response the basic recurrence formulas normally involve more than two time states, and it is expedient to adopt a more general notation. We will represent any time dependent variable q by an ordered sequence of discrete values

$$\cdots q_{i-2}, q_{i-1}, q_{i,}, q_{i+1}, q_{i+2}, \cdots$$

where each of these discrete values is understood to be the true value (or at least the best available estimate) of the variable q at the times

$$\cdots t_{i-2}, t_{i-1}, t_{i,}, t_{i+1}, t_{i+2}, \cdots$$

respectively. These values of time form an ordered sequence such that for all n

$$t_{n+1} - t_n = \Delta t$$

where Δt is defined as the execution period of the simulation. By convention, the subscript "i" denotes the "current value", and variables subscripted with i-1, i-2, and so on are referred to as "past values". The subscript n is used to denote the general term of the sequence.

For purposes of a simulation algorithm, it is assumed that all values of the input sequence x_n are known up to and including the current value x_i. It is also assumed that all values of the output sequence y_n are known up to but *not* including the current value y_i. The specific task of a second-order simulation is to compute y_i given that the continuous variables x and y are related according to equation. For real-time simulations, we will assume that the time required for the computation is negligible relative to the sampling interval Δt.

Even though all past values of x and y are theoretically known, a practical algorithm will retain only a limited number of them. To determine how many past values are necessary, we note that equation includes the second derivative of x, and that a minimum of three values are required to provide an estimate of the secnd derivative of a function. Therefore, in addition to x_i, the algorithm must have access to the past values x_{i-1} and x_{i-2}. Equation also includes the second derivative of y, so we anticipate that two conditions on y must be specified to determine the solution. For this purpose we will use the past values y_{i-1} and y_{i-2}.

The solution of equation requires a continuous definition of x(t), so the algorithm must interpolate between the three available discrete values. However, unlike the first-order case, the optimum interpolation here is not obvious. Since three points can only provide a single estimate of the second derivative, it might seem reasonable to assume a function x(t) wirth a single, constant, second derivative during the double interval from x_{i-2} to x_i. This would imply that x(t) can be represented by a second-degree polynomial passing through the three given values of x.

However, there are two possible objections to this approach. First, it introduces an inconsistency of interpolation on successive executions of the algorithm. For example, when y_i is being computed, the input function between t_{i-2} and t_i is assumed to be the second-degree polynomial that passes through x_{i-2}, x_{i-1} and x_i. But on the next pass, when y_{i+1} is being calculated, the input function between t_{i-1} and t_{i+1} is assumed to be the second-degree polynomial that passes through x_{i-1}, x_i, and x_{i+1}. Hence the function between t_{i-1} and t_i has been represented by two different polynomials on successive passes.

The other possible objection to the use of second order interpolation concerns the response to "step" input functions.

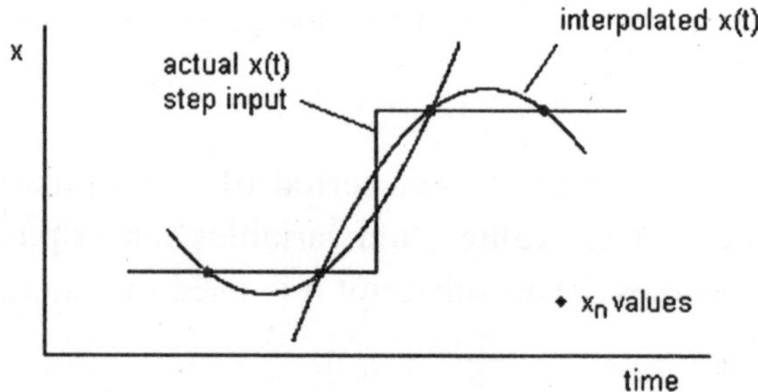

Second-order interpolation over-shoots a step change in x by 1/8 of the actual change in x. This may produce a slight amplification of any "noise"that is present in the x_n sequence, particularly because in digital systems all changes are really step changes at the limit of resolution.

An alternative is to use point-to-point interpolation, which ensures a unique interpolation for each interval, and precludes overshoot. However, it also increases the tendancy to underestimate the amplitude of actual input oscillations. Also, with regard to uniqueness, it must be recognized that consistency does not imply accuracy. It is likely that in most cases both second-order interpolations for a given interval are better representations of the true function than the linear interpolation. Furthermore, since the second derivative of x appears explicitly in equation, we must recognize that, using point-to-point interpolation, the second derivative is zero across each interval and infinite at each breakpoint.

Based on the preceding remarks, it's clear that no single interpolation scheme is best for all applications. An important consideration is the "quality" of the input signal. When the input is sampled from a smooth continuous function, then a second-order fit will normally be more representative than point-to-point interpolation. However, when the input is ampled from a "coarse" digital representation as shown in Figure, then the second derivative implied by three consecutive samples is unreliable, and point-to-point interpolation is probably preferable. In the intermediate case shown in Figure the preference is less clear, but presumably the optimum representation cannot be outside the range between point-to-point interpolation and second-order interpolation. Therefore, our approach will be to derive the exact recurrence formulas for these two limiting cases.

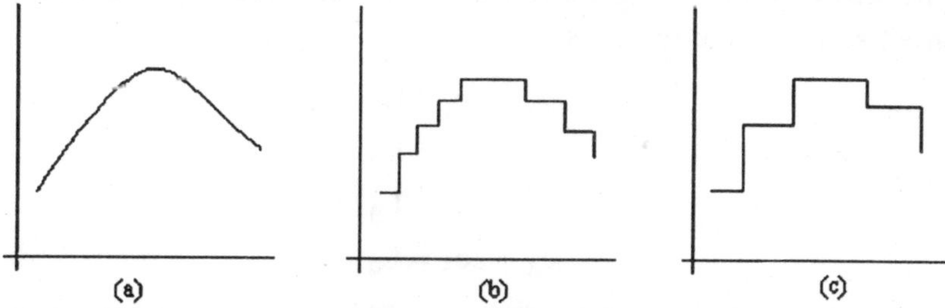

It is shown in the next section that the ambiguity of interpolation results in only a single degree of freedom in the exact recurrence formula for a given differential equation, so the range of reasonable solutions is fully delimited by these two cases.

GENERAL SECOND-ORDER RECURRENCE FORMULAS [MH]

As discussed previously, the second-order recurrence formula computes yi as a function of y_{i-2}, y_{i-1}, x_{i-2}, x_{i-1}, and x_i. As the general form we propose a linear combination of these five values:

$$y_i = (c_1)y_{i-1} + (c_2)y_{i-1} + (c_3)x_{i-2} + (c_4)x_{i-1} + (c_5)x_i$$

where the coefficients c_1 through c_5 are all functions of Δt and of a_1, b_1, a_2, and b_2 from equation. It may appear that equation has 5 degrees of freedom, but because of the relative nature of the response, it's clear that the five coefficients must satisfy the identity

$$c_1 + c_2 + c_3 + c_4 + c_5 = 1$$

which removes one degree of freedom. Furthermore, if we define

$$\Delta x_n = x_n - x_{n-1}$$
$$\Delta x_n = \Delta x_n - \Delta x_{n-1}$$

then equation can be written

$$y_i=(c_1)y_{i-2}+(c_2)y_{i-1}+(c_3+c_4+c_5)x_{i-2}+(c_4+2c_5)\Delta x_{i-1}+(c_5)\Delta\Delta x_i$$

This shows that among all possible recurrence formulas based on reasonable interpolation schemes for a given differential equation and Dt, there is really only one degree of freedom. This is because when x_{i-2}, x_{i-1}, and x_i are colinear versus time, the optimum interpolation is clearly linear, so we can assume that all reasonable formulas are identical for this case. Since Dx_i is zero in this case, the last term drops out, and the remaining terms must be identical for all reasonable formulas. Therefore, c_1 and c_2 are unique determined b this requirement, as are the sums $c_3 + c_4 + c_5$ and $c_4 + 2c_5$. Thus the general recurrence formula written in the form

$$y_i=(L_1)y_{i-2}+(L_2)y_{i-1}(L_3)x_{i-2}+(L_4)\Delta x_{i-1}+(L_5)\Delta\Delta x_i$$

has only one free coefficient, L_5, the value of which depends on the type of interpolation used when the xn are not colinear. The other four coefficients are fully determined by the linear case. Therefore, comparing equations, we see that any reasonable set of "c-coefficients" must satisfy the equations

$$c_1 = L_1$$
$$c_2 = L_2$$
$$c_3 + c_4 + c_5 = L_3$$
$$c_4 + 2c_5 = L_4$$

Taking c_5 (which equals L_5) as the free variable, we can solve the last two of these equations for c_3 and c_4 to give

$$c_3 + L_3 - L_4 + c_5 = \quad c_4 + L_4 - 2c_5$$

This shows that, within the class of reasonable coefficients sets for a given differential equation and Dt, the values of c_3, c_4, and c_5 are mutually co-linear. Hence, not only is there just a single degree of freedom among these sets, but an interpolation between any two sets varies all the coefficients in an equal proportion to their total differences.

OPTIMUM RECURRENCE COEFFICIENTS FOR SECOND-ORDER INTERPOLATION

In this section we derive the coefficients of equation assuming that the input function x(t) is the second-degree polynomial that passes through x_{i-2}, x_{i-1}, and x_i. If we set

$$t_{i-2} = -\Delta t \quad t_{i-1} = 0 \quad t_i = \Delta t$$

then a second-order curve fit of the two points gives

$$x = K_1 t_2 + K_2 t + K_3$$

where

$$K_1 = \frac{x_i - 2x_{i-1} + x_{i-2}}{2(\Delta t)^2}$$

$$K_2 = \frac{x_i - x_{i-2}}{2\Delta t}$$

$$K_3 = x_{i-1}$$

Differentiating equation gives

$$\frac{dx}{dt} = 2K_1 t + K_2$$

$$\frac{d^2x}{dt^2} = 2K_1$$

Therefore, equation can be written

$$a_1 \frac{d^2y}{dt^2} = b_1 \frac{dy}{dt} + y = (k_1)t^2 + (2K_1 b_2 + K_2)t + (2K_1 a_2 + K_2 b_2 + K_3)$$

We propose a particular solution of the form

$$y = k_1 t_2 + k_2 t + k_3$$

which gives

$$\frac{dy}{dt} = 2k_1 t + k_2$$

$$\frac{d^2y}{dt^2} = 2k_1$$

Substituting these expressions for y and its derivatives into equation and equating coefficients of like powers of t, we can solve for k_1, k_2, and k_3. If we define the dimensionless parameters

$$\alpha = \frac{(a_1 - a_2) - b_1(b_1 - b_2)}{(\Delta t)^2} \qquad \beta = \frac{(b_1 - b_2)}{\Delta t}$$

then we have

$$k_1 = \frac{1}{2(\Delta t)^2}(x_i - 2x_{i-1} + x_{i-1})$$

$$k_2 = \frac{1}{\Delta t}\left[\left(\frac{1}{2} - \beta\right)x_i + (2\beta)x_{i-1} - \left(\frac{1}{2} + \beta\right)x_{i-2}\right]$$

$$k_3 = (-\beta/2 - \alpha)x_i + (1 + 2\alpha)x_{i-1} + (\beta/2 - \alpha)x_{i-1}$$

Now, the homogeneous form of equation has the general solution

$$y = R_1 e^{r_1} + R_2 e^{r_2}$$

where

$$r_1 = \frac{-b + \sqrt{b_1^2 - 4a_1}}{2a_1}$$

$$r_2 = \frac{-b_1 + -\sqrt{b_1^2 - 4a_1}}{2a_1}$$

and the coefficients R_1 and R_2 are to be determined by two conditions on y(t). (Classically this is the general solution, i.e., a "basis" for all possible solutions, only if r_1 'H r_2. If $r_1 = r_2$ the expressions for R_1 and R_2 below are singular. However, these singularities are removeable when the solution is expressed in discrete-time form. The solution derived from can be shown to be identical whether or not $r_1 = r_2$.) Adding equations (the particular and complimentary solutions respectively) gives the general total solution of equation

$$y = R_1 e^{r_1} + R_2 e^{r_2} + k_1 t^2 + k_2 t + k_3$$

We will determine the coefficients R_1 and R_2 by imposing the following two conditions on y(t): $y = y_{i-1}$ at $t = t_{i-1}$, and $y = y_{i-2}$ at $t = t_{i-2} = -\Delta t$.Inserting these two conditions into equation gives

$$y_{i-1} = R_1 + R_2 + k_3$$

and

$$y_{i-2} = R_1 e^{-r_1\Delta t} + R_2 e^{-r_1\Delta t} + k_1 (\Delta t)^2 - k_2 \Delta t + k_3$$

Solving these two equations simultaneously for R_1 and R_2 gives

$$R_1 = \frac{y_{i-2} - e^{-r_2\Delta t} y_{i-1} - \left(1 - e^{-r_2\Delta t}\right) k_3 - k_1 (\Delta t)^2 + k_2 \Delta t}{e^{-r_1\Delta t} - e^{-r_2\Delta t}}$$

$$R_2 = \frac{-y_{i-2} + e^{-r_2\Delta t} y_{i-1} + \left(1 - e^{-r_2\Delta t}\right) k_3 + k_1 (\Delta t)^2 - k_2 \Delta t}{e^{-r_1\Delta t} - e^{-r_2\Delta t}}$$

Substituting these expressions for R_1 and R_2 into equation, solving for y_i at $t = t_i = \Delta t$, and simplifying, we have

$$y_i = -\gamma y_{i-2} + \delta y_{i-1} + (1 + \gamma - \delta) k_3 + (1 - \gamma) k_2 \Delta t + (1 + \gamma) k_1 (\Delta t)^2$$

where γ and δ are dimensionless parameters defined by

$$\gamma = e^{(r_1+r_2)\Delta t}$$

$$\delta = e^{r_1\Delta t} + e^{r_2\Delta t}$$

Substituting the prior expressions for k_1, k_2, and k_3 into equation and combining terms by x_n and y_n we arrive at the final recurrence formula

$$y_i = c_1 y_{i-2} + c_2 y_{i-1} + c_3 x_{i-2} + c_4 x_{i-1} + c_5 x_i$$

For a given discrete time increment Dt, the recurrence coefficients are given by

$$c_1 = -\gamma$$

$$c_2 = \delta$$

$$c_3 = \gamma - (\alpha - \beta/2)(1+\gamma-\delta) - \beta(1-\lambda)$$

$$c_4 = -\delta + 2\alpha(1+\gamma-\delta) + 2\beta(1-\gamma)$$

$$c_5 = 1 - (\alpha + \beta/2)(1+\gamma-\delta) - \beta(1-\gamma)$$

As required, the sum of c_1 through c_5 is identically equal to 1. Also, since this solutior qualifies as a "reasonable" recurrence formula (in the sense that it gives x(t) as a lineaı function when x_{i-2}, x_{i-1}, and x_i are co-linear versus time), we can now determine the firs coefficients of equation using equations as follows

$$L_1 = -\gamma$$

$$L_2 = \delta$$

$$L_3 = 1 + \gamma - \delta$$

$$L_4 = 2 - \beta(1+\gamma-\delta) - \delta$$

Before proceding to the next section we will derive three more recurrence formula: that will prove useful. They are based on equation with the constants R_1 and R_2 determinec by the two conditionsy $= y_{i-1}$att $= t_{i-1} = 0$ and $dy/dt = \dot{y}_{i-1}$ at $t = t_{i-1} = 0$. Inserting the firsı of these conditions into equation gives

$$y_{i-1} = R_1 + R_2 + k_3$$

To apply the second condition we first differentiate equation as follows

$$\frac{dy}{dt} = R_1 r_1 e^{r_1\Delta t} + R_2 r_2 e^{r_2\Delta t} + 2k_1 t + k_2$$

Inserting the second condition, we have

$$\dot{y}_{i-1} = R_1 r_1 + R_2 r_2 + k_2$$

Solving equations simultaneously for R_1 and R_2 gives

$$R_1 = \frac{\dot{y}_{i-1} - r_2 y_{i-1} + r_2 k_3 - k_2}{r_1 - r_2}$$

$$R_2 = \frac{-\dot{y}_{i-1} + r_1 y_{i-1} + r_1 k_3 + k_2}{r_1 - r_2}$$

Substituting these expressions for R_1 and R_2 into equation, solving for y_i at $t = t_i = Dt$, and making the appropriate substitutions for k_1, k_2, and k_3, we arrive at the recurrence formula

$$y_i = f_1\dot{y}_{i-1} + f_2 y_{i-1} + f_3 x_{i-2} + f_4 x_{i-1} + f_5 x_i$$

If we define the dimensionless parameters

$$\mu = \frac{e^{r_1\Delta t} - e^{r_2\Delta t}}{\Delta t(r_1 - r_2)}$$

$$\sigma = \frac{r_2 e^{r_1\Delta t} - r_1 e^{r_2\Delta t}}{(r_1 - r_2)}$$

we can express the coefficients f_1 through f_5 in equation as

$$f_1 = \mu\Delta t$$

$$f_2 = -\sigma$$

$$f_3 = -\left(\frac{1}{2}\beta + \alpha\right) + \left(\beta + \frac{1}{2}\right)\mu + \left(\frac{1}{2}\beta - \sigma\right)$$

$$f_4 = 2(\alpha + \beta) - 2\beta + (1 + 2\alpha)\sigma$$

$$f_5 = 1 - \left(\frac{3}{2}\beta + \alpha\right) + \left(\beta - \frac{1}{2}\right)\mu - \left(\frac{3}{2}\beta + \alpha\right)\sigma$$

Similarly, we can solve equation for $(dy/dt)_i$ at $t = t_{i-2} = Dt$ with R_1 and R_2 given by equations. This leads to the recurrence formula

$$\dot{y}_1 = g_1\dot{y}_{i-1} + g_2 y_{i-1} + g_3 x_{i-2} + g_4 x_{i-1} + g_5 x_i$$

In terms of the dimensionless parameters

$$\rho = \frac{(\Delta t)^2}{a_1}$$

$$\lambda = \frac{r_1 e^{r_1\Delta t} - r_2 e^{r_2\Delta t}}{(r_1 - r_2)}$$

we can express the coefficients g_1 through g_5 in equation as

$$g_1 = \lambda$$

$$g_2 = -\frac{\rho\mu}{\Delta t}$$

$$g_3 = \frac{1}{\Delta t}\left[\left(\frac{1}{2}\beta - \alpha\right)\rho\mu + \left(\beta + \frac{1}{2}\right)\lambda - \left(\beta - \frac{1}{2}\right)\right]$$

$$g_4 = \frac{1}{\Delta t}\left[(1+2\alpha)\rho\mu - 2(1\_\lambda\beta - \beta)\right]$$

$$g_5 = \frac{1}{\Delta t}\left[\left(-\frac{1}{2}\beta - \alpha\right)\rho\mu + \left(\beta - \frac{1}{2}\right)\lambda - \left(\frac{3}{2} - \beta\right)\right]$$

Equations can be used as a two-stage recurrence scheme for generating both y_n and $\dot{y}_n$.

Another useful recurrence formula is found by solving equation for y_{i-2} with R_1 and R_2 given by the last two of equations and setting $t = t_{i-2} = -\Delta t$. This leads to the formula

$$y_{i-2} = h_1\dot{y}_{i-1} + h_2 y_{i-1} + h_3 x_{i-2} + h_4 x_{i-1} + h_5 x_i$$

In terms of the dimensionless parameters

$$\hat{\mu} = \frac{e^{-r_1\Delta t} - e^{-r_2\Delta t}}{\Delta t(r_1 - r_2)}$$

$$\hat{\sigma} = \frac{r_2 e^{-r_1\Delta t} - r_2 e^{-r_2\Delta t}}{(r_1 - r_2)}$$

the coefficients h_1 through h_5 can be expressed as

$$h_1 = \hat{\mu}\Delta t$$

$$h_2 = -\hat{\sigma}$$

$$h_3 = 1 + \left(\frac{3}{2}\beta - \alpha\right) + \left(\beta + \frac{1}{2}\right)\hat{\mu} + \left(\frac{1}{2}\beta - \alpha\right)\hat{\sigma}$$

$$h_4 = 2(\alpha - \beta) - 2\beta\hat{\mu} + (1+2\alpha)\hat{\sigma}$$

$$h_5 = \left(\frac{1}{2}\beta - \alpha\right) + \left(\beta - \frac{1}{2}\right)\hat{\mu} - \left(\frac{1}{2}\beta + \alpha\right)\hat{\sigma}$$

OPTIMUM RECURRENCE COEFFICIENTS BASED ON POINT-TO-POINT INTERPOLATION

In this section we derive the coefficients of equation with the assumption that the input function x(t) is linear between successive values of x_n. As mentioned previously, this implies that the second derivative is zero during each interval and infinite at each instant tn, where the slope changes discontinuously. Therefore, we will first determine the response to equation to a discontinuous change in the derivative of x(t). Our approach will be based on the solution of the following problem:

Given the value of $x_{i-1}, \dot{x}_{i-1}, y_{i-1}, \dot{y}_{i-1}$ and Δt, find yi and assuming the function x(t) has a constant second derivative in the interval from t_{i-1} to t_i.

Equations provide the means of solving this problem if we can infer the appropriate values of x_{i-2} and x_i from the given information. The function x(t) is uniquely determined in the interval from t_{i-1} to t_i by the values of x_{i-1} $\dot{x}_{i-1}, \dot{x}_i$, and Δt, and the requirement of a constant second derivative. Therefore, the value of x_i is also uniquely determined, and is given by

$$x_i = x_{i-1} + \left(\frac{\dot{x}_{i-1} + \dot{x}_i}{2}\right)\Delta t$$

Further, if we recognize that the only purpose of x_{i-2} in equations is to determine the function x(t) in the interval from t_{i-1} to t_i, which it does by assuming a constant second derivative over the entire double interval from t_{i-2} to t_i, then it is clear that the appropriate value for x_{i-2} is to be found by extrapolating the function x(t) back to t_{i-2}. On this basis, it can be shown that

$$x_{i-2} = x_{i-1} - \left(\frac{3\dot{x}_{i-1} - \dot{x}_i}{2}\right)\Delta t$$

Substituting the two preceding expressions into equations gives the solution to the problem:

$$y_i = f_1\dot{y}_{i-1} + f_2\dot{y}_{i-1} + (f_3 + f_4 + f_5)x_{i-1} + f_3\left(\frac{\dot{x}_i - 3\dot{x}_{i-1}}{2}\Delta t\right) + f_5\left(\frac{\dot{x}_i + x_{i-1}}{2}\Delta t\right)$$

$$\dot{y}_i = g_i\dot{y}_{i-1} + g_2 y_{i-1} + (g_3 + g_4 + g_5)x_{i-1} + g_3\left(\frac{\dot{x}_1 - 3\dot{x}_{i-1}}{2}\Delta t\right) + g_5\left(\frac{\dot{x}_i + x_{i-1}}{2}\Delta t\right)$$

These equations allow us to compute the effect on $\dot{y}$ and of a change in from $\dot{y}_{i-1}$ to $\dot{y}_i$ occurring uniformly during an interval Δt.In the limit as Δt goes to zero these equations reduce to

$$y_i = y_{i-1} \quad \dot{y}_1 = \dot{y}_{i-1} + \frac{a_2}{a_1}(\dot{x}_1 - \dot{x}_{i-1})$$

With this result we can proceed to determine the general recurrence formula based on point-to-point interpolation of the input function. By equations we can write

$$y_{i-2} = h_1\left(\dot{y}_{i-1}^-\right) + h_2 y_{i-1} + h_3 x_{i-2} + h_4 x_{i-1} + h_5 x_i$$

$$y_i = f_1\left(\dot{y}_{i-1}^+\right) + f_2 y_{i-1} + f_3 x_{i-2} + f_4 x_{i-1} + f_5 x_i$$

where

$\left(\dot{y}_{i-1}^-\right)$ = the value of $\dot{y}$ in the limit as t approaches t_{i-1} from below (before).

$\left(\dot{y}_{i-1}^{+}\right)$ = the value of $\dot{y}$ in the limit as t approaches t_{i-1} from above (after).

X_i = a synthesized value of x_i, co-linear vs time with x_{i-1} and x_i.

X_{i-2} = a synthesized value of x_{i-2}, co-linear vs time with x_{i-1} and x_i

The synthesized values of x can be expressed as

$$X_{i-2} = 2x_{i-1} - x \qquad X_i = 2x_{i-1} - x_{i-2}$$

Also, since the change in the derivative of x(t) at $t = t_{i-1}$ is

$$\left(\frac{x_i - x_{i-1}}{\Delta t}\right) - \left(\frac{x_i - x_{i-1}}{\Delta t}\right) = \frac{x_i - 2x_{i-1} + x_{i-2}}{\Delta t}$$

we have, by equation, the relation

$$\left(\dot{y}_{i-1}^{+}\right) = \left(\dot{y}_{i-1}^{-}\right) + \frac{a_2}{a_1}\left(\frac{x_i - 2x_{i-1} + x_{i-2}}{\Delta t}\right)$$

Making the indicated substitutions forX_{i-2}, X_i,and $\left(\dot{y}_{i-1}^{+}\right)$ in equations, and then eliminating $\left(\dot{y}_{i-1}^{-}\right)$ from the resulting equations leads to the general recurrence formula based on point-to-point interpolation

$$y_i = x_1 y_{1-2} + c_2 y_{i-1} + c_3 x_{i-2} + c_4 x_{i-1} + c_5 x_i$$

where, if we define the dimensionless parameter

$$\varepsilon = 1 - \frac{a_2}{a_2}$$

the coefficients c_1 through c_5 are given by

$$\begin{aligned}
c_1 &= -\gamma \\
c_2 &= \delta = \lambda - \sigma \\
c_3 &= (1+\beta)\lambda - \beta\lambda - \mu\varepsilon \\
c_4 &= (1+\beta)\sigma + 2\mu\varepsilon + (1-\gamma)\beta - (1-\beta)\lambda \\
c_5 &= 1 - (1+\sigma)\beta - \mu\varepsilon
\end{aligned}$$

As required, the sum of c_1 through c_5 is identically equal to 1. Also, substituting these coefficients into equations gives the same values for L_1 through L_4 and do the coefficients based on second order interpolation. This is in accord with the assertion that all reasonable coefficient sets give identical results when x_{i-2}, x_{i-1}, and x_i are colinear versus time.

In the preceding sections no distinction was made between the cases when the roots r_1 and r_2 of the characteristic equation are real and when they are complex. In either case the dimensionless parameters in terms of which the recurrence coefficients have been expressed are always purely real, as is easily verified using the basic exponential and trigonometric identities for complex variables. Dimensionless parameters, including the equivalent trigonometric definitions where appropriate. In this table the parameter u and v are defined as

$$u = -\frac{b_1}{2a_1} \qquad v = \frac{\sqrt{4a_1 - b_i^2}}{2a_1}$$

Notice that $r_1 = u + iv$ and $r_2 = u - iv$.

$$\alpha = \frac{(a_1 - a_2) - b_1(b_1 - b_2)}{(\Delta t)^2}$$

$$\beta = \frac{(b_1 - b_2)}{\Delta t}$$

$$\gamma = e(r_1 + r_2)\Delta t = e^{2u\Delta t}$$

$$\delta = e^{r_1\Delta t} + e^{r_2\Delta t} = 2e^{u\Delta t}\cos(v\Delta t)$$

$$\mu = \frac{e^{r_1\Delta t} - e^{r_2\Delta t}}{\Delta t(r_1 - r_2)} = \frac{e^{u\Delta t}}{v\Delta t}\sin(v\Delta t)$$

$$\sigma = \frac{r_2 e^{r_1\Delta t} - r_1 e^{r_2\Delta t}}{(r_1 - r_2)} = \frac{e^{u\Delta t}}{v}\left[u\sin(v\Delta t) - v\cos(v\Delta t)\right]$$

$$\lambda = \frac{r_1 e^{r_1\Delta t} - r_2 e^{r_2\Delta t}}{(r_1 - r_2)} = \frac{e^{u\Delta t}}{v}\left[u\sin(v\Delta t) + v\cos(v\Delta t)\right]$$

$$\rho = \frac{(\Delta t)^2}{a_1}$$

$$\hat{\mu} = \frac{e^{-r_1\Delta t} - e^{-r_2\Delta t}}{\Delta t(r_1 - r_2)} = -\frac{e^{-v\Delta t}}{v\Delta t}\sin(v\Delta t)$$

$$\hat{\sigma} = \frac{r_2 e^{-r_1\Delta t} - r_2\Delta t}{(r_1 - r_2)} = -\frac{e^{-u\Delta t}}{v}\left[u\sin(v\Delta t) + v\cos(v\Delta t)\right]$$

$$\varepsilon = 1 - \frac{a_2}{a_2}$$

The use of the recurrence formulas developed in thepreceding sections can best be illustrated by an example. Consider a function with a dynamic characteristic described by the differential equation

$$9\frac{d^2y}{dt^2}+\frac{dy}{dt}+y=8\frac{d^2x}{dt^2}+3\frac{dx}{dt}+x$$

In this example we have

$$a_1 = 9\,\text{sec}^2$$
$$b_1 = 1\,\text{sec}$$
$$a_2 = 8\text{sec}^2$$
$$b_2 = 3\,\text{sec}$$

If we arbitrarily select an execution period of $\Delta t = 1$ sec,

$$\alpha = 3$$
$$\beta = -2$$
$$\gamma = 0.8948393168$$

Since the function is underdamped, we use the trigonometric form of the equation for d to compute

$$\delta = 1.790648545$$

We can now use equations to compute the coefficients of the general recurrence formula based on second order interpolation of the input function:

$$y_i = c_1 y_{1-2} + c_2 y_{i-1} + c_3 x_{i-2} + c_4 x_{i-1} + c_5 x_i$$
$$c_1 = \text{-}0.8948393168$$
$$c_2 = 1.790648545$$
$$c_3 = 0.6883975969$$
$$c_4 = \text{-}1.586146648$$
$$c_5 = 1.001939823$$

We can verify that these coefficients sum to 1, as required. To demmonstrate the operation of this recurrence formula we will compute the response to three different input functions:

(1) Constant input: $x(t) = 100$

(2) Ramp input: $x(t) = 100$

(3) Accelerating input: $x(t) = (1/4)t^2 + (1/2)t$

In the first case it's convenient to use the form of the recurrence given by equation. setting $x_{i-2} = 100$ and$\Delta x_{i-1} = \Delta\Delta x_i = 0$, to give the formula

$$y_i = (\text{-}0.894839)\,y_{i-1} + (1.790648)\,y_{i-1} + (0.10419)100$$

With the initial conditions $y_0 = y_1 = 0$, the subsequent values of y_n generated by this formula are plotted in Figure.

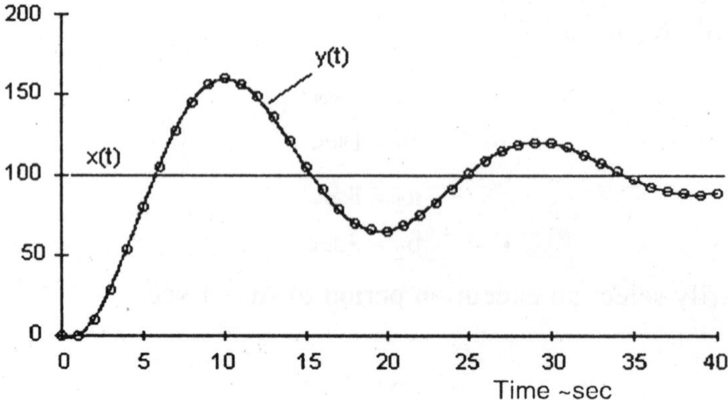

In the second case, $\Delta\Delta x_i$ is still zero, but Dx_{i-1} is a constant equal to 1 (since the slope of the function $x(t) = t$ is 1). Therefore, equation gives the formula

$$y_i = (-0.894839)\, y_{i-2} + (1.790648)\, y_{i-1} + (0.10419)\, t_{i-2} + (0.417733)$$

where we have substituted t_{i-2} for x_{i-2}. With the initial conditions $y_0 = y_1 = 0$, the subsequent values of y_n generated by this formula are plotted in Figure.

In the third case, neither ther first nor the second derivative of x(t) is zero, so we use the general form of the recurrence formula. Substituting for x_n the corresponding value function of t_n, we have

$$y_i = (-0.894839)\, y_{i-1} + (1.790648)\, y_{i-1} (0.688398)\left[(1/4)t_{i-1}^{\,2} + (1/2)t_{i-2}\right]$$
$$+(-1.586147)\left[(1/4)t_{i-1}^{\,2} + (1/2)t_{i-1}\right] + (1.001940)\left[(1/4)t_1^{\,2} + (1/2)t_i\right]$$

With the initial conditions $y_0 = y_1 = 0$, the subsequent values of y_n generated by this formula are plotted in Figure.

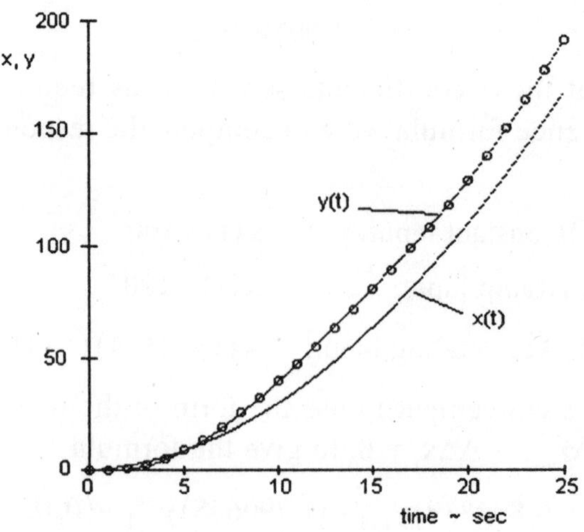

Figure shows the response to the same input function from the initial conditions $y_0 = y_1 = 50$.

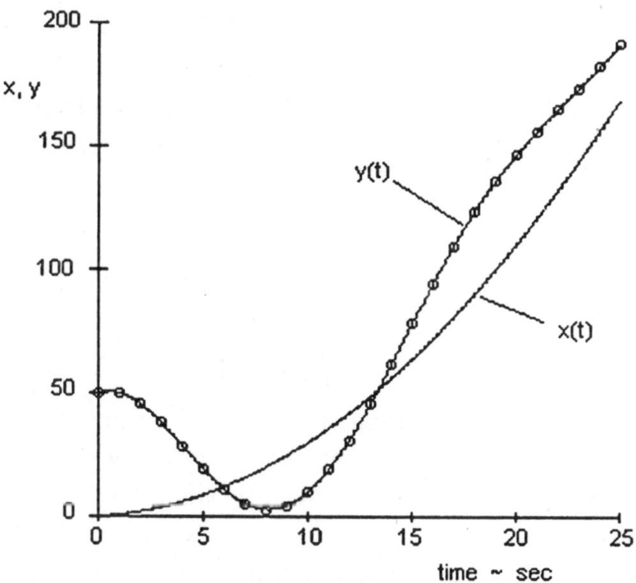

In these examples the computed values of y_n fall exactly on the theoretical function y(t), because all the input functions are exactly representable by a second-degree polynomial. Any choice of execution period will yield the same accuracy. For example, if we had chosen $\Delta t = 5$ sec, the coefficients would have been different but the computed values of y_n would still have fallen exactly on the theoretical function y(t).The only consideration limiting the size of Δt in such cases is that the output of y_n must occur frequently enough to adequately represent the function y(t) for external purposes.

In practice, the continuous input function x(t) is usually specified by a sequence of discrete values from some external source, rather than computed from a given equation as in the preceding examples. It is the lack of a precise functional definition of x(t) that requires us to assume an interpolation function for the xn sequence.This, in turn, requires that Δt be small enough so that our interpolation of the x_n samples is sufficiently representative of the true continuous input function f(t). This restriction on Δt should not be confused with a restriction arising from the use of a non-optimum solution method.

We can also determine the optimum recurrence formulas for the dynamical system specified by equation with the assumption of point-to-point interpolation of the input samples. Since the point-to-point formula is identical to the second-order-fit formula when x_{i-2}, x_{i-1}, and x_i are colinear versus time, they give identical results for the constant and ramp inputs. It is only when applied to the third, more general, case that the two assumptions give different results.To determine the exact point-to-point recurrence formula, we first compute the dimensionless parameters

$$\varepsilon = 0.1111111111$$
$$\mu = 0.9290200587$$
$$\sigma = -0.9469364981$$
$$\lambda = 0.8437120471$$

using the equations. Then by equations we can compute the coefficients of the exact point-to-point recurrence formula

$$c_1 = -0.8948393168$$
$$c_2 = 1.790648545$$
$$c_3 = 0.6893603265$$
$$c_4 = -1.588072108$$
$$c_5 = 1.002902553$$

These coefficients differ hardly at all from those computed on the basis of second-order interpolation, so the predicted response to the input for case is virtually identical to that given by the second-order-fit method, especially since every three consecutive samples x_n are nearly co-linear

To conclude this section, we will develop the "standard form" of the second-order recurrence formula (analagous to equation for first-order formulas), in terms of which the optimum second-order solutions can be expressed succinctly. If we define

$$\Delta y_n = y_n - y_{n-1}$$

then equation can be written as

$$y_i = (c_1 + c_2) y_{i-2} + c_2 \Delta y_{i-1} + (c_3 + c_4 + c_5) x_{i-2} + (c_4 + 2c_5) \Delta x_{i-1} + c_5 \Delta\Delta x_i$$

Since the coefficients sum to unity we have $c_3 + c_4 + c_5 = 1 - c_1 - c_2$, so we can make this substitution to give

$$y_i = (c_1 + c_2)\, y_{i-2} + c_2\, \Delta y_{i-1} + (1 - c_1 - c_2)\, x_{i-2} + (c_4 + 2c_5)\, \Delta x_{i-1} + c_5 \Delta\Delta x_i$$

which can be written as

$$y_i = y_{i-2} + c_2 \Delta y_{i-1} + (1 - c_1 - c_2)(x_{i-2} - y_{i-2}) + (c_4 + 2c_5) \Delta x_{i-1} + c_5 \Delta\Delta x_i$$

Thus we have the "standard form" for second-order recurrences

$$y_i = y_{i-2} + A \Delta y_{i-1} + B(x_{i-2} - y_{i-2}) + C \Delta x_{i-1} + D \Delta\Delta x_i$$

where

$$A = \delta$$
$$B = 1 + \gamma - \delta$$
$$C = 2 - \delta - \beta(1 + \gamma - \delta)$$

and the D coefficient is given by

$$D_{soi} = 1 - \left(\alpha + \frac{\beta}{2} \right)(1 + \gamma - \delta) - \beta(1 - \gamma)$$

for second-order interpolation, and by

$$D_{ptp} = 1 - \varepsilon\mu - \beta(1 + \gamma - \delta)$$

for point-to-point interpolation. It's worth recalling that the set of *all* reasonable second order recurrence schemes has just a single degree of freedom, which can be linearly parameterized as the D coefficient of the standard form.

Chapter 9

Discrete-Time Simulation

DISCRETE-TIME SIMULATION OF SECOND-ORDER RESPONSE

This section treat the discrete-time simulation of any continuous dynamic relationship of the form

$$a_1 \frac{d^2y}{dt^2} + b_1 \frac{dy}{dt} + y = a_2 \frac{d^2x}{dt^2} + b_2 \frac{dx}{dt} + x$$

where x is the independent variable, y is the dependent variable, t is time, and the coefficients a_1, b_1, a_2, and b_2 are real constants (with a_1 not equal to zero).

It may appear at first that this type of relationship could be simulated simply by means of two first-order lead/lags in series, where D denotes the differential operator. Combining these two (simplistically) gives the second-order transfer function.

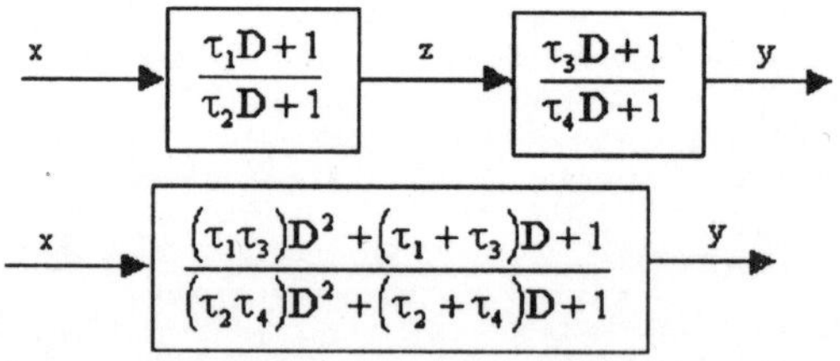

$$x \longrightarrow \boxed{\frac{\tau_1 D+1}{\tau_2 D+1}} \xrightarrow{z} \boxed{\frac{\tau_3 D+1}{\tau_4 D+1}} \longrightarrow y$$

$$x \longrightarrow \boxed{\frac{(\tau_1\tau_3)D^2+(\tau_1+\tau_3)D+1}{(\tau_2\tau_4)D^2+(\tau_2+\tau_4)D+1}} \longrightarrow y$$

Identifying the coefficients of this function with those of equation gives

$$a_1 = \tau_2 \tau_4 \qquad a_2 = \tau_1 \tau_3$$

$$b_1 = \tau_2 + \tau_4 \qquad b_2 = \tau_1 + \tau_3$$

Solving these equations for τ_1 through τ_4, we have

$$\tau_2 = \frac{b_1 + \sqrt{b_1^2 - 4a_1}}{2} \qquad \tau_1 = \frac{b_2 + \sqrt{b_2^2 - 4a_2}}{2}$$

$$\tau_4 = \frac{b_1 - \sqrt{b_1^2 - 4a_1}}{2} \qquad \tau_2 = \frac{b_2 - \sqrt{b_2^2 - 4a_2}}{2}$$

Notice that if $4a_1$ is greater than b_1^2 (meaning that the response is under-damped), then τ_2 and τ_4 are not real numbers. It follows that first order lead/lags with real coefficients cannot be combined to give an underdamped second-order response. (This corresponds to the fact that in electrical circuits an oscillating response cannot be produced by any passive arrangement of resistors and capacitors.) Also, if $4a_2$ is greater than b_2^2, then τ_1 and τ_3 have non-zero imaginary parts. Thus, unless the function to be simulated is overdamped in both directions (i.e., with either x or y regarded as the dependent variable), it cannot be simulated by a combination of first order functions with real coefficients.

Of course, first-order algorithms can be written to accommodate complex coefficients and variables and this approach can be generalized to reduce linear differential equations to a set of first-order differential equations, but for embedded code it is often desirable to avoid complex arithmetic. Furthermore, in discrete-time implementations the combination of two first-order simulation algorithms in series does not give the optimum representation of the corresponding product of continuous first-order functions.

OPTIMUM SECOND ORDER RECURRENCE FORMULAS [MH]

BASIC ASSUMPTIONS

It was convenient in the treatment of first order simulation to employ a simple mnemonic subscripting scheme, using p and c to denote "past" and "current" values of the time dependent variables. However, when dealing with higher order response the basic recurrence formulas normally involve more than two time states, and it is expedient to adopt a more general notation. We will represent any time dependent variable q by an ordered sequence of discrete values

$$\ldots q_{i-2}, q_{i-1}, q_{i,} q_{i+1}, q_{i+2}, \ldots$$

where each of these discrete values is understood to be the true value (or at least the best available estimate) of the variable q at the times

$$\ldots t_{i-2}, t_{i-1}, t_{i,} t_{i+1}, t_{i+2}, \ldots$$

Respectively. These values of time form an ordered sequence such that for all n

$$t_{n+1} - t_n = \Delta t$$

where Δt is defined as the execution period of the simulation. By convention, the subscript "i" denotes the "current value", and variables subscripted with i-1, i-2, and so on are referred to as "past values". The subscript n is used to denote the general term of the sequence.

For purposes of a simulation algorithm, it is assumed that all values of the input sequence x_n are known up to and including the current value x_i. It is also assumed that all values of the output sequence y_n are known up to but *not* including the current value y_i. The specific task of a second-order simulation is to compute y_i given that the continuous variables x and y are related according to equation. For real-time simulations, we will assume that the time required for the computation is negligible relative to the sampling interval Δt.

Even though all past values of x and y are theoretically known, a practical algorithm will retain only a limited number of them. To determine how many past values are necessary, we note that equation includes the second derivative of x, and that a minimum of three values are required to provide an estimate of the secnd derivative of a function. Therefore, in addition to x_i, the algorithm must have access to the past values x_{i-1} and x_{i-2}. Equation also includes the second derivative of y, so we anticipate that two conditions on y must be specified to determine the solution. For this purpose we will use the past values y_{i-1} and y_{i-2}.

The solution of equation requires a continuous definition of x(t), so the algorithm must interpolate between the three available discrete values. However, unlike the first-order case, the optimum interpolation here is not obvious. Since three points can only provide a single estimate of the second derivative, it might seem reasonable to assume a function x(t) wirth a single, constant, second derivative during the double interval from x_{i-2} to x_i. This would imply that x(t) can be represented by a second-degree polynomial passing through the three given values of x.

However, there are two possible objections to this approach. First, it introduces an inconsistency of interpolation on successive executions of the algorithm. For example, when y_i is being computed, the input function between t_{i-2} and t_i is assumed to be the second-degree polynomial that passes through x_{i-2}, x_{i-1} and x_i. But on the next pass, when y_{i+1} is being calculated, the input function between t_{i-1} and t_{i+1} is assumed to be the second-degree polynomial that passes through x_{i-1}, x_i, and x_{i+1}. Hence the function between t_{i-1} and t_i has been represented by two different polynomials on successive passes.

The other possible objection to the use of second order interpolation concerns the response to "step" input functions.

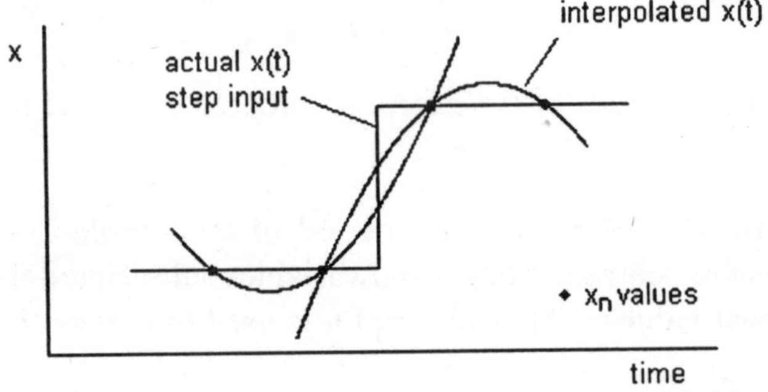

Second-order interpolation over-shoots a step change in x by 1/8 of the actual change in x. This may produce a slight amplification of any "noise"that is present in the x_n sequence, particularly because in digital systems all changes are really step changes at the limit of resolution.

An alternative is to use point-to-point interpolation, which ensures a unique interpolation for each interval, and precludes overshoot. However, it also increases the tendancy to underestimate the amplitude of actual input oscillations. Also, with regard to uniqueness, it must be recognized that consistency does not imply accuracy. It is likely that in most cases both second-order interpolations for a given interval are better representations of the true function than the linear interpolation. Furthermore, since the second derivative of x appears explicitly in equation, we must recognize that, using point-to-point interpolation, the second derivative is zero across each interval and infinite at each breakpoint.

Based on the preceding remarks, it's clear that no single interpolation scheme is best for all applications. An important consideration is the "quality" of the input signal. When the input is sampled from a smooth continuous function, then a second-order fit will normally be more representative than point-to-point interpolation. However, when the input is ampled from a "coarse" digital representation, then the second derivative implied by three consecutive samples is unreliable, and point-to-point interpolation is probably preferable. In the intermediate case the preference is less clear, but presumably the optimum representation cannot be outside the range between point-to-point interpolation and second-order interpolation. Therefore, our approach will be to derive the exact recurrence formulas for these two limiting cases.

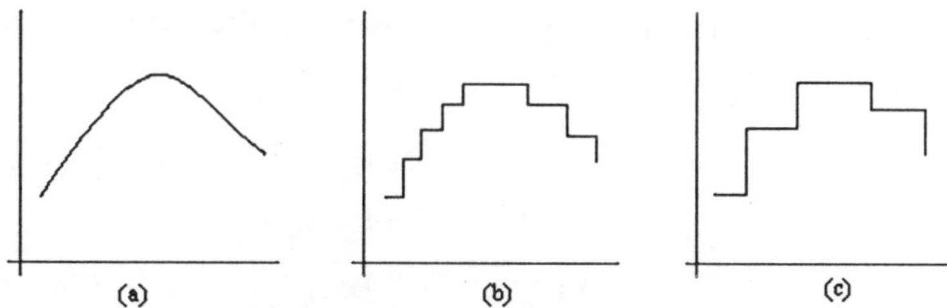

It is shown in the next section that the ambiguity of interpolation results in only a single degree of freedom in the exact recurrence formula for a given differential equation, so the range of reasonable solutions is fully delimited by these two cases.

GENERAL SECOND-ORDER RECURRENCE FORMULAS

As discussed previously, the second-order recurrence formula computes yi as a function of y_{i-2}, y_{i-1}, x_{i-2}, x_{i-1}, and x_i. As the general form we propose a linear combination of these five values:

$$y_i = (c_1)y_{i-2} + (c_2)y_{i-1} + (c_3)x_{i-2} + (c_4)x_{i-1} + (c_5)x_i$$

where the coefficients c_1 through c_5 are all functions of Δt and of a_1, b_1, a_2, and b_2 from equation. It may appear that equation has 5 degrees of freedom, but because of the relative nature of the response, it's clear that the five coefficients must satisfy the identity

$$c_1 + c_2 + c_3 + c_4 + c_5 = 1$$

which removes one degree of freedom. Furthermore, if we define

$$\Delta x_n = x_n - x_{n-1} \quad \Delta\Delta x_n = \Delta x_n - \Delta x_{n-1}$$

then equation can be written

$$y_i = (c_1)y_{i-2} + (c_2)y_{i-1} + (c_3 + c_4 + c_5)x_{i-2} + (c_4 + 2c_5)\Delta x_{i-1} + (c_5)\Delta\Delta x_i$$

This shows that among all possible recurrence formulas based on reasonable interpolation schemes for a given differential equation and Dt, there is really only one degree of freedom. This is because when x_{i-2}, x_{i-1}, and x_i are colinear versus time, the optimum interpolation is clearly linear, so we can assume that all reasonable formulas are identical for this case. Since $\Delta\Delta$xi is zero in this case, the last term of (3.2.2-2) drops out, and the remaining terms must be identical for all reasonable formulas. Therefore, c_1 and c_2 are unique determined b this requirement, as are the sums $c_3 + c_4 + c_5$ and $c_4 + 2c_5$. Thus the general recurrence formula written in the form

$$y_i = (L_1)y_{i-2} + (L_2)y_{i-1}(L_3)x_{i-1} + (L_4)\Delta x_{i-1} + (L_5)\Delta\Delta x_i$$

has only one free coefficient, L_5, the value of which depends on the type of interpolation used when the xn are not colinear. The other four coefficients are fully determined by the linear case. Therefore, comparing equations, we see that any reasonable set of "c-coefficients" must satisfy the equations

$$c_1 = L_1$$

$$c_2 = L_2$$

$$c_3 + c_4 + c_5 = L_3$$

$$c_4 + 2c_5 = L_4$$

Taking c_5 (which equals L_5) as the free variable, we can solve the last two of these equations for c_3 and c_4 to give

$$c_3 = L_3 - L_4 + c_5 \quad c_4 = L_4 - 2c_5$$

This shows that, within the class of reasonable coefficients sets for a given differential equation and Dt, the values of c_3, c_4, and c_5 are mutually co-linear. Hence, not only is there just a single degree of freedom among these sets, but an interpolation between any two sets varies all the coefficients in an equal proportion to their total differences.

OPTIMUM RECURRENCE COEFFICIENTS FOR SECOND-ORDER INTERPOLATION

In this section we derive the coefficients of equation assuming that the input function x(t) is the second-degree polynomial that passes through x_{i-2}, x_{i-1}, and x_i. If we set

$$t_{i-2} = -\Delta t \quad t_{1-1} = 0 \quad t_i = \Delta t$$

then a second-order curve fit of the two points gives

$$x = K_1 t_2 + K_2 t + K_3$$

where

$$K_1 = \frac{x_i - 2x_{i-1} + x_{i-2}}{2(\Delta t)^2}$$

$$K_2 = \frac{x_i - x_{i-2}}{2\Delta t}$$

$$K_3 = x_{i-1}$$

Differentiating equation gives

$$\frac{dx}{dt} = 2K_1 t + K_2$$

$$\frac{d^2x}{dt^2} = 2K_1$$

Therefore, equation can be written

$$a_1 \frac{d^2y}{dt^2} + b_1 \frac{dy}{dt} + y = (K_1)t^2 + (2K_1 b_2 + K_2)t + (2K_1 a_2 + K_2 b_2 + K_3)$$

We propose a particular solution of the form

$$y = k_1 t_2 + k_2 t + k_3$$

which gives

$$\frac{dy}{dt} = 2k_1 t + k_2 \quad \frac{d^2y}{dt^2} = 2k_1$$

Substituting these expressions for y and its derivatives into equation and equating coefficients of like powers of t, we can solve for k_1, k_2, and k_3. If we define the dimensionless parameters

$$\alpha = \frac{(a_1 + a_2) - b_1(b_1 - b_2)}{(\Delta t)^2}$$

$$\beta = \frac{(b_1 - b_2)}{\Delta t}$$

then we have

$$k_1 = \frac{1}{2(\Delta t)^2}(x_i - 2x_{i-1} + x_{i-2})$$

$$k_2 = \frac{1}{\Delta t}\left[\left(\frac{1}{2} - \beta\right)x_i + (2\beta)x_{i-1} - \left(\frac{1}{2} + \beta\right)x_{i-2}\right]$$

$$k_3 = (-\beta/2 - \alpha)x_i + (1 + 2\alpha)x_{i-1} + (\beta/2 - \alpha)x_{i-2}$$

Now, the homogeneous form of equation has the general solution

$$y = R_1 e^{r_1} + R_2 e^{r_2}$$

where

$$r_1 = \frac{-b_1 + \sqrt{b_1^2 - 4a_1}}{2a_1}$$

$$r_2 = \frac{-b_1 - \sqrt{b_1^2 - 4a_1}}{2a_1}$$

and the coefficients R_1 and R_2 are to be determined by two conditions on y(t). (Classically this is the general solution, i.e., a "basis" for all possible solutions, only if r_1 'H r_2. If $r_1 = r_2$ the expressions for R_1 and R_2 below are singular. However, these singularities are removeable when the solution is expressed in discrete-time form. The solution derived from can be shown to be identical whether or not $r_1 = r_2$.) Adding equations (the particular and complimentary solutions respectively) gives the general total solution of equation.

$$y = R_1 e^{r_1} + R_2 e^{r_2} + k_1 t^2 + k_2 t + k_3$$

We will determine the coefficients R_1 and R_2 by imposing the following two conditions on y(t): $y = y_{i-1}$ at $t = t_{i-1}$, and $y = y_{i-2}$ at $t = t_{i-2} = -\Delta t$. Inserting these two conditions into equation gives

$$y_{1+1} = R_1 + R_2 + k_3$$

and

$$y_{i-2} = R_1 e^{-r_1 \Delta t} + R_2 e^{-r_2 \Delta} + k_1 (\Delta t)^2 - k_2 \Delta t + k_3$$

Solving these two equations simultaneously for R_1 and R_2 gives

$$R_1 = \frac{y_{i-2} - e^{-r_2 \Delta t} y_{i-1} - \left(1 - e^{-r_2 \Delta t}\right)k_3 - k_1 (\Delta t)^2 + k_2 \Delta t}{e^{-r_1 \Delta t} - e^{-r_2 \Delta t}}$$

$$R_2 = \frac{-y_{i-2} + e^{-r_1 \Delta t} y_{i-1} + \left(1 - e^{-r_1 \Delta t}\right) k_3 + k_1 (\Delta t)^2 - k_2 \Delta t}{e^{-r_1 \Delta t} - e^{-r_2 \Delta t}}$$

Substituting these expressions for R_1 and R_2 into equation, solving for y_i at $t = t_i = \Delta t$, and simplifying, we have

$$y_i = -\gamma y_{i-2} + \delta y_{i-1} + (1 + y - \delta) k_3 + (1 - \gamma) k_2 \Delta t + (1 + \gamma) k_1 (\Delta t)^2$$

where γ and δ are dimensionless parameters defined by

$$\gamma = e^{(r_1 + r_2)\Delta t}$$
$$\delta = e^{r_1 \Delta t} + r^{r_2 \Delta t}$$

Substituting the prior expressions for k_1, k_2, and k_3 into equation and combining terms by x_n and y_n we arrive at the final recurrence formula

$$y_i = c_1 y_{i-2} + c_2 y_{i-1} + c_3 x_{i-2} + c_4 x_{i-1} + c_5 x_i$$

For a given discrete time increment Dt, the recurrence coefficients are given by

$$c_1 = -\gamma$$
$$c_2 = \delta$$
$$c_3 = \gamma - (\alpha - \beta/2)(1 + \gamma - \delta) - \beta(1 - \gamma)$$
$$c_4 = -\delta + 2\alpha(1 + y - \delta) + 2\beta(1 - \gamma)$$
$$c_5 = 1 - (\alpha + \beta/2)(1 + \gamma - \delta) - \beta(1 - \gamma)$$

As required, the sum of c_1 through c_5 is identically equal to 1. Also, since this solution qualifies as a "reasonable" recurrence formula (in the sense that it gives x(t) as a linear function when x_{i-2}, x_{i-1}, and x_i are co-linear versus time), we can now determine the first coefficients of equation using equations as follows

$$L_1 = -\gamma$$
$$L_2 = \delta$$
$$L_3 = 1 + \gamma - \delta$$
$$L_4 = 2 - \beta(1 + \gamma - \delta)$$

Before proceding to the next section we will derive three more recurrence formulas that will prove useful. They are based on equation with the constants R_1 and R_2 determined by the two conditions $y = y_{i-1}$ att $= t_{i-1} = 0$ and $dy/dt = \dot{y}_{i-1}$ at $t = t_{i-1} = 0$.Inserting the first of these conditions into equation gives

$$y_{i-1} = R_1 + R_2 + k_3$$

To apply the second condition we first differentiate equation as follows

$$\frac{dy}{dt} = R_{1r1e^{r_1 \Delta t}} + R_2 r_2 e^{r_2 \Delta t} + 2k_1 t + k_2$$

Inserting the second condition, we have

$$\dot{y}_{i-1} = R_1 r_1 + R_2 r_2 + k_2$$

Solving equations simultaneously for R_1 and R_2 gives

$$R_1 = \frac{\dot{y}_{i-1} - r_2 y_{i-1} + r_2 k_3 - k_2}{r_1 - r_2}$$

$$R_2 = \frac{-\dot{y}_{i-1} + r_1 y_{i-1} - r_1 k_3 - k_2}{r_1 - r_2}$$

Substituting these expressions for R_1 and R_2 into equation, solving for y_i at $t = t_i = Dt$, and making the appropriate substitutions for k_1, k_2, and k_3, we arrive at the recurrence formula

$$y_i = f_1 \dot{y}_{i-1} + f_2 y_{i-1} + f_3 x_{i-2} + f_4 x_{i-1} + f_5 x_i$$

If we define the dimensionless parameters

$$\mu = \frac{e^{r_1 \Delta t} - e^{r_2 \Delta t}}{\Delta t (r_1 - r_2)}$$

$$\sigma = \frac{r_2 e^{r_1 \Delta t} - r_1 e^{r_2 \Delta t}}{(r_1 - r_2)}$$

we can express the coefficients f_1 through f_5 in equation as

$$f_1 = \mu \, \Delta t$$

$$f_2 = -\sigma$$

$$f_3 = -\left(\frac{1}{2}\beta + \alpha\right) + \left(\beta + \frac{1}{2}\right)\mu + \left(\frac{1}{2}\beta - \alpha\right)\sigma$$

$$f_4 = 2(\alpha + \beta) - 2\beta\mu + (1 + 2\alpha)\sigma$$

$$f_5 = 1 - \left(\frac{3}{2}\beta + \alpha\right) + \left(\beta - \frac{1}{2}\right)\mu - \left(\frac{3}{2}\beta + \alpha\right)\sigma$$

Similarly, we can solve equation for $(dy/dt)_i$ at $t = t_{i-2} = Dt$ with R_1 and R_2 given by equations. This leads to the recurrence formula

$$\dot{y}_1 = g_1 \dot{y}_{1-1} + g_2 y_{i-1} + g_3 x_{i-2} + g_4 x_{i-1} + g_5 x_1$$

In terms of the dimensionless parameters

$$\rho = \frac{(\Delta t)^2}{a_1}$$

$$\lambda = \frac{r_1 e^{r_1 \Delta t} - r_2 e^{r_2 \Delta t}}{(r_1 - r_2)}$$

we can express the coefficients g_1 through g_5 in equation as

$$g_1 = \lambda$$

$$g_2 = -\frac{\rho\mu}{\Delta t}$$

$$g_3 = \frac{1}{\Delta t}\left[\left(\frac{1}{2}\beta - \alpha\right)\rho\mu + \left(\beta + \frac{1}{2}\right)\lambda - \left(\beta - \frac{1}{2}\right)\right]$$

$$g_4 = \frac{1}{\Delta t}\left[(1 + 2\alpha)\rho\mu - 2(1 + \lambda\beta - \beta)\right]$$

$$g_5 = \frac{1}{\Delta t}\left[\left(-\frac{1}{2}\beta - \alpha\right)\rho\mu + \left(\beta - \frac{1}{2}\right)\lambda - \left(\frac{3}{2} - \beta\right)\right]$$

Equations can be used as a two-stage recurrence scheme for generating both y_n and $\dot{y}_n$.

Another useful recurrence formula is found by solving equation for y_{i-2} with R_1 and R_2 given by the last two of equations and setting $t = t_{i-2} = -\Delta t$. This leads to the formula

$$\dot{y}_{n-2} = h_1 \dot{y}_{i-1} + h_2 y_{i-1} + h_3 x_{i-2} + h_4 x_{i-1} + h_5 x_i$$

In terms of the dimensionless parameters

$$\hat{\mu} = \frac{e^{-r_1 \Delta t} - e^{-r_2 \Delta t}}{\Delta t (r_1 - r_2)}$$

$$\hat{\sigma} = \frac{r_2 e^{-r_1 \Delta t} - r_1 e^{-r_2 \Delta t}}{(r_1 - r_2)}$$

the coefficients h_1 through h_5 can be expressed as

$$h_1 = \hat{\mu}\,\Delta t$$

$$h_2 = -\hat{\sigma}$$

$$h_3 = 1 + \left(\frac{3}{2}\beta - \alpha\right) + \left(\beta + \frac{1}{2}\right)\hat{\mu} + \left(\frac{1}{2}\beta - \alpha\right)\hat{\sigma}$$

$$h_4 = 2(\alpha - \beta) - 2\beta\hat{\mu} + (1 + 2\alpha)\hat{\sigma}$$

$$h_5 = \left(\frac{1}{2}\beta - \alpha\right) + \left(\beta - \frac{1}{2}\right)\hat{\mu} - \left(\frac{1}{2}\beta + \alpha\right)\hat{\sigma}$$

OPTIMUM RECURRENCE COEFFICIENTS BASED ON POINT-TO-POINT INTERPOLATION

In this section we derive the coefficients of equation with the assumption that the input function x(t) is linear between successive values of x_n. As mentioned previously, this implies that the second derivative is zero during each interval and infinite at each instant tn, where the slope changes discontinuously. Therefore, we will first determine the response to equation to a discontinuous change in the derivative of x(t). Our approach will be based on the solution of the following problem:

Given the value of $x_{i-1}, \dot{x}_{i-1}, y_{i-1}, \dot{y}_{i-1},$ and Δt, find y_i assuming the function x (t) has a constant second derivative in the interval from t_{i-1} to t_i

Equations provide the means of solving this problem if we can infer the appropriate values of x_{i-2} and x_i from the given information. The function x(t) is uniquely determined in the interval from t_{i-1} to t_i by the values of $x_{i-1}, \dot{x}_{i-1}, \dot{x}_i$, and Δt, and the requirement of a constant second derivative. Therefore, the value of x_i is also uniquely determined, and is given by

$$x_i = x_{i-1} + \left(\frac{\dot{x}_{i-1} + \dot{x}}{2}\right)\Delta t$$

Further, if we recognize that the only purpose of x_{i-2} in equations is to determine the function x(t) in the interval from t_{i-1} to t_i, which it does by assuming a constant second derivative over the entire double interval from t_{i-2} to t_i, then it is clear that the appropriate value for x_{i-2} is to be found by extrapolating the function x(t) back to t_{i-2}. On this basis, it can be shown that

$$x_{i-2} = x_{i-1} - \left(\frac{3\dot{x}_{i-1} + \dot{x}_i}{2}\right)\Delta t$$

Substituting the two preceding expressions into equations gives the solution to the problem:

$$y_i = f_i\dot{y}_{i-1} + f_2 y_{i-1} + (f_3 + f_4 + f_5)x_{i-1} + f_3\left(\frac{\dot{x}_i - 3\dot{x}_{i-1}}{2}\Delta t\right) + f_5\left(\frac{\dot{x}_i - \dot{x}_{i-1}}{2}\Delta t\right)$$

$$\dot{y}_i = g_1\dot{y}_{i-1} + g_2 y_{i-1} + (g_3 + g_4 + g_5)x_{i-1} + g_3\left(\frac{\dot{x}_i - 3\dot{x}_{i-1}}{2}\Delta t\right) + g_5\left(\frac{\dot{x}_i - \dot{x}_{i-1}}{2}\Delta t\right)$$

These equations allow us to compute the effect on y and $\dot{y}$ of a change in $\dot{x}$ from $\dot{y}_{i-1}$ to $\dot{y}_i$ occurring uniformly during an interval Δt. In the limit as Δt goes to zero these equations reduce to

$$\dot{y}_i = y_{i-1} \quad \dot{y}_i = \dot{y}_{i-1} + \frac{a_2}{a_2}(\dot{x}_i - \dot{x}_{i-1})$$

With this result we can proceed to determine the general recurrence formula based on point-to-point interpolation of the input function. By equations we can write

$$y_{i-2} = h_1\left(\dot{y}_{i-1}^{-}\right) + h_2 y_{i-1} + h_3 x_{i-2} + h_4 x_{i-1} + h_5 X_i$$

$$y_i = f_1\left(\dot{y}_{i-1}^{+}\right) + f_2 y_{i-1} + f_3 X_{i-2} + f_4 x_{i-1} + f_5 x_i$$

where

$\left(\dot{y}_{i-1}^{-}\right)$ = the value of $\dot{y}$ in the limit as t approaches t_{i-1} from below (before).

$\left(\dot{y}_{i-1}^{+}\right)$ = the value of $\dot{y}$ in the limit as t approaches t_{i-1} from above (after).

X_i = a synthesized value of x_i, co-linear vs time with x_{i-2} and x_i.

X_{i-2} = a synthesized value of x_{i-2}, co-linear vs time with x_{i-1} and x_i.

The synthesized values of x can be expressed as

$$X_{i-1} = 2x_{i-1} - x_i$$

$$X_{i-1} = 2x_{i-1} - x_{i-2}$$

Also, since the change in the derivative of x(t) at $t = t_{i-1}$ is

$$\left(\frac{x_i - x_{i\ 1}}{\Delta t}\right) - \left(\frac{x_i - x_{i-1}}{\Delta t}\right) = \frac{x_i - 2x_{i-1} + x_{1-2}}{\Delta t}$$

we have, by equation, the relation

$$\left(\dot{y}_{i-1}^{+}\right) = \left(\dot{y}_{i-1}^{-}\right) + \frac{a_2}{a_1}\left(\frac{x_i - 2x_{i-1} + x_{i-2}}{\Delta t}\right)$$

Making the indicated substitutions for X_{i-2}, X_i, and $\left(\dot{y}_{i-1}^{+}\right)$ in equations, and then eliminating $\left(\dot{y}_{i-1}^{-}\right)$ from the resulting equations leads to the general recurrence formula based on point-to-point interpolation

$$y_i = c_i y_{i-2} + c_2 y_{i-1} + c_3 x_{i-2} + c_4 x_{i-1} + c_5 x_i$$

where, if we define the dimensionless parameter

$$\varepsilon = 1 - \frac{a_2}{a_1}$$

the coefficients c_1 through c_5 are given by

$$c_1 = -\gamma$$

$$c_2 = \delta = \lambda - \sigma$$

$$c_3 = (1+\beta)\gamma - \beta\lambda - \mu\varepsilon$$
$$c_4 = (1+\beta)\sigma + 2\mu\varepsilon - (1-\gamma)\beta - (1-\beta)\lambda$$
$$c_5 = 1 - (1+\sigma)\beta - \mu\varepsilon$$

As required, the sum of c_1 through c_5 is identically equal to 1. Also, substituting these coefficients into equations gives the same values for L_1 through L_4 and do the coefficients based on second order interpolation. This is in accord with the assertion that all reasonable coefficient sets give identical results when x_{i-2}, x_{i-1}, and x_i are colinear versus time.

In the preceding sections no distinction was made between the cases when the roots r_1 and r_2 of the characteristic equation are real and when they are complex. In either case the dimensionless parameters in terms of which the recurrence coefficients have been expressed are always purely real, as is easily verified using the basic exponential and trigonometric identities for complex variables. Summary of these dimensionless parameters, including the equivalent trigonometric definitions where appropriate. In this table the parameter u and v are defined as

$$u = -\frac{b_1}{2a_1} \qquad v = \frac{\sqrt{4a_1 - b_1^2}}{2a_1}$$

Notice that $r_1 = u + iv$ and $r_2 = u - iv$.

$$\alpha = \frac{(a_1 - a_2) - b_1(b_1 - b_2)}{(\Delta t)^2}$$

$$\beta = \frac{(b_1 - b_2)}{\Delta t}$$

$$\gamma = e^{(r_1 + r_2)\Delta t} = e^{2u\Delta t}$$

$$\delta = e^{r_1\Delta t} + e^{r_2\Delta t} = 2e^{u\Delta t}\cos(v\Delta t)$$

$$\mu = \frac{e^{r_1\Delta t} - e^{r_2\Delta t}}{\Delta t(r_1 - r_2)} = \frac{e^{u\Delta t}}{v\Delta t}\sin(v\Delta t)$$

$$\sigma = \frac{r_2 e^{r_1\Delta t} - -r_1 e^{r_2\Delta t}}{(r_1 - r_2)} = \frac{e^{u\Delta t}}{v}\left[u\sin(v\Delta t) - v\cos(v\Delta t)\right]$$

$$\lambda = \frac{r_1 e^{r_1\Delta t} - r_2 e^{r_2\Delta t}}{(r_1 - r_2)} = \frac{e^{u\Delta t}}{v}\left[u\sin(v\Delta t) + v\cos(v\Delta t)\right]$$

$$\rho = \frac{(\Delta t)^2}{a_1}$$

$$\mu = \frac{e^{-r_1\Delta t} - e^{-r_2\Delta t}}{\Delta t(r_1 - r_2)} = -\frac{e^{-u\Delta t}}{v\Delta t}\sin(v\Delta t)$$

$$\hat{\sigma} = \frac{r_2 e^{-r_1 \Delta t} - r_1 e^{-r_2 \Delta t}}{(r_1 - r_2)} = -\frac{e^{-u\Delta t}}{v}[u \sin(v\Delta t) + v \cos(v\Delta t)]$$

$$\varepsilon = 1 - \frac{a_2}{a_1}$$

The use of the recurrence formulas developed in thepreceding sections can best be illustrated by an example. Consider a function with a dynamic characteristic described by the differential equation

$$9\frac{d^2y}{dt^2} + \frac{dy}{dt} + y = 8\frac{d^2x}{dt^2} + 3\frac{dx}{dt} + x$$

In this example we have

$$a_1 = 9\sec^2 \quad b_1 = 1\sec \quad a_2 = 8\sec^2 \quad b_2 = 3\sec$$

If we arbitrarily select an execution period of $\Delta t = 1$ sec,

$$\alpha = 3 \quad \beta = -2 \quad \gamma = 0.8948393168$$

Since the function is underdamped, we use the trigonometric form of the equation for d to compute

$$\delta = 1.790648545$$

We can now use equations to compute the coefficients of the general recurrence formula based on second order interpolation of the input function:

$$y_i = c_1 y_{i-2} + c_2 y_{i-1} + c_3 x_{i-2} + c_4 x_{i-1} + c_5 x_i$$

$$c_1 = -0.8948393168$$

$$c_2 = 1.790648545$$

$$c_3 = 0.6883975969$$

$$c_4 = 1.001939823$$

We can verify that these coefficients sum to 1, as required. To demmonstrate the operation of this recurrence formula we will compute the response to three different input functions:

$$(1)\,\text{Cons tan t input}: \quad x(t) = 100$$

$$(2)\,\text{Ramp input}: \quad x(t) = t$$

$$(1)\,\text{Accelerating input}: \quad x(t) = (1/4)t^2 + (1/2)t$$

In the first case it's convenient to use the form of the recurrence given by equation. setting $x_{i-2} = 100$ and $\Delta x_{i-1} = \Delta\Delta x_i = 0$, to give the formula

$$y_i = (-0.894839)y_{i-1} + (1.790648)y_{i-1} + (0.10419)100$$

With the initial conditions $y_0 = y_1 = 0$, the subsequent values of y_n generated by this formula are plotted in Figure.

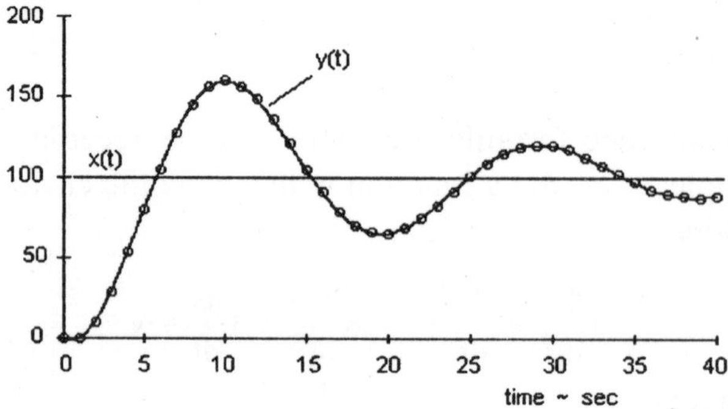

In the second case, $\Delta\Delta x_i$ is still zero, but Δx_{i-1} is a constant equal to 1 (since the slope of the function $x(t) = t$ is 1). Therefore, equation gives the formula

$$y_i = (-0.894839)y_{i-2} + (1.790648)y_{i-1} + (0.10419)t_{i-2} + (0.417733)$$

where we have substituted t_{i-2} for x_{i-2}. With the initial conditions $y_0 = y_1 = 0$, the subsequent values of y_n generated.

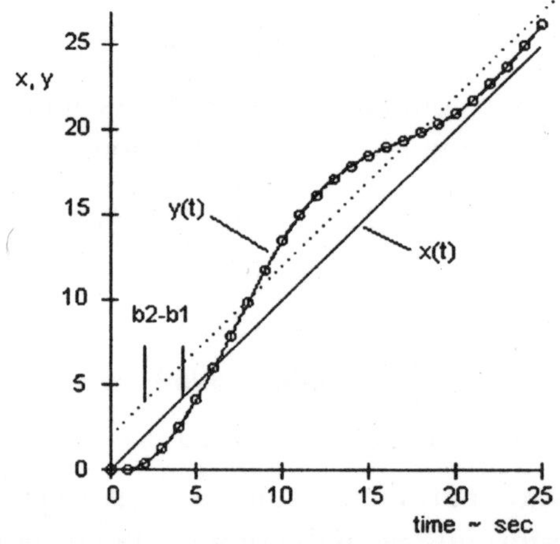

In the third case, neither ther first nor the second derivative of x(t) is zero, so we use the general form of the recurrence formula. Substituting for x_n the corresponding value function of t_n, we have

$$y_i = (-0.894839)y_{i-2} + (1.790648)y_{i-1} + (0.688398)\left[(1/4)t_{i-2}^{\ 2} + (1/2)t_{i-2}\right]$$
$$+(-1.586147)\left[(1/4)t_{i-1}^{\ 2} + (1/2)t_{i-1}\right] + (1.001940)\left[(1/4)t_i^{\ 2} + (1/2)t_i\right]$$

With the initial conditions $y_0 = y_1 = 0$, the subsequent values of y_n generated by this formula are plotted in Figure.

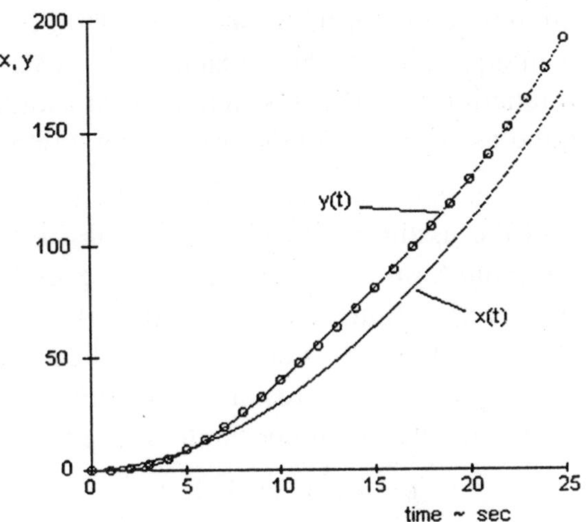

Figure shows the response to the same input function from the initial conditions $y_0 = y_1 = 50$.

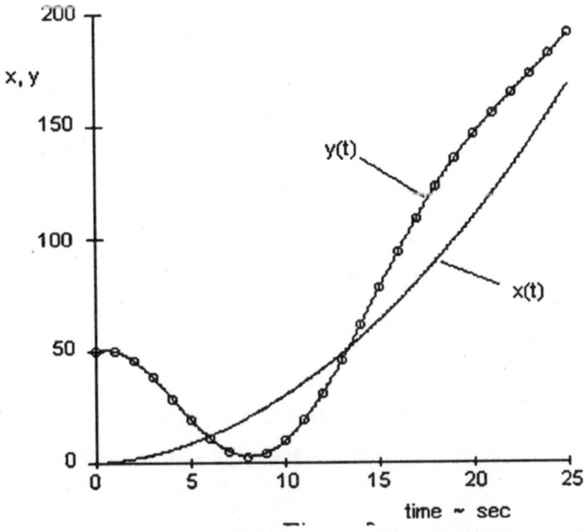

In these examples the computed values of y_n fall exactly on the theoretical function y(t), because all the input functions are exactly representable by a second-degree polynomial. Any choice of execution period will yield the same accuracy. For example, if we had chosen $\Delta t = 5$ sec, the coefficients would have been different but the computed values of y_n would still have fallen exactly on the theoretical function y(t). The only consideration limiting the size of Δt in such cases is that the output of y_n must occur frequently enough to adequately represent the function y(t) for external purposes.

In practice, the continuous input function x(t) is usually specified by a sequence of discrete values from some external source, rather than computed from a given equation as in the preceding examples. It is the lack of a precise functional definition of x(t) that requires us to assume an interpolation function for the x_n sequence. This, in turn, requires that Δt be small enough so that our interpolation of the x_n samples is sufficiently representative of the true continuous input function f(t). This restriction on Δt should not be confused with a restriction arising from the use of a non-optimum solution method.

We can also determine the optimum recurrence formulas for the dynamical system specified by equation with the assumption of point-to-point interpolation of the input samples. Since the point-to-point formula is identical to the second-order-fit formula when x_{i-2}, x_{i-1}, and x_i are colinear versus time, they give identical results for the constant and ramp inputs i.e., cases (1) and (2) above. It is only when applied to the third, more general, case that the two assumptions give different results. To determine the exact point-to-point recurrence formula, we first compute the dimensionless parameters

$$\varepsilon = 0.1111111111$$
$$\mu = 0.9290200587$$
$$\sigma = -0.9469364981$$
$$\lambda = 0.8437120471$$

using the equations. Then by equations we can compute the coefficients of the exact point-to-point recurrence formula

$$y_i = c_1 y_{i-2} + c_2 y_{i-1} + c_3 x_{i-2} + c_4 x_{i-1} + c_5 x_i$$
$$c_1 = -0.8948393168$$
$$c_2 = 1.790648545$$
$$c_3 = 0.6893603265$$
$$c_4 = 1.002902553$$

These coefficients differ hardly at all from those computed on the basis of second-order interpolation, so the predicted response to the input for case is virtually identical to that given by the second-order-fit method, especially since every three consecutive samples x_n are nearly co-linear.

To conclude this section, we will develop the "standard form" of the second-order recurrence formula (analagous to equation for first-order formulas), in terms of which the optimum second-order solutions can be expressed succinctly. If we define

$$\Delta y_n = y_n - y_{n-1}$$

then equation can be written as

$$y_i = (c_1 + c_2)\, y_{i-2} + c_2\, \Delta y_{i-1} + (c_3 + c_4 + c_5)\, x_{i-2} + (c_4 + 2c_5)\, \Delta x_{i-1} + c_5\, \Delta\Delta x_i$$

Since the coefficients sum to unity we have $c_3 + c_4 + c_5 = 1 - c_1 - c_2$, so we can make this substitution to give

$$y_i = (c_1 + c_2)\, y_{i-2} + c_2\, \Delta y_{i-1} + (1 - c_1 - c_2)\, x_{i-2} + (c_4 + 2c_5)\, \Delta x_{i-1} + c_5\, \Delta\Delta x_i$$

which can be written as

Thus we have the "standard form" for second-order recurrences

$$y_i = y_{i-2} + A\Delta y_{i-1} + B\,(x_{1-2} - y_{1-2}) + C\Delta x_{i-1} + D\Delta\Delta x_i$$

where

$$A = \delta$$
$$B = 1 + \gamma - \delta$$
$$C = 2 - \delta - B(1 + \gamma - \delta)$$

and the D coefficient is given by

$$D_{soi} = 1 - \left(\alpha + \frac{\beta}{2}\right)(1 + \gamma - \delta) - \beta(1 - \gamma)$$

for second-order interpolation, and by

$$D_{ptp} = 1 - \varepsilon\mu - \beta(1 + \gamma - \delta)$$

for point-to-point interpolation. It's worth recalling that the set of *all* reasonable second order recurrence schemes has just a single degree of freedom, which can be linearly parameterized as the D coefficient of the standard form.